Applied and Numerical Harmonic Analysis

Applied and Numerical Harmonic Analysis

Published titles

J.M. Cooper: *Introduction to Partial Differential Equations with MATLAB* (ISBN 0-8176-3967-5)

C.E. D'Attellis and E.M. Fernández-Berdaguer: *Wavelet Theory and Harmonic Analysis in Applied Sciences* (ISBN 0-8176-3953-5)

H.G. Feichtinger and T. Strohmer: *Gabor Analysis and Algorithms* (ISBN 0-8176-3959-4)

T.M. Peters, J.H.T. Bates, G.B. Pike, P. Munger, and J.C. Williams: *Fourier Transforms and Biomedical Engineering* (ISBN 0-8176-3941-1)

A.I. Saichev and W.A. Woyczyński: *Distributions in the Physical and Engineering Sciences* (ISBN 0-8176-3924-1)

R. Tolimierei and M. An: *Time-Frequency Representations* (ISBN 0-8176-3918-7)

G.T. Herman: *Geometry of Digital Spaces* (ISBN 0-8176-3897-0)

A. Procházka, J. Uhlíř, P.J.W. Rayner, and N.G. Kingsbury: *Signal Analysis and Prediction* (ISBN 0-8176-4042-8)

J. Ramanathan: *Methods of Applied Fourier Analysis* (ISBN 0-8176-3963-2)

A. Teolis: *Computational Signal Processing with Wavelets* (ISBN 0-8176-3909-8)

W.O. Bray and Č.V. Stanojević: *Analysis of Divergence* (ISBN 0-8176-4058-4)

G.T. Herman and A. Kuba: *Discrete Tomography* (ISBN 0-8176-4101-7)

J.J. Benedetto and P.J.S.G. Ferreira: *Modern Sampling Theory* (ISBN 0-8176-4023-1)

A. Abbate, C.M. DeCusatis, and P.K. Das: *Wavelets and Subbands* (ISBN 0-8176-4136-X)

L. Debnath: *Wavelet Transforms and Time-Frequency Signal Analysis* (ISBN 0-8176-4104-1)

K. Gröchenig: *Foundations of Time-Frequency Analysis* (ISBN 0-8176-4022-3)

D.F. Walnut: *An Introduction to Wavelet Analysis* (ISBN 0-8176-3962-4)

O. Bratelli and P. Jorgensen: *Wavelets through a Looking Glass* (ISBN 0-8176-4280-3)

H. Feichtinger and T. Strohmer: *Advances in Gabor Analysis* (ISBN 0-8176-4239-0)

O. Christensen: *An Introduction to Frames and Riesz Bases* (ISBN 0-8176-4295-1)

L. Debnath: *Wavelets and Signal Processing* (ISBN 0-8176-4235-8)

Forthcoming Titles

E. Prestini: *The Evolution of Applied Harmonic Analysis* (ISBN 0-8176-4125-4)

G. Bi and Y.H. Zeng: *Transforms and Fast Algorithms for Signal Analysis and Representations* (ISBN 0-8176-4279-X)

J. Benedetto and A. Zayed: *Sampling, Wavelets, and Tomography* (0-8176-4304-4)

Wavelets and Signal Processing

Lokenath Debnath
Editor

Birkhäuser
Boston • Basel • Berlin

Lokenath Debnath
The University of Texas-Pan American
Department of Mathematics
Edinburg, TX 78539-2999
USA

Library of Congress Cataloging-in-Publication Data

Wavelets and signal processing / Lokenath Debnath, editor.
p. cm. – (Applied and numerical harmonic analysis)
Includes bibliographical references and index.
ISBN 0-8176-4235-8 (alk. paper) – ISBN 3-7643-4235-8 (alk. paper)
1. Signal processing–Mathematics 2. Wavelets (Mathematics) I. Debnath, Lokenath.
II. Series.

TK5102.9.W3797 2002
621.382'2–dc21 2002074762
CIP

Printed on acid-free paper

Birkhäuser 

ISBN 0-8176-4235-8 SPIN 10832344
ISBN 3-7643-4235-8

Typeset by Larisa Martin, Lawrence, KS.
Printed in the United States of America.

9 8 7 6 5 4 3 2 1

Birkhäuser Boston • Basel • Berlin
A member of BertelsmannSpringer Science+Business Media GmbH

Series Preface

The *Applied and Numerical Harmonic Analysis (ANHA)* book series aims to provide the engineering, mathematical, and scientific communities with significant developments in harmonic analysis, ranging from abstract harmonic analysis to basic applications. The title of the series reflects the importance of applications and numerical implementation, but richness and relevance of applications and implementation depend fundamentally on the structure and depth of theoretical underpinnings. Thus, from our point of view, the interleaving of theory and applications and their creative symbiotic evolution is axiomatic.

Harmonic analysis is a wellspring of ideas and applicability that has flourished, developed, and deepened over time within many disciplines and by means of creative cross-fertilization with diverse areas. The intricate and fundamental relationship between harmonic analysis and fields such as signal processing, partial differential equations (PDEs), and image processing is reflected in our state of the art *ANHA* series.

Our vision of modern harmonic analysis includes mathematical areas such as wavelet theory, Banach algebras, classical Fourier analysis, time-frequency analysis, and fractal geometry, as well as the diverse topics that impinge on them.

For example, wavelet theory can be considered an appropriate tool to deal with some basic problems in digital signal processing, speech and image processing, geophysics, pattern recognition, biomedical engineering, and turbulence. These areas implement the latest technology from sampling methods on surfaces to fast algorithms and computer vision methods. The underlying mathematics of wavelet theory depends not only on classical Fourier analysis, but also on ideas from abstract harmonic analysis, including von Neumann algebras and the affine group. This leads to a study of the Heisenberg group and its relationship to Gabor systems, and of the metaplectic group for a meaningful interaction of signal decomposition methods. The unifying influence of wavelet theory in the aforementioned topics illustrates the justification for providing a means for centralizing and disseminating information from the broader, but still focused, area of harmonic analysis. This will be a key role of *ANHA*. We intend to publish with the scope and interaction that such a host of issues demands.

Along with our commitment to publish mathematically significant works at the frontiers of harmonic analysis, we have a comparably strong commitment to publish major advances in the following applicable topics in which harmonic analysis plays a substantial role:

Antenna theory	*Prediction theory*
Biomedical signal processing	*Radar applications*
Digital signal processing	*Sampling theory*
Fast algorithms	*Spectral estimation*
Gabor theory and applications	*Speech processing*
Image processing	*Time-frequency and time-scale analysis*
Numerical partial differential equations	*Wavelet theory*

The above point of view for the *ANHA* book series is inspired by the history of Fourier analysis itself, whose tentacles reach into so many fields.

In the last two centuries Fourier analysis has had a major impact on the development of mathematics, on the understanding of many engineering and scientific phenomena, and on the solution of some of the most important problems in mathematics and the sciences. Historically, Fourier series were developed in the analysis of some of the classical PDEs of mathematical physics; these series were used to solve such equations. In order to understand Fourier series and the kinds of solutions they could represent, some of the most basic notions of analysis were defined, e.g., the concept of "function." Since the coefficients of Fourier series are integrals, it is no surprise that Riemann integrals were conceived to deal with uniqueness properties of trigonometric series. Cantor's set theory was also developed because of such uniqueness questions.

A basic problem in Fourier analysis is to show how complicated phenomena, such as sound waves, can be described in terms of elementary harmonics. There are two aspects of this problem: first, to find, or even define properly, the harmonics or spectrum of a given phenomenon, e.g., the spectroscopy problem in optics; second, to determine which phenomena can be constructed from given classes of harmonics, as done, for example, by the mechanical synthesizers in tidal analysis.

Fourier analysis is also the natural setting for many other problems in engineering, mathematics, and the sciences. For example, Wiener's Tauberian theorem in Fourier analysis not only characterizes the behavior of the prime numbers, but also provides the proper notion of spectrum for phenomena such as white light; this latter process leads to the Fourier analysis associated with correlation functions in filtering and prediction problems, and these problems, in turn, deal naturally with Hardy spaces in the theory of complex variables.

Nowadays, some of the theory of PDEs has given way to the study of Fourier integral operators. Problems in antenna theory are studied in terms of unimodular trigonometric polynomials. Applications of Fourier analysis abound in signal processing, whether with the Fast Fourier transform (FFT), or filter design, or the adaptive modeling inherent in time-frequency-scale methods such as wavelet theory. The coherent states of mathematical physics are translated and modulated Fourier transforms, and these are used, in conjunction with the uncertainty principle, for dealing with signal reconstruction in communications theory. We are back to the raison d'être of the ANHA series!

John J. Benedetto
Series Editor
University of Maryland
College Park

Contents

Series Preface **v**
Preface **xvii**
Contributors **xxi**

I Wavelets and Wavelet Transforms **1**

1 The Wavelet Transform and Time-Frequency Analysis
Leon Cohen **3**
1.1 Introduction 3
1.2 The wavelet approach to time-frequency analysis 4
1.2.1 Density of time frequency 5
1.2.2 Marginals 5
1.2.3 Moments 6
1.2.4 Mean time and mean frequency 7
1.2.5 Conditional moments 9
1.3 Examples 11
1.3.1 Morlet wavelet 11
1.3.2 Modified Morlet wavelet 12
1.3.3 Examples 12
1.4 Conclusion 16
1.5 Appendix: Moments 18
1.5.1 The "a" moments 18
1.5.2 The time moments 19
1.5.3 The frequency moments 21
References 21

2 Multiwavelets in $\mathbb{R}^n$ with an Arbitrary Dilation Matrix
Carlos A. Cabrelli, Christopher Heil, and Ursula M. Molter **23**
2.1 Introduction 23
2.2 Self-similarity 24
2.3 Refinement equations 25

2.4 Existence of MRAs ... 35
References ... 38
3 **On the Nonexistence of Certain Divergence-free Multiwavelets**
J. D. Lakey and M. C. Pereyra **41**
3.1 Introduction ... 41
3.2 The main theorem ... 42
3.3 Shift invariant subspaces of $\mathcal{H}^2(\mathbf{R}^n, \mathbf{R}^n)$ and ω-localized projections ... 43
3.3.1 The correlation function ... 44
3.3.2 Proof of Theorem 3.3.1 ... 48
3.3.3 Proof of Theorem 3.3.2 ... 52
3.4 Nonexistence ... 53
References ... 54
4 **Osiris Wavelets and Isis Variables**
Guy Battle **55**
4.1 Introduction ... 55
4.2 The Gaussian fixed point (configuration picture) ... 66
4.3 The recursion formula for the dipole perturbation ... 73
4.4 The Gaussian fixed point (wavelet picture) ... 77
4.5 Elements of the A^∞ matrix ... 85
4.6 Linearized RG transformation for the dipole perturbation ... 90
Appendix ... 97
References ... 101
5 **Wavelets for Statistical Estimation and Detection**
R. Chandramouli and K. M. Ramachandran **103**
5.1 Introduction ... 103
5.2 Wavelet analysis ... 106
5.3 Statistical estimation ... 107
5.3.1 Popular wavelet-based signal estimation schemes ... 109
5.3.2 MPE ... 112
5.3.3 Performance analysis ... 117
5.4 Hypothesis testing ... 123
5.4.1 OR decision fusion ... 125
5.4.2 AND decision fusion ... 127
5.4.3 Asymptotic analysis ... 128
5.5 Conclusion ... 129
References ... 130

II Signal Processing and Time-Frequency Signal Analysis 133

6 High Performance Time-Frequency Distributions for Practical Applications
Boualem Boashash and Victor Sucic **135**
6.1 The core concepts of time-frequency signal processing . . . 135
6.1.1 Rationale for using time-frequency distributions . . . 135
6.1.2 Limitations of traditional signal representations . . . 136
6.1.3 Positive frequencies and the analytic signal . . . 138
6.1.4 Joint time-frequency representations . . . 139
6.1.5 Heuristic formulation of the class of quadratic TFDs . . . 144
6.2 Performance assessment for quadratic TFDs . . . 151
6.2.1 Visual comparison of TFDs . . . 151
6.2.2 Performance criteria for TFDs . . . 153
6.2.3 Performance specifications for the design of high resolution TFDs . . . 160
6.3 Selecting the best TFD for a given practical application . . . 161
6.3.1 Assessment of performance for selection of the best TFD . . . 161
6.3.2 Selecting the optimal TFD for real-life signals under given constraints . . . 169
6.4 Discussion and conclusions . . . 172
References . . . 174

7 Covariant Time-Frequency Analysis
Franz Hlawatsch, Georg Tauböck, and Teresa Twaroch **177**
7.1 Introduction . . . 177
7.2 Groups and group representations . . . 180
7.2.1 Groups . . . 180
7.2.2 Unitary and projective group representations . . . 181
7.2.3 Modulation operators . . . 183
7.2.4 Warping operators . . . 183
7.3 Dual operators . . . 187
7.3.1 Properties of dual operators . . . 187
7.3.2 Modulation and warping operators as dual operators . . . 189
7.4 Affine operators . . . 190
7.4.1 Properties of affine operators . . . 190
7.4.2 Modulation and warping operators as affine operators . . . 192
7.5 Covariant signal representations: group domain . . . 193
7.5.1 Displacement operators . . . 193
7.5.2 Covariant linear θ-representations . . . 195
7.5.3 Covariant bilinear θ-representations . . . 198
7.6 The displacement function: theory . . . 201
7.6.1 Axioms of time-frequency displacement . . . 201
7.6.2 Lie group structure and consequences . . . 202
7.6.3 Unitary equivalence results . . . 205
7.6.4 Relation between extended displacement functions . . . 206

7.7 The displacement function: construction 208
7.7.1 General construction 208
7.7.2 Warping operator and modulation operator 211
7.7.3 Dual modulation and warping operators 212
7.7.4 Affine modulation and warping operators 212
7.8 Covariant signal representations: time-frequency domain 213
7.8.1 Covariant linear time-frequency representations 213
7.8.2 Covariant bilinear time-frequency representations 215
7.8.3 Dual and affine modulation and warping operators 217
7.9 The characteristic function method 219
7.9.1 (a, b)-energy distributions 220
7.9.2 Time-frequency energy distributions 221
7.9.3 Equivalence results for dual operators 222
7.10 Extension to general LCA groups 226
7.11 Conclusion 227
References 228

8 Time-Frequency/Time-Scale Reassignment
Eric Chassande-Mottin, Francois Auger, and Patrick Flandrin **233**
8.1 Introduction 233
8.2 The reassignment principle 234
8.2.1 Localization trade-offs in classical time-frequency and time-scale analysis 234
8.2.2 Spectrograms/scalograms revisited from a mechanical analogy 236
8.2.3 Reassignment as a general principle 237
8.2.4 Connections with related approaches 241
8.3 Reassignment in action 241
8.3.1 Efficient algorithms 242
8.3.2 Analysis of AM-FM signals 243
8.3.3 Singularity characterization from reassigned scalograms 247
8.4 More on spectrogram reassignment vector fields 251
8.4.1 Statistics of spectrogram reassignment vectors 251
8.4.2 Geometric phase and level curves 253
8.5 Two variations 255
8.5.1 Supervised reassignment 255
8.5.2 Differential reassignment 256
8.6 Two signal processing applications 257
8.6.1 Time-frequency partitioning and filtering 257
8.6.2 Chirp detection 259
8.7 Conclusion 263
References 264

9 Spatial Time-Frequency Distributions: Theory and Applications
Moeness G. Amin, Yimin Zhang, Gordon J. Frazer, and Alan R. Lindsey **269**
9.1 Introduction ... 269
9.2 STFDs ... 270
9.2.1 Signal model ... 270
9.2.2 STFDs ... 271
9.2.3 Joint diagonalization and time-frequency averaging ... 272
9.3 Properties of STFDs ... 273
9.3.1 Subspace analysis for FM signals ... 274
9.3.2 Signal-to-noise ratio enhancement ... 276
9.3.3 Signal and noise subspaces using STFDs ... 278
9.4 Blind source separation ... 280
9.4.1 Source separation based on STFDs ... 280
9.4.2 Source separation based on spatial averaging ... 281
9.4.3 Effect of cross-terms between source signals ... 284
9.4.4 Source separation based on joint diagonalization and joint anti-diagonalization ... 286
9.5 Direction finding ... 288
9.5.1 Time-frequency MUSIC ... 288
9.5.2 Time-frequency maximum likelihood method ... 290
9.5.3 Effect of cross-terms ... 292
9.6 Spatial ambiguity functions ... 296
9.7 Sensor-angle distributions ... 297
9.7.1 Signal model ... 298
9.7.2 SADs ... 301
9.7.3 Characterizing local scatter ... 302
9.7.4 Simulations and examples ... 302
9.8 Conclusion ... 307
References ... 308
10 Time-Frequency Processing of Time-Varying Signals with Nonlinear Group Delay
Antonia Papandreou-Suppappola **311**
10.1 Introduction ... 311
10.2 Constant time-shift covariant QTFRs ... 313
10.2.1 Cohen's class ... 314
10.2.2 Affine class ... 315
10.3 Group delay shift covariance property ... 316
10.3.1 Definition and special cases ... 316
10.3.2 QTFR time shift and signal group delay ... 319
10.3.3 Generalized impulse and signal expansion ... 319
10.4 Group delay shift covariant QTFRs ... 322
10.4.1 Dispersively warped Cohen's class ... 323
10.4.2 Dispersively warped affine class ... 330
10.4.3 Other dispersively warped classes ... 336
10.4.4 Group delay shift covariant intersections ... 339

10.5 Simulation examples ... 342
10.5.1 Impulse example ... 343
10.5.2 Hyperbolic impulse example ... 343
10.5.3 Power impulse example ... 346
10.5.4 Power exponential impulse example ... 347
10.5.5 Hyperbolic impulse and Gaussian example ... 350
10.5.6 Impulse response of a steel beam example ... 351
10.6 Conclusion ... 353
References ... 354

11 Self-Similarity and Intermittency
Albert Benassi, Serge Cohen, Sébastien Deguy, and Jacques Istas **361**
11.1 Introduction ... 361
11.2 Wavelet decomposition of self-similar Gaussian processes ... 362
11.2.1 Self-similar Gaussian processes ... 362
11.2.2 Lemarié–Meyer basis ... 364
11.2.3 Wavelet decomposition ... 364
11.2.4 Estimation of H ... 364
11.2.5 A generalized model ... 365
11.3 Intermittency for stochastic processes ... 366
11.3.1 Encoding the dyadic cells of $\mathbb{R}^d$ with a tree ... 366
11.3.2 Intermittency ... 368
11.3.3 Geometry of trees ... 368
11.3.4 Model and main result ... 368
11.4 Appendix ... 370
11.4.1 Simulations ... 370
11.4.2 Estimations ... 371
References ... 374

12 Selective Thresholding in Wavelet Image Compression
Prashant Sansgiry and Ioana Mihaila **377**
12.1 Introduction ... 377
12.2 Selective thresholding ... 378
12.3 Results ... 379
12.4 Conclusions ... 381
References ... 381

13 Coherent Vortex Simulation of Two- and Three-Dimensional Turbulent Flows Using Orthogonal Wavelets
Alexandre Azzalini, Giulio Pellegrino, and Jörg Ziuber **383**
13.1 Introduction ... 383
13.2 Wavelet filtering for coherent vortex extraction ... 385
13.2.1 Wavelet projection ... 385
13.2.2 Multiresolution analysis ... 387
13.2.3 Donoho's theorem ... 388
13.2.4 Iterative threshold estimation ... 391
13.3 Application of wavelet filtering to 2D and 3D data fields ... 396
13.3.1 Artificial academic 2D field ... 396

13.3.2 2D turbulent field ... 399
13.3.3 Wavelet filtering of 3D turbulent flows ... 403
13.4 CVS of evolving turbulent flows ... 411
13.4.1 Implementation of CVS: adaptive basis selection ... 411
13.4.2 Simulation of 2D turbulent flows ... 414
13.4.3 Simulation of 3D flows ... 422
13.5 Summary and conclusion ... 428
References ... 431
Index **433**

Preface

The mathematical theory of wavelet analysis and its applications have been a rich area of research over the past twenty years. Nowadays, wavelet analysis is an integral part of the solution to difficult problems in fields as diverse as signal processing, computer vision, data compression, pattern recognition, image processing, computer graphics, medical imaging, and defense technology. In recognition of the many new and exciting results in wavelet analysis and these related disciplines, this volume consists of invited articles by mathematicians and engineering scientists who have made important contributions to the study of wavelets, wavelet transforms, time-frequency signal analysis, turbulence, and signal and image processing. Like its predecessor *Wavelet Transforms and Time-Frequency Signal Analysis*, (Birkhäuser, 0-8176-4104-1), this work emphasizes both theory and practice, and provides directions for future research.

Given that the literature in wavelet analysis must continually change to accompany the discovery and implementation of new ideas and techniques, it has become increasingly important for a broad audience of researchers to keep pace with new developments in the subject. Much of the latest research, however, is widely scattered across journals and books and is often inaccessible or difficult to locate. This work addresses that difficulty by bringing together original research articles in key areas, as well as expository/survey articles, open questions, unsolved problems, and updated bibliographies. Designed for pure and applied mathematicians, engineers, physicists, and graduate students, this reference text brings these diverse groups to the forefront of current research.

The book is divided into two main parts: Part I, on Wavelets and Wavelet Transforms; and Part II, on Signal Processing and Time-Frequency Signal Analysis. The opening chapter by Leon Cohen deals with a general method for calculating relevant time-frequency moments for the wavelet transform. Explicit results are given for the time, frequency, and scale moments. It is shown that the wavelet transform has unique characteristics that are not possessed by previous known methods. Special attention is given to the fact as to whether these unusual characteristics are mathematically or physically important. Exactly solvable examples are given and the results are contrasted with the standard methods, such as the spectrogram and the Wigner–Ville distribution.

In Chapter 2, Carlos A. Cabrelli, Christopher Heil, and Ursula M. Molter outline some recent developments in the construction of higher-dimensional wavelet bases exploiting the fact that the refinement equation is a statement that the scaling function satisfies some kind of self-similarity. Special focus is on the notion of self-similarity, as applied to wavelet theory, so that wavelets associated with a multiresolution analysis of R^n allow arbitrary dilation matrices without any restrictions on the number of scaling functions.

Chapter 3, by J.D. Lakey and M.C. Pereyra, deals with the nonexistence of biorthogonal multiwavelets having some regularity, such that both biorthogonal families have compactly supported and divergence-free generators. The authors generalize Lamarie's bivariate result based on vector-valued multiresolution analyses.

Osiris wavelets and Isis variables are the topics covered in Chapter 4 by Guy Battle. Based on his new hierarchical approximations, the author discusses the kinematic coupling of wavelet amplitudes which arises in various models based on Osiris wavelets. The variables associated with different length scales are no longer independent with respect to the massless continuum Gaussian. This creates undefined Gaussian states that are fixed with respect to the renormalization group iteration—in addition to the usual Gaussian fixed points which influence the flow of the iteration. This is followed by a derivation of the recursion formula—and its linearization—for dipole perturbations of the Gaussian fixed point. Such a formula is best described by Isis variables.

In Chapter 5, R. Chandramouli and K.M. Ramachandran discuss two important application of wavelets: (a) signal estimation and (b) signal detection. Several wavelets based on signal estimator/denoising schemes are reviewed. This is followed by a new signal estimator which attempts to estimate and preserve some moments of the underlying original signal. A relationship between signal estimation and data compression is also discussed. Analytical and experimental results are presented to explain the performance of these signal estimation schemes. A wavelet decomposition level-dependent binary hypothesis test is investigated for signal detection. Outputs of likelihood ratio test-based local detectors are combined by a fusion rule to obtain a global decision about the true hypothesis. Due to the combinatorial nature of this detection fusion problem, two specific global detection fusion rules—OR and AND—are discussed in some detail. A method to compute the global thresholds and the asymptotic properties of the hypothesis test is also derived.

Part II deals with the study of signal processing and time-frequency signal analysis. It begins with Chapter 6 by Boualem Boashash and Victor Sucic who present three interrelated sections dealing with key concepts and techniques needed to design and use high performance time-frequency distributions (TFDs) in real-world applications. They discuss the core concepts of time-frequency signal processing (TFSP) in recent developments, including the design of high resolution quadratic TFDs for multicomponent signal analysis. This is followed by methods of assessment for the performance of time-frequency techniques in terms of the resolution performance of TFDs. Finally, a methodology for selecting the most suitable TFD for a given real-life signal is given.

Chapter 7 by Franz Hlawatsch, Georg Tauböck and Teresa Twaroch is concerned with a general theory of linear and bilinear quadratic time-frequency (TF) representations based on the covariance principle. The covariance theory provides a unified framework for important TF representations. It yields a theoretical basis for TF analysis and allows for the systematic construction of covariant TF representations. Special attention is given to unitary and projective group representations that provide mathematical models for displacement operators. This is followed by a discussion of important classes of operator families which include modulation and warping operators as well as pairs of dual and affine operators. The results of the covariance theory are then applied to these operator classes. Also studied are the fundamental properties of the displacement function connecting the group domain and the TF domain with their far-reaching consequences. A systematic method for constructing the displacement function is also presented. Finally, the characteristic function method is considered for construction of bilinear TF representations. It is shown that for dual operators the characteristic function method is equivalent to the covariance method.

In Chapter 8, Eric Chassande-Mottin, Francois Auger and Patrick Flandrin introduce the idea of reassignment in order to move computed values of time-frequency (or time-scale) distributions to increase their localization properties. Attention is given to the practical implementation of the reassignment operators and to the resulting computational effectiveness with analytical examples of both chirp-like signals and isolated singularities. An in-depth study of (spectrogram) reassignment vector fields is presented from both statistical and geometric viewpoints. Two variations of the initial method, including supervised reassignment and differential reassignment, are discussed. This is followed by two examples of applications of reassignment methods to signal processing. The first one is devoted to the partitioning of the time-frequency plane, with time-varying filtering, denoising and signal components extraction as natural byproducts. The second example covers chirp detection with special emphasis on power-law chirps, which is associated with gravitational waves emitted from the coalescence of astrophysical compact bodies.

Chapter 9, by Moeness G. Amin, Yimin Zhang, Gordon J. Frazer and Alan R. Lindsey, is devoted to a comprehensive treatment of the hybrid area of time-frequency distribution (TFD) and array signal processing. Two basic formulations are presented to incorporate the spatial information into quadratic distributions. Special attention is given to applications of these formulations with advances made through time-frequency analysis in the direction-of-arrival estimation, signal synthesis, and near-field source characterization.

In Chapter 10, A. Papandreou-Suppappola discusses quadratic time-frequency representations (QTFRs) preserving constant and dispersive time shifts. Under study are QTFR classes that always preserve nonlinear, frequency-dependent time shifts on the analysis signal which may be the result of a signal propagating through systems with dispersive time-frequency characteristics. It is shown that these generalized time-shift covariant QTFR classes are related to the constant time-shift covariant QTFR classes through a unitary warping transformation that is fixed by the dispersive system characteristics. Various examples to demonstrate the importance of matching the group delay of the analysis signal with the generalized time shift of the QTFR are included.

In Chapter 11, a class of stochastic processes with an extended self-similarity property and intermittency is discussed by Albert Benassi, Serge Cohen, Sebastian Deguy, and Jacques Istas. Presented are several simulations of such processes as well as the estimation of their characteristic parameters.

Selective thresholding in wavelet image compression is the topic of Chapter 12 by Prashant Sansgiry and Ioana Mihaila. Image compression can be achieved by using several types of wavelets. Lossy compression is obtained by applying various levels of thresholds to the compressed matrix of the image. The authors explore non-uniform ways to threshold this matrix and discuss the effects of these thresholding techniques on the quality of the image and the compression ratio.

Chapter 13 is devoted to coherent vortex simulation (CVS) of two- and three-dimensional turbulent flows based on the orthogonal wavelet decomposition of the vorticity field which can be highly compressed using wavelet filters. Direct numerical simulations (DNS) using CVS are carried out for turbulent flows that deal with the integration of the Navier–Stokes equations in a compressed adaptive basis. In order to check the potential of the CVS approach, DNS and CVS are compared in both two and three dimensions. This leads to an analysis of the effect of the filtering procedure onto the dynamical evolution of the flow.

I express my grateful thanks to the authors for their cooperation and their excellent chapters. Some of these chapters grew out of lectures originally presented at the NSF-CBMS conference at the University of Central Florida in May 1998. I hope the reader will share in the excitement of present-day research and become stimulated to explore wavelets, wavelet transforms, and their diverse applications. I also hope this volume will not only generate new, useful leads for those engaged in advanced study and research, but will also attract newcomers to these fields. Finally, I would like to thank Tom Grasso and the staff of Birkhäuser Boston for their constant help throughout the many stages of publishing this volume.

Lokenath Debnath
University of Texas-Pan American
Edinburg, Texas

Contributors

Moeness G. Amin
Center for Advanced Communications
Villanova University
Villanova, PA, USA
E-mail: moeness@ece.vill.edu

Francois Auger
C.R.T.T.
Boulevard de l'Université
Boite postale 406
44602 Saint-Nazaire Cedex, France
E-mail: auger@ge44.univ-nantes.fr

Alexandre Azzalini
Laboratoire de Météorologie Dynamique du CNRS
Ecole Normale Supérieure
24, rue Lhomond
F-75231 Paris Cedex 9, France
E-mail: azzalini@lmd.ens.fr

Guy Battle
Department of Mathematics
Texas A&M University
College Station, TX 77843, USA
E-mail: battle@math.tamu.edu

Albert Benassi
Université de Blaise Pascal
(Clermont–Ferrand II)
LAMP, CNRS UPRESA 6016
63177 Aubiera Cedex, France
E-mail: benassi@opgc.univ-bpclermont.fr

Boualem Boashash
Signal Processing Research Centre
School of Electrical Engineering
Queensland University of Technology
2 George St., Brisbane QLD 4000, Australia
E-mail: b.boashash@qut.edu.au

Carlos A. Cabrelli
Departamento de Matemática
Facultad de Ciencias Exactas y Naturales
Universidad de Buenos Aires
Cuidad Universitaria
Pabellón I
1428 Capital Federal
Buenos Aires, Argentina
E-mail: ccabrelli@dm.uba.ar

R. Chandramouli
207 Burchard Building
Department of Electrical and Computer Engineering
Stevens Institute of Technology
Hoboken, NJ 07030, USA
E-mail: mouli@stevens-tech.edu

Eric Chassande-Mottin
Albert-Einstein-Institut, MPI für Grav. Physik
Am Mühlenberg, 5
D-14476 Golm, Germany
E-mail: ecm@obs-nice.fr

Leon Cohen
Department of Physics and Astronomy
Hunter College
City University of New York
695 Park Avenue
New York, NY 10021, USA
E-mail: leon.cohen@hunter.cuny.edu

Serge Cohen
Laboratoire de Statistique et de Probabilities
Université Paul Sabatier
118 route de Noarbonne
31062 Toulouse Cedex, France
E-mail: Serge.cohen@math.ups-tlse.fr

Sébastien Deguy
Université Blaise Pascal (Clermont-Ferrand II)
Laboratoire de Logique
Algorithmique et Informatique de Clermont 1
63177 Aubere Cedex, France
E-mail: deguy@llaic.u-clermont1.fr

Patrick Flandrin
Ecole Normale Supérieure de Lyon
46 Allée d'Italie
69364 Lyon Cedex 07, France
E-mail: flandrin@ens-lyon.fr

Gordon J. Frazer
Surveillance Systems Division
Defence Science and Technology Organisation
P.O. Box 1500
Edinburgh, SA 5111, Australia
E-mail: frazer@ieee.org

Christopher E. Heil
School of Mathematics
Georgia Institute of Technology
Atlanta, GA 30332-0160, USA
E-mail: heil@math.gatech.edu

Franz Hlawatsch
Institut für Nachrichtentechnik und Hochfrequenztechnik
Technische Universität Wien
Gusshausstrasse 25/389
A- 1040 Vienna, Austria
E-mail: fhlawats@pop.tuwien.ac.at

Jacques Istas
Département IMSS–BSHM
Université Pierre Mendes–France
38000 Grenoble, France
E-mail: Jacques.istas@upmf-grenoble.fr

J. D. Lakey
Department of Mathematical Sciences
New Mexico State University
Las Cruces, NM 88003-8001, USA
E-mail: jlakey@nmsu.edu

Alan R. Lindsey
Air Force Research Laboratory
525 Brooks Road
Rome, NY 13441, USA
E-mail: alan.lindsey@rl.af.mil

Ursula M. Molter
Departamento de Matemática
Facultad de Ciencias Exactas y Naturales
Universidad de Buenos Aires
Cuidad Universitaria
Pabellón I
1428 Capital Federal
Buenos Aires, Argentina
E-mail: umolter@dm.uba.ar

Ioana Mihaila
Department of Mathematics
California State Polytechnic University, Pomona
Pomona, CA 91768, USA
Email: imihaila@csupomona.edu

Antonia Papandreou-Suppappola
Department of Electrical Engineering
Arizona State University
P.O. Box 877206
Tempe, AZ 85287-7206, USA
E-mail: papandreou@asu.edu

Giulio Pellegrino
Laboratoire de Modélisation et Simulation Numérique en Mécanique
CNRS et Université d'Aix-Marseille
38, rue F. Joliot-Curie
F-13451 Marseille Cedex 13, France
Email: pellegri@13m.univ-mrs.fr

M. C. Pereyra
Department of Mathematics and Statistics
University of New Mexico
Albuquerque, NM 87131-1141, USA
E-mail: crisp@math.unm.edu

K. M. Ramachandran
Department of Mathematics
University of South Florida
Tampa, FL 33620-5700, USA
E-mail: ram@chuma.cas.usf.edu

Prashant Sansgiry
Department of Mathematics and Statistics
Coastal Carolina University
Conway, SC 29526, USA
E-mail: sansgirp@coastal.edu

Victor Sucic
Signal Processing Research Centre
School of Electrical Engineering
Queensland University of Technology
2 George St., Brisbane QLD 4000, Australia
E-mail: v. sucic@qut.edu.au

Georg Tauböck
Forschungszentrum Telekommunikation Wien
Tech Gate Vienna
Donau-City-Straße 1
A-1220 Wien, Austria
E-mail: tauboeck@ftw.at

Teresa Twaroch
Institut für Analysis und Technische Mathematik
Vienna University of Technology
Wiedner Hauptstrasse 8-10
A-1040 Vienna, Austria
E-mail: teresa.twaroch@ci.tuwien.ac.at

Yimin Zhang
Center for Advanced Communications
Villanova University
Villanova, PA, 19085-1603, USA
E-mail: yimin@ieee.org

Jörg Ziuber
Institut für Chemische Technik
Universität Karlsruhe
Kaiserstrasse 12
D-76128 Karlsruhe, Germany
E-mail: ziuber@ict.uni-karlsruhe.de

Wavelets and Signal Processing

Part I

Wavelets and Wavelet Transforms

1

The Wavelet Transform and Time-Frequency Analysis

Leon Cohen

ABSTRACT We present a general method for calculating relevant moments of the wavelet transform. Explicit results are given for the time, frequency, and scale moments. Using the results obtained, we show that the wavelet transform has unique characteristics that are not possessed by other methods. We discuss whether these unusual characteristics are physically or mathematically important. Exactly solvable examples are given, and the results are contrasted to those of the standard methods such as the spectrogram and the Wigner distribution.

1.1 Introduction

The aim of time-frequency analysis is to understand from both a physical and mathematical point of view how the spectral properties of a signal change with time [1]–[3]. While many methods have been proposed and developed, no complete general theory exists. With the advent of wavelets, there have been suggestions that their study may shed some light on the time-frequency properties of signals and may lead toward a useful theory. However, very few relevant mathematical calculations have been reported that are aimed toward understanding the time-frequency structure of signals, as most of the results have been of a pictorial nature. The pictures are curious and generally unlike the ones usually obtained using conventional methods of time-frequency analysis such as the spectrogram or the Wigner distribution. These unusual results may be good or bad: good if indeed we are learning something new, bad if the results are confusing and of no physical or mathematical relevance. Also, we point out that many early papers developed the wavelet approach in isolation, in that they did not compare the wavelet approach with the standard ones or calculate the relevant physical quantity predicted by the wavelet approach. Recently, there has been a more balanced presentation, for example, in the book by Mallat [4], where one of the chapters presents nonwavelet-based methods, like the Wigner distribution [5] and Choi–Williams distribution [6], [7].

The wavelet approach may be seen as a method involving variable windowed spectrograms which has a long history [8]. Jeong and Williams [9], [10] and Williams et al. [11] devised a general approach to variable window spectrograms and showed that the wavelet transform is a special case. Also, Rioul and Flandrin derived interesting relationships with other time-frequency representations [12]. However, the wavelet transform has a particular mathematical form, and this form forces certain results that differ from other methods.

In this article, we develop some of the mathematical concepts needed to understand the contribution of the wavelet approach to time-frequency analysis. In particular, we explicitly calculate simple moments. We point out that the answers to these simple moments are known, since they depend only on the signal or spectrum, and therefore we have something to compare to the results of the wavelet approach and other approaches. We also emphasize that while the wavelet transform has found many applications in almost all aspects of science, we limit ourselves to its possible application to time-frequency analysis [4], [13]– [20].

We use the standard definition of the wavelet transform,

$$WT(t,a) = \frac{1}{\sqrt{a}} \int s(u) \psi^* \left(\frac{u-t}{a} \right) du \qquad (1.1.1)$$

$$= \sqrt{a} \int e^{j\omega t} \hat{s}(\omega) \hat{\psi}^*(a\omega) d\omega \qquad (1.1.2)$$

where $s(t)$ is the signal and $\psi(t)$ is the wavelet and where $\hat{s}(\omega)$ and $\hat{\psi}(\omega)$ are, respectively, the Fourier transforms

$$\hat{s}(\omega) = \frac{1}{\sqrt{2\pi}} \int s(t) e^{-j\omega t} dt \qquad (1.1.3)$$

$$\hat{\psi}(\omega) = \frac{1}{\sqrt{2\pi}} \int \psi(t) e^{-j\omega t} dt \qquad (1.1.4)$$

1.2 The wavelet approach to time-frequency analysis

The wavelet transform is not a function of time and frequency but time and "a" and hence one must somehow relate "a" to frequency. What is typically done is to first define a joint density of time and "a", $P_{WT}(t,a)$, which is called the scalogram,

$$P_{WT}(t,a) = \frac{1}{2\pi C a^2} |WT(t,a)|^2. \qquad (1.2.1)$$

The factor $1/2\pi C a^2$ is inserted for proper normalization. The constant C is chosen so that the energy obtained by using the wavelet transform is the same as the energy of the signal,

$$\int |s(t)|^2 dt = \int \int P_{WT}(t,a) dt da = 1. \qquad (1.2.2)$$

Imposing Eq. (1.2.2) one obtains that [12], [13]

$$C = \int \frac{|\hat{\psi}(\omega)|^2}{|\omega|} d\omega. \tag{1.2.3}$$

It is very interesting that this constant is the same constant that is usually called the admissibility condition. The admissibility condition is a requirement that one may uniquely obtain the signal from the wavelet transform.

1.2.1 *Density of time frequency*

To obtain a time-frequency density, one chooses a reference frequency ω_r and defines the frequency, ω, from

$$a = \frac{\omega_r}{\omega} \qquad \omega = \frac{\omega_r}{a}. \tag{1.2.4}$$

The density of time and frequency, $P_{WT}(t, \omega)$, is then obtained from

$$P(t, \omega) d\omega dt = P(t, a) da dt. \tag{1.2.5}$$

Using

$$da = -\frac{\omega_r}{\omega^2} d\omega = -\frac{a^2}{\omega_r} d\omega \qquad \frac{da}{|a|} = \frac{d\omega}{|\omega|}, \tag{1.2.6}$$

we have

$$P_{WT}(t, \omega) d\omega dt = P_{WT}(t, a = \omega_r/\omega) \frac{\omega_r}{\omega^2} d\omega dt \tag{1.2.7}$$

and therefore

$$P_{WT}(t, \omega) = \frac{\omega_r}{\omega^2} P_{WT}(t, a = \omega_r/\omega) \tag{1.2.8}$$

$$= \frac{1}{2\pi C \omega_r} |WT(t, \omega_r/\omega)|^2. \tag{1.2.9}$$

1.2.2 *Marginals*

From the joint distribution one obtains the individual densities of time and frequency by integrating over the other variables. The resulting densities are called the marginals. In the case of time-frequency analysis we know what the marginals should be, namely, the signal and Fourier transform squared, respectively. That is, if we have a joint density $P(t, \omega)$, ideally one should have

$$P(t) = \int P(t, \omega) d\omega = |s(t)|^2 \tag{1.2.10}$$

$$P(\omega) = \int P(t,\omega)dt = |\hat{s}(\omega)|^2. \qquad (1.2.11)$$

Also, if Eq. (1.2.11) is taken as the marginal for frequency, then the marginal of a is given by

$$P(a) = P(\omega)\frac{d\omega}{da}. \qquad (1.2.12)$$

These are the correct marginals, and we point out that the Wigner distribution satisfies them exactly. Let us now see what they are for the wavelet transform. For the time marginal, we have

$$P(t) = \int P_{WT}(t,a)da \qquad (1.2.13)$$

$$= \frac{1}{2\pi C}\int\int \frac{1}{|a|} e^{j(\omega-\omega')t}\hat{s}(\omega)\hat{s}^*(\omega')\hat{\psi}^*(\omega a)\hat{\psi}(\omega' a)d\omega d\omega' da$$

$$= \frac{1}{2\pi C}\int\int e^{j(\omega-\omega')t}\hat{s}(\omega)\hat{s}^*(\omega')L(\omega,\omega')d\omega d\omega' \qquad (1.2.14)$$

with

$$L(\omega,\omega') = \int \frac{\hat{\psi}^*(\omega a)\hat{\psi}(\omega' a)}{|a|}da. \qquad (1.2.15)$$

Note that

$$L(\omega,\omega) = C. \qquad (1.2.16)$$

The frequency marginal is

$$P_{WT}(\omega) = \int P_{WT}(t,\omega)dt \qquad (1.2.17)$$

$$= \frac{1}{|\omega| C}\int |\hat{s}(\omega')|^2 \left|\hat{\psi}\left(\frac{\omega_r}{\omega}\omega'\right)\right|^2 d\omega'. \qquad (1.2.18)$$

The marginal for scale a is then

$$P_{WT}(a) = \int P_{WT}(t,a)dt \qquad (1.2.19)$$

$$= \frac{1}{|a| C}\int |\hat{s}(\omega)|^2 |\hat{\psi}(\omega a)|^2 d\omega. \qquad (1.2.20)$$

Thus, we see that the marginals are combinations of the signal and wavelet.

1.2.3 Moments

In the appendix, we explicitly derive the global moments for time, frequency, and scale using the wavelet approach. Here we just discuss the low-order moments and, in particular, the mean time and frequency. We also contrast the results with other distributions. We first point out that for all distributions the mean time, frequency, duration, and bandwidth are defined in the same way, namely,

$$\langle t \rangle = \int\int tP(t,\omega)dtd\omega \tag{1.2.21}$$

$$\langle \omega \rangle = \int\int \omega P(t,\omega)dtd\omega \tag{1.2.22}$$

$$T^2 = \int\int (t - \langle t \rangle)^2 P(t,\omega)dtd\omega \tag{1.2.23}$$

$$B^2 = \int\int (\omega - \langle \omega \rangle)^2 P(t,\omega)dtd\omega. \tag{1.2.24}$$

The correct answers for these quantities are

$$\langle t \rangle = \int t|s(t)|^2 dt \tag{1.2.25}$$

$$\langle \omega \rangle = \int \omega \left|S(\omega)\right|^2 d\omega \tag{1.2.26}$$

$$T^2 = \int (t - \langle t \rangle)^2 |s(t)|^2 dt \tag{1.2.27}$$

$$B^2 = \int (\omega - \langle \omega \rangle)^2 \left|S(\omega)\right|^2 d\omega. \tag{1.2.28}$$

As we will be contrasting the wavelet approach with the spectrogram and Wigner distribution, we give them here. For the spectrogram, we denote the window by $h(t)$. The short-time Fourier transform, $S_t(\omega)$, is

$$S_t(\omega) = \frac{1}{\sqrt{2\pi}} \int e^{-j\omega\tau} s(\tau)h(\tau - t)d\tau, \tag{1.2.29}$$

and the spectrogram is

$$P(t,\omega) = \left|S_t(\omega)\right|^2, \tag{1.2.30}$$

which is a two-dimensional density in time and frequency.

The Wigner distribution is

$$W(t,\omega) = \frac{1}{2\pi} \int s^* \left(t - \frac{1}{2}\tau\right) s\left(t + \frac{1}{2}\tau\right) e^{-j\tau\omega} d\tau. \tag{1.2.31}$$

1.2.4 *Mean time and mean frequency*

From the appendix, the mean time using wavelets is given by

$$\langle t \rangle_{WT} = \langle t \rangle_s - \frac{1}{C} \left\langle \frac{1}{\omega} \right\rangle_s \langle t \rangle_\psi, \tag{1.2.32}$$

where

$$\langle t \rangle_s = \int t|s(t)|^2 dt \tag{1.2.33}$$

$$\left\langle \frac{1}{\omega} \right\rangle_s = \int \frac{1}{\omega} |\hat{s}(\omega)|^2 d\omega \tag{1.2.34}$$

$$\langle t \rangle_\psi = \int t|\psi(t)|^2 dt. \tag{1.2.35}$$

The correct answer is the first term in Eq. (1.2.32) and, is given by many of the distributions that have been developed over the past 60 years. That is, for many distributions, e.g., the Wigner distribution, one has that

$$\langle t \rangle_{\text{Wigner}} = \langle t \rangle_s \,. \tag{1.2.36}$$

The spectrogram, which is perhaps the most used time-frequency distribution, does not satisfy Eq. (1.2.33), and therefore it is interesting to contrast the wavelet approach to the spectrogram. The first moment in time for the spectrogram is given by

$$\langle t \rangle = \langle t \rangle_s - \langle t \rangle_h. \tag{1.2.37}$$

That is, the mean time for the spectrogram is the mean time of the signal minus the mean time of the window. Now let us contrast the two results. While the spectrogram does not give the exact answer, it gives an answer that is readily understood and hence can be taken into account. Note that for the mean time it involves only time quantities of the signal and window. Also, if we take a window such that $\langle t \rangle_h = 0$, then we get the correct answer. On the other hand, the wavelet transform approach involves frequency, more specifically, the inverse frequency of the signal. Therefore, the wavelet transform approach mixes time and frequency properties of the signal even though we are only considering a time quantity. For the spectrogram, the mean time will exist if the mean time of the signal and window exist. However, for the wavelet transform, the mean time does not exist if the average of $1/\omega$ of the signal does not exist. This is "unphysical" in that there are many signals that clearly have a reasonable mean time and yet for the wavelet approach the answer would be that they do not exist!

For the mean frequency, one has that

$$\langle \omega \rangle = \frac{\omega_r}{C} \langle \omega \rangle_s \left\langle \frac{1}{\omega|\omega|} \right\rangle_\psi . \tag{1.2.38}$$

Thus, for the mean frequency to exist, $\left\langle \frac{1}{\omega|\omega|} \right\rangle_\psi$ must exist. In contrast, for the spectrogram, we have that

$$\langle \omega \rangle = \langle \omega \rangle_{(s)} + \langle \omega \rangle_{(h)}, \tag{1.2.39}$$

which again is easily understood and shows that only the moment of the window has to exist. Also, if we chose a window so that $\langle \omega \rangle_{(h)} = 0$, then we get the right answer. Notice also that the properties of the window are additive, while in the wavelet approach they are multiplicative.

1.2.5 Conditional moments

Important moments in time-frequency analysis are the conditional moments. The reason for that is the following. Suppose we consider the average frequency defined in the usual way

$$\langle \omega \rangle = \int \omega |S(\omega)|^2 d\omega \tag{1.2.40}$$

and write it in terms of the phase and amplitude of the signal,

$$s(t) = A(t)e^{j\varphi(t)}. \tag{1.2.41}$$

One obtains that

$$\langle \omega \rangle = \int \varphi'(t) A^2(t) dt, \tag{1.2.42}$$

which says that the average frequency may be obtained by integrating "something" with the density over all time. Hence, the instantaneous frequency, the frequency at each moment, $\omega_i(t)$, is

$$\omega_i(t) = \varphi'(t). \tag{1.2.43}$$

Now, it was Ville who argued that instantaneous frequency should be the first conditional moment of frequency for a given time. The conditional distribution of frequency for a given time is

$$P(\omega \mid t) = \frac{P(t,\omega)}{P(t)}, \tag{1.2.44}$$

where $P(t)$ is the marginal distribution,

$$P(t) = \int P(t,\omega) d\omega. \tag{1.2.45}$$

We have used $P(t)$ rather than $|s(t)|^2$ to allow for the possibility that the marginals are not satisfied. Now we would like the conditional frequency to be the derivative of the phase,

$$\langle \omega \rangle_t = \frac{1}{P(t)} \int \omega P(t,\omega) d\omega = \varphi'(t). \tag{1.2.46}$$

Are there distributions that satisfy Eq. (1.2.46)? Yes, there are an infinite number, and the Wigner distribution is one of them,

$$\frac{1}{|s(t)|^2} \int \omega W(t,\omega) d\omega = \varphi'(t). \tag{1.2.47}$$

However, the spectrogram is not one of them. For the spectrogram,

$$\langle\omega\rangle_t = \frac{1}{P(t)} \int A^2(\tau) A_h^2(\tau - t)\varphi'(\tau) d\tau, \tag{1.2.48}$$

where $A_h(t)$ is the window function which we take to be real. Even though the right-hand side does not equal $\varphi'(t)$, it does approach it in some sense. If we narrow the window so that $A_h^2(t)$ approaches a delta function,

$$A_h^2(t) \to \delta(t), \tag{1.2.49}$$

then

$$\langle\omega\rangle_t \to \varphi'(t). \tag{1.2.50}$$

That is, if the window is narrowed to get increasing time resolution, the limiting value of the estimated instantaneous frequency is the derivative of the phase of the signal.

Let us now see what one obtains for the wavelet transform. First consider the conditional average of the scale, a,

$$\langle a\rangle_t = \frac{1}{P(t)} \int aP(t,a) da. \tag{1.2.51}$$

We have

$$\langle a\rangle_t = \frac{1}{P(t)} \int a \frac{1}{2\pi C a^2} |WT(t,a)|^2 da \tag{1.2.52}$$

$$= \frac{1}{P(t)} \frac{1}{2\pi C} \int\int e^{j(\omega-\omega')t} \hat{s}(\omega)\hat{s}^*(\omega')\hat{\psi}^*(\omega a)\hat{\psi}(\omega' a) d\omega d\omega' da$$

$$= \frac{1}{P(t)} \frac{1}{2\pi C} \int\int e^{j(\omega-\omega')t} \hat{s}(\omega)\hat{s}^*(\omega') L_1(\omega,\omega') d\omega d\omega' \tag{1.2.53}$$

with

$$L_1(\omega,\omega') = \int \hat{\psi}^*(\omega a)\hat{\psi}(\omega' a) da. \tag{1.2.54}$$

Similarly,

$$\langle 1/a\rangle_t = \frac{1}{2\pi C} \int\int e^{j(\omega-\omega')t} \hat{s}(\omega)\hat{s}^*(\omega') L_2(\omega,\omega') d\omega d\omega', \tag{1.2.55}$$

where

$$L_2(\omega,\omega') = \int \frac{\hat{\psi}^*(\omega a)\hat{\psi}(\omega' a)}{a^2} da. \tag{1.2.56}$$

Now consider the conditional frequency

$$\langle\omega\rangle_t = \frac{1}{P(t)} \int \omega P(t,\omega) d\omega \tag{1.2.57}$$

$$= \frac{1}{P(t)} \int \frac{\omega_r}{\omega} P\left(t, a = \frac{\omega_r}{\omega}\right) d\omega \tag{1.2.58}$$

$$= \frac{1}{P(t)} \int \frac{\omega_r}{a} P(t, a) da, \tag{1.2.59}$$

and therefore

$$\langle \omega \rangle_t = \omega_r \left\langle \frac{1}{a} \right\rangle_t, \tag{1.2.60}$$

where $\langle \frac{1}{a} \rangle_t$ is given by Eq. (1.2.55).

1.3 Examples

We now consider some exactly solvable problems. We use two wavelets, the Morlet wavelet and the modified Morlet wavelet. We have chosen these two wavelets because they are commonly used to calculate time-frequency properties of signals. We first discuss the two wavelets and their properties.

1.3.1 Morlet wavelet

The Morlet wavelet and its transform are

$$\psi(t) = A_\psi e^{-\eta t^2/2} \left[e^{j\omega_m t} - e^{-\omega_m^2/(2\eta)} \right] \tag{1.3.1}$$

$$\hat{\psi}(\omega) = \frac{A_\psi}{\sqrt{\eta}} \left[e^{-(\omega-\omega_m)^2/(2\eta)} - e^{-(\omega^2+\omega_m^2)/(2\eta)} \right], \tag{1.3.2}$$

where A_ψ is chosen so that

$$\int |\psi(t)|^2 dt = 1. \tag{1.3.3}$$

Explicitly,

$$A_\psi^2 = \sqrt{\frac{\eta}{\pi}} \frac{1}{1 + e^{-\omega_m^2/\eta} - 2e^{-3\omega_m^2/(4\eta)}}. \tag{1.3.4}$$

This wavelet satisfies

$$\int \psi(t) dt = 0, \tag{1.3.5}$$

a condition typically chosen in wavelet analysis so that the wavelet has "wave-like" properties, although for real waves, e.g., electromagnetic or acoustic ones, Eq. (1.3.5) has no meaning. The Morlet wavelet does not satisfy the admissibility condition: it does so approximately if ω_m is chosen appropriately, typically $\omega_m \sim$ 5–6.

1.3.2 Modified Morlet wavelet

The modified Morlet wavelet is

$$\psi(t) = A_\psi e^{-\eta t^2/2} e^{j\omega_m t} \tag{1.3.6}$$

$$\hat{\psi}(\omega) = \frac{A_\psi}{\sqrt{\eta}} e^{-(\omega-\omega_m)^2/(2\eta)} \tag{1.3.7}$$

with

$$A_\psi^2 = \sqrt{\frac{\eta}{\pi}} \tag{1.3.8}$$

This wavelet does not satisfy the admissibility condition but is nontheless commonly used. Sometimes this wavelet is called the "Gabor wavelet," but that term is improper because Gabor had nothing to do with wavelets. He was one of the great scientists of the century, having been awarded the Nobel prize for holography, and was one of the founders of time-frequency analysis [21].

1.3.3 Examples

For each example we just list the final results for the relevant quantities. As is often the case, nonnormalizable densities arise, and the way to use them is to only consider relative densities, that is, how intense one region is relative to another. Hence, in comparing ratios of densities, C, although infinite, "drops out" for those situations. Also, when the signal is not normalizable, then delta function normalization is used.

Example 1

$$s(t) = A_s e^{j\omega_0 t}; \quad \hat{s}(\omega) = \sqrt{2\pi} A_s \delta(\omega - \omega_0) \tag{1.3.9}$$

$$\psi(t) = A_\psi e^{-\eta t^2/2} \left[e^{j\omega_m t} - e^{-\omega_m^2/(2\eta)} \right]$$

$$\hat{\psi}(\omega) = \frac{A_\psi}{\sqrt{\eta}} \left[e^{-(\omega-\omega_m)^2/(2\eta)} - e^{-(\omega^2+\omega_m^2)/(2\eta)} \right] \tag{1.3.10}$$

$$\begin{aligned} WT(t,a) &= \sqrt{2\pi} A_s \sqrt{a} e^{j\omega_0 t} \hat{\psi}^*(a\omega_0) \qquad (1.3.11) \\ &= A_s A_\psi \sqrt{\frac{2\pi a}{\eta}} e^{j\omega_0 t} \left[e^{-(a\omega_0-\omega_m)^2/(2\eta)} - e^{-(a^2\omega_0^2+\omega_m^2)/(2\eta)} \right] \end{aligned}$$

$$|WT(t,a)|^2 = 2\pi A_s^2 A_\psi^2 \frac{|a|}{\eta} \left[e^{-(a\omega_0-\omega_m)^2/(2\eta)} - e^{-(a^2\omega_0^2+\omega_m^2)/(2\eta)} \right]^2 \tag{1.3.12}$$

$$\begin{aligned} |WT(t,\omega_r/\omega)|^2 &= 2\pi A_s^2 A_\psi^2 \frac{\omega_r}{\eta|\omega|} \\ &\quad \times \left[e^{-(\omega_r\omega_0/\omega-\omega_m)^2/(2\eta)} - e^{-(\omega_r^2\omega_0^2/\omega^2+\omega_m^2)/(2\eta)} \right]^2 \qquad (1.3.13) \end{aligned}$$

$$|WT(t,\omega_r/\omega)|^2 = 2\pi A_s^2 A_\psi^2 \frac{\omega_r}{\eta|\omega|} \times \left\{ \exp\left[-\omega_r^2 \frac{(\omega-\omega_0)^2}{2\eta\omega^2}\right] - \exp\left[-\omega_r^2 \frac{\omega^2+\omega_0^2}{2\eta\omega^2}\right]\right\}^2 \quad \text{for } \omega_r = \omega_m \qquad (1.3.14)$$

$$P_{WT}(t,a) = \frac{A_\psi^2 A_s^2}{C|a|\eta}\left\{ \exp\left[-\omega_r^2 \frac{(\omega-\omega_0)^2}{2\eta\omega^2}\right] - \exp\left[-\omega_r^2 \frac{\omega^2+\omega_0^2}{2\eta\omega^2}\right]\right\}^2 \qquad (1.3.15)$$

$$P_{WT}(t,\omega) = \frac{A_\psi^2 A_s^2}{C\eta|\omega|}\left[e^{-(\omega_r\omega_0/\omega-\omega_m)^2/(2\eta)} - e^{-(\omega_r^2\omega_0^2/\omega^2+\omega_m^2)/(2\eta)}\right]^2 \qquad (1.3.16)$$

Example 2

$$s(t) = A_s e^{j\omega_0 t}; \quad \hat{s}(\omega) = \sqrt{2\pi} A_s \delta(\omega-\omega_0) \qquad (1.3.17)$$

$$\psi(t) = A_\psi e^{-\eta t^2/2} e^{j\omega_m t}$$

$$\hat{\psi}(\omega) = \frac{A_\psi}{\sqrt{\eta}} e^{-(\omega-\omega_m)^2/(2\eta)} \qquad A_\psi^2 = \sqrt{\frac{\eta}{\pi}} \qquad (1.3.18)$$

$$WT(t,a) = \sqrt{2\pi}A_s\sqrt{2\pi}A_s\sqrt{a}e^{j\omega_0 t}\hat{\psi}^*(a\omega_0) \qquad (1.3.19)$$

$$= A_\psi\sqrt{\frac{a}{\eta}}e^{j\omega_0 t}e^{-(a\omega_0-\omega_m)^2/(2\eta)} \qquad (1.3.20)$$

$$|WT(t,a)|^2 = 2\pi A_s^2 A_\psi^2 \frac{a}{\eta} e^{-(a\omega_0-\omega_m)^2/\eta} \qquad (1.3.21)$$

$$|WT(t,\omega_r/\omega)|^2 = 2\pi A_s^2 A_\psi^2 \frac{\omega_r}{\eta\omega} e^{-(\omega_r\omega_0/\omega-\omega_m)^2/\eta} \qquad (1.3.22)$$

$$|WT(t,\omega_r/\omega)|^2 = 2\pi A_s^2 A_\psi^2 \frac{\omega_r}{\eta|\omega|} \exp\left[-\omega_r^2\frac{(\omega-\omega_0)^2}{\eta\omega^2}\right] \qquad \omega_r = \omega_m \qquad (1.3.23)$$

$$P_{WT}(t,a) = \frac{A_s^2 A_\psi^2}{Ca\eta} e^{-(a\omega_0-\omega_m)^2/\eta} \qquad (1.3.24)$$

$$P_{WT}(t,\omega) = \frac{A_\psi^2 A_s^2}{\eta|\omega|} \exp\left[-\omega_r^2\frac{(\omega-\omega_0)^2}{\eta\omega^2}\right] \qquad (1.3.25)$$

Example 3

$$s(t) = A_s\delta(t-t_0) \qquad \hat{s}(\omega) = \frac{A_s}{\sqrt{2\pi}} e^{-jt_0\omega} \qquad (1.3.26)$$

$$\psi(t) = A_\psi e^{-\eta t^2/2}\left[e^{j\omega_m t} - e^{-\omega_m^2/(2\eta)}\right]$$

$$\hat{\psi}(\omega) = \frac{A_\psi}{\sqrt{\eta}} \left[e^{-(\omega-\omega_m)^2/(2\eta)} - e^{-(\omega^2+\omega_m^2)/(2\eta)} \right] \tag{1.3.27}$$

$$WT(t,a) = \frac{A_s}{\sqrt{a}} \psi^*(\frac{t_0 - t}{a}) \tag{1.3.28}$$

$$= \frac{A_\psi A_s}{\sqrt{a}} e^{-\eta(t_0-t)^2/(2a^2)} \left[e^{-j\omega_m(t_0-t)/a} - e^{-\omega_m^2/(2\eta)} \right] \tag{1.3.29}$$

$$|WT(t,a)|^2 = \frac{A_\psi^2 A_s^2}{a} e^{-\eta(t_0-t)^2/a^2} \left| e^{-j\omega_m(t_0-t)/a} - e^{-\omega_m^2/(2\eta)} \right|^2 \tag{1.3.30}$$

$$WT(t,\omega_r/\omega) = A_\psi^2 A_s^2 \sqrt{\frac{\omega}{\omega_r}} e^{-\eta\omega^2(t_0-t)^2/(2\omega_r^2)} \times \left[e^{-j\omega_m\omega(t_0-t)/\omega_r} - e^{-\omega_m^2/(2\eta)} \right] \tag{1.3.31}$$

$$|WT(t,\omega_r/\omega)|^2 = \frac{A_\psi^2 A_s^2 |\omega|}{\omega_r} e^{-\eta\omega^2(t_0-t)^2/\omega_r^2} \left| e^{-j\omega_m\omega(t_0-t)/\omega_r} - e^{-\omega_m^2/(2\eta)} \right|^2$$

$$WT(t,\omega_r/\omega) = A_\psi A_s \sqrt{\frac{\omega}{\omega_r}} e^{-\eta\omega^2(t_0-t)^2/(2\omega_r^2)} \times \left[e^{-j\omega(t_0-t)} - e^{-\omega_r^2/(2\eta)} \right] \qquad \omega_r = \omega_m \tag{1.3.32}$$

$$|WT(t,\omega_r/\omega)|^2 = \frac{A_\psi^2 A_s^2 |\omega|}{\omega_r} e^{-\eta\omega^2(t_0-t)^2/\omega_r^2} \left| e^{-j\omega(t_0-t)} - e^{-\omega_r^2/(2\eta)} \right| \tag{1.3.33}$$

$$P_{WT}(t,a) = \frac{A_\psi^2 A_s^2}{2\pi C a^3} e^{-\eta(t_0-t)^2/a^2} \left| e^{-j\omega_m(t_0-t)/a} - e^{-\omega_m^2/(2\eta)} \right| \tag{1.3.34}$$

$$P_{WT}(t,\omega) = \frac{A_\psi^2 |\omega|}{2\pi C \omega_r^2} e^{-\eta\omega^2(t_0-t)^2/\omega_r^2} \left| e^{-j\omega(t_0-t)} - e^{-\omega_r^2/(2\eta)} \right|^2 \tag{1.3.35}$$

Example 4

$$s(t) = A_s \delta(t - t_0); \quad \hat{s}(\omega) = \frac{A_s}{\sqrt{2\pi}} e^{jt_0\omega} \tag{1.3.36}$$

$$\psi(t) = A_\psi e^{-\eta t^2/2} e^{j\omega_m t}$$

$$\hat{\psi}(\omega) = \frac{A_\psi}{\sqrt{\eta}} e^{-(\omega-\omega_m)^2/(2\eta)} \qquad A_\psi^2 = \sqrt{\frac{\eta}{\pi}} \tag{1.3.37}$$

$$WT(t,a) = \frac{A_s}{\sqrt{a}} \psi^*(\frac{t_0 - t}{a}) \tag{1.3.38}$$

$$= \frac{A_\psi A_s}{\sqrt{a}} e^{-\eta(t_0-t)^2/(2a^2)} e^{-j\omega_m(t_0-t)/a} \tag{1.3.39}$$

$$|WT(t,a)|^2 = \frac{A_\psi^2 A_s^2}{a} e^{-\eta(t_0-t)^2/a^2} \tag{1.3.40}$$

$$WT(t, \omega_r/\omega) = A_\psi^2 A_s^2 \sqrt{\frac{\omega}{\omega_r}} e^{-\eta\omega^2(t_0-t)^2/(2\omega_r^2) - j\omega_m\omega(t_0-t)/\omega_r} \tag{1.3.41}$$

$$|WT(t, \omega_r/\omega)|^2 = \frac{A_\psi^2 A_s^2 \omega}{\omega_r} e^{-\eta\omega^2(t_0-t)^2/\omega_r^2} \tag{1.3.42}$$

$$WT(t, \omega_r/\omega) = A_\psi A_s \sqrt{\frac{\omega}{\omega_r}} e^{-\eta\omega^2(t_0-t)^2/(2\omega_r^2)} e^{j\omega(t_0-t)} \quad \text{for } \omega_r = \omega_m \tag{1.3.43}$$

$$|WT(t, \omega_r/\omega)|^2 = \frac{A_\psi^2 A_s^2 \omega}{\omega_r} e^{-\eta\omega^2(t_0-t)^2/\omega_r^2} \tag{1.3.44}$$

$$P_{WT}(t, a) = \frac{A_\psi^2 A_s^2}{2\pi C a^3} e^{-\eta(t_0-t)^2/a^2} \tag{1.3.45}$$

$$P_{WT}(t, \omega) = \frac{A_\psi^2 A_s^2 |\omega|}{2\pi C \omega_r^2} e^{-\eta\omega^2(t_0-t)^2/\omega_r^2} \tag{1.3.46}$$

Example 5

$$s(t) = A_s e^{-\alpha(t-t_0)^2/2 + j\omega_0 t};$$

$$A_s = (\pi/\alpha)^{1/4} \qquad \hat{s}(\omega) = \frac{A_s}{\sqrt{\alpha}} e^{-(\omega-\omega_0)^2/(2\alpha) - j(\omega-\omega_0)t_0} \tag{1.3.47}$$

$$\psi(t) = A_\psi e^{-\eta t^2/2} e^{j\omega_m t}$$

$$\hat{\psi}(\omega) = \frac{A_\psi}{\sqrt{\eta}} e^{-(\omega-\omega_m)^2/(2\eta)} \qquad A_\psi^2 = \sqrt{\frac{\eta}{\pi}} \tag{1.3.48}$$

$$\begin{aligned} WT(t, a) &= \frac{\sqrt{2\pi}\sqrt{a} A_\psi A_s}{\sqrt{\eta + \alpha a^2}} e^{j\omega_0 t_0} \\ &\quad \times \exp\left[-\frac{1}{2} \frac{\alpha\eta(t-t_0)^2 + (a\omega_0 - \omega_m)^2}{\eta + \alpha a^2} \right. \\ &\qquad \left. + j(t-t_0) \frac{\eta\omega_0 + \alpha a \omega_m}{\eta + \alpha a^2} \right] \end{aligned} \tag{1.3.49}$$

$$\begin{aligned} WT(t, \omega_r/\omega) &= \frac{\sqrt{2\pi}\sqrt{\omega_r|\omega|} A_\psi A_s}{\sqrt{\eta\omega^2 + \alpha\omega_r^2}} e^{j\omega_0 t_0} \\ &\quad \times \exp\left[-\frac{1}{2}\omega^2 \frac{\alpha\eta(t-t_0)^2}{\eta\omega^2 + \alpha\omega_r^2} - \frac{1}{2} \frac{(\omega_r\omega_0 - \omega_m\omega)^2}{\eta\omega^2 + \alpha\omega_r^2} \right. \\ &\qquad \left. + j(t-t_0)\omega \frac{\eta\omega_0\omega + \alpha\omega_r\omega_m}{\eta\omega^2 + \alpha\omega_r^2} \right] \end{aligned} \tag{1.3.50}$$

$$|WT(t, a)|^2 = 2\pi A_\psi^2 A_s^2 \frac{a}{\eta + \alpha a^2} \exp\left[-\frac{\alpha\eta(t-t_0)^2 + (a\omega_0 - \omega_m)^2}{\eta + \alpha a^2} \right] \tag{1.3.51}$$

$$|WT(t,\omega_r/\omega)|^2 = 2\pi A_\psi^2 A_s^2 \frac{\omega_r|\omega|}{\eta\omega^2+\alpha\omega_r^2} \times \exp\left[-\frac{\alpha\eta\omega^2(t-t_0)^2}{\eta\omega^2+\alpha\omega_r^2} - \frac{(\omega_m\omega-\omega_0\omega_r)^2}{\eta\omega^2+\alpha\omega_r^2}\right] \tag{1.3.52}$$

$$WT(t,\omega_r/\omega) = \frac{\sqrt{2\pi}\sqrt{\omega_r|\omega|}A_\psi A_s}{\sqrt{\eta\omega^2+\alpha\omega_r^2}} e^{j\omega_0 t_0} \times \exp\left[-\frac{1}{2}\omega^2\frac{\alpha\eta(t-t_0)^2}{\eta\omega^2+\alpha\omega_r^2} - \frac{1}{2}\frac{(\omega_r^2(\omega_0-\omega))^2}{\eta\omega^2+\alpha\omega_r^2} + j(t-t_0)\omega\frac{\eta\omega_0\omega+\alpha\omega_r^2}{\eta\omega^2+\alpha\omega_r^2}\right], \quad \omega_m=\omega_r \tag{1.3.53}$$

$$|WT(t,\omega_r/\omega)|^2 = 2\pi A_\psi^2 A_s^2 \frac{\omega_r\omega}{\eta\omega^2+\alpha\omega_r^2}\exp\left[-\frac{\alpha\eta\omega^2(t-t_0)^2}{\eta\omega^2+\alpha\omega_r^2} - \frac{\omega_r^2(\omega-\omega_0)^2}{\eta\omega^2+\alpha\omega_r^2}\right] \tag{1.3.54}$$

$$P_{WT}(t,a) = \frac{A_\psi^2 A_s^2}{|a|C}\frac{1}{\eta+\alpha a^2}\exp\left[-\frac{\alpha\eta(t-t_0)^2+(a\omega_0-\omega_r)^2}{\eta+\alpha a^2}\right] \tag{1.3.55}$$

$$P_{WT}(t,\omega) = \frac{A_\psi^2 A_s^2}{C}\frac{|\omega|}{\eta\omega^2+\alpha\omega_r^2}\exp\left[-\frac{\alpha\eta\omega^2(t-t_0)^2}{\eta\omega^2+\alpha\omega_r^2} - \frac{\omega_r^2(\omega-\omega_0)^2}{\eta\omega^2+\alpha\omega_r^2}\right] \tag{1.3.56}$$

$$P(a) = \frac{A_\psi^2 A_s^2}{|a|C\sqrt{\alpha\eta}}\sqrt{\frac{\pi}{\eta+\alpha a^2}}\exp\left[-\frac{(a\omega_0-\omega_r)^2}{\eta+\alpha a^2}\right] \tag{1.3.57}$$

$$P(\omega) = \frac{A_\psi^2 A_s^2}{C\sqrt{\alpha\eta}}\sqrt{\frac{\pi}{\eta\omega^2+\alpha\omega_r^2}}\exp\left[-\frac{\omega_r^2(\omega-\omega_0)^2}{\eta\omega^2+\alpha\omega_r^2}\right] \tag{1.3.58}$$

1.4 Conclusion

There are a number of fundamental problems with the wavelet approach to time-frequency analysis. The most serious can be seen from the following consideration. Let us consider signals that have the same amplitude but arbitrary phase. In particular, to illustrate our point, we take a Gaussian amplitude centered at t_0 and phase $\varphi(t)$:

$$s(t) = (\alpha/\pi)^{1/4} e^{-\alpha(t-t_0)^2/2+j\varphi(t)}. \tag{1.4.1}$$

The mean time, mean time squared, and duration are readily obtained:

$$\langle t\rangle = \sqrt{\frac{\alpha}{\pi}}\int t e^{-\alpha(t-t_0)^2}dt = t_0 \tag{1.4.2}$$

$$\langle t^2\rangle = \sqrt{\frac{\alpha}{\pi}}\int t^2 e^{-\alpha(t-t_0)^2}dt = \frac{1}{2\alpha}+t_0^2 \tag{1.4.3}$$

$$\sigma_t^2 = \langle t^2\rangle - \langle t\rangle^2 = \frac{1}{2\alpha}. \tag{1.4.4}$$

These results are reasonable, and it is particularly important to note that they do not depend on the phase of the signal. That is, all signals with that envelope have the same mean time and duration. Now, for the wavelet approach, the answer for the mean time is given by

$$\langle t \rangle_{WT} = t_0 - \frac{1}{C} \left\langle \frac{1}{\omega} \right\rangle_s \langle t \rangle_\psi \,. \tag{1.4.5}$$

Thus, the mean time depends on $\left\langle \frac{1}{\omega} \right\rangle_s$, which has to do with spectral properties of the signal and which in turn depends on the phase. Therefore, different signals with the same amplitude may have different mean times. More seriously, it says that for many signals the mean time does not exist, since for many well-behaved signals $\left\langle \frac{1}{\omega} \right\rangle_s$ does not exist. There may be a deep meaning to this, but it does not seem clear. If something so simple and so fundamental as the mean time has been misunderstood for all these years, and the wavelet result sheds new light and is indeed plausible, then there would be a new world according to wavelets.

Like the spectrogram, the wavelet approach mixes properties of the window with the signal. But in the wavelet case the mixing is unusual because it mixes time and frequency properties in a way that seems to have no physical relevance and gives results that are not readily interpretable. Moreover, for most situations it predicts the nonexistence of simple physical quantities. The nonexistence of simple physical quantities, as predicted by the wavelet approach, certainly makes the approach problematic.

We now address the issue of "tiling of the time-frequency plane," which is often presented as a reasonable visual picture of the power of variable windowing. The basic idea of tiling is to give a picturesque notion for capturing the time-frequency properties of a signal. It focuses on the properties of the window and imagines the window to be a net that captures the local properties around a particular time-frequency point. It is commonly argued that the spectrogram, having a fixed window, tiles the time-frequency plane with constant rectangles, while the wavelet transform, having a variable window, tiles the time-frequency plane differently in different regions and hence is more powerful. However, this picture of nature is misleading because it confuses the uncertainty principle of the window with the uncertainty principle of the signal [22], [23]. Windowing, which is something we impose, distorts the frequency content of the original signal, sometimes to such an extent that it has no relation to the original signal! The tiling concept as commonly used does not face the issue as it disregards the intertwining of window and signal. Whether we like it or not, using a very small duration window takes a reasonable signal and produces a gigantic glob of frequencies that have nothing or almost nothing to do with the original signal. We point out that the tiling concept does not arise in methods such as the Wigner or the positive distributions [2], [24].

1.5 Appendix: Moments

We present the results separately for the scale, time, and frequency moments [23].

1.5.1 The "a" moments

The "a" moments are defined by

$$\langle a^n \rangle = \int a^n P(t,a)dtda \tag{1.5.1}$$

$$= \int a^n P(a)da, \tag{1.5.2}$$

and substituting Eq. (1.2.20) we have

$$\langle a^n \rangle = \int\int a^n \frac{1}{|a|C} |\hat{s}(\omega)|^2 |\hat{\psi}(\omega a)|^2 d\omega da \tag{1.5.3}$$

$$= \frac{1}{C}\int\int a^n \frac{1}{|a|} |\hat{s}(\omega)|^2 |\hat{\psi}(\omega a)|^2 d\omega da \tag{1.5.4}$$

$$= \frac{1}{C}\int\int \frac{1}{\omega^n} \frac{\omega'^n}{|\omega'|} |\hat{s}(\omega)|^2 |\hat{\psi}(\omega')|^2 d\omega d\omega' \tag{1.5.5}$$

or

$$\langle a^n \rangle = \frac{1}{C} \left\langle \frac{1}{\omega^n} \right\rangle_s \left\langle \frac{\omega^n}{|\omega|} \right\rangle_\psi , \tag{1.5.6}$$

where

$$\left\langle \frac{1}{\omega^n} \right\rangle_s = \int \left\langle \frac{1}{\omega^n} \right\rangle_s |\hat{s}(\omega)|^2 d\omega \tag{1.5.7}$$

$$\left\langle \frac{\omega^n}{|\omega|} \right\rangle_\psi = \int \left\langle \frac{\omega^n}{|\omega|} \right\rangle_\psi |\hat{\psi}(\omega)|^2 d\omega \tag{1.5.8}$$

The first two moments and standard deviation are

$$\langle a \rangle = \frac{1}{C} \left\langle \frac{1}{\omega} \right\rangle_s \left\langle \frac{\omega}{|\omega|} \right\rangle_\psi \tag{1.5.9}$$

$$\langle a^2 \rangle = \frac{1}{C} \left\langle \frac{1}{\omega^2} \right\rangle_s \left\langle \frac{\omega^2}{|\omega|} \right\rangle_\psi \tag{1.5.10}$$

$$\sigma_a^2 = \frac{1}{C} \left\langle \frac{1}{\omega^2} \right\rangle_s \left\langle \frac{\omega^2}{|\omega|} \right\rangle_\psi - \frac{1}{C^2} \left\langle \frac{1}{\omega} \right\rangle_s^2 \left\langle \frac{\omega}{|\omega|} \right\rangle_\psi^2 . \tag{1.5.11}$$

1.5.2 The time moments

The expectation values of the time function, $g(t)$, with respect to the signal and wavelet are defined, respectively, as

$$\langle g(t)\rangle_s = \int g(t)|s(t)|^2 dt \tag{1.5.12}$$

$$\langle g(t)\rangle_\psi = \int g(t)|\psi(t)|^2 dt. \tag{1.5.13}$$

It is useful to use and manipulate with the time operator [2]

$$\mathcal{T} = -\frac{1}{j}\frac{\partial}{\partial\omega}. \tag{1.5.14}$$

We note that for any signal $f(t)$ whose Fourier transform is $\hat{f}(\omega)$ we have that

$$\langle g(t)\rangle = \int g(t)|f(t)|^2 dt = \int \hat{f}^*(\omega) g(\mathcal{T})\hat{f}(\omega)d\omega. \tag{1.5.15}$$

In particular,

$$\int t|f(t)|^2 dt = \int \hat{f}^*(\omega)\mathcal{T}\hat{f}(\omega)d\omega \tag{1.5.16}$$

$$\int t^2|f(t)|^2 dt = \int \hat{f}^*(\omega)\mathcal{T}^2\hat{f}(\omega)d\omega = \int \left|\mathcal{T}\hat{f}(\omega)\right|^2 d\omega. \tag{1.5.17}$$

In general,

$$\langle t^n\rangle = \int\int t^n P(t,a)dtda \tag{1.5.18}$$

$$= \int t^n P(t)dt \tag{1.5.19}$$

$$= \frac{1}{2\pi C}\int\int t^n e^{j(\omega-\omega')t}\hat{s}(\omega)\hat{s}^*(\omega')L(\omega,\omega')d\omega d\omega' dt \tag{1.5.20}$$

$$= \frac{1}{2\pi C}\int\int\left\{\left(\frac{d}{jd\omega}\right)^n e^{j(\omega-\omega')t}\right\}\hat{s}(\omega)\hat{s}^*(\omega')L(\omega,\omega')d\omega d\omega' dt \tag{1.5.21}$$

$$= \frac{1}{C}\int\int\left\{\left(\frac{d}{jd\omega}\right)^n \delta(\omega-\omega')\right\}\hat{s}(\omega)\hat{s}^*(\omega')L(\omega,\omega')d\omega d\omega' \tag{1.5.22}$$

$$= \frac{1}{C}\int\left(\frac{d}{-jd\omega}\right)^n \hat{s}(\omega)\hat{s}^*(\omega')L(\omega,\omega')|_{\omega=\omega'}d\omega' \tag{1.5.23}$$

$$= \frac{1}{C}\int\int\frac{1}{|a|}\hat{s}^*(\omega')\hat{\psi}(\omega' a)\left(\frac{d}{-jd\omega}\right)^n \hat{s}(\omega)\hat{\psi}^*(\omega a)|_{\omega=\omega'}dad\omega' \tag{1.5.24}$$

$$= \frac{1}{C}\int\int\frac{1}{|a|}\hat{s}^*(\omega)\hat{\psi}(\omega a)\left(\frac{d}{-jd\omega}\right)^n \hat{s}(\omega)\hat{\psi}^*(\omega a)dad\omega \tag{1.5.25}$$

$$= \frac{1}{C}\int\int \frac{1}{|a|}\hat{s}^*(\omega)\hat{\psi}(\omega a)\mathcal{T}^n \hat{s}(\omega)\hat{\psi}^*(\omega a)dad\omega. \tag{1.5.26}$$

For the first moment,

$$\langle t\rangle = \frac{1}{C}\int\int \frac{1}{|a|}\hat{s}^*(\omega)\hat{\psi}(\omega a)\mathcal{T}\hat{s}(\omega)\hat{\psi}^*(\omega a)dad\omega \tag{1.5.27}$$

$$= \frac{1}{C}\int\int \frac{1}{|a|}\hat{s}^*(\omega)\hat{\psi}(\omega a)\left[\hat{\psi}^*(\omega a)\mathcal{T}\hat{s}(\omega) + \hat{s}(\omega)\mathcal{T}\hat{\psi}^*(\omega a)\right] dad\omega \tag{1.5.28}$$

$$= \int\int \hat{s}^*(\omega)\mathcal{T}\hat{s}(\omega)d\omega + \frac{1}{C}\int\int \frac{1}{|a|}\hat{s}^*(\omega)\hat{\psi}(\omega a)\left[\hat{s}(\omega)\mathcal{T}\hat{\psi}^*(\omega a)\right] dad\omega$$

$$= \langle t\rangle_s + \frac{1}{C}\int\int \frac{1}{|a|}|\hat{s}(\omega)|^2\hat{\psi}(\omega a)\mathcal{T}\hat{\psi}^*(\omega a)dad\omega \tag{1.5.29}$$

$$= \langle t\rangle_s + \frac{1}{C}\int \frac{1}{\omega}|\hat{s}(\omega)|^2\hat{\psi}(x)\mathcal{T}_x\hat{\psi}^*(x)dxd\omega \tag{1.5.30}$$

$$= \langle t\rangle_s - \frac{1}{C}\left\langle\frac{1}{\omega}\right\rangle_s \langle t\rangle_\psi . \tag{1.5.31}$$

For the t^2 moment we have

$$\left\langle t^2\right\rangle = \frac{1}{C}\int\int \frac{1}{|a|}\hat{s}^*(\omega)\hat{\psi}(\omega a)\mathcal{T}^2\hat{s}(\omega)\hat{\psi}^*(\omega a)dad\omega \tag{1.5.32}$$

$$= \frac{1}{C}\int\int \frac{1}{|a|}\left|\mathcal{T}\hat{s}(\omega)\hat{\psi}^*(\omega a)\right|^2 dad\omega \tag{1.5.33}$$

$$= \frac{1}{C}\int\int \frac{1}{|a|}\left|\hat{\psi}^*(\omega a)\mathcal{T}\hat{s}(\omega) + \hat{s}(\omega)\mathcal{T}\hat{\psi}^*(\omega a)\right|^2 dad\omega \tag{1.5.34}$$

$$\begin{aligned} &= \frac{1}{C}\int\int \frac{1}{|a|}\left|\mathcal{T}\hat{s}(\omega)\right|^2\left|\hat{\psi}(\omega a)\right|^2 d\omega da \\ &\quad + \frac{1}{C}\int\int \frac{1}{|a|}\left|\hat{s}(\omega)\right|^2\left|\mathcal{T}\hat{\psi}^*(\omega a)\right|^2 d\omega da \\ &\quad + \frac{1}{C}\int \frac{1}{|a|}(\hat{s}(\omega)\mathcal{T}\hat{s}^*(\omega))(\hat{\psi}(\omega a)\mathcal{T}\hat{\psi}^*(\omega a)) \\ &\qquad - (\hat{s}^*(\omega)\mathcal{T}\hat{s}(\omega))(\hat{\psi}^*(\omega a)\mathcal{T}\hat{\psi}(\omega a)) \end{aligned} \tag{1.5.35}$$

$$= \left\langle t^2\right\rangle_s + \frac{1}{C}\int\int \frac{1}{\omega^2}\left|\hat{s}(\omega)\right|^2 x\left|\mathcal{T}_x\hat{\psi}(x)\right|^2 d\omega dx \tag{1.5.36}$$

$$+ \frac{1}{C}\int\int \frac{1}{\omega}(\hat{\psi}^*(x)\mathcal{T}_x\hat{\psi}(x))\left\{(\hat{s}(\omega)\mathcal{T}\hat{s}^*(\omega)) - (\hat{s}^*(\omega)\mathcal{T}\hat{s}(\omega))\right\} dxd\omega. \tag{1.5.37}$$

Hence,

$$\left\langle t^2\right\rangle = \left\langle t^2\right\rangle_s + \frac{1}{C}\left\langle\frac{1}{\omega^2}\right\rangle_s \int \omega|\mathcal{T}\psi(\omega)|^2 d\omega - \frac{1}{C}\langle t\rangle_\psi \left\langle[1/\omega, \mathcal{T}]_+\right\rangle_s , \tag{1.5.38}$$

where the symbol $[1/\omega, \mathcal{T}]_+$ is the anticommutator of two operators

$$[1/\omega, \mathcal{T}]_+ = \frac{1}{\omega}\mathcal{T} + \mathcal{T}\frac{1}{\omega}. \tag{1.5.39}$$

1.5.3 The frequency moments

The frequency moments are

$$\langle \omega^n \rangle = \int\int \omega^n P(t,\omega) dt d\omega \tag{1.5.40}$$

$$= \int \omega^n P(\omega) d\omega \tag{1.5.41}$$

$$= \frac{1}{C}\int\int \frac{\omega^n}{|\omega|} |\hat{s}(\omega')|^2 \left|\hat{\psi}\left(\frac{\omega_r}{\omega}\omega'\right)\right|^2 d\omega' d\omega \tag{1.5.42}$$

$$= \frac{1}{C}\omega_r^n \int \omega^n \frac{1}{\omega'^n|\omega'|} |\hat{s}(\omega)|^2 |\hat{\psi}(\omega')|^2 d\omega' d\omega. \tag{1.5.43}$$

That is,

$$\langle \omega^n \rangle = \frac{1}{C}\omega_r^n \langle \omega^n \rangle_s \left\langle \frac{1}{\omega^n|\omega|} \right\rangle_\psi . \tag{1.5.44}$$

The first two frequency moments are

$$\langle \omega \rangle = \frac{\omega_r}{C} \langle \omega \rangle_s \left\langle \frac{1}{\omega|\omega|} \right\rangle_\psi \tag{1.5.45}$$

$$\langle \omega^2 \rangle = \frac{\omega_r^2}{C} \langle \omega^2 \rangle_s \left\langle \frac{1}{\omega^2|\omega|} \right\rangle_\psi , \tag{1.5.46}$$

and the standard deviation is given by

$$\sigma_\omega^2 = \omega_r^2 \left[\frac{1}{C} \langle \omega^2 \rangle_s \left\langle \frac{1}{\omega^2|\omega|} \right\rangle_s - \frac{1}{C^2} \langle \omega \rangle_s^2 \left\langle \frac{1}{\omega|\omega|} \right\rangle_\psi^2 \right]. \tag{1.5.47}$$

Acknowledgment. Work supported by the Air Force Information Institute Research Program (Rome, NY) and the NSA HBCU/MI program.

References

[1] L. Cohen, Time-Frequency Distributions—A Review, *Proc. of the IEEE*, **77** (1989), 941–981.

[2] L. Cohen, *Time-Frequency Analysis*, Prentice-Hall, Englewood Cliffs, NJ, 1995.
[3] P. Flandrin, *Time-Frequency and Time-Scale Analysis*, Academic Press, New York, 1999.
[4] S. Mallat, *A Wavelet Tour of Signal Processing*, Academic Press, 1998.
[5] E. P. Wigner, On the quantum correction for thermodynamic equilibrium, *Physical Review*, **40** (1932), 749–759.
[6] H. Choi and W. Williams, Improved Time-Frequency Representation of Multicomponent Signals Using Exponential Kernels, *IEEE Trans. on Acoust., Speech, Sig. Proc.,* **37** (1989), 862–871.
[7] J. Jeong and W. Williams, Kernel Design for Reduced Interference Distributions, *IEEE Trans. on Sig. Proc.,* **40** (1992), 402–412.
[8] R. Koenig, H. K Dunn, and L. Y. Lacy, The sound spectrograph, *J. Acoust. Soc. Am.*, **18** (1946), 19–49.
[9] J. Jeong and W. J. Williams. Variable windowed spectrograms: connecting Cohen's class and the wavelet transform, in *IEEE ASSP Workshop on Spectrum Estimation and Modeling*, (1990), 270–273.
[10] J. Jeong, *Time-Frequency Signal Analysis and Synthesis Algorithms*, Thesis, The University of Michigan, Ann Arbor, MI, 1990.
[11] W. J. Williams, T.-H. Sang, J. C. O'Neill, and E. J. Zalubas, Wavelet windowed time-frequency distribution decompositions, in *Advanced Signal Processing Architectures and Implementations*, Proc. SPIE, volume **3162** (1997), 149–160.
[12] O. Rioul and P. Flandrin, Time-scale energy distributions: a general class extending wavelet transforms, *IEEE Trans. on Signal Processing*, **40** (1992), 1746–1757.
[13] Ali N. Akansu and R. A. Haddad, *Multiresolution Signal Analysis*, Academic Press, New York, 1992.
[14] G. Strang and T. Nguyen, *Wavelets and Filter Banks*, Wellesley-Cambridge Press, Wellesley, MA, 1996.
[15] C. H. Chui, *An Introduction to Wavelets*, Academic Press, New York, 1992.
[16] I. Daubechies, *Ten Lectures on Wavelets*, SIAM, Philadelphia, PA, 1992.
[17] J. C. Goswami and A. K. Chan, *Fundamentals of Wavelets*, Wiley, New York, 1999.
[18] A. Grossmann, R. Kronland-Martinet, and J. Morlet, Reading and Understanding the Continuous Wavelet Transform, in *Wavelets: Time–Frequency Methods and Phase Space*, J. M. Combes, A. Grossman, and P. Tchamitchian, eds., Springer-Verlag, New York, 1989, pp. 2–20.
[19] Y. Meyer, *Wavelets*, SIAM, Philadelphia, PA, 1993.
[20] B. B. Hubbard, *The World According to Wavelets*, A K Peters, Natick, MA, 1998.
[21] D. Gabor, Theory of communication, *IEE J. Comm. Engrng.,* **93** (1946), 429–441.
[22] L. Cohen, "The Uncertainly Principle for the Short-Time Fourier Transform and Wavelet Transform", in *Wavelet Transforms and Time-Frequency Analysis*, Lokenath Debnath (editor), Birkhäuser, Boston, MA, 2001, pp. 217–232.
[23] L. Cohen, Wavelet Moments and Time-Frequency Analysis, *Proc. SPIE,* **3807** (1999), 434–445.
[24] P. Loughlin, J. Pitton, and L. Atlas, Construction of positive time-frequency distributions, *IEEE Trans. Sig. Proc.*, **42** (1994), 2697–2705.

2

Multiwavelets in $\mathbb{R}^n$ with an Arbitrary Dilation Matrix

Carlos A. Cabrelli
Christopher Heil
Ursula M. Molter

ABSTRACT We present an outline of how the ideas of self-similarity can be applied to wavelet theory, especially in connection to wavelets associated with a multiresolution analysis of $\mathbb{R}^n$ allowing arbitrary dilation matrices and no restrictions on the number of scaling functions.

2.1 Introduction

Wavelet bases have proved highly useful in many areas of mathematics, science, and engineering. One of the most successful approaches for the construction of such a basis begins with a special functional equation, the *refinement equation*. The solution to this refinement equation, called the scaling function, then determines a multiresolution analysis, which in turn determines the wavelet and the wavelet basis. In order to construct wavelet bases with prescribed properties, we must characterize those particular refinement equations which yield scaling functions that possess some specific desirable property. Much literature has been written on this topic for the classical one-dimensional, single-function, two-scale refinement equation, but when we move from the one-dimensional to the higher-dimensional setting or from the single wavelet to the multiwavelet setting, it becomes increasingly difficult to find and apply such characterizations.

Our goal in this paper is to outline some recent developments in the construction of higher-dimensional wavelet bases that exploit the fact that the refinement equation is a statement that the scaling function satisfies a certain kind of self-similarity. In the classical one-dimensional case with dilation factor two, there are a variety of tools in addition to self-similarity which can be used to analyze the refinement equation. However, many of these tools become difficult or impossible to apply in the multidimensional setting with a general dilation matrix, whereas self-similarity becomes an even more natural and important tool in this setting. By viewing scaling functions as particular cases of "generalized self-similar functions," we showed in [5] that the tools of functional analysis can be applied to analyze refinement equations in the general higher-dimensional and multifunction setting. We derived conditions for the existence of continuous or L^p solutions to the refinement equation in this general setting, and showed how these conditions can be combined with the analysis of the accuracy of scaling functions from [4], [3] to construct new examples of nonseparable (nontensor product) two-dimensional multiwavelets using a quincunx dilation matrix.

We will sketch some of the ideas and results from [5] in this paper, attempting to provide some insights into the techniques without dwelling on the mass of technical details that this generality necessitates. We emphasize that this work is intimately tied and connected to the vast literature on wavelets and refinement equations, and while we cannot trace those connections here, a full discussion with extensive references is presented in [5]. In particular, the important and fundamental contributions of Daubechies, Lagarias, Wang, Jia, Jiang, Shen, Plonka, Strela, and many others are discussed in [5].

2.2 Self-similarity

The seed for this approach can be traced back to Bajraktarevic [1], who in 1957 studied solutions to equations of the form

$$\mathbf{u}(x) = \mathcal{O}(x, (\mathbf{u} \circ g_1)(x), \dots, (\mathbf{u} \circ g_m)(x)), \tag{2.2.1}$$

where $g_i \colon X \to X$ and $\mathcal{O} \colon X \times E^m \to E$, and the solution $\mathbf{u} \colon X \to E$ lies in some function space $\mathcal{F}$. Bajraktarevic proved that, under mild conditions on $\mathcal{O}$ and the g_i, there is a unique solution to (2.2.1). (See also [9].) A generalized version of this equation of the form

$$\mathbf{u}(x) = \mathcal{O}(x, \varphi_1(x, (\mathbf{u} \circ g_1)(x)), \dots, \varphi_m(x, (\mathbf{u} \circ g_m)(x))), \tag{2.2.2}$$

where $\varphi_i \colon X \times E \to E$, was studied in [6]. We will state one uniqueness result below, and then in later sections demonstrate the fundamental connection between (2.2.2) and wavelets. If there exists a set B that is *self-similar* with respect to the functions g_i, i.e., if $B = \bigcup_{i=1}^m g_i^{-1}(B)$, then we refer to the solution $\mathbf{u}$ of (2.2.2) as a *generalized self-similar function*. This is because at a given point $x \in B$, the value of $\mathbf{u}(x)$ is obtained by combining the values of $\mathbf{u}(g_i(x))$ through the action of the operator $\mathcal{O}$, with each $g_i(x)$ lying in B.

In order to state the uniqueness result, we require the following notation. Let X be a closed subset of $\mathbb{R}^n$, and let $\|\cdot\|$ be any fixed norm on $\mathbb{C}^r$. Then we define $L^\infty(X, \mathbb{C}^r)$ to be the Banach space of all mappings $\mathbf{u} \colon X \to \mathbb{C}^r$ such that

$$\|\mathbf{u}\|_{L^\infty} = \sup_{x \in X} \|g(x)\| < \infty.$$

This definition is independent of the choice of norm on $\mathbb{C}^r$ in the sense that each choice of norm for $\mathbb{C}^r$ yields an equivalent norm for $L^\infty(X, \mathbb{C}^r)$. If E is a nonempty closed subset of $\mathbb{C}^r$, then $L^\infty(X, E)$ will denote the closed subset of $L^\infty(X, \mathbb{C}^r)$ consisting of functions that take values in E. We say that a function $\mathbf{u}\colon X \to E$ is *stable* if $\mathbf{u}(B)$ is a bounded subset of E whenever B is a bounded subset of X.

The following result is a special case of more general results proved in [6]. In particular, we will consider here only uniform versions of this result; it is possible to formulate L^p and other versions as well.

Theorem 2.2.1. *Let X be a compact subset of $\mathbb{R}^n$, and let E be a closed subset of $\mathbb{C}^r$. Let $\|\cdot\|$ be any norm on $\mathbb{C}^r$. Let $m \geq 1$, and assume that functions w_i, φ_i, and $\mathcal{O}$ are chosen with the following properties.*

- *For each $i = 1, \ldots, m$, let $w_i\colon X \to X$ be continuously differentiable, injective maps.*
- *Let $\varphi_i\colon X \times E \to E$ for $i = 1, \ldots, m$ satisfy the Lipschitz-like condition*

$$\max_{1 \leq i \leq m} \|\varphi_i(x, u) - \varphi_i(x, v)\| \leq C\|u - v\|. \tag{2.2.3}$$

- *Let $\mathcal{O}\colon X \times E^m \to E$ be nonexpansive for each $x \in X$, i.e.,*

$$\|\mathcal{O}(x, u_1, \ldots, u_m) - \mathcal{O}(x, v_1, \ldots, v_m)\| \leq \max_{1 \leq i \leq m} \|u_i - v_i\|. \tag{2.2.4}$$

Let t_0 be an arbitrary point in E. For $u \in L^\infty(X, E)$, define

$$T\mathbf{u}(x) = \mathcal{O}(x, \varphi_1(x, \mathbf{u}(w_1^{-1}(x))), \ldots, \varphi_m(x, \mathbf{u}(w_m^{-1}(x)))),$$

where we interpret

$$\mathbf{u}(w_i^{-1}(x)) = t_0 \qquad \text{if } x \notin w_i(X).$$

If $\mathcal{O}$ and the φ_i are stable, then T maps $L^\infty(X, E)$ into itself and satisfies

$$\|T\mathbf{u} - T\mathbf{v}\|_{L^\infty} \leq C\|\mathbf{u} - \mathbf{v}\|_{L^\infty}.$$

In particular, if $C < 1$, then T is contractive, and there exists a unique function $\mathbf{v}^* \in L^\infty(X, E)$ such that $T\mathbf{v}^* = \mathbf{v}^*$, and, moreover, $\mathbf{v}^*$ is continuous. Further, if $C < 1$ and $\mathbf{v}^{(0)}$ is any function in $L^\infty(X, E)$, then the iteration $\mathbf{v}^{(i+1)} = T\mathbf{v}^{(i)}$ converges to $\mathbf{v}^*$ in $L^\infty(X, E)$.

2.3 Refinement equations

The connection between Theorem 2.2.1 and wavelets is provided by the now-classical concept of multiresolution analysis (MRA). To construct an MRA in $\mathbb{R}^n$, one begins with a *refinement equation* of the form

$$f(x) = \sum_{k \in \Lambda} c_k f(Ax - k), \qquad x \in \mathbb{R}^n, \tag{2.3.1}$$

where Λ is a subset of the lattice $\mathbb{Z}^n$ and A is a *dilation matrix*, i.e., $A(\mathbb{Z}^n) \subset \mathbb{Z}^n$ and every eigenvalue λ of A satisfies $|\lambda| > 1$. We assume now that A, Λ, and c_k are fixed for the remainder of this paper.

A solution of the refinement equation is called a *scaling function* or a *refinable function*. If f is scalar valued, then the coefficients c_k are scalars, while if one allows vector-valued ($f\colon \mathbb{R}^n \to \mathbb{C}^r$) or matrix-valued ($f\colon \mathbb{R}^n \to \mathbb{C}^{r\times\ell}$) functions, then the c_k are $r \times r$ matrices. We will consider the case $f\colon \mathbb{R}^n \to \mathbb{C}^r$ in this paper. We say that the number r is the *multiplicity* of the scaling function f.

The fact that A can be any dilation matrix (instead of just a "uniform" dilation such as $2I$) means that the geometry of $\mathbb{R}^n$ must be carefully considered with respect to the action of A. Note that since $A(\mathbb{Z}^n) \subset \mathbb{Z}^n$, the dilation matrix A necessarily has an integer determinant. We define

$$m = |\det(A)|.$$

By [13], to each scaling function that generates a MRA there will be associated $(m-1)$ "mother wavelets," so it is desirable for some applications to consider "small" m.

The *refinement operator* associated with the refinement equation is the mapping S acting on vector functions $\mathbf{u}\colon \mathbb{R}^n \to \mathbb{C}^r$ defined by

$$S\mathbf{u}(x) = \sum_{k\in\Lambda} c_k \mathbf{u}(Ax - k), \qquad x \in \mathbb{R}^n. \tag{2.3.2}$$

A scaling function is thus a fixed point of S.

We will focus on compactly supported solutions of the refinement equation and therefore will require that the subset Λ be finite. Let us consider the support of a solution to the refinement equation in this case. For each $k \in \mathbb{Z}^n$, let $w_k\colon \mathbb{R}^n \to \mathbb{R}^n$ denote the contractive map

$$w_k(x) = A^{-1}(x + k). \tag{2.3.3}$$

Now let $\mathcal{H}(\mathbb{R}^n)$ denote the set of all nonempty, compact subsets of $\mathbb{R}^n$ equipped with the Hausdorff metric. Then it can be shown that the mapping w on $\mathcal{H}(\mathbb{R}^n)$ defined by

$$w(B) = \textstyle\bigcup_{k\in\Lambda} w_k(B) = A^{-1}(B + \Lambda)$$

is a contractive mapping of $\mathcal{H}(\mathbb{R}^n)$ into itself [11]. Hence, there is a unique compact set K_Λ such that

$$K_\Lambda = w(K_\Lambda) = \textstyle\bigcup_{k\in\Lambda} A^{-1}(K_\Lambda + k).$$

In the terminology of iterated function systems (IFSs), the set K_Λ is the attractor of the IFS generated by the collection $\{w_k\}_{k\in K}$. It can be shown that if f is a compactly supported solution of the refinement equation, then necessarily $\operatorname{supp}(f) \subset K_\Lambda$ [5].

Let

$$D = \{d_1, \dots, d_m\}$$

be a *full set of digits* with respect to A and $\mathbb{Z}^n$, i.e., a complete set of representatives of the order-m group $\mathbb{Z}^n/A(\mathbb{Z}^n)$. Because D is a full set of digits, the lattice $\mathbb{Z}^n$ is partitioned into the m disjoint cosets

$$\Gamma_d = A(\mathbb{Z}^n) - d = \{Ak - d : k \in \mathbb{Z}^n\}, \qquad d \in D.$$

Let Q be the attractor of the IFS generated by $\{w_d\}_{d\in D}$, i.e., Q is the unique nonempty compact set satisfying

$$Q = K_D = \bigcup_{d\in D} A^{-1}(Q + d).$$

We will say that Q is a *tile* if its $\mathbb{Z}^n$ translates cover $\mathbb{R}^n$ with overlaps of measure 0. In that case, the Lebesgue measure of Q is 1 [2], and the characteristic function of Q generates an MRA in $\mathbb{R}^n$ [10]. This MRA is the n-dimensional analogue of the Haar MRA in $\mathbb{R}$, because if we consider dilation by 2 in $\mathbb{R}$ with digit set $D = \{0, 1\}$, and set

$$w_0(x) = \frac{1}{2}x \quad \text{and} \quad w_1(x) = \frac{1}{2}x + \frac{1}{2},$$

then the set $[0, 1]$ satisfies

$$[0, 1] = w_0([0, 1]) \bigcup w_1([0, 1])$$

and therefore is the attractor for the IFS $\{w_0, w_1\}$. Note that $[0, 1]$ is a tile, and that the Lebesgue measure of $[0, 1]$ is 1.

Example 2.3.1. Tiles may have fractal boundaries. For example, if we consider the dilation matrix

$$A_1 = \begin{bmatrix} 1 & -1 \\ 1 & 1 \end{bmatrix}$$

and digit set $D = \{(0, 0), (1, 0)\}$, then the tile Q is the celebrated "twin dragon" fractal shown on the left in Figure 2.1. On the other hand, if

$$A_2 = \begin{bmatrix} 1 & 1 \\ 1 & -1 \end{bmatrix}$$

and $D = \{(0, 0), (1, 0)\}$, then the tile Q is the parallelogram with vertices $\{(0, 0),$ $(1, 0)$, $(2, 1)$, $(1, 1)\}$ pictured on the right in Figure 2.1. For these two matrices A_1 and A_2, the sublattices $A_1(\mathbb{Z}^2)$ and $A_2(\mathbb{Z}^2)$ coincide. This sublattice is called the *quincunx sublattice* of $\mathbb{Z}^2$. As a consequence, these two matrices A_1, A_2 are often referred to as *quincunx dilation matrices*.

It is not always the case that, given an arbitrary dilation matrix A, there exists a set of digits such that the associated attractor of $\{w_d\}_{d\in D}$ is a tile [14], [12]. We will not address this question here and will only consider dilation matrices for which a tile Q exists, and we assume that the digit set D has been chosen in such a way that Q is a tile. Without loss of generality, we can assume that $0 \in D$, and therefore the tile Q will contain the origin [5].

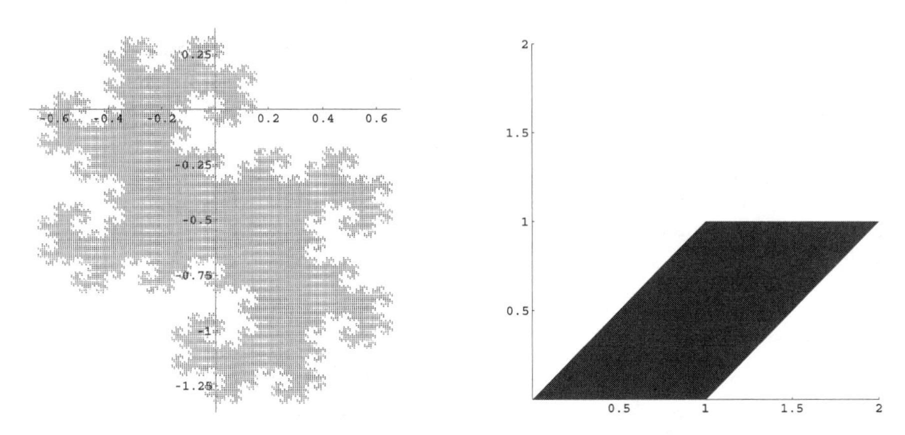

FIGURE 2.1. Twin dragon and parallelogram attractors.

Let us now return to setting up the notation required to connect the refinement equation (2.3.1) to Theorem 2.2.1.

Since $\mathrm{supp}(f) \subset K_\Lambda$, which is compact, and since Q is a tile and therefore covers $\mathbb{R}^n$ by translates, there exists a finite subset $\Omega \subset \mathbb{Z}^n$ such that

$$K_\Lambda \subset Q + \Omega = \bigcup_{\omega\in\Omega}(Q+\omega) = \{q+\omega : q \in Q, \omega \in \Omega\}.$$

Consider now any function $g\colon \mathbb{R}^n \to \mathbb{C}^r$ such that $\mathrm{supp}(g) \subset K_\Lambda$. Define the *folding* of g to be the function $\Phi g\colon Q \to (\mathbb{C}^r)^\Omega$ given by

$$\Phi g(x) = [g(x+k)]_{k\in\Omega}, \qquad x \in Q.$$

If for $k \in \Omega$ we write $(\Phi g)_k(x) = g(x+k)$ for the kth component of $\Phi g(x)$, then this folding has the property that $(\Phi g)_{k_1}(x_1) = (\Phi g)_{k_2}(x_2)$ whenever $x_1, x_2 \in Q$ and $k_1, k_2 \in \Omega$ are such that $x_1 + k_1 = x_2 + k_2$ (it can be shown that such points x_1, x_2 would necessarily have to lie on the boundary of Q [5]).

Here (and whenever we deal with vectors indexed by general sets) we consider that Ω has been ordered in some way; the choice of ordering is not important as long as the same ordering is used throughout. We use square brackets, e.g., $[u_k]_{k\in\Omega}$, to denote column vectors, and round brackets, e.g., $(u_k)_{k\in\Omega}$, to denote row vectors.

Since Q is the attractor of the IFS $\{w_d\}_{d\in D}$, it satisfies

$$Q = \bigcup_{d\in D} A^{-1}(Q+d).$$

Moreover, since Q is a tile, if $d_1 \neq d_2$, then $A^{-1}(Q+d_1) \cap A^{-1}(Q+d_2)$ has measure zero, and in fact it can be shown that these sets can intersect only along their boundaries. We will require subsets Q_d of $A^{-1}(Q+d)$ whose union is Q but which have *disjoint* intersections, i.e., such that

$$\bigcup_{d\in D} Q_d = Q \quad \text{and} \quad Q_{d_1} \cap Q_{d_2} = \emptyset \qquad \text{if } d_1 \neq d_2.$$

A precise method for creating these sets Q_d is given in [5].

For each $d \in D$, define a matrix T_d by

$$T_d = [c_{Aj-k+d}]_{j,k\in\Omega}.$$

Note that T_d consists of an $\Omega \times \Omega$ collection of $r \times r$ blocks, i.e., $T_d \in (\mathbb{C}^{r\times r})^{\Omega\times\Omega}$. Assume that E is a subset (but not necessarily a subspace) of $(\mathbb{C}^r)^\Omega$ that is invariant under each matrix T_d (we will specify E precisely later). Then for each $d \in D$ we can define $\varphi_d \colon Q \times E \to E$ by

$$\varphi_d(x, e) = T_d e, \tag{2.3.4}$$

and define $\mathcal{O} \colon Q \times E^D \to E$ by

$$\mathcal{O}(x, \{e_d\}_{d\in D}) = \sum_{d\in D} \chi_{Q_d}(x) \cdot e_d. \tag{2.3.5}$$

That is, $\mathcal{O}(x, \{e_d\}_{d\in D}) = e_d$ if $x \in Q_d$. It is easy to see that this operator $\mathcal{O}$ is stable and satisfies the nonexpansivity condition (2.2.4). Now define an operator T acting on vector functions $\mathbf{u} \colon Q \to E$ by

$$\begin{aligned} T\mathbf{u}(x) &= \mathcal{O}(x, \{\varphi_d(x, \mathbf{u}(w_d^{-1}(x)))\}_{d\in D}) \\ &= \sum_{d\in D} \chi_{Q_d}(x) \cdot T_d \mathbf{u}(Ax - d). \end{aligned} \tag{2.3.6}$$

Or, equivalently, T can be defined by

$$T\mathbf{u}(x) = T_d \mathbf{u}(Ax - d) \qquad \text{if } x \in Q_d.$$

This operator T is connected to the refinement operator S defined by (2.3.2) as follows [5].

Proposition 2.3.1. *Let $\Omega \subset \mathbb{Z}^n$ be such that $K_\Lambda \subset Q + \Omega$. If $g \colon \mathbb{R}^n \to \mathbb{C}^r$ satisfies* $\operatorname{supp}(g) \subset K_\Lambda$, *then*

$$\Phi S g = T \Phi g \ \text{a.e.} \tag{2.3.7}$$

If the function g satisfies $\operatorname{supp}(g) \subset K_\Lambda$ and additionally vanishes on the boundary of K_Λ, then the equality in (2.3.7) holds everywhere and not merely almost everywhere. This is the case, for example, if g is continuous and supported in K_Λ.

In light of Proposition 2.3.1, in order to solve the refinement equation (2.3.1), we need to find a solution to the equation

$$\mathbf{u} = T\mathbf{u},$$

and this is precisely the type of generalized self-similarity that is defined in (2.2.2).

To do this, we apply Theorem 2.2.1. The operator $\mathcal{O}$ is nonexpansive, the w_k are affine maps, and the functions φ_d are linear. Hence, if there exists a constant C with $0 < C < 1$ and a norm $\|\cdot\|$ on $(\mathbb{C}^r)^\Omega$ such that

$$\forall d \in D, \quad \forall x \in Q, \quad \forall e \in E, \qquad \|\varphi_d(x, e)\| \leq C\|e\|,$$

then T will have a unique fixed point. Considering the definition of φ_d, this means that there must exist a norm in $(\mathbb{C}^r)^\Omega$ such that

$$\forall d \in D, \quad \forall e \in E, \qquad \|T_d e\| \leq C\|e\|.$$

In other words, there must exist a norm under which all the matrices T_d are *simultaneously* contractive on some set. This leads naturally to the definition of the *joint spectral radius* of a set of matrices. Here we will focus only on the uniform joint spectral radius; it is possible to consider various generalizations as well. The uniform joint spectral radius was first introduced in [15] and was rediscovered and applied to refinement equations by Daubechies and Lagarias in [8].

If $\mathcal{M} = \{M_1, \dots, M_m\}$ is a finite collection of $s \times s$ matrices, then the *uniform joint spectral radius* of $\mathcal{M}$ is

$$\hat{\rho}(\mathcal{M}) = \lim_{\ell \to \infty} \max_{\Pi \in \mathcal{P}_\ell} \|\Pi\|^{1/\ell}, \tag{2.3.8}$$

where

$$\mathcal{P}_0 = \{I\} \quad \text{and} \quad \mathcal{P}_\ell = \{M_{j_1} \cdots M_{j_\ell} : 1 \leq j_i \leq m\}.$$

It is easy to see that the limit in (2.3.8) exists and is independent of the choice of norm $\|\cdot\|$ on $\mathbb{C}^{s \times s}$.

Note that if there is a norm such that $\max_j \|M_j\| \leq \delta$, then $\hat{\rho}(\mathcal{M}) \leq \delta$. Rota and Strang [15] proved the following converse result.

Proposition 2.3.2. *Assume that* $\mathcal{M} = \{M_1, \dots, M_m\}$ *is a finite collection of* $s \times s$ *matrices. If* $\hat{\rho}(\mathcal{M}) < \delta$, *then there exists a vector norm* $\|\cdot\|$ *on* $\mathbb{C}^s$ *such that* $\max_j \|M_j\| \leq \delta$.

Consequently, a given set of matrices is simultaneously contractive (i.e., there exists a norm such that $\max_j \|M_j\| < 1$) if and only if the uniform joint spectral radius of $\mathcal{M}$ satisfies $\hat{\rho}(\mathcal{M}) < 1$.

We can now state the main theorem relating generalized self-similarity to the existence of a continuous solution to the refinement equation.

Theorem 2.3.1. *Let* $\Omega \subset \mathbb{Z}^n$ *be a finite set such that* $K_\Lambda \subset Q + \Omega$. *Let* E *be a nonempty closed subset of* $(\mathbb{C}^r)^\Omega$ *such that* $T_d(E) \subset E$ *for each* $d \in D$. *Let* V *be a subspace of* $(\mathbb{C}^r)^\Omega$ *which contains* $E - E$ *and which is right invariant under each* T_d. *Define*

$$\mathcal{F} = \{g \in L^\infty(\mathbb{R}^n, \mathbb{C}^r) : \operatorname{supp}(g) \subset K_\Lambda \text{ and } \Phi g(Q) \subset E\}. \tag{2.3.9}$$

If $\mathcal{F} \neq \emptyset$ *and* $\hat{\rho}(\{T_d|_V\}_{d \in D}) < 1$, *then there exists a function* $f \in \mathcal{F}$ *that is a solution to the refinement equation* (2.3.1), *and the cascade algorithm* $f^{(i+1)} = Sf^{(i)}$ *converges uniformly to* f *for each starting function* $f^{(0)} \in \mathcal{F}$. *Furthermore, if there exists any continuous function* $f^{(0)} \in \mathcal{F}$, *then* f *is continuous.*

Proof: We will apply Theorem 2.2.1 with $X = Q$ and with E the specified subset of $(\mathbb{C}^r)^\Omega$. We let w_d, φ_d, $\mathcal{O}$, and T be as defined above, specifically, by equations (2.3.3), (2.3.4), (2.3.5), and (2.3.6).

We will show that the hypotheses of Theorem 2.2.1 are satisfied. First, $w_d(x) = A^{-1}(x + d)$ is clearly injective and continuously differentiable.

Second, let δ be any number such that

$$\hat{\rho}(\{T_d|_V\}_{d \in D}) < \delta < 1.$$

Then by Proposition 2.3.2 applied to the matrices $T_d|_V$, there exists a vector norm $\|\cdot\|_V$ on V such that

$$\max_{d \in D} \|T_d w\|_V \leq \delta \|w\|_V, \qquad \text{all } w \in V.$$

Let $\|\cdot\|$ denote any extension of this norm to all of $(\mathbb{C}^r)^\Omega$. Recall that $\varphi_d(x, e) = T_d e$. Since $E - E \subset V$, we therefore have for each $x \in Q$ and $u, v \in E$ that

$$\max_{d \in D} \|\varphi_d(x, u) - \varphi_d(x, v)\| = \max_{d \in D} \|T_d(u - v)\| \leq \delta \|u - v\|.$$

Therefore, the functions φ_d satisfy the condition (2.2.3) with constant $C = \delta$. It is easy to check that each φ_d is stable.

Finally, $\mathcal{O}$ is nonexpansive. Thus, the hypotheses of Theorem 2.2.1 are satisfied. Since $C < 1$, Theorem 2.2.1 implies that T maps $L^\infty(Q, E)$ into itself and satisfies

$$\|T\mathbf{u} - T\mathbf{v}\| \leq C \|\mathbf{u} - \mathbf{v}\|.$$

It follows that T is contractive on $L^\infty(Q, E)$ and there exists a unique function $\mathbf{v}^* \in L^\infty(Q, E)$ such that $T\mathbf{v}^* = \mathbf{v}^*$. Further, the iteration $\mathbf{v}^{(i+1)} = T\mathbf{v}^{(i)}$ converges in $L^\infty(Q, E)$ to $\mathbf{v}^*$ for each function $\mathbf{v}^{(0)} \in L^\infty(Q, E)$.

We want to relate now the fixed point $\mathbf{v}^*$ for T to a solution to the refinement equation. First, it can be shown that the space $\mathcal{F}$ is invariant under the refinement operator S. Hence, by Proposition 2.3.1, the following diagram commutes, with T in particular being a contraction:

$$\begin{array}{ccc} \mathcal{F} & \xrightarrow{\Phi} & L^\infty(Q, E) \\ {\scriptstyle S}\downarrow & & \downarrow{\scriptstyle T} \\ \mathcal{F} & \xrightarrow[\Phi]{} & L^\infty(Q, E). \end{array}$$

Now suppose that $f^{(0)}$ is any function in $\mathcal{F}$, and define $f^{(i+1)} = Sf^{(i)}$. Then $f^{(i)} \in \mathcal{F}$ for each i, and if we set $\mathbf{v}^{(i)} = \Phi f^{(i)}$, then

$$\mathbf{v}^{(i+1)} = \Phi f^{(i+1)} = \Phi S f^{(i)} = T \Phi f^{(i)} = T\mathbf{v}^{(i)},$$

so $\mathbf{v}^{(i)}$ must converge uniformly to $\mathbf{v}^*$. By choosing an appropriate choice of norm on $\mathcal{F}$ (see [5]), it follows that $f^{(i)}$ converges uniformly to some function

$f \in L^\infty(\mathbb{R}^n, \mathbb{C}^r)$. We must have $f \in \mathcal{F}$ since $\mathcal{F}$ is a closed subset of $L^\infty(\mathbb{R}^n, \mathbb{C}^r)$. Further,

$$\Phi f = \mathbf{v}^* = T\mathbf{v}^* = T\Phi f = \Phi S f \text{ a.e.}$$

Therefore, f satisfies the refinement equation (2.3.1) almost everywhere. Since $\mathbf{v}^*$ is unique, the cascade algorithm must converge to this particular f for any starting function $f^{(0)} \in \mathcal{F}$. It only remains to observe that if any $f^{(0)} \in \mathcal{F}$ is continuous, then the iterates $f^{(i)}$ obtained from $f^{(0)}$ are continuous and converge uniformly to f, so f must itself be continuous. □

From the proof of Theorem 2.3.1, it is clear that the rate of convergence of the cascade algorithm is geometric and can be specified explicitly if desired.

The preceding theorem immediately suggests two questions:

- Does there always exist a space E that is invariant for all T_d?
- Does $\mathcal{F}$ always contain a continuous function?

The answer to both of these questions is yes, under some mild additional hypotheses.

To answer the question of the existence of the space E, let us recall the one-dimensional, single-function case. In this setting, if we impose the standard "minimal accuracy condition"

$$\sum_{k\in\mathbb{Z}} c_{2k} = \sum_{k\in\mathbb{Z}} c_{2k+1} = 1, \tag{2.3.10}$$

then E is the hyperplane through $(1, 0, \dots, 0)$ that is orthogonal to the row vector $(1, 1, \dots, 1)$. This vector is a common left eigenvector to all of the matrices T_d [8]. The minimal accuracy condition is so-called because it is directly related to the *accuracy* of the solution f. In n dimensions with multiplicity r, i.e., with $f\colon \mathbb{R}^n \to \mathbb{C}^r$, the accuracy of f is defined to be the largest integer $\kappa > 0$ such that every polynomial $q(x) = q(x_1, \dots, x_n)$ with $\deg(q) < \kappa$ can be written

$$q(x) = \sum_{k\in\mathbb{Z}^n} a_k f(x+k) = \sum_{k\in\mathbb{Z}^n} \sum_{i=1}^{r} a_{k,i} f_i(x+k) \text{ a.e.}, \qquad x \in \mathbb{R}^n,$$

for some row vectors $a_k = (a_{k,1}, \dots, a_{k,r}) \in \mathbb{C}^{1\times r}$. If no polynomials are reproducible from translates of f, then we set $\kappa = 0$. We say that f has at least *minimal accuracy* if the constant polynomial is reproducible from translates of f, i.e., if $\kappa \geq 1$. We say that translates of f along $\mathbb{Z}^n$ are *linearly independent* if $\sum_{k\in\mathbb{Z}^n} a_k f(x+k) = 0$ implies $a_k = 0$ for each k. In one dimension, under the hypotheses of linear independence of translates, the minimal accuracy condition (2.3.10) implies that f has at least minimal accuracy. In the general setting of n dimensions and multiplicity r, the minimal accuracy condition is more complicated to formulate than (2.3.10). However, this condition is still the appropriate tool to construct an appropriate set E. We present here a weak form of the minimal accuracy condition and refer to [4] for a general result.

Theorem 2.3.2. *Let $f \colon \mathbb{R}^n \to \mathbb{C}^r$ be an integrable, compactly supported solution of the refinement equation* (2.3.1)*, such that translates of f along $\mathbb{Z}^n$ are linearly independent. Then the following statements are equivalent.*

(a) *f has accuracy $\kappa \geq 1$.*
(b) *There exists a row vector $u_0 \in \mathbb{C}^{1\times r}$ such that $u_0 \hat{f}(0) \neq 0$ and*

$$u_0 = \sum_{k \in \Gamma_d} u_0 c_k \qquad \textit{for each } d \in D.$$

In the case that either statement holds, we have

$$\sum_{k \in \Gamma} u_0 f(x+k) = 1 \textit{ a.e.}$$

Assume now that the minimal accuracy condition given in Theorem 2.3.2 is satisfied, and let u_0 be the row vector such that $\sum_{k\in\mathbb{Z}^n} u_0 f(x+k) = 1$ a.e. It can be shown that the inclusions $\operatorname{supp}(f) \subset K_\Lambda \subset Q + \Omega$ imply that if $x \in Q$, then the only nonzero terms in the series $\sum_{k\in\mathbb{Z}^n} u_0 f(x+k) = 1$ occur when $k \in \Omega$. Hence, if we set $e_0 = (u_0)_{k\in\Omega}$, i.e., e_0 is the row vector obtained by repeating the block u_0 once for each $k \in \Omega$, then

$$e_0 \Phi f(x) = \sum_{k\in\Omega} u_0 f(x+k) = \sum_{k\in\mathbb{Z}^n} u_0 f(x+k) = 1 \text{ a.e.}, \qquad \text{for } x \in Q.$$

Thus, the values of $\Phi f(x)$ are constrained to lie in a particular hyperplane E_0 in $(\mathbb{C}^r)^\Omega$, namely, the collection of column vectors $v = [v_k]_{k\in\Omega}$ such that $e_0 v = \sum_{k\in\Omega} u_0 v_k = 1$. This hyperplane E_0 is a canonical choice for the set E appearing in the hypotheses of Theorem 2.3.1. In order to invoke Theorem 2.3.1, the starting functions $f^{(0)}$ for the cascade algorithm should therefore also have the property that $\Phi f^{(0)}(x)$ always lies in this hyperplane E_0. Note that with this definition of E_0, the set of differences $V_0 = E_0 - E_0$ is the subspace consisting of vectors $v = [v_k]_{k\in\Omega}$ such that $e_0 v = \sum_{k\in\Omega} u_0 v_k = 0$. Hence, the minimal accuracy condition immediately provides an appropriate choice for the space E, namely, we take $E = E_0$.

Now, having defined $E = E_0$, we are ready to address the second question, whether the set $\mathcal{F}$ defined by (2.3.9) always contains a continuous function. First we rewrite $\mathcal{F}$ as

$$\mathcal{F} = \left\{ g \in L^\infty(\mathbb{R}^n, \mathbb{C}^r) : \operatorname{supp}(g) \subset K_\Lambda \text{ and } \sum_{k\in\mathbb{Z}^n} u_0 g(x+k) = 1 \right\},$$

and note that this set is determined by two quantities: the set Λ and the row vector u_0. The set Λ is the support of the set of coefficients c_k in the refinement equation and is determined only by the location of the c_k and not by their values. The vector u_0, on the other hand, is determined by the values of the c_k as well as their locations. However, it can be shown that the question of whether $\mathcal{F}$ contains a continuous function is determined solely by Λ and not by u_0. Thus, only the location of the coefficients c_k is important for this question, and not their actual values. This is made precise in the following result [5].

Lemma 2.3.1. *Let $\Lambda \subset \mathbb{Z}^n$ be finite, and let u_0 be a nonzero row vector in $\mathbb{C}^{1\times r}$. Then the following statements are equivalent.*

(a) $\mathcal{F} \neq \emptyset$.
(b) $\mathcal{F}$ *contains a continuous function.*
(c) $K_\Lambda^\circ + \mathbb{Z}^n = \mathbb{R}^n$, *i.e., lattice translates of the interior K_Λ° of K_Λ cover $\mathbb{R}^n$.*

Thus, in designing a multiwavelet system, after choosing the dilation matrix A and digit set D, the next step is to choose a set Λ that fulfills the requirements of Lemma 2.3.1. Small Λ are preferable, since the larger Λ is, the larger the matrices T_d will be, and the more computationally difficult the computation of the joint spectral radius becomes. While we expect that some "small" Λ may fail the requirement $K_\Lambda^\circ + \mathbb{Z}^n = \mathbb{R}^n$, it is not true that all "large" Λ will necessarily satisfy this requirement (see [5] for an example).

In summary, once we impose the minimal accuracy condition and choose an appropriate set Λ, in order to check for the existence of a continuous scaling function, we must evaluate the uniform joint spectral radius $\hat{\rho}(\{T_d|_{V_0}\}_{d\in D})$. Unfortunately, this might involve the computation of products of large matrices. It can be shown that if the coefficients c_k satisfy the conditions for higher-order accuracy, then V_0 is only the largest of a decreasing chain of common invariant subspaces

$$V_0 \supset V_1 \supset \cdots \supset V_{\kappa-1}$$

of the matrices T_d, and that, as a consequence, the value of $\hat{\rho}(\{T_d|_{V_0}\}_{d\in D})$ is determined by the value of $\hat{\rho}(\{T_d|_{V_{\kappa-1}}\}_{d\in D})$ [5]. This reduction in dimension can ease the computational burden of approximating the joint spectral radius. Moreover, these invariant spaces V_s are directly determined from the coefficients c_k via the accuracy conditions, which are a system of linear equations. Hence, it is a simple matter to compute the matrices $T_d|_{V_{\kappa-1}}$. Additionally, the fact that accuracy implies such specific structure in the matrices T_d suggests that this structure could potentially be used to develop theoretical design criteria for multiwavelet systems.

A final question concerns the converse of Theorem 2.3.1. What can we say if after choosing coefficients c_k that satisfy the minimal accuracy condition, the joint spectral radius of $\hat{\rho}(\{T_d|_{V_0}\}_{d\in D})$ exceeds 1? The following theorem answers this question. It is somewhat surprising, because it essentially says that if a given operator has a fixed point, then that operator must *necessarily* be contractive. This theorem is proved in this generality in [5] but is inspired by a one-dimensional theorem of Wang [17].

Theorem 2.3.3. *Let f be a continuous, compactly supported solution to the refinement equation* (2.3.1) *such that f has L^∞-stable translates* (*defined below*). *Assume that there exists a row vector $u_0 \in \mathbb{C}^{1\times r}$ such that*

$$u_0 \hat{f}(0) \neq 0 \quad \text{and} \quad u_0 = \sum_{k\in\Gamma_d} u_0 c_k \qquad \text{for } d \in D.$$

If $\Omega \subset \mathbb{Z}^n$ is any set such that

$$K_\Lambda \subset Q + \Omega \quad and \quad A^{-1}(\Omega + \Lambda - D) \cap \mathbb{Z}^n \subset \Omega,$$

then

$$\hat{\rho}(\{T_d|_{V_0}\}_{d \in D}) < 1.$$

Here, we say that a vector function $g \in L^\infty(\mathbb{R}^n, \mathbb{C}^r)$ has L^∞-*stable translates* if there exist constants $C_1, C_2 > 0$ such that

$$C_1 \sup_{k \in \Gamma} \max_i |a_{k,i}| \leq \left\| \sum_{k \in \Gamma} a_k g(x+k) \right\|_{L^\infty} \leq C_2 \sup_{k \in \Gamma} \max_i |a_{k,i}|$$

for all sequences of row vectors $a_k = (a_{k,1}, \ldots, a_{k,r})$ with only finitely many a_k nonzero.

2.4 Existence of MRAs

In this section, we turn to the problem of using the existence of a solution to the refinement equation to construct orthonormal multiwavelet bases for $L^2(\mathbb{R}^n)$. As in the classical one-dimensional, single-function theory, the key point is that a vector scaling function that has orthonormal lattice translates determines an MRA for $\mathbb{R}^n$. The MRA then, in turn, determines a wavelet basis for $L^2(\mathbb{R}^n)$.

The main novelty here, more than allowing more than one scaling function or working in arbitrary dimensions, is the result of having an arbitrary dilation matrix. The viewpoint of self-similarity and IFSs still leads naturally to the correct decompositions [5].

Definition 2.4.1. *An MRA of multiplicity* r associated with a dilation matrix A is a sequence of closed subspaces $\{\mathcal{V}_j\}_{j \in \mathbb{Z}}$ of $L^2(\mathbb{R}^n)$ which satisfy the properties:

P1. $\mathcal{V}_j \subset \mathcal{V}_{j+1}$ for each $j \in \mathbb{Z}$,
P2. $g(x) \in \mathcal{V}_j \iff g(Ax) \in \mathcal{V}_{j+1}$ for each $j \in \mathbb{Z}$,
P3. $\bigcap_{j \in \mathbb{Z}} \mathcal{V}_j = \{0\}$,
P4. $\bigcup_{j \in \mathbb{Z}} \mathcal{V}_j$ is dense in $L^2(\mathbb{R}^n)$, and
P5. there exist functions $\varphi_1, \ldots, \varphi_r \in L^2(\mathbb{R}^n)$ such that the collection of lattice translates

$$\{\varphi_i(x-k)\}_{k \in \mathbb{Z}^n, i=1,\ldots,r}$$

forms an orthonormal basis for $\mathcal{V}_0$.

If these conditions are satisfied, then the vector function $\varphi = (\varphi_1, \ldots, \varphi_r)^{\mathrm{T}}$ is referred to as a *vector scaling function* for the MRA.

The usual technique for constructing an MRA is to start from a vector function $\varphi = (\varphi_1, \ldots, \varphi_r)^{\mathrm{T}}$ such that $\{\varphi_i(x-k)\}_{k \in \mathbb{Z}^n, i=1,\ldots,r}$ is an orthonormal system in

$L^2(\mathbb{R}^n)$, and then to construct the subspaces $\mathcal{V}_j \subset L^2(\mathbb{R}^n)$ as follows. First, let $\mathcal{V}_0$ be the closed linear span of the translates of the component functions φ_i, i.e.,

$$\mathcal{V}_0 = \overline{\text{span}}\{\varphi_i(x-k)\}_{k\in\mathbb{Z}^n, i=1,\dots,r}. \tag{2.4.1}$$

Then, for each $j \in \mathbb{Z}$, define $\mathcal{V}_j$ to be the set of all dilations of functions in $\mathcal{V}_0$ by A^j, i.e.,

$$\mathcal{V}_j = \{g(A^j x) : g \in \mathcal{V}_0\}. \tag{2.4.2}$$

If $\{\mathcal{V}_j\}_{j\in\mathbb{Z}}$ defined in this way forms an MRA for $L^2(\mathbb{R}^n)$, then we say that it is the *MRA generated by* φ.

Example 2.4.1. In one dimension, the box function $\varphi = \chi_{[0,1)}$ generates an MRA for $L^2(\mathbb{R})$. This MRA is usually referred to as the *Haar MRA*, because the wavelet basis it determines is the classical Haar system $\{2^{n/2}\psi(2^n x - k)\}_{n,k\in\mathbb{Z}}$, where $\psi = \chi_{[0,1/2)} - \chi_{[1/2,1)}$.

Gröchenig and Madych [10] proved that there is a Haar-like MRA associated to each choice of dilation matrix A and digit set D for which the attractor $Q = K_D$ is a tile. In particular, they proved that if Q is a tile, then the scalar-valued function χ_Q generates an MRA of $L^2(\mathbb{R}^n)$ of multiplicity 1. By extension of the one-dimensional terminology, this MRA is called the *Haar MRA associated with A and D*. Note that the fact that $\{\chi_Q(x-k)\}_{k\in\Gamma}$ forms an orthonormal basis for $\mathcal{V}_0$ is a restatement of the assumption that the lattice translates of the tile Q have overlaps of measure zero. Further, χ_Q is refinable because Q is self-similar and because the lattice translates of Q have overlaps of measure zero.

We will characterize those φ that generate MRAs in the following theorem. To motivate this result, note that property P2 is achieved trivially when $\mathcal{V}_j$ is defined by (2.4.2). Moreover, property P5 is simply a statement that lattice translates of φ are orthonormal. It can be seen [5] that the fact that φ has orthonormal lattice translates implies that property P3 is also automatically satisfied. Thus, the main problem in determining whether φ generates an MRA is the question of when properties P1 and P4 are satisfied. One necessary requirement for P1 is clear. If φ does generate an MRA, then P1 implies that $\varphi_i \in \mathcal{V}_0 \subset \mathcal{V}_1$ for $i = 1, \dots, r$. Since P2 and P5 together imply that $\{m^{1/2}\varphi_j(Ax-k)\}_{k\in\mathbb{Z}^n, j=1,\dots,r}$ forms an orthonormal basis for $\mathcal{V}_1$, each function φ_i must therefore equal some (possibly infinite) linear combination of the functions $\varphi_j(Ax-k)$. Consequently, the vector function φ must satisfy a refinement equation of the form

$$\varphi(x) = \sum_{k\in\mathbb{Z}^n} c_k \varphi(Ax-k) \tag{2.4.3}$$

for some choice of $r \times r$ matrices c_k. Since we only consider the case where the functions φ_i have compact support and since φ has orthonormal lattice translates, this implies that only finitely many of the matrices c_k in (2.4.3) can be nonzero. Hence, in this case the refinement equation in (2.4.3) has the same form as the refinement equation (2.3.1).

Theorem 2.4.1. *Assume that* $\varphi = (\varphi_1, \ldots, \varphi_r)^{\mathrm{T}} \in L^2(\mathbb{R}^n, \mathbb{C}^r)$ *is compactly supported and has orthonormal lattice translates, i.e.,*

$$\langle \varphi_i(x-k) \rangle \, \varphi_j(x-\ell) = \int \varphi_i(x-k)\overline{\varphi_j(x-\ell)}dx = \delta_{i,j}\delta_{k,\ell}.$$

Let $\mathbf{V}_j \subset L^2(\mathbb{R}^n)$ *for* $j \in \mathbb{Z}$ *be defined by* (2.4.1) *and* (2.4.2). *Then the following statements hold.*

(a) *Properties P2, P3, and P5 are satisfied.*

(b) *Property P1 is satisfied if and only if* φ *satisfies a refinement equation of the form*

$$\varphi(x) = \sum_{k \in \Lambda} c_k \varphi(Ax - k) \tag{2.4.4}$$

for some $r \times r$ *matrices* c_k *and some finite set* $\Lambda \subset \mathbb{Z}^n$.

(c) *If*

$$\sum_{i=1}^{r} |\hat{\varphi}_i(0)|^2 = \sum_{i=1}^{r} \left| \int \varphi_i(x)dx \right|^2 = |Q| = 1, \tag{2.4.5}$$

then Property P4 is satisfied. If φ *is refinable, i.e., if* (2.4.4) *holds, then Property P4 is satisfied if and only if* (2.4.5) *holds.*

Note that the assumption that φ_i is square integrable and compactly supported implies that $\varphi_i \in L^1(\mathbb{R}^n)$, so $\hat{\varphi}_i(0) = \int \varphi_i(x)dx$ is well defined.

Theorem 2.4.1 generalizes a result of Cohen [7], which applied specifically to the case of multiplicity 1 and dilation $A = 2I$. Cohen's estimates used a decomposition of $\mathbb{R}^n$ into dyadic cubes, making essential use of the fact that the uniform dilation $A = 2I$ maps dyadic cubes into dyadic cubes. However, this need not be true for an arbitrary dilation matrix A, so this particular decomposition is no longer feasible. Instead, the proof in [5] uses a decomposition based on the tile Q and the associated Haar MRA discussed in Example 2.4.1. One of the key observations lies in counting the number of lattice translates of Q which lie in the interior of a dilated tile $A^jQ, j \geq 1$. The fact that Q is self-similar combined with the fact that translates of Q tile $\mathbb{R}^n$ with overlaps with measure zero implies that A^jQ is a union of exactly m^j translates of Q, with each such translate lying entirely inside A^jQ (but not necessarily in the *interior* of A^jQ). It can be shown that the ratio of the number of those translates $Q + k$ that intersect the boundary of A^jQ to the total number lying inside A^jQ converges to zero.

We conclude by showing in Figure 2.2 a pair of wavelets associated to an MRA obtained by numerically solving the "accuracy 2" conditions given in [4] to obtain the coefficients c_k for a scaling vector $\varphi\colon \mathbb{R}^2 \to \mathbb{R}^2$ with orthonormal lattice translates that is refinable with respect to a quincunx dilation matrix (these numerical estimates were obtained by A. Ruedin, see [16] for related results). Using the results outlined in this paper, one can prove that these coefficients yield a continuous scaling vector that generates an MRA whose "mother wavelets" are those pictured in Figure 2.2.

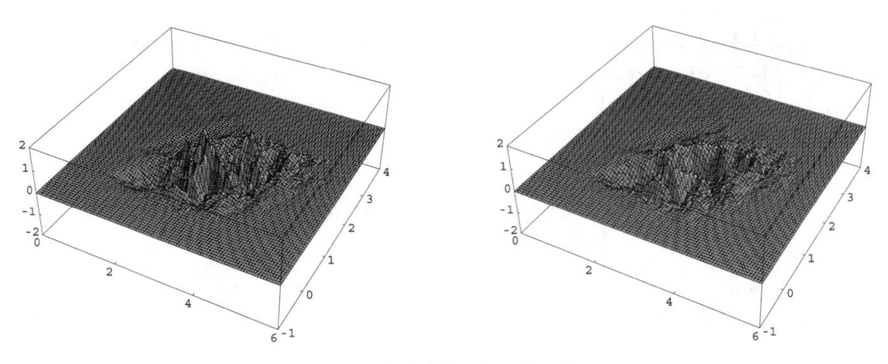

FIGURE 2.2. Wavelets ψ_1, ψ_2.

Acknowledgments. The authors thank Professor Lokenath Debnath for organizing the NSF-CBMS conference in Orlando (1998) and for inviting them to write this article.

Cabrelli and Molter also thank the School of Mathematics of the Georgia Institute of Technology for inviting them to visit several times, during which time they performed part of the research for the results behind this paper.

The research of Cabrelli and Molter was partially supported by grants UBACyT TW84 and CONICET, PIP456/98. The research of Heil was partially supported by NSF Grant DMS-9970524.

References

[1] M. Bajraktarevic, Sur une équation fonctionnelle, *Glasnik Mat.-Fiz. Astr. Društvo Mat. Fiz. Hrvatske Ser. II.*, **12** (1957), 201–205.

[2] C. Bandt, Self-similar sets, V. Integer matrices and fractal tilings of $\mathbb{R}^n$, *Proc. Amer. Math. Soc.*, **112** (1991), 549–562.

[3] C. Cabrelli, C. Heil, and U. Molter, Accuracy of several multidimensional refinable distributions, *J. Fourier Anal. Appl.*, **6** (2000), 483–502.

[4] C. Cabrelli, C. Heil, and U. Molter, Accuracy of lattice translates of several multidimensional refinable functions, *J. Approx. Theory*, **95** (1998), 5–52.

[5] C. Cabrelli, C. Heil, and U. Molter, Self-similarity and multiwavelets in higher dimensions, preprint (1999).

[6] C. A. Cabrelli and U. M. Molter, Generalized self-similarity, *J. Math. Anal. Appl.*, **230** (1999), 251–260.

[7] A. Cohen, Ondelettes, analyses multirésolutions et filtres miroirs en quadrature, *Ann. Inst. H. Poincaré Anal. Non Linéaire*, **7** (1990), 439–459.

[8] I. Daubechies and J. C. Lagarias, Two-scale difference equations: II. Local regularity, infinite products and fractals, *SIAM J. Math. Anal.*, **23** (1992), 1031–1079.

[9] G. de Rahm, Sur quelques courbes définies par des équations fonctionnelles, *Rend. Sem. Mat. dell'Univ. e del Politecnico di Torino*, **16** (1957), 101–113.

[10] K. Gröchenig and W. R. Madych, Multiresolution analysis, Haar bases, and self-similar tilings of $\mathbb{R}^n$, *IEEE Trans. Inform. Theory*, **38** (1992), 556–568.

[11] J. E. Hutchinson, Fractals and self-similarity, *Indiana Univ. Math. J.*, **30** (1981), 713–747.

[12] J. C. Lagarias and Y. Wang, Corrigendum and addendum to: Haar bases for $L^2(\mathbb{R}^n)$ and algebraic number theory, *J. Number Theory*, **76** (1999), 330–336.

[13] Y. Meyer, *Wavelets and Operators*, Cambridge University Press, Cambridge, 1992.

[14] A. Potiopa, A problem of Lagarias and Wang, Master's Thesis, Siedlce University, Siedlce, Poland (Polish), 1997.

[15] G. C. Rota and G. Strang, A note on the joint spectral radius, *Indagationes Mathematicae*, **22** (1960), 379–381.

[16] A. M. C. Ruedin, Construction of nonseparable multiwavelets for nonlinear image compression, *EURASIP J. Appl. Signal Process.*, **2002** (2002), 73–79.

[17] Y. Wang, Two-scale dilation equations and the cascade algorithm, *Random Comput. Dynamics*, **3** (1995), 289–307.

3

On the Nonexistence of Certain Divergence-free Multiwavelets

J. D. Lakey
M. C. Pereyra

ABSTRACT We show that there are no biorthogonal pairs of divergence-free multiwavelet families on $\mathbf{R}^n$, having any regularity, such that both biorthogonal families have compactly supported, divergence-free generators. This main result generalizes Lemarié's bivariate result. In particular, our method is based on vector-valued multiresolution analyses.

3.1 Introduction

We will demonstrate the nonexistence of two-scale biorthogonal (multi)wavelets having some regularity, such that both biorthogonal families have compactly supported, divergence-free generators. This gives a negative answer to a question posed to us concerning the possible use of multiwavelets to circumvent the nonexistence result of Lemarié [3]. Actually, this fact does not impart severe obstacles to the application of wavelet-Galerkin methods for Navier–Stokes systems [5]. Nevertheless, it clarifies the sort of limitations that the wavelets must have, cf., [4].

The method is based on Lemarié's bivariate result [3], which relies on a dimensionality trick: divergence-free vector fields in two dimensions have the form $(\partial f/\partial x_2, -\partial f/\partial x_1)$, where $f : \mathbf{R}^2 \to \mathbf{R}$ is a scalar-valued function. This trick goes a long way toward making the analysis essentially scalar valued and it is not available in higher dimensions. Lemarié's method is based, in turn, on his characterization of projectors onto shift invariant spaces [2] together with a key result of DeVore, de Boor, and Ron [1]. To extend his technique requires extending results in [2] and [1] to a subspace $\mathcal{H}^2 = \mathcal{H}^2(\mathbf{R}^n, \mathbf{R}^n)$ of $L^2(\mathbf{R}^n, \mathbf{R}^n)$ consisting of distributionally divergence-free vector fields. Here $\mathcal{H}$ stands for Hardy! The required extensions are done here. One of the main factors that allows the analysis to go through is the nontrivial fact that $\mathcal{H}^2$ is preserved under scalar-valued Fourier multipliers of L^2. We restrict our wavelet analysis to the case of two-scale dilations: more general dilations that map integers to integers can be considered as in [3].

3.2 The main theorem

First, we need some conventions. Unless explicitly stated, all functions of x will map $\mathbf{R}^n \to \mathbf{R}^n$, and their values will be regarded as column vectors, their transposed values as row vectors. Then $f^\top g$ will be scalar valued; in particular, the L^2 inner product is $\langle f, g\rangle = \int f^\top g$. But $fg^\top$ will take values in the $n\times n$ matrices $\mathcal{M}_n(\mathbf{R}^n)$. The Fourier transform is defined on integrable functions by the formula $\widehat{f}(\xi) = \int_{\mathbf{R}} e^{-2\pi i x\cdot\xi} f(x)dx \in \mathbf{C}^n$. The integration in this case is carried out componentwise on f. However, when we represent an operator K by integration against a kernel $k(x,y)$, the kernel will typically have values in $\mathcal{M}_n(\mathbf{R}^n)$.

The divergence-free wavelet families considered here come from two-scale dilations and will be indexed by sets of the form

$$\Lambda = \{\lambda = (\varepsilon, \alpha, i, Q) : \varepsilon \in \{0,1\}^n \setminus \{0,0\}, i = 1, \dots, n-1, Q \in \mathcal{Q}, \alpha \in \mathcal{A}\},$$

where $\mathcal{Q}$ denotes the collection of dyadic cubes, and $\mathcal{A}$ is some finite index that is used to indicate that the wavelets are allowed to be "multi"wavelets. We will denote by $Q(\lambda)$ the dyadic cube belonging to the index λ. By a biorthogonal pair of wavelet bases for the space $\mathcal{H}^2(\mathbf{R}^n, \mathbf{R}^n)$ we mean a pair $\{\psi_\lambda, \psi^*_{\lambda'}\}_{\lambda,\lambda'\in\Lambda}$ such that any $f \in \mathcal{H}^2(\mathbf{R}^n, \mathbf{R}^n)$ possesses an expansion of the form

$$f = \sum_{\lambda\in\Lambda} \langle f, \psi^*_\lambda\rangle\, \psi_\lambda; \qquad \|f\|_2^2 \sim \sum_{\lambda\in\Lambda} |\langle f, \psi^*_\lambda\rangle|^2 .$$

The square-integrable vector fields ψ_λ are divergence-free in the sense of distributions modulo polynomials, but the ψ^*_λ need not be. In fact, the main result is nothing but the statement that under certain niceness conditions they *cannot* be.

Theorem 3.2.1. *Biorthogonal wavelet families $\{\psi_\lambda, \psi^*_{\lambda'}\}$ defined as above cannot be simultaneously divergence-free, compactly supported, and Hölder continuous of some positive order.*

The result does not pose obstacles to wavelet-Galerkin methods because the *synthesizing* wavelets have the differential property built in. The only property required of the *analyzing* wavelets is that they allow wavelet coefficients to be computed rapidly.

Here is the idea of the proof: Gradients of harmonic, scalar-valued functions h are locally square integrable. In particular, if the ψ^*_λ have compact support, then the coefficients $\langle \nabla h, \psi^*_\lambda\rangle$ are well defined and, when $\nabla\cdot\psi^*_\lambda = 0$, they vanish in view of integration by parts. As it turns out, this allows one to identify the restriction of a certain cutoff of ∇h with the restriction of its "V_0" multiresolution projection to a neighborhood of the origin. If the ψ_λ have compact support, then the space V_0 turns out to be locally finite dimensional. Most of the work lies in verifying this. In contrast, harmonic functions form a locally infinite-dimensional space. This contradicts the local equality between ∇h and its V_0-projection. Regularity and compact support of the ψ^*_λ are required to apply integration by parts; they are required of the ψ_λ to conclude that V_0 is locally finite dimensional.

3.3 Shift invariant subspaces of $\mathcal{H}^2(\mathbf{R}^n, \mathbf{R}^n)$ and ω-localized projections

Much of this section amounts to lifting known facts about $L^2(\mathbf{R}^n)$ to the case of $\mathcal{H}^2(\mathbf{R}^n, \mathbf{R}^n)$. First, a closed subspace V_0 of $\mathcal{H}^2(\mathbf{R}^n, \mathbf{R}^n)$ is said to be *shift invariant* if, whenever $f \in V_0$ and $k \in \mathbf{Z}^n$, one has $f(x-k) \in V_0$. In what follows, V_0 will always be shift invariant. The operator P_0 that projects onto V_0 is shift invariant if for all $f \in \mathcal{H}^2(\mathbf{R}^n, \mathbf{R}^n)$ one has $(P_0 f)(x-k) = P_0(f(\cdot - k))(x)$. When is V_0 generated by the translates $\phi_\alpha(x-k)$ of a finite family $\{\phi_\alpha\}_{\alpha \in A}$? Such a space is said to be a *finite shift invariant* (FSI) space. More particularly, when does $\{\phi_\alpha(x-k)\}_{\alpha \in A}$ form a Riesz basis for V_0?

The answer can be phrased in terms of regularity of the kernel of P_0. This was formulated for $L^2(\mathbf{R}^n)$ by Lemarié. We say that P_0 has an *ω-localized kernel for $\mathcal{H}^2$* provided its action is given by integration against a locally integrable kernel function $p(x,y) \in \mathcal{M}_n$ such that $\nabla_x^\top \cdot p(x,y) = 0$ in the sense that

$$\langle P_0 f, g\rangle = \int\int \overline{g}^\top(x) p(x,y) f(y)\, dy\, dx; \quad \langle P_0 f, \nabla_x h\rangle = 0, \qquad h \in C_c^\infty(\mathbf{R}^n),$$

where

$$\int_{x\in[0,1)^n} \int_{y \in \mathbf{R}^n} \omega(x-y)|p(x,y)|^2\, dx\, dy < \infty$$
$$\int_{y\in[0,1)^n} \int_{x \in \mathbf{R}^n} \omega(x-y)|p(x,y)|^2\, dx\, dy < \infty.$$

Here, $|p(x,y)| = [\sum_{i,j=1}^n |p_{ij}(x,y)|^2]^{1/2}$. We will need to assume that ω is a symmetric *Beurling weight*. This means that ω is a nonnegative valued function that is strictly bounded below, has growth of at most polynomial order, $\omega(x) = \omega(-x)$, $\omega(x+y) \le \omega(x)\omega(y)$, and $\omega^{-1} \in L^1(\mathbf{R}^n)$ and $\omega^{-1} * \omega^{-1} \le C\omega^{-1}$. For a review of the properties of Beurling weights, refer to Annexe A in [2]. We have the following theorem.

Theorem 3.3.1. *If P_0 has an ω-localized kernel for $\mathcal{H}^2$, then V_0 has a Riesz basis of the form $\{\phi_\alpha(x-k)\}_{\alpha\in\mathcal{A}, k\in\mathbf{Z}^n}$. The cardinality $A = |\mathcal{A}|$ does not depend on the choice of the basis. In fact,*

$$A = \operatorname{tr} \int_{[0,1)^n} \int_{[0,1)^n} \widetilde{p}(x,y)\widetilde{p}(y,x)\, dy\, dx,$$

where

$$\widetilde{p}(x,y) = \sum_{k \in \mathbf{Z}^n} p(x, y-k).$$

Note that A is well defined because $\widetilde{p}$ is periodic and locally square integrable. The kernel of P_0 can be defined *a posteriori* by the basis expansion

$$p(x,y) = \sum_k \sum_\alpha \phi_\alpha(x-k)\phi_\alpha(y-k)^\top,$$

which is $\mathcal{M}_n$-valued when ϕ is $\mathbf{R}^n$-valued. But we must construct the basis given the ω-localized kernel.

The appearance of Beurling weights in this theory might look a bit peculiar, so we will briefly remark on their role here. The basis functions ϕ_α will be assembled from certain functions in the weighted space L^2_ω, which is a subspace of the Beurling algebra. The crucial step in the assembly is a local inversion at the Fourier level, which is made possible because elements of the Beurling algebra have absolutely convergent Fourier transforms. In the case of Theorem 3.2.1, the Beurling weight that arises is just $(1+|x|)^{n+\rho}$, where ρ depends on the Hölder regularity.

Before proceeding, it makes sense to introduce here the analogue of an additional key idea due to DeVore, de Boor, and Ron [1] which addresses the remaining crucial ingredient for Theorem 3.2.1: the role of compact support. In [1] an FSI space is called *local* if it has compactly supported generators $\{\phi_\alpha\}$. Local FSI spaces are locally finite dimensional. That is, given any compact set, there is a finite family of functions in the space whose restrictions to that set span the restrictions of all elements of the space to that set. The following corresponds to Corollary 3.36 of [1].

Theorem 3.3.2. *Let V_0 be an FSI subspace of $\mathcal{H}^2$ and suppose that the compactly supported elements of V_0 are dense in V_0. Then V_0 is local.*

Note that in the case of a multiresolution analysis and, in particular, in the setup of Theorem 3.3.1, V_0 can be identified with the closed linear span of the wavelets living on long scales. In this case, V_0 is local when the wavelets have compact support.

3.3.1 The correlation function

Given $f, g \in \mathcal{H}^2(\mathbf{R}^n, \mathbf{R}^n)$, we define their *correlation function*

$$C(f,g)(\xi) = \sum_{k\in\mathbf{Z}^n} \widehat{f}(\xi+k)^\top \overline{\widehat{g}}(\xi+k),$$

which is a periodic, $\mathbf{C}$-valued function. We call $C(f,f) = C(f)$ the *autocorrelation function* of f. When f, g are square integrable, their autocorrelation functions are integrable over the unit cube $[0,1)^n$. Hence, by Cauchy–Schwarz, $C(f,g) \in L^1([0,1)^n)$. The next three lemmas are just $\mathbf{R}^n$-valued versions of Lemmas 1–3 in section I.1 of [2].

Lemma 3.3.1. *The family $\{f(\cdot - k)\}$ is a Bessel family if and only if $C(f) \in L^\infty(\mathbf{T}^n)$.*

Proof: The point is that $C(f)$ can be regarded as a multiplier of $L^2(\mathbf{T}^n)$. Thus, if $\{\alpha_k\} \in l^2(\mathbf{Z}^n)$, Plancherel's theorem gives

$$\left\|\sum \alpha_k f(\cdot - k)\right\|_2^2 = \int_{[0,1)^n} \left|\sum \alpha_k e^{-2\pi i k\cdot\xi}\right|^2 C(f)(\xi)d\xi \leq C \left\|\{\alpha_k\}\right\|_{l^2}^2$$

with C independent of $\{\alpha_k\}$ if and only if $C(f) \in L^\infty(\mathbf{T}^n)$. □

Recall that a family of functions $\{f_\kappa\}_{\kappa\in K}$ is a Riesz family if whenever $\{\alpha_\kappa\} \in l^2(K)$ one has $\|\{\alpha_\kappa\}\|_{l^2} \leq C\|\sum \alpha_\kappa f_\kappa\|_2$. When $K = \mathcal{A} \times \mathbf{Z}^n$ for a fixed, finite set $\mathcal{A}$, $\{f_\kappa\}_{\kappa\in K} = \{f_\alpha(\cdot - k)\}_{\alpha\in\mathcal{A}, k\in\mathbf{Z}^n} \equiv \{f_\alpha(\cdot - k)\}$. Then one can form the *correlation matrix* $M(\xi) = \{C(f_\alpha, f_{\alpha'})\}_{\alpha,\alpha'} \in \mathcal{M}_A$, where $A = |\mathcal{A}|$. We shall write $M(\xi) = M_{f_\alpha}(\xi)$ when we wish to distinguish the correlations of a specific family $\{f_\alpha\}_{\alpha\in\mathcal{A}}$ from those of another family. We want to address the issue of when $\{f_\alpha(\cdot - k)\}$ is a Riesz basis (i.e., Bessel family and Riesz family) for V_0. For any such basis there exists a basis $\{f_\kappa^*\}_{\kappa\in K}$ for $V_0^* = (\ker P_0)^\perp$ that is biorthogonal to $\{f_\alpha(\cdot - k)\}$ in the sense that $\langle f_{\alpha',k'}^*, f_\alpha(\cdot - k)\rangle = \delta_{\alpha\alpha'}\delta_{kk'}$. Actually, the uniqueness of basis expansions implies that $f_{\alpha,k}^*(x) \equiv f_\alpha^*(x - k)$. We shall determine the functions $\{f_\alpha^*\}$ explicitly in terms of $\{f_\alpha\}$, but first we shall use their existence to formulate a useful condition for determining whether $\{f_\alpha(\cdot - k)\}$ is a Riesz basis.

Lemma 3.3.2. *The family $\{f_\alpha(\cdot - k)\}$ is a Riesz basis for its span if and only if $M(\xi) \in \mathcal{M}_A(L^\infty)$ and $1/\det M(\xi) \in L^\infty$.*

Proof: First assume that $\{f_\alpha(\cdot - k)\}$ is a Riesz basis for its span and let $\{f_\alpha^*(\cdot - k)\}$ be a basis of $\{f_\alpha(\cdot - k)\}$. Since

$$f_\alpha^* = \sum_k \sum_{\alpha'} \langle f_\alpha^*, f_{\alpha'}^*(\cdot - k)\rangle\, f_{\alpha'}(\cdot - k),$$

one has

$$\widehat{f_\alpha^*}(\xi) = \sum_k \sum_{\alpha'} \langle f_\alpha^*, f_{\alpha'}^*(\cdot - k)\rangle\, e^{-2\pi i k\cdot\xi} \widehat{f_{\alpha'}}(\xi).$$

But, $\langle f_\alpha^*, f_{\alpha'}^*(\cdot - k)\rangle$ is the kth Fourier coefficient of $C(f_\alpha^*, f_{\alpha'}^*)$, whence, by Parseval's formula,

$$\widehat{f_\alpha^*}(\xi) = \sum_{\alpha'} C(f_\alpha^*, f_{\alpha'}^*)(\xi)\widehat{f_{\alpha'}}(\xi).$$

By a similar computation, one concludes that

$$C(f_\alpha^*, f_{\alpha'})(\xi) = \sum_{\alpha''} C(f_\alpha^*, f_{\alpha''}^*)(\xi) C(f_{\alpha''}, f_{\alpha'})(\xi).$$

On the other hand, biorthogonality implies that

$$C(f_\alpha^*, f_{\alpha'})(\xi) = \sum_k \langle f_\alpha^*, f_{\alpha'}(\cdot - k)\rangle\, e^{-2\pi i k\cdot\xi} = \delta_{\alpha\alpha'}.$$

Therefore, $(M_{f_\alpha}(\xi))^{-1} = M_{f_\alpha^*}(\xi)$, where the latter matrix is the correlation matrix for the dual generators. Now the $\{f_\alpha^*(x-k)\}$ form a Bessel family because they are also a Riesz basis for their span. By the previous lemma, together with Cauchy–Schwarz, it follows that both $M_{f_\alpha}(\xi) \in \mathcal{M}_A(L^\infty)$ and $M_{f_\alpha^*}(\xi) \in \mathcal{M}_A(L^\infty)$. In particular,

$$1 = (\det M_{f_\alpha}(\xi))\left(\det M_{f_\alpha^*}(\xi)\right) \le \det M_{f_\alpha}(\xi)\left\|\det M_{f_\alpha^*}(\xi)\right\|_\infty$$

so that

$$\det M_{f_\alpha}(\xi) \ge \frac{1}{\left\|\det M_{f_\alpha^*}(\xi)\right\|_\infty}.$$

As the right-hand side is independent of ξ, the "only if" follows.

To prove the "if" part, assume that $M(\xi) \in \mathcal{M}_A(L^\infty)$ and that $\det M(\xi)$ is essentially bounded below. Since $(f,g) \mapsto C(f,g)(\xi)$ is Hermitian, it follows that $M(\xi)$ is a positive definite Hermitian matrix almost everywhere. One concludes that there exists a function $\gamma(\xi)$ at least as large as the *smallest* eigenvalue of $M(\xi)$ such that, for any vector $\lambda = \{\lambda_\alpha\}_{\alpha\in\mathcal{A}}$, one has $\lambda^\top M(\xi)\overline{\lambda} \ge \gamma(\xi)\sum_\alpha |\lambda_\alpha|^2$. On the other hand, if $\Lambda(\xi)$ is the *largest* eigenvalue of $M(\xi)$, then

$$\gamma(\xi) \ge \frac{\det M(\xi)}{\Lambda(\xi)^{A-1}}.$$

Since $\Lambda(\xi) \le [A\sum_\alpha C(f_\alpha)(\xi)]^{1/2}$, $\gamma(\xi)$ is minorized by some fixed constant $\gamma > 0$. Thus, setting $\nu_\alpha(\xi) = \sum_k \lambda_{k,\alpha} e^{-2\pi i k\cdot\xi}$ and $\nu(\xi) = \overrightarrow{\nu_\alpha}(\xi) \in \mathbf{C}^A$ and applying Plancherel yields

$$\begin{aligned}\left\|\sum_k\sum_\alpha \lambda_{k,\alpha} f_\alpha(x-k)\right\|_2^2 &= \int_{[0,1)^n} \nu^\top(\xi)M(\xi)\overline{\nu}(\xi)\\ &\ge \gamma\int_{[0,1)^n}\sum_\alpha |\nu_\alpha(\xi)|^2\,d\xi\\ &= \gamma\sum_\alpha\sum_k |\lambda_{k,\alpha}|^2,\end{aligned}$$

which shows that the family $\{f_\alpha(x-k)\}$ forms a Riesz basis for its span. □

In general, one can define the correlation matrix $M(\{f_\alpha\},\{g_\beta\})(\xi)$ between two different families of generators $\{f_\alpha\}_{\alpha\in\mathcal{A}}$ and $\{g_\beta\}_{\beta\in\mathcal{B}}$: once again the entries are the correlations between the various generators. We take this approach presently because it will give a precise description of V_0 as a complemented subspace.

We say that two subspaces V, W of $\mathcal{H}^2$ are in duality, provided $\mathcal{H}^2 = V \oplus W^\perp$.

Lemma 3.3.3. *Let V and W be two closed subspaces of $\mathcal{H}^2$ such that V has a Riesz basis of the form $\{f_\alpha(x-k)\}$ and W has a Riesz basis of the form $\{g_\beta(x-k)\}$. Also let $\{\phi_\delta\}_{\delta\in\mathcal{D}} \subseteq V$ and $\{\psi_\tau\}_{\tau\in\mathcal{T}} \subseteq W$. Then:*

(a) $M(\{\phi_\delta\},\{\psi_\tau\}) = M(\{\phi_\delta\},\{f^*_\alpha\})M(\{f_\alpha\},\{g_\beta\})M(\{g^*_\beta\},\{\psi_\tau\})$, *where* $\{f^*_\alpha\}$ *and* $\{g^*_\beta\}$ *denote the bases dual to* $\{f_\alpha\},\{g_\beta\}$ *in* V, W, *resp.*

(b) $\{\phi_\delta(x-k)\}$ *is a Riesz basis of* V *if and only if* $|\mathcal{D}| = |\mathcal{A}|$ *and* $M(\{\phi_\delta\},\{f^*_\alpha\})$ *belongs to* $\mathcal{M}_A(L^\infty)$ *with inverse in* $\mathcal{M}_A(L^\infty)$.

(c) *If* $N(\xi) = [M(\{f_\alpha\},\{f_\alpha\})(\xi)]^{-1/2}$ *and* $\widehat{\phi_\alpha}(\xi) = \sum_{\alpha'} N_{\alpha,\alpha'}(\xi)\widehat{f_{\alpha'}}(\xi)$, *then* $\{\phi_\alpha(x-k)\}$ *forms an orthonormal basis of* V.

(d) V, W *are in duality if and only if* $|\mathcal{A}| = |\mathcal{B}|$ *and* $M(\{f_\alpha\},\{g_\beta\}) \in \mathcal{M}_A(L^\infty)$. *Moreover, if the matrix* D *satisfies* $M(\{f_\alpha\},\{g_\beta\})^\top \overline{D} = I$, *then the functions* $\{\gamma^*_\alpha\}$ *defined by* $\widehat{\gamma^*_\alpha}(\xi) = \sum_\beta D_{\alpha,\beta}(\xi)\widehat{g_\beta}(\xi)$ *form a basis of* W *dual to* $\{f_\alpha\}$.

Proof: The proof of (a) follows from the fact that $\widehat{\phi_\eta} = \sum_\alpha C(\phi_\eta, f^*_\alpha)\widehat{f_\alpha}$ and a similar identity represents ψ_τ. If $\{\phi_\delta(x-k)\}$ forms a Riesz basis of V, then $M_{\phi_\delta} = M(\{\phi_\delta\},\{\phi_\delta\})$ has maximal rank $|\mathcal{D}|$ almost everywhere. Because one can factorize $M_{f_\alpha} = M(\{f_\alpha\},\{\phi^*_\delta\})M_{\phi_\delta}M(\{\phi^*_\delta\},\{f_\alpha\})$ and one can factorize M_{ϕ_δ} in terms of $\{f_\alpha\}$ similarly, maximal rank of each implies $|\mathcal{D}| = |\mathcal{A}| = A$. Using factorization to compute determinants yields

$$\det M_{\phi_\delta} = \det M_{f_\alpha} \left|\det M(\{\phi_\delta\},\{f^*_\alpha\})\right|^2.$$

By Lemma 3.3.2, $|\det M(\{\phi_\delta\},\{f^*_\alpha\})|$ is minorized a.e. by a fixed constant; in particular, $M(\{\phi_\delta\},\{f^*_\alpha\})$ is uniformly invertible. The converse is immediate from Lemma 3.3.2. This proves (b).

To verify (c), first we show that $\{\phi_\alpha(x-k)\}$ forms a Riesz basis of V. By (a) and the biorthogonality of $\{f_\alpha\},\{f^*_\alpha\}$, $M(\{\phi_\alpha\},\{f^*_\alpha\})(\xi) = N(\xi)$; hence, by multiplicativity of determinants,

$$\det M(\{\phi_\alpha\},\{f^*_\alpha\}) = \frac{1}{\sqrt{\det M(\{f_\alpha\},\{f_\alpha\})}}.$$

Next we wish to show that

$$N(\xi) = \frac{2}{\pi}\int_0^\infty (I + t^2 M(\xi))^{-1} dt$$

has bounded coefficients. Let $\lambda(\xi)$ and $\Lambda(\xi)$ denote the smallest and largest eigenvalues of $M(\xi)$. Then

$$(1 + t^2\lambda(\xi))^A \le \det(I + t^2 M(\xi)) \le (1 + t^2\Lambda(\xi))^A$$

and, consequently,

$$\|(I + t^2 M(\xi))^{-1}\|_{M_A(L^\infty)} \le \frac{C}{1+t^2},$$

hence $N(\xi) \in M_A(L^\infty)$. This proves that $\{\phi_\alpha(x-k)\}$ is a Riesz basis of V. To show that $\{\phi_\alpha(x-k)\}$ is an orthonormal family, it is enough to show that $M(\{\phi_\delta\},\{\phi_\delta\}) = I$. But, by (a),

$$M(\{\phi_\delta\},\{\phi_\delta\}) = N(\xi)M(\{f_\alpha\},\{f_\alpha\})N(\xi) = I.$$

This proves (c).

Finally, (d) is clear since V, W are in duality if and only if W has a Riesz basis $\{\gamma_\alpha^*(x-k)\}_\alpha$ such that $M(\{f_\alpha\},\{\gamma_\alpha^*\}) = I_A$. □

3.3.2 *Proof of Theorem 3.3.1*

We assume that P_0 has an ω-localized kernel $p(x,y)$. The estimate just below and the proof of the next lemma will rely, in turn, on properties of Beurling weights described in Lemma 12 of [2]. First, recall that $\widetilde{p}(x,y) = \sum_k p(x, y-k)$, is periodic in both x, y because of shift covariance: $p(x, y-k) = p(x+k, y)$. To verify local square integrability, taking p in $\mathbf{C}^{n\times n}$, Cauchy–Schwarz gives

$$\begin{aligned}\int_{[0,1)^{2n}} \left|\sum_k p(x,y-k)\right|^2 dxdy &\le \int_{[0,1)^{2n}} \sum_k |p(x,y-k)|^2\, \omega(x-y+k)dxdy \\ &\quad\times \sup_{(x,y)\in[0,1)^{2n}} \left|\sum_k \frac{1}{\omega(x-y+k)}\right| \\ &= C\int_{[0,1)^n}\int_{\mathbf{R}^n} |p(x,y)|^2\,\omega(x-y)\,dxdy < \infty\end{aligned}$$

by hypothesis on P_0 and the definition of ω. We conclude, in particular, that the operator $\widetilde{P}$ on $L^2([0,1)^n, \mathbf{R}^n)$ defined by $f \mapsto \int_{[0,1)^n} \widetilde{p}(x,y) f(y)dy$ is bounded.

The next four lemmas are extensions of corresponding Lemmas 4–7 in section I.2 of [2] to the setting of $\mathcal{H}^2_\omega$ versus L^2_ω. In the next lemma, even though the result will be applied to functions in $\mathcal{H}^2$, the estimates just depend on the size of the functions. We set $\widetilde{\mathcal{H}^2} = \{\widetilde{f} = \sum_k f(x-k) : f \in \mathcal{H}^2\}$. Clearly, this defines a subspace of $L^2([0,1)^n, \mathbf{R}^n)$. We also set $\mathcal{H}^2_\omega = \mathcal{H}^2 \cap L^2_\omega$, where $L^2_\omega = \{f : \mathbf{R}^n \to \mathbf{R}^n : |f(x)|^2\omega(x) \in L^1(\mathbf{R}^n)\}$. We have the following.

Lemma 3.3.4. (i) *If* $f \in \mathcal{H}^2_\omega$, *then* $\widetilde{f} \in \widetilde{\mathcal{H}^2} \cap L^2([0,1)^n, \mathbf{R}^n)$.

(ii) *If* $f \in \mathcal{H}^2_\omega$, *then* $P_0 f \in \mathcal{H}^2_\omega$.

(iii) *If* $g \in \mathcal{H}^2$ *is supported in* $[0,1]^n$, *then* $\widetilde{P_0 g} = \widetilde{P}g$.

(iv) $\widetilde{P}$ *projects* $\widetilde{\mathcal{H}^2}$ *onto* $\widetilde{V_0} = \{\widetilde{f} : f \in V_0 \cap \mathcal{H}^2_\omega\}$.

Proof: To prove (i), we just note that, because ω is a Beurling weight,

$$\begin{aligned}\int_{[0,1)^n} \left|\sum_k f(x-k)\right|^2 dx &\le \int_{[0,1)^n} \sum_k |f(x-k)|^2\,\omega(x-k)\sum_k \frac{1}{\omega(x-k)}dx \\ &\le C\int_{\mathbf{R}^n} |f(x)|^2\,\omega(x)dx.\end{aligned}$$

To verify (ii), we note that $P_0 f \in \mathcal{H}^2$ because of the divergence-free condition on the kernel. To verify that $P_0 f \in L^2_\omega$, using properties of Beurling weights again and the shift invariance of p gives

$$\begin{aligned}
&\int \omega(x) \left| \int p(x,y) f(y) dy \right|^2 dx \\
&\quad \leq \int \omega(x) \left[\sum_k \left(\int_{k+[0,1)^n} |p(x,y)|^2 \, dy \right)^{1/2} \left(\int_{k+[0,1)^n} |f(y)|^2 \, dy \right)^{1/2} \right]^2 dx \\
&\quad \leq \int \sum_k \left[\int_{k+[0,1)^n} |p(x,y)|^2 \, \omega(x-k) \, dy \right] \left[\int_{k+[0,1)^n} |f(y)|^2 \, dy \, \omega(k) \right] \\
&\qquad \times \frac{\omega(x)}{\omega(x-k)\omega(k)} dx \\
&\leq C \sum_k \left[\int_{k+[0,1)^n} |f(y)|^2 \, dy \, \omega(k) \right] \\
&\qquad \times \int_{\mathbf{R}^n} \int_{k+[0,1)^n} |p(x-k, y-k)|^2 \, \omega(x-k) \, dy \, dx \\
&\leq C \int_{\mathbf{R}^n} \int_{[0,1)^n} |p(x,y)|^2 \, \omega(x-y) \, dy \, dx \int_{\mathbf{R}^n} |f(y)|^2 \, \omega(y) dy,
\end{aligned}$$

whence the claim follows from the definition of an ω-localized kernel.

To prove (iii), we note that if $g \in \mathcal{H}^2$ is supported in $[0,1]^n$, then $\widetilde{g} = g$ almost everywhere on $[0,1)^n$. To show that $\widetilde{P_0 g} = \widetilde{P} g$:

$$\begin{aligned}
\widetilde{P} g(x) &= \int_{[0,1)^n} \widetilde{p}(x,y) g(y) \, dy = \sum_l \int_{[0,1)^n} \widetilde{p}(x,y) g(y-l) \, dy \\
&= \sum_l \int_{[0,1)^n} \widetilde{p}(x+l,y) g(y-l) \, dy, \qquad \text{since } \widetilde{p}(\cdot, y) \text{ is periodic} \\
&= \sum_k \sum_l \int_{[0,1)^n} p(x-k+l, y) g(y-l) dy \\
&= \sum_k \sum_l \int_{[0,1)^n} p(x-k, y-l) g(y-l) dy \qquad \text{by shift covariance} \\
&= \sum_k \int_{\mathbf{R}^n} p(x-k, y) g(y) \, dy = \sum_k (P_0 g)(x-k) = \widetilde{P_0 g}(x).
\end{aligned}$$

Therefore, it remains only to prove (iv). First we note that any element of $\widetilde{\mathcal{H}^2}$ defines a periodic, locally square-integrable divergence-free distribution. Hence, any such element g can be identified with an element of $\mathcal{H}^2$ supported in $[0,1]^n$, which automatically belongs to $\mathcal{H}^2_\omega$. In view of (iii), therefore, it is enough to show that if $f \in V_0 \cap \mathcal{H}^2_\omega$, then $\widetilde{P}(\widetilde{f}) = \widetilde{f}$. But shift covariance gives

$$
\begin{aligned}
\widetilde{P}\widetilde{f}(x) &= \int_{[0,1)^n} \sum_k p(x, y-k) \sum_l f(y-l)\, dy \\
&= \int_{[0,1)^n} \sum_k p(x+k, y) \sum_l f(y-l)\, dy \\
&= \sum_l \int_{[0,1)^n} \sum_k p(x+k-l, y-l) f(y-l)\, dy \\
&= \sum_k \int_{\mathbf{R}^n} p(x+k, y) f(y)\, dy = \sum_k (P_0 f)(x+k) = \widetilde{f}(x) \qquad \square
\end{aligned}
$$

Lemma 3.3.5. $\dim \widetilde{V_0} = A$*, where* A *is defined as in Theorem 3.3.1.*

Proof: Since $\widetilde{p} \in L^2([0,1)^{2n}, \mathcal{M}_n)$, the operator $\widetilde{P}$ is Hilbert–Schmidt, hence compact. Since the unit ball $\widetilde{B}$ of $\widetilde{V_0}$ is bounded, it follows that $\widetilde{P}(\widetilde{B})$ is relatively compact, so that $\dim \widetilde{V_0}$ is finite. Now if $\{\widetilde{f}_\alpha\}$ is a basis of $\widetilde{V_0}$ and $\{\widetilde{f}^*_\alpha\}$ is a dual basis of $\{\widetilde{f}_\alpha\}$ in $(\ker \widetilde{P_0})^\perp$, then

$$
\widetilde{p}(x, y) = \sum_{\alpha \in \mathcal{A}} \widetilde{f}_\alpha(x) \overline{\widetilde{f}_\alpha}^{*\top}(y) \text{ a.e.}
$$

so that

$$
\operatorname{tr} \int_{[0,1)^n} \int_{[0,1)^n} [\widetilde{p}(x,y)\widetilde{p}(y,x)]\, dy\, dx = \sum_{\alpha \in \mathcal{A}} \left\langle \widetilde{f}_\alpha, \widetilde{f}^*_\alpha \right\rangle = A.
$$

This proves the lemma. $\square$

It still needs to be shown that V_0 *has* a Riesz basis of the desired form. We begin with the following.

Lemma 3.3.6. *Let* $\{f_\alpha\}_{\alpha \in A} \subset L^2(\mathbf{R}^n, \mathbf{R}^n)$*. Then the Gram matrix of* L^2 *inner products of the vector functions* $\{\widetilde{f}_\alpha\}$ *is the autocorrelation matrix* $M_{f_\alpha}(\xi)|_{\xi=0}$.

Proof: A simple computation shows that $\langle \widetilde{f_\alpha}, \widetilde{f_{\alpha'}} \rangle_{L^2([0,1)^n)} = C(f_\alpha, f_{\alpha'})(0)$. $\square$

In view of the previous lemma, the main trick in recovering properties of V_0 from those of $\widetilde{V_0}$ is to replace the role of $\xi = 0$ by other ξ. In particular, abusing the meaning of the zero in V_0, we define $V_\xi = \{e^{-2\pi i x\cdot\xi} f : f \in V_0\}$. Similarly, we set $V^*_\xi = \{e^{-2\pi i x\cdot\xi} f : f \in V^*_0\}$. Then let P_ξ be the projector onto V_ξ parallel to $(V^*_\xi)^\perp$. Its kernel $p_\xi(x,y) = p(x,y)e^{-2\pi i\xi(x-y)}$ is ω-localized. Finally, set $\widetilde{V_\xi} = \{\widetilde{f} : f \in V_\xi \cap e^{-2\pi i x\cdot\xi}\mathcal{H}^2_\omega\}$.

Lemma 3.3.7. $\dim \widetilde{V_\xi} = A$ *for all* ξ.

Proof: As P_ξ commutes with integer translations and has an ω-localized kernel, just as before,

$$\dim \widetilde{V_\xi} = \operatorname{tr} \int_{[0,1)^n} \int_{[0,1)^n} \widetilde{p}_\xi(x,y)\widetilde{p}_\xi(y,x)\, dy\, dx = A.$$

This can be seen by direct expansion in terms of p or else from dominated convergence, from which we can infer that the integral varies continuously in the parameter ξ. Since it takes integer values, it must be constant. □

Now we construct a Riesz basis of V_0 of the form $\{\phi_\alpha(x-k)\}_{\alpha\in\mathcal{A}}$. By the previous lemma, for each $\xi \in [0,1)^n$ there exist A functions $\{f_\alpha^\xi\}_{\alpha\in\mathcal{A}}$ in $V_0 \cap H^2_\omega$ such that $\{(e^{-2\pi i x\xi} f_\alpha^\xi)^\sim\}_{\alpha\in\mathcal{A}}$ is a basis of $\widetilde{V_\xi}$. But the Gram matrix of $\{(e^{-2\pi i x\xi} f_\alpha^\xi)^\sim\}_{\alpha\in\mathcal{A}}$ at $\eta \in [0,1)^n$ is nothing other than the autocorrelation matrix $M_{f_\alpha^\xi}$ evaluated at η. Since $f_\alpha^\xi \in L^2_\omega$, the coefficients of $M_{f_\alpha^\xi}$ have absolutely convergent Fourier series, hence, so does its determinant. Therefore, if $\det M_{f_\alpha^\xi}$ does not vanish at $\eta = \xi$, then it remains minorized by some $\gamma(\xi) > 0$, and the coefficients remain majorized in modulus by $1/\gamma(\xi)$ in a neighborhood $B(\xi, r(\xi))$ of ξ. Then, by the compactness of $[0,1]^n$, one can extract a finite family $B_\nu = B(\xi_\nu, r_\nu(\xi)) \cap [0,1)^n$, $1 \le \nu \le N$ that still covers $[0,1)^n$. Set $C_\nu = B_\nu \setminus \cup_{\mu<\nu} B_\mu$ and let D_ν be the union of all integer shifts of C_ν. Now define

$$\widehat{\phi}_\alpha(\xi) = \sum_{\nu=1}^{N} \widehat{f_\alpha^{\xi_\nu}} \chi_{D_\nu}(\xi).$$

We must show that $\phi_\alpha \in V_0$. First, $\widehat{f_\alpha^{\xi_\nu}} \chi_{D_\nu} \in \widehat{\mathcal{H}^2}$ because χ_{D_ν} is bounded and scalar valued. It follows that if $g \in (V_0)^\perp$, then, by periodicity of χ_{D_ν},

$$\begin{aligned} C(\left(\widehat{f_\alpha^{\xi_\nu}} \chi_{D_\nu}\right)^\vee, g)(\xi) &= \chi_{D_\nu}(\xi) C(f_\alpha^{\xi_\nu}, g)(\xi) \\ &= \chi_{D_\nu}(\xi) \sum_{k\in\mathbf{Z}^n} \langle f_\alpha^{\xi_\nu}(\cdot + k), g\rangle\, e^{-2\pi i k\cdot\xi} = 0 \end{aligned}$$

because $\langle f_\alpha^{\xi_\nu}(\cdot + k), g\rangle = 0$ for all k. One concludes that $\langle (\widehat{f_\alpha^{\xi_\nu}} \chi_{D_\nu})^\vee, g\rangle = 0$ and, therefore, each $\phi_\alpha \in V_0$. Next, since the sets D_ν are disjoint,

$$C(\phi_\alpha, \phi_{\alpha'}) = \sum_{\nu,\nu'=1}^{N} \chi_{D_\nu}\chi_{D_{\nu'}} C(f_\alpha^{\xi_\nu}, f_{\alpha'}^{\xi_{\nu'}}) = \sum_{\nu=1}^{N} \chi_{D_\nu} C(f_\alpha^{\xi_\nu}, f_{\alpha'}^{\xi_\nu}).$$

Hence, the coefficients of the autocorrelation matrix of the $\{\phi_\alpha\}$ remain majorized by $\max_{1\le\nu\le N}\{1/\gamma(\xi_\nu)\}$, and the determinant remains bounded below by $\min_{1\le\nu\le N}\{\gamma(\xi_\nu)\}$. By Lemma 3.3.2, therefore, the $\{\phi_\alpha(\cdot - k)\}$, $\alpha \in \mathcal{A}$, $k \in \mathbf{Z}^n$ form a Riesz family in V_0.

It just remains to show that the $\{\phi_\alpha\}$ generate all of V_0. But, if $f \in V_0 \cap \mathcal{H}^2_\omega$, then $(e^{-2\pi i x\cdot\xi} f)^\sim$ can be expressed as a linear combination of $(e^{-2\pi i x\cdot\xi} f^{\xi_\nu}_\alpha)^\sim$ whenever $\xi \in C_\nu$, where the coefficients are bounded. In fact, if we call these coefficients $r_{\alpha,\nu}(\xi)$, they satisfy

$$M_{f^{\xi_\nu}_\alpha}|_\xi \cdot \overrightarrow{r}_{\alpha,\nu}(\xi) = \overrightarrow{C}(f^{\xi_\nu}_\alpha, f),$$

where the vector index runs over $\mathcal{A}$. Now if $R_{\alpha,\nu}(\xi) = \sum_k r_{\alpha,\nu}(\xi+k)$, then $\widehat{f}\chi_{D_\nu} = \sum_\alpha R_{\alpha,\nu}(\xi)\chi_{D_\nu}\widehat{f^{\xi_\nu}_\alpha}$ so that, finally,

$$\widehat{f} = \sum_\alpha \left(\sum_\nu R_{\alpha,\nu}(\xi)\chi_{D_\nu}(\xi) \right) \widehat{\phi_\alpha},$$

which proves that the $\{\phi_\alpha(x-k)\}$ are complete in V_0. Since $V_0 \cap \mathcal{H}^2_\omega = P_0(\mathcal{H}^2_\omega)$ and $\mathcal{H}^2_\omega$ is dense in $\mathcal{H}^2$, it follows that $V_0 \cap \mathcal{H}^2_\omega$ is dense in V_0. Hence, $\{\phi_\alpha(x-k)\}$ form a Riesz basis for V_0. This completes the proof of Theorem 3.3.1.

3.3.3 Proof of Theorem 3.3.2

Lemma 3.3.8. *If the finite collection $\{\phi_\alpha\}_{\alpha\in\mathcal{A}}$ has compact support, then the correlation matrix $M_{\phi_\alpha}(\xi)$ of $\{\phi_\alpha\}$ takes values in trigonometric polynomials.*

Proof: Given any pair of square-integrable compactly supported functions, their correlation function is a trigonometric polynomial. This follows from expanding the correlation function in a Fourier series and using Parseval's formula. □

Before getting to the final point, we need a little more notation. Let V be an FSI subspace of $\mathcal{H}^2$ that possesses an ω-localized kernel so that, in particular, Theorem 3.3.1 holds. Let $\{\varphi_\beta\}_{\beta\in\mathcal{B}}$ be any finite subset of V, with $|\mathcal{B}| = B$.

Lemma 3.3.9. *The collection of vectors $\{\widehat{\varphi}_\beta(\xi+k)\}_k \subset l^2(\mathbf{Z}^n, \mathbf{C}^n)$ is linearly independent for $\beta \in \mathcal{B}$ if and only if the correlation matrix $M_{\varphi_\beta}(\xi)$ is nonsingular at ξ.*

Proof: Suppose that $\{\widehat{\varphi}_\beta(\xi+k)\}_k$, $\beta \in \mathcal{B}$ are linearly dependent. Then there are constants a_β, not all zero, such that $\sum_\beta a_\beta \widehat{\varphi}_\beta(\xi+k) = 0$. This implies that $0 = C(\sum_\beta a_\beta\varphi_\beta, \varphi_\alpha)(\xi) = \sum_\beta a_\beta C(\varphi_\beta, \varphi_\alpha)(\xi)$. Therefore, the rows of $M_{\varphi_\beta}(\xi)$ are linearly dependent. Conversely, suppose that the rows of $M_{\varphi_\beta}(\xi)$ are linearly dependent. Then, as before, there is a nontrivial solution of $0 = C(\sum_\beta a_\beta\varphi_\beta, \varphi_\alpha)(\xi)$. Therefore, the vector $\{\sum_\beta a_\beta\widehat{\varphi}_\beta(\xi+k)\}_k$ is orthogonal to each $\{\widehat{\varphi}_\beta(\xi+k)\}_k$. This can only happen if $\{\sum_\beta a_\beta\widehat{\varphi}_\beta(\xi+k)\}_k$ is the zero vector. In particular, the vectors $\{\widehat{\varphi}_\beta(\xi+k)\}_k$ are linearly dependent. □

The proof of Theorem 3.3.2 now is simple. Let $\{\varphi_\beta\}_{\beta\in\mathcal{B}}$ be a compactly supported family in V_0 that is maximal with respect to the property that $\det M_{\varphi_\beta} =$

0 only on a set of measure zero. Then $\{\varphi_\beta\}_{\beta\in\mathcal{B}}$ must be a finite set. In fact, $\{\varphi_\beta(\cdot - k)\}_{\beta\in\mathcal{B}}$ forms a basis for the compactly supported elements of V_0. Otherwise, there would be a compactly supported element $\varphi_0 \in V_0$ independent of these. Look at the correlation matrix of $\{\varphi_0, \varphi_\beta\}_{\beta\in\mathcal{B}}$. Its determinant is a trigonometric polynomial, and hence either vanishes identically or vanishes on a set of measure zero. The latter case would contradict the maximality of $\{\varphi_\beta\}_{\beta\in\mathcal{B}}$. The former case implies that the rows of the correlation matrix are linearly dependent, which is tantamount to saying that $C(\varphi_0, \varphi_\beta)(\xi) = \sum a_{\beta'}(\xi) C(\varphi_{\beta'}, \varphi_\beta)(\xi)$, where the functions $a_{\beta'}(\xi)$ are trigonometric polynomials as well. As we saw above, this amounts to saying that $\{\widehat{\varphi}_0(\xi + k)\}_k \equiv \sum_{\beta,l} a_{\beta,l} e^{-2\pi i l\cdot\xi}\{\widehat{\varphi}_\beta(\xi + k)\}_k$, where $a_{\beta,l} \neq 0$ for only finitely many $l \in \mathbf{Z}^n$. But this implies that $\varphi_0(x) = \sum_{\beta,l} a_{\beta,l}\varphi_\beta(x - l)$, which contradicts the independence of $\{\varphi_0, \varphi_\beta(\cdot - k)\}_{\beta\in\mathcal{B}}$. Therefore, the compactly supported elements of V_0 are contained in the span of $\{\varphi_\beta(\cdot - k)\}_{\beta\in\mathcal{B}}$. Since the compactly supported elements of V_0 are dense in V_0, it follows that V_0 is the closure of the span of $\{\varphi_\beta(\cdot - k\}_{\beta\in\mathcal{B}}$. This completes the proof of Theorem 3.3.2. □

3.4 Nonexistence

To complete the proof that no compactly supported divergence-free wavelet bases exist, we need to apply the preceding results to the special case where the space in question is generated by the long-scale wavelets. In particular, the fact that the kernel is ω-localized should be a simple corollary of the compact support of the wavelets.

Lemma 3.4.1. *Under the hypotheses of Theorem 3.2.1, the projector P_0 onto the subspace V_0 of $\mathcal{H}^2$ generated by $\{\psi_\lambda\}_{\lambda\in\Lambda,|Q(\lambda)|>1}$ has an ω-localized kernel.*

Proof: First, because of biorthogonality, it is clear that the kernel of P_0 should have the form

$$p(x, y) = \sum_{\lambda\in\Lambda,|Q(\lambda)|>1} \psi_\lambda(x)\overline{\psi_\lambda^{*\top}}(y).$$

The fact that the components of this matrix-valued kernel are ω-localized follows from precisely the same argument as in the scalar case. In particular, one can show that $|p(x, y)|_{\mathcal{M}_n} \leq C(1 + |x - y|)^{-n-\gamma}$, where $\gamma > 0$ depends on the Hölder regularity of the wavelets. But then p is ω-localized with respect to $\omega = (1 + |x|)^{n+\rho}$ whenever $0 < \rho < 2\gamma$. □

Corollary 3.4.1. *Under the same hypotheses, the compactly supported elements of V_0 are dense in V_0.*

Proof: Clearly, finite linear combinations of $\{\psi_\lambda\}_{\lambda\in\Lambda,|Q(\lambda)|>1}$ are compactly supported. If $f \in V_0$, then its wavelet coefficient sequence $\{\langle f, \psi_\lambda^*\rangle\}_{\lambda\in\Lambda,|Q(\lambda)|>1}$ is square summable, which implies that f can be approximated in V_0 by compactly supported elements. □

Now, applying Theorem 3.3.2 gives the following.

Corollary 3.4.2. *V_0 is locally finite dimensional.*

Proof: By Theorem 3.3.2, there is a finite collection of compactly supported functions $\{\phi_\beta\} \subset V_0$ such that V_0 is the l^2-closed span of $\{\phi_\beta(x-k)\}$. In particular, the restriction of any $f \in V_0$ to a compact set K is a linear combination of the restrictions to K of the $\{\phi_\beta(x-k)\}$, and only finitely many of these do not vanish identically on K. □

Now we can use the corollary to obtain a contradiction that proves Theorem 3.2.1. Assume that $V_0 = \oplus_{-\infty}^{-1} W_j = \mathcal{H}^2 \ominus (\oplus_0^\infty W_j)$, where W_j is generated by the families $\{\psi_\lambda : |Q(\lambda)| = 2^{-nj}\}$, where the ψ_λ and the dual wavelets $\psi_{\lambda'}^*$ are both divergence-free and compactly supported. Let h be a harmonic function on $\mathbf{R}^n$ so that ∇h is divergence-free. Set $h_M = \nabla h \chi_{\{|x|<M\}}$, where M is chosen such that if ψ_λ^* has support intersecting $[0,1)^n$ and $|Q(\lambda)| \leq 1$, then ψ_λ^* is supported in $\{|x| < M\}$. For such ψ_λ^* it follows that

$$\int_{\mathbf{R}^n} h_M \cdot \psi_\lambda^* = \int_{\mathbf{R}^n} \nabla h \cdot \psi_\lambda^* = -\int_{\mathbf{R}^n} h\,(\nabla \cdot \psi_\lambda^*) = 0$$

by integration by parts together with the fact that $\nabla \cdot \psi_\lambda^* = 0$.

This computation shows that all wavelets ψ_λ^* such that $|Q(\lambda)| \leq 1$ having support intersecting $[0,1)^n$ vanish as linear functionals against ∇h. We conclude that the restriction of $P_0(\nabla h \chi_{\{|x|<M\}})$ to $[0,1)^n$ agrees with the restriction of ∇h to $[0,1)^n$. But the previous results show that restrictions of V_0 to $[0,1)^n$ define a finite-dimensional space, whereas restrictions of ∇h to $[0,1)^n$, where h is harmonic, form an infinite-dimensional space. This contradiction proves the main theorem. □

Acknowledgment. Both authors were supported by Sandia National Laboratories' SURP program. They would like to thank Jeff Geronimo, Doug Hardin, Peter Massopust, and David Roach for important insights pertaining to this work.

References

[1] de Boor, R., DeVore, R., and Ron, A., The structure of finitely generated shift invariant spaces in $L^2(\mathbf{R}^d)$, *J. Functional Anal.*, **119** (1994), 37–78.

[2] Lemarié-Rieusset, P. G., Projecteurs invariants, matrices de dilatation, ondelettes et analyses multi-résolution, *Rev. Mat. Iberoamericana*, **8** (1992), 221–237.

[3] Lemarié-Rieusset, P. G., Un théorème d'inexistence pour les ondelettes vecteurs à divergence nulle, *C. R. Acad. Sci. Paris Sér. I Math.*, **319** (1994), 811–813.

[4] Lakey, J., Massopust, P., and Pereyra, M.C., Divergence-free multiwavelets, in Proc. Approximation Theory IX. **2**:161–168 (1998), C. K. Chui and L. L. Schumaker (eds).

[5] Urban, K., On divergence-free wavelets, *Adv. Comput. Math.*, **4** (1995), 51–81.

4

Osiris Wavelets and Isis Variables

Guy Battle

ABSTRACT In the familiar formulations of the renormalization group in equilibrium statistical mechanics, the variables associated with different length scales are independent random variables with respect to the massless Gaussian continuum measures (or its equivalent for the given model). We have recently introduced a new type of modeling based on *Osiris wavelets*. While such models represent a new hierarchical approximation that is expected to have important advantages over the familiar hierarchical approximations, the focus of this paper is on the *kinematic coupling* of the wavelet amplitudes that arises for these models. The variables associated with different length scales are no longer independent with respect to the massless Gaussian. This creates undefined Gaussian states that are fixed with respect to the renormalization group iteration—in addition to the usual Gaussian fixed point, which is a well-defined state. We show that these spurious Gaussian fixed points influence the flow of the iteration. We also derive the recursion formula—and its linearization—for dipole perturbations of the Gaussian fixed point. Such a formula is best described by *Isis variables*. This paper deals with two dimensions only.

4.1 Introduction

In the analysis of critical phenomena—and our program [1–3] is no exception—the application of the renormalization group always involves a formulation based on some notion of a fluctuation. Even when a renormalization group transformation is initially defined by the innocent-looking formula involving the covariant action of an averaging transformation on expectations of random variables, the multiscale analysis of non-Gaussian probability measures on the space of configurations calls for an alternative formula that integrates fluctuations out of the functional integral. In such a case, the fluctuations are the configurations that are annihilated by the averaging transformation, while the minimizer associated with the averaging transformation is applied to the new variables to realize the nonfluctuating contributions to the old variables [4–12]. Another standard formulation is to simply decompose an arbitrary configuration into its contributions to different scales of momentum. The renormalization group transformation integrates out the variables associated with the maximum momentum scale, which corresponds to the minimum length scale. This approach is often referred to as *momentum slicing* [13–16], and it is more commonly used in Euclidean field theory. All of the standard approaches share one characteristic: variables associated with different length scales are independent with respect to that Gaussian measure which is a fixed point of the renormalization group transformation—i.e., the Gaussian fixed point factors over the scales of fluctuations. This Gaussian also corresponds to the massless Gaussian in the continuum.

We have recently embarked on a program to analyze the critical behavior of classical systems in equilibrium by introducing a new hierarchical approximation [1–3]. One advantage that our approximation has over the familiar hierarchical approximations [17–22] is the expectation of obtaining a nonzero value of the critical exponent η for the Ginzburg–Landau ferromagnet. The idea is to define a multiscale set of expansion functions that are continuous and piecewise linear, and the most important property of this set is that the localization block of each function *supports* that function. The price we pay is that the set is not complete, but this kind of sacrifice is not new. In an effort to calculate the critical exponent η, Golner [23] based his renormalization group analysis on fictitious expansion functions having more variation in their localization blocks than those used by Wilson [11] to obtain the first hierarchical approximation. From a qualitative point of view, it is easy to see that Golner's set is incomplete. Our program follows Golner in spirit, but with mathematically concrete expansion functions whose description has suggested the name *Osiris wavelets* [1–3]. In addition to the lack of completeness, we must also deal with a certain lack of Sobolev orthogonality between the Osiris wavelets of different length scales. This means that the fluctuations (amplitudes for the wavelets) on different length scales are no longer independent with respect to the massless Gaussian. We refer to this as *kinematic coupling*, and it has an important consequence: in addition to the Gaussian fixed point, there are undefined Gaussian fixed points that influence the flow of the renormalization group iteration.

Actually, this phenomenon is not peculiar to Osiris wavelets at all, but can be understood in a very elementary way that does not depend on any knowledge of the renormalization group. Consider a discrete Gaussian process with *an initial random variable*. Let each single-variable distribution be the unit Gaussian, except that of the initial variable, which has a weight parameter. Suppose further that the process involves couplings between successive random variables only. Let $\alpha_1, \alpha_2, \alpha_3, \ldots$ be the sequence of variables. The two-point correlation for the Gaussian is given by

$$\langle \alpha_j \alpha_k \rangle_{\mu,\lambda} = \lim_{N \to \infty} \left[Z_N(\mu,\lambda)^{-1} \left(\prod_{\ell=1}^{N} \int_{-\infty}^{\infty} d\alpha_\ell \right) \alpha_j \alpha_k \right. \\ \left. \times \exp\left(-\frac{1}{2}\mu\alpha_1^2 - \frac{1}{2}\sum_{\ell=2}^{N} \alpha_\ell^2 + \lambda \sum_{\ell=2}^{N} \alpha_\ell \alpha_{\ell-1} \right) \right], \quad (4.1.1)$$

$$Z_N(\mu,\lambda) = \left(\prod_{\ell=1}^{N} \int_{-\infty}^{\infty} d\alpha_\ell\right) \exp\left(-\frac{1}{2}\mu\alpha_1^2 - \frac{1}{2}\sum_{\ell=2}^{N}\alpha_\ell^2 + \lambda\sum_{\ell=2}^{N}\alpha_\ell\alpha_{\ell-1}\right), \tag{4.1.2}$$

with $\mu, \lambda > 0$. By performing the α_1-integration and making the substitutions $\alpha_k \to \alpha_{k-1}$ for the other integration variables, one can easily establish that

$$\langle\alpha_j\alpha_k\rangle_{\mu,\lambda} = \langle\alpha_{j-1}\alpha_{k-1}\rangle_{R(\mu),\lambda}, \tag{4.1.3}$$

$$R(\mu) = 1 - \frac{\lambda^2}{\mu}. \tag{4.1.4}$$

This observation has been widely known for a long time. The fixed points of the transformation R are the translation-invariant states in this context. To find such states, one needs only to solve the equation

$$\mu = 1 - \frac{\lambda^2}{\mu}, \tag{4.1.5}$$

or rather

$$\mu^2 - \mu + \lambda^2 = 0. \tag{4.1.6}$$

The fixed-point values of μ are

$$\mu = \mu_\pm \equiv \frac{1}{2} \pm \frac{1}{2}\sqrt{1-4\lambda^2}, \tag{4.1.7}$$

but only one of the corresponding Gaussian processes is defined. The expectation functional $\langle\cdot\rangle_{\mu_-,\lambda}$ is not defined. Indeed, integrating by parts with respect to α_1 yields the relation

$$\langle\alpha_1^2\rangle_{\mu_-,\lambda} = \frac{\lambda^2}{\mu_-^2}\langle\alpha_2^2\rangle_{\mu_-,\lambda} + \frac{1}{\mu_-}. \tag{4.1.8}$$

On the other hand, the fixed-point property implies that

$$\langle\alpha_2^2\rangle_{\mu_-,\lambda} = \langle\alpha_1^2\rangle_{\mu_-,\lambda}. \tag{4.1.9}$$

Hence,

$$\langle\alpha_1^2\rangle_{\mu_-,\lambda} = \frac{\mu_-}{\mu_-^2 - \lambda^2}, \tag{4.1.10}$$

but $\mu_-^2 < \lambda^2$, so this contradicts the positivity of $\langle\alpha_1^2\rangle_{\mu_-,\lambda}$.

If we iterate the transformation R, the flow is easy to analyze. Clearly, if $\mu > \mu_+$, then $\mu^2 - \mu + \lambda^2 > 0$, and so $R(\mu) < \mu$. However, if $\mu_- < \mu < \mu_+$, then $\mu^2 - \mu + \lambda^2 < 0$, and so $R(\mu) > \mu$. This means that μ_+ is an attractive fixed point with respect to the flow of the iteration, while μ_- is repelling. This also means that

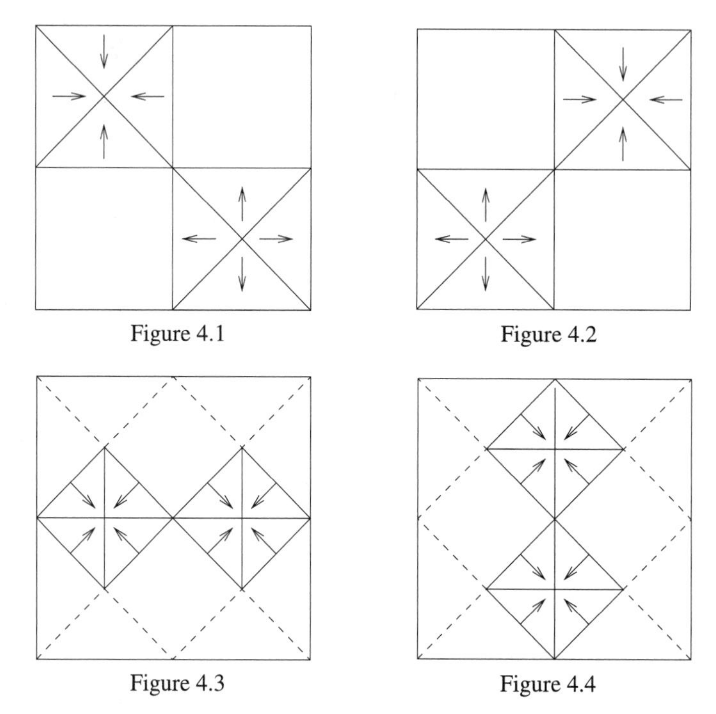

FIGURE 4.1–4.4. Diagrammatical description of the mother wavelets.

the Gaussian functional $\langle\cdot\rangle_{\mu,\lambda}$ exists for $\mu > \mu_-$. A more complicated version of this phenomenon occurs in the renormalization group analysis of the Osiris models. The occurrence of undefined Gaussian fixed points in addition to the well-defined Gaussian fixed point is not a characteristic of the more standard models. The reason is that there is no *kinematic* coupling in those models—all interscale coupling is due to the interaction (non-Gaussian perturbation). In the elementary example here, the analogue would be to set $\lambda = 0$, in which case $R(\mu) = 1$, and so there would be only one Gaussian fixed point, defined or otherwise, in this elementary image of the familiar modeling.

In this paper, we address the issue for Osiris wavelet modeling in two dimensions. In [1] we introduced four mother wavelets defined schematically by the diagrams shown in Figures 4.1–4.4. In these figures, the arrows represent constant gradients on their respective triangles, and these continuous, piecewise-linear functions vanish on the blank regions. For these wavelets $\Psi_1, \Psi_2, \Psi_3, \Psi_4$, we have chosen the normalization

$$\int (\nabla\Psi_\iota)^2 = 1, \tag{4.1.11}$$

which is preserved under the scalings and translations

$$\Psi_{\iota r \vec{n}}(\vec{x}) = \Psi_\iota(2^{-r}\,\vec{x} - \vec{n}), \qquad r \in \mathbb{Z}, \quad \vec{n} \in \mathbb{Z}^2. \tag{4.1.12}$$

For the Sobolev norm, the amplitude scaling factor is unity in two dimensions. The overlap matrix for this scale-coherent wavelet set is given by

$$S_{rr';\vec{n}\,\vec{n}';\iota\iota'} = \int \nabla\Psi_{\iota r\vec{n}} \cdot \nabla\Psi_{\iota' r'\vec{n}'}. \tag{4.1.13}$$

Since the wavelet set is not Sobolev orthonormal, this matrix is not the identity matrix. However, in [1] we proved that

$$S \geq \frac{3}{4}(2-\sqrt{3}). \tag{4.1.14}$$

The most convenient property of S in our search for Gaussian fixed points is the property

$$S_{rr'} = 0, \qquad |r-r'| > 1, \tag{4.1.15}$$

which follows from the fact that Osiris wavelets with length scales differing by more than one factor of 2 are Sobolev orthogonal. For this reason, we are dealing with a class of Gaussian states that is quite analogous to the class of discrete Gaussian processes discussed above.

Remark. For the Osiris wavelets constructed in three dimensions [2], the multiscale Sobolev overlap matrix does *not* satisfy (4.1.15). The investigation of kinematic coupling will be more complicated in three dimensions than it is here, but the principle illustrated by the discrete Gaussian process will be the same.

In the wavelet amplitude picture, the quadratic form corresponding to an arbitrary Gaussian state is given by

$$\begin{aligned}(S_G\alpha,\alpha) = &\sum_{\vec{n}\in\mathbb{Z}^2}\sum_{\iota,\kappa=1}^{4} G_{\iota\kappa}\alpha^{(0)}_{\iota\vec{n}}\alpha^{(0)}_{\kappa\vec{n}} \\ &+\left(\sum_{r=1}^{\infty}\sum_{r'=0}^{\infty}+\sum_{r=0}^{\infty}\sum_{r'=1}^{\infty}\right)\sum_{\vec{n},\vec{n}'\in\mathbb{Z}^2}\sum_{\iota,\iota=1}^{4} S_{rr';\vec{n}\,\vec{n}';\iota\iota'}\alpha^{(r)}_{\iota\vec{n}}\alpha^{(r')}_{\iota'\vec{n}'},\end{aligned} \tag{4.1.16}$$

where $\alpha^{(r)}_{\iota\vec{n}}$ denotes the amplitude of the contribution of $\Psi_{\iota r\vec{n}}$ to the continuum configuration, which is regularized by the cutoff $r \geq 0$ as well as conditioned by the incompleteness of the wavelet set. The parametrizing matrix has the form

$$G = \begin{bmatrix} 1 & 0 & 0 & 0 \\ 0 & 1 & 0 & 0 \\ 0 & 0 & g_{11} & g_{12} \\ 0 & 0 & g_{21} & g_{22} \end{bmatrix}, \tag{4.1.17}$$

while (4.1.15) together with $S_{rr} = 1$ and the symmetry $S_{r-1,r} = S_{r,r-1}{}^*$ implies

$$(S_G\alpha,\alpha) = \sum_{\vec{n}\in\mathbb{Z}^2}\sum_{\iota,\kappa=1}^{4} G_{\iota\kappa}\alpha^{(0)}_{\iota\vec{n}}\alpha^{(0)}_{\kappa\vec{n}} + \sum_{r=1}^{\infty}\sum_{\vec{n}\in\mathbb{Z}^2}\sum_{\iota=1}^{4}\alpha^{(r)2}_{\iota\vec{n}}$$
$$+ 2\sum_{r=1}^{\infty}\sum_{\vec{n},\vec{n}'\in\mathbb{Z}^2}\sum_{\iota,\iota'=1}^{4} S_{r,r-1;\vec{n}\,\vec{n}';\iota\iota'}\alpha^{(r)}_{\iota\vec{n}}\alpha^{(r-1)}_{\iota'\vec{n}'}. \tag{4.1.18}$$

In Section 4.4 we reduce the renormalization group transformation to a nonlinear operator on 2×2 matrices given by

$$\mathcal{R}\left(\begin{bmatrix} g_{11} & g_{12} \\ g_{21} & g_{22} \end{bmatrix}\right) = \begin{bmatrix} 7/8 & 0 \\ 0 & 7/8 \end{bmatrix} - \frac{1}{16\mathfrak{g}}\begin{bmatrix} g_{11} & -g_{21} \\ -g_{21} & g_{22} \end{bmatrix}, \tag{4.1.19}$$

$$\mathfrak{g} = g_{11}g_{22} - g_{12}g_{21}. \tag{4.1.20}$$

The fixed points of $\mathcal{R}$ are the four matrices

$$\frac{1}{16}(7 \pm \sqrt{33})\begin{bmatrix} 1 & 0 \\ 0 & 1 \end{bmatrix}, \qquad \frac{1}{16}\begin{bmatrix} 7 & \pm\sqrt{33} \\ \pm\sqrt{33} & 7 \end{bmatrix},$$

and the influence of each fixed point on the iteration of $\mathcal{R}$ has to be investigated.

To this end, we identify each 2×2 matrix with a 4×1 matrix and calculate the 4×4 Jacobian matrix of $\mathcal{R}$. Since

$$\frac{\partial}{\partial g_{11}}\left(\frac{1}{\mathfrak{g}}\right) = -\frac{1}{\mathfrak{g}^2}g_{22}, \tag{4.1.21}$$

$$\frac{\partial}{\partial g_{22}}\left(\frac{1}{\mathfrak{g}}\right) = -\frac{1}{\mathfrak{g}^2}g_{11}, \tag{4.1.22}$$

$$\frac{\partial}{\partial g_{12}}\left(\frac{1}{\mathfrak{g}}\right) = -\frac{1}{\mathfrak{g}^2}g_{21}, \tag{4.1.23}$$

$$\frac{\partial}{\partial g_{21}}\left(\frac{1}{\mathfrak{g}}\right) = -\frac{1}{\mathfrak{g}^2}g_{12}, \tag{4.1.24}$$

it follows from the choice

$$\begin{bmatrix} g_{11} & g_{12} \\ g_{21} & g_{22} \end{bmatrix} \longrightarrow \begin{bmatrix} g_{11} \\ g_{22} \\ g_{12} \\ g_{21} \end{bmatrix}$$

that

$$J(\mathcal{R})\left(\begin{bmatrix} g_{11} \\ g_{22} \\ g_{12} \\ g_{21} \end{bmatrix}\right) = \frac{1}{16\mathfrak{g}^2}\begin{bmatrix} g_{12}g_{21} & g_{11}^2 & -g_{11}g_{21} & -g_{11}g_{12} \\ g_{22}^2 & g_{12}g_{21} & -g_{22}g_{21} & -g_{22}g_{12} \\ -g_{21}g_{22} & -g_{21}g_{11} & g_{21}^2 & g_{11}g_{22} \\ -g_{12}g_{22} & -g_{12}g_{11} & g_{11}g_{22} & g_{12}^2 \end{bmatrix}. \tag{4.1.25}$$

Clearly,

$$J(\mathcal{R})\left(\frac{1}{16}(7\pm\sqrt{33})\begin{bmatrix}1\\1\\0\\0\end{bmatrix}\right)=\frac{(7\mp\sqrt{33})^2}{16}\begin{bmatrix}0&1&0&0\\1&0&0&0\\0&0&0&1\\0&0&1&0\end{bmatrix}, \tag{4.1.26}$$

$$J(\mathcal{R})\left(\frac{1}{16}\begin{bmatrix}7\\7\\\pm\sqrt{33}\\\pm\sqrt{33}\end{bmatrix}\right)=\frac{1}{256}\begin{bmatrix}33&49&\mp7\sqrt{33}&\mp7\sqrt{33}\\49&33&\mp7\sqrt{33}&\mp7\sqrt{33}\\\mp7\sqrt{33}&\mp7\sqrt{33}&33&49\\\mp7\sqrt{33}&\mp7\sqrt{33}&49&33\end{bmatrix}. \tag{4.1.27}$$

The eigenvalues of the Jacobian matrix at

$$\frac{1}{16}(7+\sqrt{33})\begin{bmatrix}1\\1\\0\\0\end{bmatrix}$$

are $\pm(1/16)(7-\sqrt{33})^2$, each with multiplicity two. Since these eigenvalues have absolute value less than unity, this fixed point is stable. We shall see that it corresponds to a well-defined Gaussian state. The eigenvalues of the Jacobian matrix at

$$\frac{1}{16}(7-\sqrt{33})\begin{bmatrix}1\\1\\0\\0\end{bmatrix}$$

are $\pm(1/16)(7+\sqrt{33})^2$, each with multiplicity two. These eigenvalues have absolute value greater than unity, so this fixed point is completely unstable. We shall prove in the Appendix that the corresponding Gaussian state is undefined. The eigenvalues of the Jacobian matrix at

$$\frac{1}{16}\begin{bmatrix}7\\7\\\sqrt{33}\\\sqrt{33}\end{bmatrix}$$

are -1 and $(1/16)(7\pm\sqrt{33})^2$, where the eigenvalue -1 has multiplicity two. This fixed point is semistable. One of the positive eigenvalues is greater than unity, while the other is less than unity. We shall prove that the Gaussian state associated with this fixed point is undefined as well. Finally, the eigenvalues of the Jacobian matrix at

$$\frac{1}{16}\begin{bmatrix}7\\7\\-\sqrt{33}\\-\sqrt{33}\end{bmatrix}$$

are -1 and $(1/16)(7\pm\sqrt{33})^2$, where, again, the eigenvalue -1 has multiplicity two. This fixed point shares the properties of its twin.

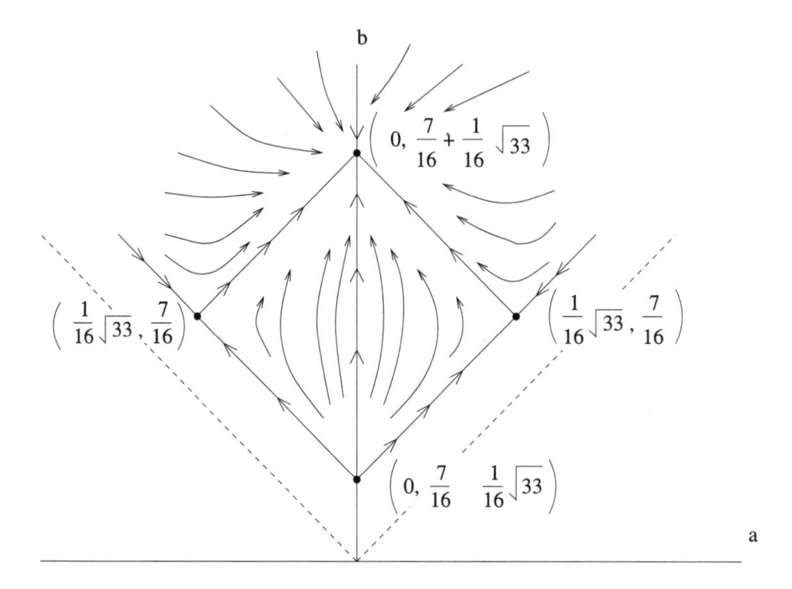

FIGURE 4.5. Flow diagram for RG iteration on the submanifold.

Naturally, these calculations determine the flow of the renormalization group (RG) iteration only in the neighborhoods of the fixed points. The flow in this four-dimensional space is not so easy to determine globally. However, the two-dimensional manifold of matrices of the form

$$G = \begin{bmatrix} b & a \\ a & b \end{bmatrix}, \qquad a \neq \pm b, \tag{4.1.28}$$

is preserved by the transformation and contains all of the fixed points. In this submanifold, the flow of the RG iteration is given globally by the diagram of Figure 4.5, for which we give the argument in Section 4.4. In the Appendix we prove that neither the unstable fixed point nor the semistable fixed points correspond to states at all, but they obviously influence the iteration of the transformation.

Another major objective is to derive the Hamiltonian for the stable Gaussian fixed point. In this case, the desired Hamiltonian is a quadratic form in the continuous, piecewise-linear configurations based on a unit-scale triangulation. The *canonical* Hamiltonian is given by

$$H_0(\phi) = \frac{1}{2} \int (\nabla \phi)^2 \tag{4.1.29}$$

with ϕ ranging over such configurations. The triangulation of the unit square is given by the diagram in Figure 4.6. The effect of the RG transformation on the triangulation is to delete certain lines to obtain the diagram shown in Figure 4.7 and then to scale the unit square down to the square $[0, 1/2]^2$. Clearly, Figure 4.7 gives the

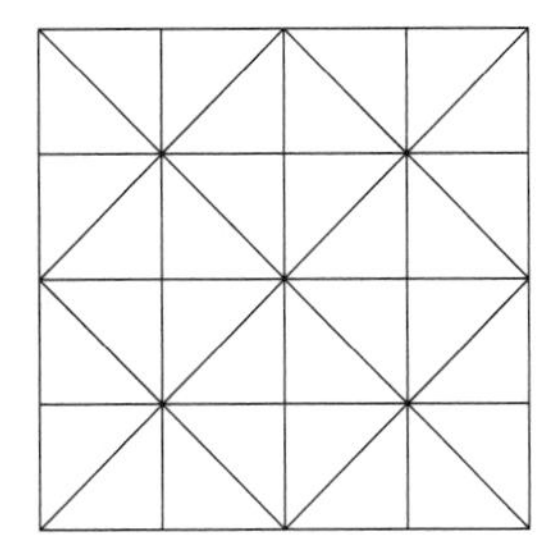

FIGURE 4.6. Isis triangulation of unit square.

basic triangulation, while Figure 4.6 gives the four copies of this basic triangulation needed for the mother wavelets. It is also clear that the variation of a continuous, piecewise-linear configuration based on Figure 4.7 is determined by the directional derivatives $\sigma_1, \ldots, \sigma_8$ along the radiating line segments labeled $1, \ldots, 8$, respectively. We have chosen to call these directional derivatives *Isis variables*. Actually, when we introduced these variables in [1], we defined them as

$$\begin{aligned}
\sigma_1 &= \text{outward directional derivative along 1st line segment,} \\
\sigma_2 &= \sqrt{2} \text{ times outward directional derivative along 2nd line segment,} \\
\sigma_3 &= \text{outward directional derivative along 3rd line segment,} \\
\sigma_4 &= \sqrt{2} \text{ times outward directional derivative along 4th line segment,} \\
\sigma_5 &= \text{ inward directional derivative along 5th line segment,} \\
\sigma_6 &= \sqrt{2} \text{ times inward directional derivative along 6th line segment,} \\
\sigma_7 &= \text{ inward directional derivative along 7th line segment,} \\
\sigma_8 &= \sqrt{2} \text{ times inward directional derivative along 8th line segment.}
\end{aligned}$$

We shall adhere to this convention for consistency.

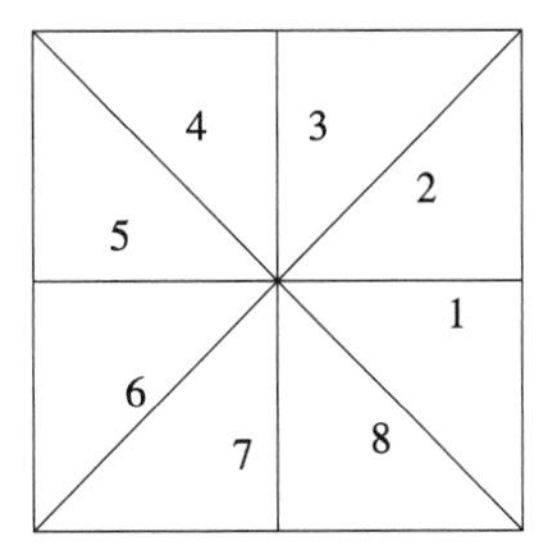

FIGURE 4.7. Basic triangulation.

Let $(A^{(0)}\sigma, \sigma)$ be the quadratic form in the Isis variables that corresponds to the canonical Hamiltonian. Since Figure 4.7 represents the standard $1/2 \times 1/2$ square, we have

$$(A^{(0)}\sigma, \sigma) = \int_{[0,\frac{1}{2}]^2} (\nabla\phi)^2 \tag{4.1.30}$$

for the association $\phi \mapsto (\sigma_1, \dots, \sigma_8)$. We shall calculate

$$\begin{aligned}(A^{(0)}\sigma, \sigma) = \frac{1}{32}[&2\sigma_1^2 + (\sigma_1 + \sigma_8)^2 + (\sigma_2 - \sigma_1)^2 + (\sigma_2 - \sigma_3)^2 \\ &+ 2\sigma_3^2 + (\sigma_3 - \sigma_4)^2 + (\sigma_4 + \sigma_5)^2 + 2\sigma_5^2 + (\sigma_6 - \sigma_5)^2 \\ &+ (\sigma_6 - \sigma_7)^2 + 2\sigma_7^2 + (\sigma_7 - \sigma_8)^2].\end{aligned} \tag{4.1.31}$$

Thus,

$$A^{(0)} = \frac{1}{32}\begin{bmatrix} 4 & -1 & 0 & 0 & 0 & 0 & 0 & 1 \\ -1 & 2 & -1 & 0 & 0 & 0 & 0 & 0 \\ 0 & -1 & 4 & -1 & 0 & 0 & 0 & 0 \\ 0 & 0 & -1 & 2 & 1 & 0 & 0 & 0 \\ 0 & 0 & 0 & 1 & 4 & -1 & 0 & 0 \\ 0 & 0 & 0 & 0 & -1 & 2 & -1 & 0 \\ 0 & 0 & 0 & 0 & 0 & -1 & 4 & -1 \\ 1 & 0 & 0 & 0 & 0 & 0 & -1 & 2 \end{bmatrix}. \tag{4.1.32}$$

We shall subsequently combine this with RG iteration to calculate the quadratic form $(A^{\infty}\sigma, \sigma)$ in the Isis variables that corresponds to the stable, well-defined Gaussian fixed point.

Now recall that in [1] we represented an arbitrary dipole Hamiltonian as a function of the Isis variables. Since the RG transformation is defined by integrating out Osiris wavelet amplitudes, the hierarchical structure enabled us to represent the transformed Hamiltonian in the same way. Indeed, the reduced RG transformation of an arbitrary function $\widehat{H}(\sigma_1, \dots, \sigma_8)$ of the Isis variables is given by

$$\begin{aligned}&\exp(-\widehat{R}(\widehat{H})(\sigma_1', \dots, \sigma_8')) \\ &= \widehat{Z}_{\widehat{H}}^{-1}\left(\prod_{\iota=1}^{4}\int_{-\infty}^{\infty} d\tau_\iota\right) \\ &\quad\times \exp\Bigg[-\widehat{H}\bigg(\frac{1}{2}\sigma_6' - \frac{1}{2}\sigma_7' + \tau_2 + \tau_4, \frac{1}{2}\sigma_6' + \tau_2, \\ &\qquad\qquad \frac{1}{2}\sigma_6' - \frac{1}{2}\sigma_5' + \tau_2 + \tau_3, \frac{1}{2}\sigma_6' - \sigma_5' + \tau_2, \\ &\qquad\qquad \frac{1}{2}\sigma_5' - \tau_2, \frac{1}{2}\sigma_6' - \tau_2, \frac{1}{2}\sigma_7' - \tau_2, \sigma_7' - \frac{1}{2}\sigma_6' - \tau_2\bigg) \\ &\qquad - \widehat{H}\bigg(\frac{1}{2}\sigma_1' + \tau_1, \sigma_1' + \frac{1}{2}\sigma_8' + \tau_1,\end{aligned}$$

$$
\begin{aligned}
&\frac{1}{2}\sigma_1' + \frac{1}{2}\sigma_8' + \tau_1 + \tau_3, \frac{1}{2}\sigma_8' + \tau_1,\\
&\frac{1}{2}\sigma_7' - \frac{1}{2}\sigma_8' - \tau_1 - \tau_4, \sigma_7' - \frac{1}{2}\sigma_8' - \tau_1,\\
&\frac{1}{2}\sigma_7' - \tau_1, \frac{1}{2}\sigma_8' - \tau_1\Big)\\
- \widehat{H}\Big(&\frac{1}{2}\sigma_3' - \frac{1}{2}\sigma_4' + \tau_4 - \tau_1, \sigma_3' - \frac{1}{2}\sigma_4' - \tau_1,\\
&\frac{1}{2}\sigma_3' - \tau_1, \frac{1}{2}\sigma_4' - \tau_1, \frac{1}{2}\sigma_5' + \tau_1,\\
&\frac{1}{2}\sigma_4' + \sigma_5' + \tau_1, \frac{1}{2}\sigma_4' + \frac{1}{2}\sigma_5' + \tau_1 - \tau_3, \frac{1}{2}\sigma_4' + \tau_1\Big)\\
- \widehat{H}\Big(&\frac{1}{2}\sigma_1' - \tau_2, \frac{1}{2}\sigma_2' - \tau_2, \frac{1}{2}\sigma_3' - \tau_2\\
&\sigma_3' - \frac{1}{2}\sigma_2' - \tau_2, \frac{1}{2}\sigma_2' - \frac{1}{2}\sigma_3' + \tau_2 - \tau_4,\\
&\frac{1}{2}\sigma_2' + \tau_2, \frac{1}{2}\sigma_2' - \frac{1}{2}\sigma_1' + \tau_2 - \tau_3, \frac{1}{2}\sigma_2' - \sigma_1' + \tau_2\Big)\Big].
\end{aligned} \tag{4.1.33}
$$

$$
\begin{aligned}
\widehat{Z}_{\widehat{H}} = &\left(\prod_{\iota=1}^{4}\int_{-\infty}^{\infty} d\tau_\iota\right)\\
&\times \exp\Big(-\widehat{H}(\tau_2 + \tau_4, \tau_2, \tau_2 + \tau_3, \tau_2, -\tau_2, -\tau_2, -\tau_2, -\tau_2)\\
&\quad - \widehat{H}(\tau_1, \tau_1, \tau_1 + \tau_3, \tau_1, -\tau_1 - \tau_4, -\tau_1, -\tau_1, -\tau_1)\\
&\quad - \widehat{H}(\tau_4 - \tau_1, -\tau_1, -\tau_1, -\tau_1, \tau_1, \tau_1, \tau_1 - \tau_3, \tau_1)\\
&\quad -\widehat{H}(-\tau_2, -\tau_2, -\tau_2, -\tau_2, \tau_2 - \tau_4, \tau_2, \tau_2 - \tau_3, \tau_2)\Big).
\end{aligned} \tag{4.1.34}
$$

A less explicit but more intelligible version of this formula is given by

$$
\begin{aligned}
&\exp(-\widehat{R}(\widehat{H})(\sigma_1', \dots, \sigma_8'))\\
&\quad = \widehat{Z}_{\widehat{H}}^{-1}\left(\prod_{\iota=1}^{4}\int_{-\infty}^{\infty} d\tau_\iota\right)\\
&\quad\quad \times \exp\left(-\sum_{\vec{\varepsilon}\in\{0,1\}^2} \widehat{H}\left(\frac{1}{2}L_{\vec{\varepsilon}}(\sigma') + \sum_{\iota=1}^{4}\tau_\iota N(\iota, \vec{\varepsilon})\right)\right),
\end{aligned} \tag{4.1.35}
$$

$$
N(\iota, \vec{\varepsilon}) = (N(\iota, \vec{\varepsilon})_1, \dots, N(\iota, \vec{\varepsilon})_8), \tag{4.1.36}
$$

$$L_{\vec{\varepsilon}}(\sigma') = (L_{\vec{\varepsilon}}(\sigma')_1, \ldots, L_{\vec{\varepsilon}}(\sigma')_8), \tag{4.1.37}$$

where $N(\iota, \vec{\varepsilon})_\mu$ is the contribution of the wavelet Ψ_ι to the μth Isis variable of the square $[0, 1/2]^2 + (1/2)\,\vec{\varepsilon}$ in Figure 4.6 and $L_{\vec{\varepsilon}}(\sigma')_\mu$ is the contribution of ϕ' to the scaling down of the same variable. Actually, ϕ' is determined by $\sigma' = (\sigma'_1, \ldots, \sigma'_8)$ only up to a constant, but $L_{\vec{\varepsilon}}(\sigma')_\mu$ is still uniquely defined because it is a directional derivative. From this formula we shall derive the recursion formula for the class of dipole perturbations of the fixed quadratic form. The canonical Hamiltonians in this class have the reduced form

$$\widehat{H}(\sigma) = \frac{1}{2}(A^\infty \sigma, \sigma) + \int_{[0,\frac{1}{2}]^2} F(\nabla\phi), \tag{4.1.38}$$

$$F(\vec{z}) = \sum_{k=2}^{\infty} c_k (\vec{z} \cdot \vec{z})^k. \tag{4.1.39}$$

We also derive the linearization of the RG transformation of these perturbations.

4.2 The Gaussian fixed point (configuration picture)

We first consider the configuration picture—i.e., the cone of positive quadratic forms on the eight-dimensional space of Isis variables. The action of the reduced RG transformation on this special set of reduced Hamiltonians

$$\widehat{H}(\sigma) = \frac{1}{2}(A\sigma, \sigma) \tag{4.2.1}$$

is given by

$$\begin{aligned}
&\exp\left(-\frac{1}{2}(\widehat{R}(A)\sigma', \sigma')\right) \\
&\quad = \exp\left(-\frac{1}{2} \sum_{\vec{\varepsilon} \in \{0,1\}^2} 2^{-2}(AL_{\vec{\varepsilon}}(\sigma'), L_{\vec{\varepsilon}}(\sigma'))\right) \\
&\quad \times \widehat{Z}_A^{-1} \left(\prod_{\iota=1}^{4} \int_{-\infty}^{\infty} d\tau_\iota\right) \\
&\quad \times \exp\left(-\sum_{\iota=1}^{4} \tau_\iota \sum_{\vec{\varepsilon} \in \{0,1\}^2} 2^{-1}(AL_{\vec{\varepsilon}}(\sigma'), N(\iota, \vec{\varepsilon}))\right. \\
&\qquad\qquad \left. -\frac{1}{2} \sum_{\iota,\iota'=1}^{4} \tau_\iota \tau_{\iota'} \sum_{\vec{\varepsilon} \in \{0,1\}^2} (AN(\iota, \vec{\varepsilon}), N(\iota', \vec{\varepsilon}))\right),
\end{aligned} \tag{4.2.2}$$

$$\widehat{Z}_A = \left(\prod_{\iota=1}^{4} \int_{-\infty}^{\infty} d\tau_\iota\right) \exp\left(-\frac{1}{2}\sum_{\iota,\iota'=1}^{4} \tau_\iota \tau_{\iota'} \sum_{\vec{\varepsilon}\in\{0,1\}^2} (AN(\iota,\vec{\varepsilon}), N(\iota',\vec{\varepsilon}))\right). \tag{4.2.3}$$

Since we are dealing with a Gaussian density, the multiple integration can be done explicitly. We have

$$\exp\left(-\frac{1}{2}(\widehat{R}(A)\sigma',\sigma')\right) = \exp\left(-\frac{1}{2}\sum_{\vec{\varepsilon}\in\{0,1\}^2} 2^{-2}(AL_{\vec{\varepsilon}}(\sigma'), L_{\vec{\varepsilon}}(\sigma'))\right) \times \exp\left(\frac{1}{2}\sum_{\iota,\iota'=1}^{4} (G_A^{-1})_{\iota\iota'} \gamma_A(\sigma')_{\iota'} \gamma_A(\sigma')_\iota\right) \tag{4.2.4}$$

with G_A the 4×4 matrix and $\gamma_A(\sigma')$ the 4-vector given by

$$(G_A)_{\iota\iota'} = \sum_{\vec{\varepsilon}\in\{0,1\}^2} (AN(\iota,\vec{\varepsilon}), N(\iota',\vec{\varepsilon})), \tag{4.2.5}$$

$$\gamma_A(\sigma')_\iota = \sum_{\vec{\varepsilon}\in\{0,1\}^2} 2^{-1}(AL_{\vec{\varepsilon}}(\sigma'), N(\iota,\vec{\varepsilon})), \tag{4.2.6}$$

respectively. Thus,

$$(\widehat{R}(A)\sigma',\sigma') = \sum_{\vec{\varepsilon}\in\{0,1\}^2} 2^{-2}(AL_{\vec{\varepsilon}}(\sigma'), L_{\vec{\varepsilon}}(\sigma')) - \sum_{\iota,\iota'=1}^{4} (G_A^{-1})_{\iota\iota'} \gamma_A(\sigma')_{\iota'} \gamma_A(\sigma')_\iota \tag{4.2.7}$$

as a quadratic form. If we introduce the 4×8 matrix $N_{\vec{\varepsilon}}$ given by

$$(N_{\vec{\varepsilon}})_{\iota\lambda} = N(\iota,\vec{\varepsilon})_\lambda \tag{4.2.8}$$

and identify the Isis-space operator $L_{\vec{\varepsilon}}$ with its 8×8 matrix, we may write

$$\widehat{R}(A) = \frac{1}{4}\sum_{\vec{\varepsilon}\in\{0,1\}^2} L_{\vec{\varepsilon}}{}^\dagger A L_{\vec{\varepsilon}} - \frac{1}{2}\sum_{\vec{\varepsilon},\vec{\varepsilon}'\in\{0,1\}^2} L_{\vec{\varepsilon}}{}^\dagger A N_{\vec{\varepsilon}}{}^\dagger G_A^{-1} N_{\vec{\varepsilon}'} A L_{\vec{\varepsilon}'}, \tag{4.2.9}$$

$$G_A = \sum_{\vec{\varepsilon}\in\{0,1\}^2} N_{\vec{\varepsilon}}{}^\dagger A N_{\vec{\varepsilon}}. \tag{4.2.10}$$

While $\widehat{R}(A)$ is nonlinear in A, note that $A \mapsto \widehat{R}(A)$ preserves scalar multiplication.

Obviously, every Gaussian fixed point for the RG transformation reduces to a fixed point of this matrix transformation. It is more practical to introduce an initial matrix $A^{(0)}$ and define

$$A^{(N)} = \widehat{R}^N(A^{(0)}). \tag{4.2.11}$$

If $A^\infty = \lim_{N\to\infty} A^{(N)}$ exists, then

$$\widehat{R}(A^\infty) = A^\infty. \tag{4.2.12}$$

Our choice of $A^{(0)}$ is given by (4.1.30)—the canonical Gaussian. It is a very natural choice because

$$\frac{1}{4}\sum_{\vec{\varepsilon}\in\{0,1\}^2} L_{\vec{\varepsilon}}^{\dagger} A^{(0)} L_{\vec{\varepsilon}} = A^{(0)}. \tag{4.2.13}$$

Indeed, if $\vec{z}_\mu$ denotes the constant value of $\nabla\phi$ on the basic triangle that succeeds the μth radiating line segment in the clockwise sense, then

$$\sigma_1 = \vec{z}_1 \cdot \hat{i} = \vec{z}_2 \cdot \hat{i}, \tag{4.2.14}$$
$$\sigma_2 = \vec{z}_2 \cdot (\hat{i} + \hat{j}) = \vec{z}_3 \cdot (\hat{i} + \hat{j}), \tag{4.2.15}$$
$$\sigma_3 = \vec{z}_3 \cdot \hat{j} = \vec{z}_4 \cdot \hat{j}, \tag{4.2.16}$$
$$\sigma_4 = \vec{z}_4 \cdot (\hat{j} - \hat{i}) = \vec{z}_5 \cdot (\hat{j} - \hat{i}), \tag{4.2.17}$$
$$\sigma_5 = \vec{z}_5 \cdot \hat{i} = \vec{z}_6 \cdot \hat{i}, \tag{4.2.18}$$
$$\sigma_6 = \vec{z}_6 \cdot (\hat{i} + \hat{j}) = \vec{z}_7 \cdot (\hat{i} + \hat{j}), \tag{4.2.19}$$
$$\sigma_7 = \vec{z}_7 \cdot \hat{j} = \vec{z}_8 \cdot \hat{j} \tag{4.2.20}$$
$$\sigma_8 = \vec{z}_8 \cdot (\hat{j} - \hat{i}) = \vec{z}_1 \cdot (\hat{j} - \hat{i}) \tag{4.2.21}$$

as a consequence of the way in which the Isis variables were defined in the Section 4.1. Hence,

$$\vec{z}_1 = \sigma_1\hat{i} + (\sigma_8 + \sigma_1)\hat{j}, \tag{4.2.22}$$
$$\vec{z}_2 = \sigma_1\hat{i} + (\sigma_2 - \sigma_1)\hat{j}, \tag{4.2.23}$$
$$\vec{z}_3 = (\sigma_2 - \sigma_3)\hat{i} + \sigma_3\hat{j}, \tag{4.2.24}$$
$$\vec{z}_4 = (\sigma_3 - \sigma_4)\hat{i} + \sigma_3\hat{j}, \tag{4.2.25}$$
$$\vec{z}_5 = \sigma_5\hat{i} + (\sigma_4 + \sigma_5)\hat{j}, \tag{4.2.26}$$
$$\vec{z}_6 = \sigma_5\hat{i} + (\sigma_6 - \sigma_5)\hat{j}, \tag{4.2.27}$$
$$\vec{z}_7 = (\sigma_6 - \sigma_7)\hat{i} + \sigma_7\hat{j}, \tag{4.2.28}$$
$$\vec{z}_8 = (\sigma_7 - \sigma_8)\hat{i} + \sigma_7\hat{j}, \tag{4.2.29}$$

and since (4.1.30) yields

$$(A^{(0)}\sigma, \sigma) = \frac{1}{32}\sum_{\mu=1}^{8} \vec{z}_\mu \cdot \vec{z}_\mu, \tag{4.2.30}$$

it follows that

$$(A^{(0)}\sigma,\sigma) = \frac{1}{32}\left[2\sigma_1^2 + (\sigma_1+\sigma_8)^2 + (\sigma_2-\sigma_1)^2 + (\sigma_2-\sigma_3)^2 + 2\sigma_3^2 \right.$$
$$+ (\sigma_3-\sigma_4)^2 + 2\sigma_5^2 + (\sigma_4+\sigma_5)^2 + (\sigma_6-\sigma_5)^2 + 2\sigma_7^2$$
$$\left. + (\sigma_6-\sigma_7)^2 + (\sigma_7-\sigma_8)^2\right]. \qquad (4.2.31)$$

This is precisely the formula (4.1.31) that we promised to derive when we introduced $A^{(0)}$. On the other hand,

$$L_{\vec{0}}(\sigma') = (\sigma_6'-\sigma_7', \sigma_6', \sigma_6'-\sigma_5', \sigma_6'-2\sigma_5', \sigma_5', \sigma_6', \sigma_7', 2\sigma_7'-\sigma_6'), \qquad (4.2.32)$$
$$L_{\hat{i}}(\sigma') = (\sigma_1', \sigma_8'+2\sigma_1', \sigma_8'+\sigma_1', \sigma_8', \sigma_7'-\sigma_8', 2\sigma_7'-\sigma_8', \sigma_7', \sigma_8'), \qquad (4.2.33)$$
$$L_{\hat{j}}(\sigma') = (\sigma_3'-\sigma_4', 2\sigma_3'-\sigma_4', \sigma_3', \sigma_4', \sigma_5', \sigma_4'+2\sigma_5', \sigma_4'+\sigma_5', \sigma_4'), \qquad (4.2.34)$$
$$L_{\hat{i}+\hat{j}}(\sigma') = (\sigma_1', \sigma_2', \sigma_3', 2\sigma_3'-\sigma_2', \sigma_2'-\sigma_3', \sigma_2', \sigma_2'-\sigma_1', \sigma_2'-2\sigma_1'), \qquad (4.2.35)$$

and therefore

$$(A^{(0)}L_{\vec{0}}(\sigma'), L_{\vec{0}}(\sigma'))$$
$$= \frac{1}{32}\left[2(\sigma_6'-\sigma_7')^2 + \sigma_7'^2 + \sigma_7'^2 + \sigma_5'^2 + 2(\sigma_6'-\sigma_5')^2\right.$$
$$+ \sigma_5'^2 + 2\sigma_5'^2 + (\sigma_6'-\sigma_5')^2 + (\sigma_6'-\sigma_5')^2 + 2\sigma_7'^2$$
$$\left. + (\sigma_6'-\sigma_7')^2 + (\sigma_6'-\sigma_7')^2\right]$$
$$= \frac{1}{8}\left[\sigma_5'^2 + \sigma_7'^2 + (\sigma_6'-\sigma_5')^2 + (\sigma_6'-\sigma_7')^2\right], \qquad (4.2.36)$$

$$(A^{(0)}L_{\hat{i}}(\sigma'), L_{\hat{i}}(\sigma'))$$
$$= \frac{1}{32}\left[2\sigma_1'^2 + (\sigma_8'+\sigma_1')^2 + (\sigma_1'+\sigma_8')^2 + \sigma_1'^2\right.$$
$$+ 2(\sigma_8'+\sigma_1')^2 + \sigma_1'^2 + 2(\sigma_7'-\sigma_8')^2 + \sigma_7'^2 + \sigma_7'^2 + 2\sigma_7'^2$$
$$\left. + (\sigma_7'-\sigma_8')^2 + (\sigma_7'-\sigma_8')^2\right]$$
$$= \frac{1}{8}\left[\sigma_1'^2 + \sigma_7'^2 + (\sigma_8'+\sigma_1')^2 + (\sigma_7'-\sigma_8')^2\right], \qquad (4.2.37)$$

$$(A^{(0)}L_{\hat{j}}(\sigma'), L_{\hat{j}}(\sigma'))$$
$$= \frac{1}{32}\left[2(\sigma_3'-\sigma_4')^2 + \sigma_3'^2 + \sigma_3'^2 + (\sigma_3'-\sigma_4')^2 + 2\sigma_3'^2\right.$$
$$+ (\sigma_3'-\sigma_4')^2 + 2\sigma_5'^2 + (\sigma_4'+\sigma_5')^2 + (\sigma_4'+\sigma_5')^2$$
$$\left. + 2(\sigma_4'+\sigma_5')^2 + \sigma_5'^2 + \sigma_5'^2\right]$$
$$= \frac{1}{8}\left[\sigma_3'^2 + \sigma_5'^2 + (\sigma_3'-\sigma_4')^2 + (\sigma_4'+\sigma_5')^2\right], \qquad (4.2.38)$$

$$
\begin{aligned}
&(A^{(0)}L_{\hat{i}+\hat{j}}(\sigma'), L_{\hat{i}+\hat{j}}(\sigma')) \\
&\quad = \frac{1}{32}\left[2\sigma'^2_1 + (\sigma'_2-\sigma'_1)^2 + (\sigma'_2-\sigma'_1)^2 + (\sigma'_2-\sigma'_3)^2\right. \\
&\qquad\quad + 2\sigma'^2_3 + (\sigma'_2-\sigma'_3)^2 + 2(\sigma'_2-\sigma'_3)^2 + \sigma'^2_3 + \sigma'^2_3 \\
&\qquad\quad \left. +2(\sigma'_2-\sigma'_1)^2 + \sigma'^2_1 + \sigma'^2_1\right] \\
&\quad = \frac{1}{8}\left[\sigma'^2_1 + \sigma'^2_3 + (\sigma'_2-\sigma'_1)^2 + (\sigma'_2-\sigma'_3)^2\right]. \qquad (4.2.39)
\end{aligned}
$$

The sum of these four equations yields the desired formula (4.2.13).

Our goal is the calculation of $A^{(N)}$, but the N-fold iteration of $\widehat{R}$ does not have an obvious simplification from the matrix-algebraic point of view. Therefore, we return to the integral representation of $\widehat{R}$ and describe its iteration from the continuum point of view, as already indicated by (4.1.30). For N iterations, the *initial* range of continuum configurations ϕ on $[0, 2^{N-1}]^2$ is given by

$$
\phi(\vec{x}) = \phi^{(N)}(2^{-N}\,\vec{x}) + \sum_{\iota=1}^{4}\sum_{r=0}^{N-1}\sum_{\vec{\ell}\in\Lambda_{N-r}} \alpha^{(r)}_{\iota\vec{\ell}}\Psi_{\iota r\vec{\ell}}(\vec{x}), \qquad (4.2.40)
$$

$$
\Lambda_s = \{0, 1, \ldots, 2^{s-1}-1\}^2. \qquad (4.2.41)
$$

These continuous, piecewise-linear configurations are based on a triangulation where each $2^{-N-1}\times 2^{-N-1}$ square is divided as in Figure 4.7. Recall from [1] that for any Hamiltonian of the form

$$
H_W(\phi) = \overline{H}_W(\nabla\phi) = \int W(\nabla\phi(\vec{x}))d\,\vec{x}, \qquad (4.2.42)
$$

the transformed Hamiltonian has the half-scale localization property—i.e.,

$$
\mathcal{R}^N(\overline{H}_W)(\nabla\phi^{(N)}) = \sum_{\vec{\varepsilon}\in\{0,1\}^2}\sum_{\vec{m}} \mathcal{R}^N(\overline{H}_W)(\chi_{[0,\frac{1}{2}]^2+\vec{m}+\frac{1}{2}\vec{\varepsilon}}\nabla\phi^{(N)}). \qquad (4.2.43)
$$

Moreover,

$$
\begin{aligned}
&\exp(-\mathcal{R}^N(\overline{H}_W)(\chi_{[0,\frac{1}{2}]^2}\nabla\phi^{(N)})) \\
&\quad = \widetilde{Z}^{(N)\ -1}_{W,\{\vec{0}\}}\left(\prod_{\iota=1}^{4}\prod_{r=0}^{N-1}\prod_{\vec{\ell}\in\Lambda_{N-r}}\int_{-\infty}^{\infty} d\alpha^{(r)}_{\iota\vec{\ell}}\right) \\
&\qquad \times\exp\left(-\int_{[0,2^{N-1}]^2} d\,\vec{x}\;W\left(2^{-N}\nabla\phi^{(N)}(2^{-N}\,\vec{\xi})\right.\right. \\
&\qquad\qquad \left.\left. +\sum_{\iota=1}^{4}\sum_{r=0}^{N-1}\sum_{\vec{\ell}\in\Lambda_{N-r}} \alpha^{(r)}_{\iota\vec{\ell}}\nabla\Psi_{\iota r\vec{\ell}}(\vec{x})\right)\right),
\end{aligned}
\qquad (4.2.44)
$$

$$\widetilde{Z}^{(N)}_{W,\{\vec{0}\}} = \left(\prod_{\iota=1}^{4} \prod_{r=0}^{N-1} \prod_{\vec{\ell} \in \Lambda_{N-r}} \int_{-\infty}^{\infty} d\alpha^{(r)}_{\iota \vec{\ell}} \right) \times \exp\left(- \int_{[0,2^{N-1}]^2} d\vec{x}\; W\left(\sum_{\iota=1}^{4} \sum_{r=0}^{N-1} \sum_{\vec{\ell} \in \Lambda_{N-r}} \alpha^{(r)}_{\iota \vec{\ell}} \Psi_{\iota r \vec{\ell}}(\vec{x}) \right) \right). \tag{4.2.45}$$

Since $A^{(0)}$ is our initial matrix, we are actually dealing with the special case

$$W(\vec{z}) = W_0(\vec{z}) \equiv \frac{1}{2} \vec{z} \cdot \vec{z} \tag{4.2.46}$$

because

$$\overline{H}_{W_0}(\chi_\Omega \nabla\phi) = \frac{1}{2} \int_\Omega (\nabla\phi)^2 \tag{4.2.47}$$

for every rectangle Ω. Clearly,

$$\begin{aligned} &\exp(-\mathcal{R}^N(\overline{H}_{W_0})(\chi_{[0,\frac{1}{2}]^2} \nabla\phi^{(N)})) \\ &= \exp\left(-\frac{1}{2} \int_{[0,\frac{1}{2}]^2} d\vec{x}^{\,(N)} (\nabla\phi^{(N)}(\vec{x}^{\,(N)}))^2 \right) \\ &\times \widetilde{Z}^{(N)}_{W_0,\{\vec{0}\}}{}^{-1} \left(\prod_{\iota=1}^{4} \prod_{r=0}^{N-1} \prod_{\vec{\ell} \in \Lambda_{N-r}} \int_{-\infty}^{\infty} d\alpha^{(r)}_{\iota \vec{\ell}} \right) \\ &\times \exp\Bigg(-\sum_{\iota=1}^{4} \prod_{r=0}^{N-1} \sum_{\vec{\ell} \in \Lambda_{N-r}} \alpha^{(r)}_{\iota \vec{\ell}} \int_{[0,2^{N-1}]^2} d\vec{x}\; 2^{-N} \nabla\phi^{(N)}(2^{-N}\vec{x}) \\ &\qquad \cdot \nabla\Psi_{\iota r \vec{\ell}}(\vec{x}) \\ &\qquad - \frac{1}{2} \sum_{\iota,\iota'=1}^{4} \sum_{r,r'=0}^{N-1} \sum_{\vec{\ell} \in \Lambda_{N-r}} \sum_{\vec{\ell}' \in \Lambda_{N-r'}} \alpha^{(r)}_{\iota \vec{\ell}} \alpha^{(r')}_{\iota' \vec{\ell}'} \int_{[0,2^{N-1}]^2} d\vec{x} \\ &\qquad \cdot \nabla\Psi_{\iota r \vec{\ell}}(\vec{x}) \nabla\Psi_{\iota' r' \vec{\ell}'}(\vec{x}) \Bigg). \end{aligned} \tag{4.2.48}$$

Once again, the integration can be done explicitly because we have a Gaussian density. If we set

$$(G_N)_{\iota\iota';rr';\vec{\ell}\,\vec{\ell}'} = \int_{[0,2^{N-1}]^2} d\vec{x}\ \nabla\Psi_{\iota r\vec{\ell}}(\vec{x})\cdot\nabla\Psi_{\iota' r'\vec{\ell}'}(\vec{x}), \tag{4.2.49}$$

$$\gamma^{(N)}(\phi^{(N)})_{\iota r\vec{\ell}} = \int_{[0,2^{N-1}]^2} d\vec{x}\ 2^{-N}\nabla\phi^{(N)}(2^{-N}\vec{x})\cdot\nabla\Psi_{\iota r\vec{\ell}}(\vec{x}), \tag{4.2.50}$$

then applying the change of variable $\vec{x}^{\,(N)} = 2^{-N}\vec{x}$ to the leading integral, we obtain

$$\begin{aligned}&\mathcal{R}^N(\overline{H}_{W_0})(\chi_{[0,\frac{1}{2}]^2}\nabla\phi^{(N)})\\ &\quad= \frac{1}{2}\int_{[0,\frac{1}{2}]^2} d\vec{x}^{\,(N)}(\nabla\phi^{(N)}(\vec{x}^{\,(N)}))^2\\ &\quad\quad-\frac{1}{2}\sum_{\iota,\iota'=1}^{4}\sum_{r,r'=0}^{N-1}\sum_{\vec{\ell}\in\Lambda_{N-r}}\sum_{\vec{\ell}'\in\Lambda_{N-r'}}(G_N^{-1})_{\iota\iota';rr';\vec{\ell}\,\vec{\ell}'}\gamma^{(N)}(\phi^{(N)})_{\iota' r'\vec{\ell}'}\\ &\quad\quad\times\gamma^{(N)}(\phi^{(N)})_{\iota r\vec{\ell}}.\end{aligned} \tag{4.2.51}$$

The remainder of the calculation depends on the determination of G_N, which we reserve for Section 4.4. In the meantime, it is worthwhile to mention a couple of basic facts. Let Δ_μ denote the μth basic triangle in $[0,1/2]^2$ corresponding to the $\vec{z}_\mu$-variable. Then

$$\int_{\Delta_\mu}\nabla\Psi_{\iota s\vec{\ell}} = 0, \qquad s<-1, \tag{4.2.52}$$

as we shall discuss in Section 4.5. For $s=-1$, it is easy enough to calculate the integrals. We have

$$\int_{\Delta_1}\nabla\Psi_{1,-1,\vec{0}} = \int_{\Delta_5}\nabla\Psi_{1,-1,\vec{0}} = \frac{1}{16\sqrt{2}}(\hat{\imath}+\hat{\jmath}), \tag{4.2.53}$$

$$\int_{\Delta_2}\nabla\Psi_{1,-1,\vec{0}} = \int_{\Delta_3}\nabla\Psi_{1,-1,\vec{0}} = \int_{\Delta_6}\nabla\Psi_{1,-1,\vec{0}} = \int_{\Delta_7}\nabla\Psi_{1,-1,\vec{0}} = \vec{0}, \tag{4.2.54}$$

$$\int_{\Delta_4}\nabla\Psi_{1,-1,\vec{0}} = \int_{\Delta_8}\nabla\Psi_{1,-1,\vec{0}} = -\frac{1}{16\sqrt{2}}(\hat{\imath}+\hat{\jmath}), \tag{4.2.55}$$

$$\int_{\Delta_1}\nabla\Psi_{2,-1,\vec{0}} = \int_{\Delta_4}\nabla\Psi_{2,-1,\vec{0}} = \int_{\Delta_5}\nabla\Psi_{2,-1,\vec{0}} = \int_{\Delta_8}\nabla\Psi_{2,-1,\vec{0}} = \vec{0}, \tag{4.2.56}$$

$$\int_{\Delta_2}\nabla\Psi_{2,-1,\vec{0}} = \int_{\Delta_6}\nabla\Psi_{2,-1,\vec{0}} = \frac{1}{16\sqrt{2}}(\hat{\jmath}-\hat{\imath}), \tag{4.2.57}$$

$$\int_{\Delta_3}\nabla\Psi_{2,-1,\vec{0}} = \int_{\Delta_7}\nabla\Psi_{2,-1,\vec{0}} = \frac{1}{16\sqrt{2}}(\hat{\imath}-\hat{\jmath}), \tag{4.2.58}$$

$$\int_{\Delta_1}\nabla\Psi_{3,-1,\vec{0}} = \int_{\Delta_6}\nabla\Psi_{3,-1,\vec{0}} = \frac{1}{16\sqrt{2}}\hat{\jmath}, \tag{4.2.59}$$

$$\int_{\Delta_2} \nabla\Psi_{3,-1,\vec{0}} = \int_{\Delta_5} \nabla\Psi_{3,-1,\vec{0}} = -\frac{1}{16\sqrt{2}}\hat{j}, \tag{4.2.60}$$

$$\int_{\Delta_3} \nabla\Psi_{3,-1,\vec{0}} = \int_{\Delta_4} \nabla\psi_{3,-1,\vec{0}} = \int_{\Delta_7} \nabla\Psi_{3,-1,\vec{0}} = \int_{\Delta_8} \nabla\Psi_{3,-1,\vec{0}} = \vec{0}, \tag{4.2.61}$$

$$\int_{\Delta_1} \nabla\Psi_{4,-1,\vec{0}} = \int_{\Delta_2} \nabla\Psi_{4,-1,\vec{0}} = \int_{\Delta_5} \nabla\Psi_{4,-1,\vec{0}} = \int_{\Delta_6} \nabla\Psi_{4,-1,\vec{0}} = \vec{0}, \tag{4.2.62}$$

$$\int_{\Delta_3} \nabla\Psi_{4,-1,\vec{0}} = \int_{\Delta_8} \nabla\Psi_{4,-1,\vec{0}} = -\frac{1}{16\sqrt{2}}\hat{i}, \tag{4.2.63}$$

$$\int_{\Delta_4} \nabla\Psi_{4,-1,\vec{0}} = \int_{\Delta_7} \nabla\Psi_{4,-1,\vec{0}} = \frac{1}{16\sqrt{2}}\hat{i}. \tag{4.2.64}$$

We clarify this in Section 4.5.

4.3 The recursion formula for the dipole perturbation

Before we complete our analysis of the Gaussian fixed point, we derive the hierarchical reduction of the RG transformation for the Osiris wavelet conditioning of the dipole perturbation in two dimensions. Our initial Hamiltonian for the RG iteration is

$$H(\phi) = \frac{1}{2} \sum_{\vec{m}\in\mathbb{Z}^2} (A^{\infty}\sigma_{\frac{1}{2}\vec{m}}, \sigma_{\frac{1}{2}\vec{m}}) + \int F(\nabla\phi), \tag{4.3.1}$$

where $F(\vec{z})$ is a real analytic function in $\vec{z} \cdot \vec{z}$ that is bounded below and vanishes to second order at zero. This Hamiltonian is not to be confused with the dipole Hamiltonian given in [1], where the quadratic form is

$$\int (\nabla\phi)^2 = \sum_{\vec{m}\in\mathbb{Z}^2} (A^{(0)}\sigma_{\frac{1}{2}\vec{m}}, \sigma_{\frac{1}{2}\vec{m}}). \tag{4.3.2}$$

Now if we apply the RG transformation to H, we must calculate (with $\alpha_{\iota\vec{m}} = \alpha^{(0)}_{\iota\vec{m}}$)

$$\begin{aligned}
&\exp(-\mathcal{R}(H)(\phi')) \\
&= \exp\left(-\frac{1}{2} \sum_{\vec{\varepsilon}\in\{0,1\}^2} \sum_{\vec{m}\in\mathbb{Z}^2} 2^{-2}(A^{\infty}L_{\vec{\varepsilon}}(\sigma'_{\frac{1}{2}\vec{m}}), L_{\vec{\varepsilon}}(\sigma'_{\frac{1}{2}\vec{m}}))\right) \\
&\quad \times \lim_{\Lambda\nearrow\mathbb{Z}^2} \widetilde{Z}^{-1}_{H,\Lambda} \left(\prod_{\iota=1}^{4} \prod_{\vec{m}\in\Lambda} \int_{-\infty}^{\infty} d\alpha_{\iota\vec{m}}\right)
\end{aligned}$$

$$\times \exp\left(-\sum_{\iota=1}^{4}\sum_{\vec{m}\in\Lambda}\alpha_{\iota\vec{m}}\sum_{\vec{\varepsilon}\in\{0,1\}^2}2^{-1}(A^{\infty}L_{\vec{\varepsilon}}(\sigma'_{\frac{1}{2}\vec{m}}),N(\iota,\vec{\varepsilon}))\right.$$

$$-\frac{1}{2}\sum_{\iota,\iota'=1}^{4}\sum_{\vec{m}\in\Lambda}\alpha_{\iota\vec{m}}\alpha_{\iota'\vec{m}}\sum_{\vec{\varepsilon}\in\{0,1\}^2}(A^{\infty}N(\iota,\vec{\varepsilon}),N(\iota',\vec{\varepsilon}))$$

$$\left.-\sum_{\vec{m}\in\Lambda}\int_{[0,1]^2}d\vec{x}F\left(\frac{1}{2}\nabla\phi'\left(\frac{1}{2}\vec{x}+\frac{1}{2}\vec{m}\right)+\sum_{\iota=1}^{4}\alpha_{\iota\vec{m}}\nabla\Psi_{\iota\vec{m}}(\vec{x})\right)\right), \tag{4.3.3}$$

$$\widetilde{Z}_{H,\Lambda}=\left(\prod_{\iota=1}^{4}\prod_{\vec{m}\in\Lambda}\int_{-\infty}^{\infty}d\alpha_{\iota\vec{m}}\right)$$

$$\times \exp\left(-\frac{1}{2}\sum_{\iota,\iota'=1}^{4}\sum_{\vec{m},\vec{m}'\in\Lambda}\alpha_{\iota\vec{m}}\alpha_{\iota'\vec{m}'}\sum_{\vec{\varepsilon}\in\{0,1\}^2}(A^{\infty}N(\iota,\vec{\varepsilon}),N(\iota',\vec{\xi}\,'))\right.$$

$$\left.-\sum_{\vec{m}\in\Lambda}\int_{[0,1]^2}d\vec{x}\;F\left(\sum_{\iota=1}^{4}\alpha_{\iota\vec{m}}\nabla\Psi_{\iota\vec{m}}(\vec{x})\right)\right), \tag{4.3.4}$$

where we have incorporated (4.2.2) and applied

$$\text{supp}\;\Psi_{\iota\vec{m}}\subset[0,1]^2+\vec{m} \tag{4.3.5}$$

as well. The support property yields the usual factorization. With the association

$$\phi\mapsto(\sigma_{\frac{1}{2}\vec{m},1},\ldots,\sigma_{\frac{1}{2}\vec{m},8})$$

of Isis variables to the field in each square $[0,1/2]^2+(1/2)\,\vec{m}$, we may write

$$H(\phi)=\sum_{\vec{m}\in\mathbb{Z}^2}\widehat{H}(\sigma_{\frac{1}{2}\vec{m}}). \tag{4.3.6}$$

The point is that

$$\mathcal{R}(H)(\phi')=\sum_{\vec{m}\in\mathbb{Z}^2}\widehat{R}(\widehat{H})(\sigma'_{\frac{1}{2}\vec{m}}), \tag{4.3.7}$$

$$\exp(-\widehat{R}(\widehat{H})(\sigma'))=\exp\left(-\frac{1}{2}\sum_{\vec{\varepsilon}\in\{0,1\}^2}2^{-2}(A^{\infty}L_{\vec{\varepsilon}}(\sigma'),L_{\vec{\varepsilon}}(\sigma'))\right)$$

$$\times\widehat{Z}_{\widehat{H}}^{-1}\left(\prod_{\iota=1}^{4}\int_{-\infty}^{\infty}d\tau_{\iota}\right)$$

$$\times\exp\left(-\sum_{\iota=1}^{4}\tau_{\iota}\sum_{\vec{\varepsilon}\in\{0,1\}^2}2^{-1}(A^{\infty}L_{\vec{\varepsilon}}(\sigma'),N(\iota,\vec{\varepsilon}))\right.$$

$$-\frac{1}{2}\sum_{\iota,\iota'=1}^{4}\tau_\iota\tau_{\iota'}\sum_{\vec{\varepsilon}\in\{0,1\}^2}(A^\infty N(\iota,\vec{\varepsilon}),N(\iota',\vec{\varepsilon}))$$
$$-\int_{[0,1]^2}d\vec{x}\,F\left(\frac{1}{2}\nabla\phi'\left(\frac{1}{2}\vec{x}\right)+\sum_{\iota=1}^{4}\tau_\iota\nabla\Psi_\iota(\vec{x})\right)\Bigg)$$

(4.3.8)

$$\widehat{Z}_{\widehat{H}}=\widetilde{Z}_{H,\{\vec{0}\}}. \qquad (4.3.9)$$

The locality of the perturbation has not been preserved, except over $1/2 \times 1/2$ squares, inside of which the transformed perturbation is nonlocal.

At this point, it is convenient to introduce a sigma-dependent probability measure on the space of τ-variables that is associated with the Gaussian fixed point. Let

$$d\mu_\infty(\sigma')(\tau)=\widehat{X}_\infty(\sigma')^{-1}$$
$$\times\exp\left(-\frac{1}{2}\sum_{\iota=1}^{4}\tau_\iota\sum_{\vec{\varepsilon}\in\{0,1\}^2}(A^\infty L_{\vec{\varepsilon}}(\sigma'),N(\iota,\vec{\varepsilon}))\right.$$
$$\left.-\frac{1}{2}\sum_{\iota,\iota'=1}^{4}\tau_\iota\tau_{\iota'}\sum_{\vec{\varepsilon}\in\{0,1\}^2}(A^\infty N(\vec{\varepsilon},\iota),N(\vec{\varepsilon},\iota'))\right)\prod_{\iota=1}^{4}d\tau_\iota,$$

(4.3.10)

$$\widehat{X}_\infty(\sigma')=\left(\prod_{\iota=1}^{4}\int_{-\infty}^{\infty}d\tau_\iota\right)$$
$$\times\exp\left(-\frac{1}{2}\sum_{\iota=1}^{4}\tau_\iota\sum_{\vec{\varepsilon}\in\{0,1\}^2}(A^\infty L_{\vec{\varepsilon}}(\sigma'),N(\iota,\vec{\varepsilon}))\right.$$
$$\left.-\frac{1}{2}\sum_{\iota,\iota'=1}^{4}\tau_\iota\tau_{\iota'}\sum_{\vec{\varepsilon}\in\{0,1\}^2}(A^\infty N(\vec{\varepsilon},\iota),N(\vec{\varepsilon},\iota'))\right).$$

(4.3.11)

Then

$$\exp(-\widehat{R}(\widehat{H})(\sigma'))$$
$$=\exp\left(-\frac{1}{2}\sum_{\vec{\varepsilon}\in\{0,1\}^2}2^{-2}(A^\infty L_{\vec{\varepsilon}}(\sigma'),L_{\vec{\varepsilon}}(\sigma'))\right)$$
$$\times\widehat{X}_\infty(\sigma')\widehat{Z}_{A^\infty}^{-1}\widehat{Z}_{A^\infty}\widehat{Z}_{\widehat{H}}^{-1}\int d\mu_\infty(\sigma')(\tau)$$
$$\times\exp\left(-\int_{[0,1]^2}d\vec{x}\;F\left(\frac{1}{2}\nabla\phi'\left(\frac{1}{2}\vec{x}\right)+\sum_{\iota=1}^{4}\tau_\iota\nabla\Psi_\iota(\vec{x})\right)\right). \qquad (4.3.12)$$

On the other hand, it is obvious that

$$\begin{aligned}
&\widehat{Z}_{A^\infty}^{-1} \exp\left(-\frac{1}{2}\sum_{\vec{\varepsilon}\in\{0,1\}^2} 2^{-2}(A^\infty L_{\vec{\varepsilon}}(\sigma'), L_{\vec{\varepsilon}}(\sigma'))\right)\widehat{X}_\infty(\sigma') \\
&\quad = \exp\left(-\frac{1}{2}(\widehat{R}(A^\infty)\sigma',\sigma')\right) \\
&\quad = \exp\left(-\frac{1}{2}(A^\infty\sigma',\sigma')\right),
\end{aligned} \tag{4.3.13}$$

while

$$\widehat{Z}_{A^\infty}^{-1}\widehat{Z}_{\widehat{H}} = \int d\mu_\infty(0)(\tau)\exp\left(-\int_{[0,1]^2} d\,\vec{x}\;\, F\left(\sum_{\iota=1}^{4}\tau_\iota\nabla\Psi_\iota(\vec{x})\right)\right). \tag{4.3.14}$$

Hence,

$$\widehat{R}(\widehat{H})(\sigma') = \frac{1}{2}(A^\infty\sigma',\sigma') + \widehat{R}_\infty(\widehat{F})(\sigma'), \tag{4.3.15}$$

$$\begin{aligned}
\exp(-\widehat{R}_\infty(\widehat{F})(\sigma')) &= \left[\int d\mu_\infty(0)(\tau)\exp\left(-\int_{[0,1]^2} d\,\vec{x}\,F\left(\sum_{\iota=1}^{4}\tau_\iota\nabla\Psi_\iota(\vec{x})\right)\right)\right]^{-1} \\
&\quad\times\int d\mu_\infty(\sigma')(\tau) \\
&\quad\times\exp\left(-\int_{[0,1]^2} d\,\vec{x}\;\, F\left(\frac{1}{2}\nabla\phi'\left(\frac{1}{2}\,\vec{x}\right) + \sum_{\iota=1}^{4}\tau_\iota\Psi_\iota(\vec{x})\right)\right),
\end{aligned} \tag{4.3.16}$$

where

$$\widehat{F}(\sigma_1,\ldots,\sigma_8) = \int_{[0,\frac{1}{2}]^2} d\,\vec{x}\;\, F(\nabla\phi(\vec{x})). \tag{4.3.17}$$

This describes the transformation of the perturbation of the $\widehat{R}$-fixed quadratic form.

The transformed perturbation is no longer local in the continuum, so in order to iterate the transformation, we must consider perturbations that are general functions $\widehat{F}(\sigma)$. The initial Hamiltonian simply involves the special case (4.3.17), so if we write

$$\begin{aligned}
&\int_{[0,1]^2} d\,\vec{x}\;\, F\left(\frac{1}{2}\nabla\phi'\left(\frac{1}{2}\,\vec{x}\right) + \sum_{\iota=1}^{4}\tau_\iota\nabla\Psi_\iota(\vec{x})\right) \\
&\quad = \sum_{\vec{\varepsilon}\in\{0,1\}^2}\int_{[0,\frac{1}{2}]^2} d\,\vec{x}\;\, F\left(\frac{1}{2}\nabla\phi'\left(\frac{1}{2}\,\vec{x} + \frac{1}{4}\,\vec{\varepsilon}\right) + \sum_{\iota=1}^{4}\tau_\iota\nabla\Psi_\iota\left(\vec{x} + \frac{1}{2}\,\vec{\epsilon}\right)\right)
\end{aligned} \tag{4.3.18}$$

and recall from Section 4.1 that $N(\iota, \vec{\varepsilon})_\mu$ is the contribution of $\nabla\Psi_\iota(\vec{x} + (1/2)\vec{\varepsilon})$ to the μth Isis variable of the square $[0, 1/2]^2$, we have

$$\begin{aligned}\int_{[0,1]^2} d\vec{x}\, F\left(\frac{1}{2}\nabla\phi'\left(\frac{1}{2}\vec{x}\right) + \sum_{\iota=1}^{4} \tau_\iota \nabla\Psi_\iota(\vec{x})\right) \\ = \sum_{\vec{\varepsilon}\in\{0,1\}^2} F\left(\frac{1}{2}L_{\vec{\varepsilon}}(\sigma') + \sum_{\iota=1}^{4} \tau_\iota N(\iota, \vec{\varepsilon})\right),\end{aligned} \tag{4.3.19}$$

where we have also recalled that $L_{\vec{\varepsilon}}(\sigma')_\mu$ is the contribution of $\nabla\phi'((1/2)\vec{x} + (1/4)\vec{\varepsilon})$ to the same variable. The right-hand side of this equation is the key to the extension of our transformation formula to one suitable for iteration. For the more general function $\widehat{F}(\sigma)$, we see that

$$\begin{aligned}\exp(-\widehat{R}_\infty(\widehat{F})(\sigma')) = \widehat{Z}_\infty(\widehat{F})^{-1} \int d\mu_\infty(\sigma')(\tau) \\ \exp\left(-\sum_{\vec{\varepsilon}\in\{0,1\}^2} \widehat{F}\left(\frac{1}{2}L_{\vec{\varepsilon}}(\sigma') + \sum_{\iota=1}^{4} \tau_\iota N(\iota, \vec{\varepsilon})\right)\right),\end{aligned} \tag{4.3.20}$$

$$\widehat{Z}_\infty(\widehat{F}) = \int d\mu_\infty(0)(\tau) \exp\left(-\sum_{\vec{\varepsilon}\in\{0,1\}^2} \widehat{F}\left(\sum_{\iota=1}^{4} \tau_\iota N(\iota, \vec{\varepsilon})\right)\right). \tag{4.3.21}$$

4.4 The Gaussian fixed point (wavelet picture)

At this point we consider the quadratic form in the wavelet amplitudes associated with the unit-scale configurations. Our starting point is the expansion [14] of the Osiris-constrained continuum scalar field with unit-scale cutoff. Thus

$$\int (\nabla\phi)^2 = \sum_{r,r'=0}^{\infty} \sum_{\vec{n},\vec{n}'\in\mathbb{Z}^2} \sum_{\iota,\iota'=1}^{4} \alpha^{(r)}_{\iota\vec{n}} \alpha^{(r')}_{\iota'\vec{n}'} S_{rr';\vec{n}\vec{n}';\iota\iota'} \equiv (S\alpha, \alpha) \tag{4.4.1}$$

is the initial quadratic form for the RG iteration. We have seen in [1] that the matrix S has the properties

$$S_{rr'} = T_{r-r'}, \tag{4.4.2}$$

$$T_r = 0, \qquad r \neq -1, 0, 1, \tag{4.4.3}$$

$$T_0 = 1, \tag{4.4.4}$$

$$T_{-1} = T_1^*, \tag{4.4.5}$$

$$T_{1;\vec{n},2\vec{m}+\vec{\varepsilon};\iota\iota'} = Q^{\vec{\varepsilon}}_{\iota\iota'}\delta_{\vec{m}\vec{n}}, \qquad \vec{\varepsilon} \in \{0,1\}^2, \tag{4.4.6}$$

where the 4×4 matrices $Q^{\vec{\varepsilon}}$ shall be recalled below. As a result, this quadratic form reduces to

$$(S\alpha,\alpha) = \sum_{r=0}^{\infty} \sum_{\vec{n}\in\mathbb{Z}^2} \sum_{\iota=1}^{4} \alpha^{(r)2}_{\iota\vec{n}} + 2\sum_{r=1}^{\infty} \sum_{\vec{m}\in\mathbb{Z}^2} \sum_{\vec{\varepsilon}\in\{0,1\}^2} \sum_{\iota,\iota'=1}^{4} Q^{\vec{\varepsilon}}_{\iota\iota'}\alpha^{(r-1)}_{\iota',2\vec{m}+\vec{\varepsilon}}\alpha^{(r)}_{\iota\vec{m}}. \tag{4.4.7}$$

Since the RG transformation is based on just integrating out the unit-scale wavelet amplitudes, we have

$$\begin{aligned}
&\exp\left(-\frac{1}{2}(\mathcal{R}(S)\alpha',\alpha')\right) \\
&= \exp\left(-\frac{1}{2}\sum_{r=1}^{\infty} \sum_{\vec{n}\in\mathbb{Z}^2} \sum_{\iota=1}^{4} \alpha^{(r)2}_{\iota\vec{n}} - \sum_{r=2}^{\infty} \sum_{\vec{m}\in\mathbb{Z}^2} \sum_{\vec{\varepsilon}\in\{0,1\}^2} \sum_{\iota,\iota'=1}^{4} Q^{\vec{\varepsilon}}_{\iota\iota'}\alpha^{(r-1)}_{\iota',2\vec{m}+\vec{\varepsilon}}\alpha^{(r)}_{\iota\vec{m}}\right) \\
&\quad\times \prod_{\vec{m}\in\mathbb{Z}^2} \prod_{\vec{\varepsilon}\in\{0,1\}^2} \left[\widehat{Z}_0^{-1}\left(\prod_{\iota=1}^{4}\int_{-\infty}^{\infty} d\tau_\iota\right) \times \exp\left(-\frac{1}{2}\sum_{\iota=1}^{4}\tau_\iota^2 - \sum_{\iota,\iota'=1}^{4} Q^{\vec{\varepsilon}}_{\iota\iota'}\tau_{\iota'}\alpha'_{\iota\vec{m}}\right)\right],
\end{aligned} \tag{4.4.8}$$

$$\widehat{Z}_0 = \left(\prod_{\iota=1}^{4}\int_{-\infty}^{\infty} d\tau_\iota\right)\exp\left(-\frac{1}{2}\sum_{\iota=1}^{4}\tau_\iota^2\right), \tag{4.4.9}$$

$$\alpha^{(r)}_{\iota\vec{n}} = \alpha'^{(r-1)}_{\iota\vec{n}}, \qquad r \geq 1, \tag{4.4.10}$$

where we have applied the factorization of the integration over unit squares labeled by $2\vec{m} + \vec{\varepsilon}$ and set

$$\alpha^{(0)}_{\iota,2\vec{m}+\vec{\varepsilon}} = \alpha_{\iota,2\vec{m}+\vec{\varepsilon}} = \tau_\iota, \tag{4.4.11}$$

$$\alpha^{(1)}_{\iota\vec{m}} = \alpha'^{(0)}_{\iota\vec{m}} = \alpha'_{\iota\vec{m}}, \tag{4.4.12}$$

for each quadruple integral. Since (4.4.8) involves a Gaussian integral, we can integrate explicitly. We obtain

$$\widehat{Z}_0^{-1}\left(\prod_{\iota=1}^{4}\int_{-\infty}^{\infty}d\tau_\iota\right)\exp\left(-\frac{1}{2}\sum_{\iota=1}^{4}\tau_\iota^2-\sum_{\iota,\iota'=1}^{4}Q^{\vec{\varepsilon}}_{\iota\iota'}\tau_{\iota'}\alpha'_{\iota\vec{m}}\right)$$
$$=\exp\left(\frac{1}{2}\sum_{\iota'=1}^{4}\left(\sum_{\iota=1}^{4}Q^{\vec{\varepsilon}}_{\iota\iota'}\alpha'_{\iota\vec{m}}\right)^2\right), \tag{4.4.13}$$

and therefore

$$(\mathcal{R}(S)\alpha',\alpha')=(S\alpha',\alpha')-\sum_{\vec{m}\in\mathbb{Z}^2}\sum_{\vec{\varepsilon}\in\{0,1\}^2}\sum_{\iota'=1}^{4}\left(\sum_{\iota=1}^{4}Q^{\vec{\varepsilon}}_{\iota\iota'}\alpha'_{\iota\vec{m}}\right)^2. \tag{4.4.14}$$

This completes the first step in the iteration.

Clearly, the expression (4.4.7) for the initial quadratic form is not quite preserved by the RG transformation. Instead, we have

$$(\mathcal{R}(S)\alpha',\alpha')=\sum_{\vec{m}\in\mathbb{Z}^2}\sum_{\iota,\kappa=1}^{4}\alpha'_{\iota\vec{m}}\alpha'_{\kappa\vec{m}}\left(\delta_{\iota\kappa}-\left(\sum_{\vec{\varepsilon}\in\{0,1\}^2}Q^{\vec{\varepsilon}}Q^{\vec{\varepsilon}*}\right)_{\iota\kappa}\right)$$
$$+\sum_{r=1}^{\infty}\sum_{\vec{n}\in\mathbb{Z}^2}\sum_{\iota=1}^{4}\alpha'^{(r)2}_{\iota\vec{n}}$$
$$+2\sum_{r=1}^{\infty}\sum_{\vec{m}\in\mathbb{Z}^2}\sum_{\vec{\varepsilon}\in\{0,1\}^2}\sum_{\iota,\iota'=1}^{4}Q^{\vec{\varepsilon}}_{\iota\iota'}\alpha'^{(r-1)}_{\iota',2\vec{m}+\vec{\varepsilon}}\alpha'^{(r)}_{\iota\vec{m}}, \tag{4.4.15}$$

and it is this expression with the generalization

$$\delta_{\iota\kappa}-\left(\sum_{\vec{\varepsilon}\in\{0,1\}^2}Q^{\vec{\varepsilon}}Q^{\vec{\varepsilon}*}\right)_{\iota\kappa}\longmapsto G_{\iota\kappa}$$

that survives the RG iteration. Indeed, if we set

$$(S_G\alpha,\alpha)=\sum_{\vec{m}\in\mathbb{Z}^2}\sum_{\iota,\kappa=1}^{4}\alpha_{\iota\vec{m}}\alpha_{\kappa\vec{m}}G_{\iota\kappa}+\sum_{r=1}^{\infty}\sum_{\vec{n}\in\mathbb{Z}^2}\sum_{\iota=1}^{4}\alpha^{(r)2}_{\iota\vec{n}}$$
$$+2\sum_{r=1}^{\infty}\sum_{\vec{m}\in\mathbb{Z}^2}\sum_{\vec{\varepsilon}\in\{0,1\}^2}\sum_{\iota,\iota'=1}^{4}Q^{\vec{\varepsilon}}_{\iota\iota'}\alpha^{(r-1)}_{\iota',2\vec{m}+\vec{\varepsilon}}\alpha^{(r)}_{\iota\vec{m}}, \tag{4.4.16}$$

then the Gaussian integration in this slightly more general case yields

$$\mathcal{R}(S_G)=S_{\mathcal{R}(G)}, \tag{4.4.17}$$
$$\mathcal{R}(G)=1-\sum_{\vec{\varepsilon}\in\{0,1\}^2}Q^{\vec{\varepsilon}}G^{-1}Q^{\vec{\varepsilon}*}. \tag{4.4.18}$$

Thus, we have described the Nth iteration of the transformation.

Actually, this iteration is even simpler than these formulas suggest. The matrices $Q^{\vec{\varepsilon}}$ were calculated in [1], and we display them here:

$$Q^{\vec{0}} = \frac{1}{8}\begin{bmatrix} 0 & 0 & 0 & 0 \\ 0 & 0 & 0 & 0 \\ -1 & -1 & 0 & 1 \\ 1 & -1 & 1 & 0 \end{bmatrix}, \tag{4.4.19}$$

$$Q^{\hat{\imath}} = \frac{1}{8}\begin{bmatrix} 0 & 0 & 0 & 0 \\ 0 & 0 & 0 & 0 \\ -1 & -1 & 0 & 1 \\ -1 & 1 & 1 & 0 \end{bmatrix}, \tag{4.4.20}$$

$$Q^{\hat{\jmath}} = \frac{1}{8}\begin{bmatrix} 0 & 0 & 0 & 0 \\ 0 & 0 & 0 & 0 \\ 1 & 1 & 0 & 1 \\ 1 & -1 & 1 & 0 \end{bmatrix}, \tag{4.4.21}$$

$$Q^{\hat{\imath}+\hat{\jmath}} = \frac{1}{8}\begin{bmatrix} 0 & 0 & 0 & 0 \\ 0 & 0 & 0 & 0 \\ 1 & 1 & 0 & 1 \\ -1 & 1 & 1 & 0 \end{bmatrix}, \tag{4.4.22}$$

For each $\vec{\varepsilon} \in \{0,1\}^3$, it is obvious that

$$Q^{\vec{\varepsilon}} Q^{\vec{\varepsilon}*} = \frac{3}{64}\begin{bmatrix} 0 & 0 & 0 & 0 \\ 0 & 0 & 0 & 0 \\ 0 & 0 & 1 & 0 \\ 0 & 0 & 0 & 1 \end{bmatrix}, \tag{4.4.23}$$

which implies

$$G_1 \equiv 1 - \sum_{\vec{\varepsilon} \in \{0,1\}^2} Q^{\vec{\varepsilon}} Q^{\vec{\varepsilon}*} = \frac{1}{16}\begin{bmatrix} 16 & 0 & 0 & 0 \\ 0 & 16 & 0 & 0 \\ 0 & 0 & 13 & 0 \\ 0 & 0 & 0 & 13 \end{bmatrix}. \tag{4.4.24}$$

A slight generalization of (4.4.23) is the calculation

$$Q^{\vec{\varepsilon}} \begin{bmatrix} 1 & 0 & 0 & 0 \\ 0 & 1 & 0 & 0 \\ 0 & 0 & c & 0 \\ 0 & 0 & 0 & c \end{bmatrix} Q^{\vec{\varepsilon}*} = \frac{1}{64}(c+2)\begin{bmatrix} 0 & 0 & 0 & 0 \\ 0 & 0 & 0 & 0 \\ 0 & 0 & 1 & 0 \\ 0 & 0 & 0 & 1 \end{bmatrix}, \tag{4.4.25}$$

from which it follows that

$$\sum_{\vec{\varepsilon} \in \{0,1\}^2} Q^{\vec{\varepsilon}} G_1^{-1} Q^{\vec{\varepsilon}*} = \frac{1}{16}\left(\frac{16}{13} + 2\right)\begin{bmatrix} 0 & 0 & 0 & 0 \\ 0 & 0 & 0 & 0 \\ 0 & 0 & 1 & 0 \\ 0 & 0 & 0 & 1 \end{bmatrix}. \tag{4.4.26}$$

Hence,

$$G_2 \equiv 1 - \sum_{\vec{\varepsilon} \in \{0,1\}^2} Q^{\vec{\varepsilon}} G_1^{-1} Q^{\vec{\varepsilon}*} = \frac{1}{104} \begin{bmatrix} 104 & 0 & 0 & 0 \\ 0 & 104 & 0 & 0 \\ 0 & 0 & 83 & 0 \\ 0 & 0 & 0 & 83 \end{bmatrix}, \tag{4.4.27}$$

which is also a diagonal matrix of the form involved in (4.4.25). The limiting matrix for this iteration has the form

$$G_\infty = \begin{bmatrix} 1 & 0 & 0 & 0 \\ 0 & 1 & 0 & 0 \\ 0 & 0 & b & 0 \\ 0 & 0 & 0 & b \end{bmatrix} \tag{4.4.28}$$

and satisfies the fixed-point equation

$$G_\infty = 1 - \sum_{\vec{\varepsilon} \in \{0,1\}^2} Q^{\vec{\varepsilon}} G_\infty^{-1} Q^{\vec{\varepsilon}*}. \tag{4.4.29}$$

More explicitly,

$$\begin{bmatrix} 1 & 0 & 0 & 0 \\ 0 & 1 & 0 & 0 \\ 0 & 0 & b & 0 \\ 0 & 0 & 0 & b \end{bmatrix} = 1 - \frac{1}{16}\left(\frac{1}{b} + 2\right) \begin{bmatrix} 0 & 0 & 0 & 0 \\ 0 & 0 & 0 & 0 \\ 0 & 0 & 1 & 0 \\ 0 & 0 & 0 & 1 \end{bmatrix}, \tag{4.4.30}$$

and so

$$16b^2 - 14b + 1 = 0. \tag{4.4.31}$$

Obviously, there are two solutions to this system; namely,

$$b = \frac{7}{16} \pm \frac{1}{16}\sqrt{33}. \tag{4.4.32}$$

We claim that G_∞ arises from the larger value of b. Thus

$$\begin{aligned} (S_{G_\infty}\alpha, \alpha) &= \sum_{\vec{m} \in \mathbb{Z}^2} \left[({\alpha_{1\vec{m}}}^2 + {\alpha_{2\vec{m}}}^2) + \left(\frac{7}{16} + \frac{1}{16}\sqrt{33}\right) ({\alpha_{3\vec{m}}}^2 + {\alpha_{4\vec{m}}}^2) \right] \\ &+ \sum_{r=1}^{\infty} \sum_{\vec{n} \in \mathbb{Z}^2} \sum_{\iota=1}^{4} \alpha_{\iota\vec{n}}^{(r)2} \\ &+ 2 \sum_{r=1}^{\infty} \sum_{\vec{m} \in \mathbb{Z}^2} \sum_{\vec{\varepsilon} \in \{0,1\}^2} \sum_{\iota,\iota'=1}^{4} Q_{\iota\iota'}^{\vec{\varepsilon}} \alpha_{\iota',2\vec{m}+\vec{\varepsilon}}^{(r-1)} \alpha_{\iota\vec{m}}^{(r)} \end{aligned} \tag{4.4.33}$$

is the quadratic form associated with the limiting Gaussian fixed point.

To see why the preceding iteration does indeed drive the quadratic form to $(S_{G_\infty}\alpha, \alpha)$, it is necessary only to note that

$$b > \frac{7}{16} + \frac{1}{16}\sqrt{33} \Rightarrow 16b^2 - 14b + 1 > 0$$
$$\Rightarrow \rho(b) = \frac{7}{8} - \frac{1}{16b} < b,$$

where we have adopted the notation

$$\mathcal{R}\left(\begin{bmatrix} 1 & 0 & 0 & 0 \\ 0 & 1 & 0 & 0 \\ 0 & 0 & b & 0 \\ 0 & 0 & 0 & b \end{bmatrix}\right) = \begin{bmatrix} 1 & 0 & 0 & 0 \\ 0 & 1 & 0 & 0 \\ 0 & 0 & \rho(b) & 0 \\ 0 & 0 & 0 & \rho(b) \end{bmatrix}. \tag{4.4.34}$$

Since

$$b > \frac{7}{16} + \frac{1}{16}\sqrt{33} \Rightarrow \rho(b) = \frac{7}{8} - \frac{1}{16b} > \frac{7}{16} + \frac{1}{16}\sqrt{33}$$

as well, it is clear that the iteration drives b in the negative direction toward $7/16 + (1/16)\sqrt{33}$. This fixed point is attractive because we also have

$$\frac{7}{16} - \frac{1}{16}\sqrt{33} < b < \frac{7}{16} + \frac{1}{16}\sqrt{33} \Rightarrow 16b^2 - 14b + 1 < 0$$
$$\Rightarrow \rho(b) = \frac{7}{8} - \frac{1}{16b} > b.$$

Obviously, in this regime the iteration drives b away from the other fixed point. Indeed, the latter is repelling because we also have

$$0 < b < \frac{7}{16} - \frac{1}{16}\sqrt{33} \Rightarrow 16b^2 - 14b + 1 > 0$$
$$\Rightarrow \rho(b) = \frac{7}{8} - \frac{1}{16b} < b.$$

Incidentally, note that if $0 < b \leq 1/14$, the corresponding state is undefined because $\rho(b) \leq 0$ in that case. Actually, we shall prove in the Appendix that even the state corresponding to $b = 7/16 - (1/16)\sqrt{33}$ does not exist.

Since our iteration is based on an initial quadratic form that is special, we have not determined whether all of the fixed points have been found. To resolve this issue, we consider the fixed-point equation

$$G = 1 - \sum_{\vec{\varepsilon} \in \{0,1\}^2} Q^{\vec{\varepsilon}} G^{-1} Q^{\vec{\varepsilon}*} \tag{4.4.35}$$

for the most general possible G. First note that each 4×4 matrix

$$Q^{\vec{\varepsilon}} G^{-1} Q^{\vec{\varepsilon}*}$$

has nonzero entries in the 2×2 lower right submatrix only. It follows from the fixed-point equation that G has the form

$$\begin{bmatrix} 1 & 0 & 0 & 0 \\ 0 & 1 & 0 & 0 \\ 0 & 0 & g_{11} & g_{12} \\ 0 & 0 & g_{21} & g_{22} \end{bmatrix},$$

and therefore

$$\begin{bmatrix} 1 & 0 & 0 & 0 \\ 0 & 1 & 0 & 0 \\ 0 & 0 & g_{11} & g_{12} \\ 0 & 0 & g_{21} & g_{22} \end{bmatrix} = 1 - \frac{1}{\mathfrak{g}} \sum_{\vec{\varepsilon} \in \{0,1\}^2} Q^{\vec{\varepsilon}} \begin{bmatrix} \mathfrak{g} & 0 & 0 & 0 \\ 0 & \mathfrak{g} & 0 & 0 \\ 0 & 0 & g_{22} & -g_{12} \\ 0 & 0 & -g_{21} & g_{11} \end{bmatrix} Q^{\vec{\varepsilon} *}, \tag{4.4.36}$$

$$\mathfrak{g} = g_{11} g_{22} - g_{12} g_{21}. \tag{4.4.37}$$

On the other hand, it is easy to calculate

$$Q^{\vec{\varepsilon}} \begin{bmatrix} \mathfrak{g} & 0 & 0 & 0 \\ 0 & \mathfrak{g} & 0 & 0 \\ 0 & 0 & g_{22} & -g_{12} \\ 0 & 0 & -g_{21} & g_{11} \end{bmatrix} Q^{\vec{\varepsilon} *} = \frac{1}{64} \begin{bmatrix} 0 & 0 & 0 & 0 \\ 0 & 0 & 0 & 0 \\ 0 & 0 & g_{11} + 2\mathfrak{g} & -g_{21} \\ 0 & 0 & -g_{12} & g_{22} + 2\mathfrak{g} \end{bmatrix}, \tag{4.4.38}$$

and therefore

$$g_{11} = 1 - \frac{1}{16\mathfrak{g}}(g_{11} + 2\mathfrak{g}) = \frac{7}{8} - \frac{1}{16\mathfrak{g}} g_{11}, \tag{4.4.39}$$

$$g_{12} = \frac{1}{16\mathfrak{g}} g_{21}, \tag{4.4.40}$$

$$g_{21} = \frac{1}{16\mathfrak{g}} g_{12}, \tag{4.4.41}$$

$$g_{22} = 1 - \frac{1}{16\mathfrak{g}}(g_{22} + 2\mathfrak{g}) = \frac{7}{8} - \frac{1}{16\mathfrak{g}} g_{22}. \tag{4.4.42}$$

Case 1. If $g_{12} = 0$, then $g_{21} = 0$, in which case $\mathfrak{g} = g_{11} g_{22}$. Thus,

$$g_{11} = 1 - \frac{1}{16}\left(\frac{1}{g_{22}} + 2\right), \tag{4.4.43}$$

$$g_{22} = 1 - \frac{1}{16}\left(\frac{1}{g_{11}} + 2\right), \tag{4.4.44}$$

which imply

$$16 g_{11} g_{22} = 14 g_{22} - 1, \tag{4.4.45}$$

$$16 g_{11} g_{22} = 14 g_{11} - 1, \tag{4.4.46}$$

respectively. Hence, $g_{22} = g_{11}$, and so we have the case already considered above.

On the other hand, if $g_{12} \neq 0$, then $\mathfrak{g} = \pm(1/16)$.

Case 2. If $\mathfrak{g} = -1/16$, then Eq. (4.4.39) is inconsistent as well as Eq. (4.4.42).

Case 3. If $\mathfrak{g} = 1/16$, then $g_{21} = g_{12}$ and

$$g_{11} = g_{22} = \frac{7}{16}. \tag{4.4.47}$$

It follows from (4.4.37) that

$$\left(\frac{7}{16}\right)^2 - g_{22}^2 = \frac{1}{16}, \tag{4.4.48}$$

and therefore

$$g_{12} = \pm\frac{1}{16}\sqrt{33}. \tag{4.4.49}$$

We have found two more fixed points; namely,

$$G = \begin{bmatrix} 1 & 0 & 0 & 0 \\ 0 & 1 & 0 & 0 \\ 0 & 0 & \frac{7}{16} & \pm\frac{1}{16}\sqrt{33} \\ 0 & 0 & \pm\frac{1}{16}\sqrt{33} & \frac{7}{16} \end{bmatrix}. \tag{4.4.50}$$

Now, the linear manifold of matrices of the form

$$G = \begin{bmatrix} 1 & 0 & 0 & 0 \\ 0 & 1 & 0 & 0 \\ 0 & 0 & b & a \\ 0 & 0 & a & b \end{bmatrix}, \qquad b \neq \pm a, \tag{4.4.51}$$

is obviously preserved by the nonlinear transformation and contains all four of the fixed points. Since $\mathfrak{g} = b^2 - a^2$ for this form, the transformed matrix

$$\mathcal{R}(G) = \begin{bmatrix} 1 & 0 & 0 & 0 \\ 0 & 1 & 0 & 0 \\ 0 & 0 & b' & a' \\ 0 & 0 & a' & b' \end{bmatrix} \tag{4.4.52}$$

is given by

$$b' = \frac{7}{8} - \frac{b}{16(b^2 - a^2)}, \tag{4.4.53}$$

$$a' = \frac{a}{16(b^2 - a^2)}. \tag{4.4.54}$$

Note that this transformation has a natural decomposition:

$$b' + a' = \frac{7}{8} + \frac{a-b}{16(b^2-a^2)}$$
$$= \frac{7}{8} - \frac{1}{16(b+a)} = \rho(b+a), \tag{4.4.55}$$
$$b' - a' = \frac{7}{8} - \frac{a+b}{16(b^2-a^2)}$$
$$= \frac{7}{8} - \frac{1}{16(b-a)} = \rho(b-a). \tag{4.4.56}$$

We have already discussed the effect of ρ. In this context, the quantity $b+a$ is driven toward the value $7/16 + (1/16)\sqrt{33}$ and away from the value $7/16 - (1/16)\sqrt{33}$ by the iteration. Moreover, the flow of this iteration is independent of the quantity $b-a$, which is driven in exactly the same way. These observations yield the flow diagram of Figure 4.5 in Section 4.1.

4.5 Elements of the $\boldsymbol{A}^{\infty}$ matrix

Our starting point in calculating the elements of the matrix $A^{(N)}$ is to recall the integral representation involving the wavelet amplitudes that have been integrated out by the N iterations of the RG transformation. We have

$$\exp\left(-\frac{1}{2}(A^{(N)}\sigma^{(N)}, \sigma^{(N)})\right)$$
$$= \widetilde{Z}^{(N)}_{W_0,\{\vec{0}\}}{}^{-1}\left(\prod_{r=0}^{N-1}\prod_{\vec{\ell}\in\Lambda_{N-r}}\prod_{\iota=1}^{4}\int_{-\infty}^{\infty} d\alpha^{(r)}_{\iota\vec{\ell}}\right)$$
$$\times \exp\Bigg(-\sum_{r=0}^{N-1}\sum_{\vec{\ell}\in\Lambda_{N-r}}\sum_{\iota=1}^{4}\alpha^{(r)}_{\iota\vec{\ell}}\int_{[0,2^{N-1}]^2} d\vec{x}\, 2^{-N}\nabla\phi^{(N)}(2^{-N}\vec{x})\cdot\nabla\Psi_{\iota r\vec{\ell}}(\vec{x})$$
$$-\frac{1}{2}\sum_{r,r'=0}^{N-1}\sum_{\vec{\ell}\in\Lambda_{N-r'}}\sum_{\vec{\ell}'\in\Lambda_{N-r'}}\sum_{\iota,\iota'=1}^{4}\alpha^{(r)}_{\iota\vec{\ell}}\alpha^{(r')}_{\iota'\vec{\ell}'}\int_{[0,2^{N-1}]^2} d\vec{x}$$
$$\times \nabla\Psi_{\iota r\vec{\ell}}(\vec{x})\cdot\nabla\Psi_{\iota' r'\vec{\ell}'}(\vec{x})\Bigg)$$
$$\times \exp\left(-\frac{1}{2}\int_{[0,\frac{1}{2}]^2} d\vec{x}^{(N)}\,(\nabla\phi^{(N)}(\vec{x}^{(N)}))^2\right), \tag{4.5.1}$$

which follows from (4.2.48) and (4.1.30). Inspection reveals that

$$\int_{[0,2^{N-1}]^2} d\vec{x}\;\nabla\phi^{(N)}(2^{-N}\vec{x})\cdot\nabla\Psi_{\iota r\vec{\ell}}(\vec{x}) = 0, \qquad r < N-1, \tag{4.5.2}$$

as the triangulation for such a scaling of $\phi^{(N)}$ is coarse, relative to the length scale of such wavelets. Moreover,

$$\int_{[0,2^{N-1}]^2} \nabla\Psi_{\iota r \vec{\ell}} \cdot \nabla\Psi_{\iota' r' \vec{\ell}'} = S_{rr';\vec{\ell}\vec{\ell}';\iota\iota'}, \qquad r, r' \leq N-1, \tag{4.5.3}$$

so by (4.4.7) we have the reduction (noting that $\Lambda_1 = \{\vec{0}\}$)

$$\begin{aligned}
&\exp\left(-\frac{1}{2}(A^{(N)}\sigma^{(N)}, \sigma^{(N)})\right) \\
&= \exp\left(-\frac{1}{2}\int_{[0,\frac{1}{2}]^2} d\,\vec{x}^{\,(N)}\,(\nabla\phi^{(N)}(\vec{x}^{\,(N)}))^2\right) \\
&\quad\times \widetilde{Z}^{(N)\,-1}_{W_0,\{\vec{0}\}}\left(\prod_{\iota=1}^{4}\int_{-\infty}^{\infty} d\alpha^{(N-1)}_{\iota\vec{\ell}}\right) \\
&\quad\times\left[\exp\left(-\sum_{\iota=1}^{4}\alpha^{(N-1)}_{\iota\vec{0}}\int_{[0,2^{N-1}]^2} 2^{-N} d\,\vec{x}\;\nabla\phi^{(N)}(2^{-N}\,\vec{x})\cdot\nabla\Psi_{\iota,N-1,\vec{0}}(\vec{x})\right)\right. \\
&\qquad\times\left(\prod_{r=0}^{N-2}\prod_{\vec{\ell}\in\Lambda_{N-r}}\prod_{\iota=1}^{4}\int_{-\infty}^{\infty} d\alpha^{(r)}_{\iota\vec{\ell}}\right) \\
&\qquad\times\exp\left(-\frac{1}{2}\sum_{r=0}^{N-1}\sum_{\vec{n}\in\Lambda_{N-r}}\sum_{\iota=1}^{4}\alpha^{(r)2}_{\iota\vec{n}}\right. \\
&\qquad\qquad\left.\left.-\sum_{r=1}^{N-1}\sum_{\vec{m}\in\Lambda_{N-r}}\sum_{\vec{\varepsilon}\in\{0,1\}^2}\sum_{\iota,\iota'=1}^{4} Q^{\vec{\varepsilon}}_{\iota\iota'}\alpha^{(r-1)}_{\iota',2\vec{m}+\varepsilon}\alpha^{(r)}_{\iota\vec{m}}\right)\right].
\end{aligned} \tag{4.5.4}$$

Since the field configuration is coupled to the largest-scale modes only, the process of integrating out the amplitudes from the smallest scale up is identical to the iterative calculation in the previous section. Thus, if we integrate out all amplitudes except the $r = N-1$ amplitudes, we obtain

$$\begin{aligned}
&\exp\left(-\frac{1}{2}(A^{(N)}\sigma^{(N)}, \sigma^{(N)})\right) \\
&= \exp\left(-\frac{1}{2}\int_{[0,\frac{1}{2}]^2} d\,\vec{x}^{\,(N)}\,(\nabla\phi^{(N)}(\vec{x}^{\,(N)}))^2\right) \\
&\quad\times \widetilde{Z}^{\,-1}_{W_0,\{\vec{0}\}}\left(\prod_{\iota=1}^{4}\int_{-\infty}^{\infty} d\alpha^{(N-1)}_{\iota\vec{0}}\right) \\
&\quad\times\exp\left(-\sum_{\iota=1}^{4}\alpha^{(N-1)}_{\iota\vec{0}}\int_{[0,2^{N-1}]^2} d\,\vec{x}\;2^{-N}\nabla\phi^{(N)}(2^{-N}\,\vec{x})\cdot\nabla\Psi_{\iota,N-1,\vec{0}}(\vec{x})\right)
\end{aligned}$$

$$\times \exp\left(-\frac{1}{2}\sum_{\iota,\kappa=1}^{4} \alpha^{(N-1)}_{\iota\,\vec{0}}\,\alpha^{(N-1)}_{\kappa\,\vec{0}}(G_N)_{\iota\kappa}\right), \tag{4.5.5}$$

$$G_{s+1} = 1 - \sum_{\vec{\varepsilon}\in\{0,1\}^2} Q^{\vec{\varepsilon}}(G_s)^{-1}Q^{\vec{\varepsilon}\,*}, \tag{4.5.6}$$

$$G_0 = 1. \tag{4.5.7}$$

On the other hand, the residual integral is just a Gaussian integral in the four remaining wavelet amplitudes, so it can be explicitly written in terms of $G^{(N)}$. Thus

$$\begin{aligned}(A^{(N)}\sigma^{(N)},\sigma^{(N)}) = &\int_{[0,\frac{1}{2}]^2} d\,\vec{x}^{\,(N)}\,(\nabla\phi^{(N)}(\vec{x}^{\,(N)}))^2 - \sum_{\iota,\kappa=1}^{4}(G_N)^{-1}_{\iota\kappa}\\ &\times\int_{[0,2^{N-1}]^2} d\,\vec{x}\;2^{-N}\,\nabla\phi^{(N)}(2^{-N}\,\vec{x})\cdot\nabla\Psi_{\iota,N-1,\vec{0}}(\vec{x})\\ &\times\int_{[0,2^{N-1}]^2} d\,\vec{x}\;2^{-N}\,\nabla\phi^{(N)}(2^{-N}\,\vec{x})\nabla\Psi_{\kappa,N-1,\vec{0}}(\vec{x}).\end{aligned} \tag{4.5.8}$$

Fixing the configuration variables before taking the $N = \infty$ limit of the quadratic form, we may write

$$\begin{aligned}(A^{(N)}\sigma,\sigma) = &\int_{[0,\frac{1}{2}]^2}(\nabla\phi)^2\\ &-\sum_{\iota,\kappa=1}^{4}(G_N)^{-1}_{\iota\kappa}\left(\int_{[0,\frac{1}{2}]^2}\nabla\phi\cdot\nabla\Psi_{\iota,-1,\vec{0}}\right)\left(\int_{[0,\frac{1}{2}]^2}\nabla\phi\cdot\nabla\Psi_{\kappa,-1,\vec{0}}\right),\end{aligned} \tag{4.5.9}$$

where we have also made the change of variable $\vec{x}\mapsto 2^N\,\vec{x}$ and applied the relation

$$\nabla\Psi_{\iota,N-1,\vec{0}}(2^N\,\vec{x}) = 2^{-N}\nabla\Psi_{\iota,-1,\vec{0}}(\vec{x}). \tag{4.5.10}$$

As we have seen in Section 4.4,

$$G_\infty = \begin{bmatrix}1 & 0 & 0 & 0\\ 0 & 1 & 0 & 0\\ 0 & 0 & \frac{7}{16}+\frac{1}{16}\sqrt{33} & 0\\ 0 & 0 & 0 & \frac{7}{16}+\frac{1}{16}\sqrt{33}\end{bmatrix} \tag{4.5.11}$$

Therefore,

$$G_\infty^{-1} = \begin{bmatrix}1 & 0 & 0 & 0\\ 0 & 1 & 0 & 0\\ 0 & 0 & 7-\sqrt{33} & 0\\ 0 & 0 & 0 & 7-\sqrt{33}\end{bmatrix}. \tag{4.5.12}$$

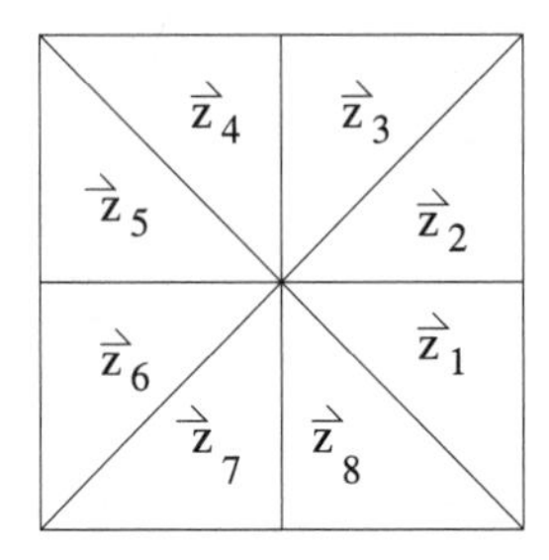

FIGURE 4.8. Triangulation of a square.

Combining this with (4.1.30) and the $N = \infty$ limit of (4.5.9), we get

$$(A^{\infty}\sigma, \sigma) = (A^{(0)}\sigma, \sigma) - \sum_{\iota=1}^{2} \left(\int_{[0,\frac{1}{2}]^2} \nabla\phi \cdot \nabla\Psi_{\iota,-1,\vec{0}} \right)^2 - (7 - \sqrt{33}) \sum_{\iota=3}^{4} \left(\int_{[0,\frac{1}{2}]^2} \nabla\phi \cdot \nabla\Psi_{\iota,-1,\vec{0}} \right)^2 . \tag{4.5.13}$$

The remaining task is to write the integrals in terms of the Isis variables.

As the triangulation for the piecewise-linear configuration ϕ is one length scale coarser than the triangulation for the wavelets $\Psi_{\iota,-1,\vec{0}}$, we are comparing the triangulation of the square $[0, 1/2]^2$ given in Figure 4.8 to the gradient representation of these four wavelets given in Figures 4.9–4.12. Equations (4.2.53)–(4.2.64) are the input here, but we have not yet examined the calculations. Visually, it is easy to calculate the $\vec{z}_\mu$-dependence of the integrals in (4.5.13).

The normalization condition on the wavelets requires each vector in Figures 4.9 and 4.10 to have magnitude equal to $2\sqrt{2}$. Similarly, each vector in Figures 4.11 and 4.12 must have magnitude equal to 4. Thus,

$$\int_{[0,\frac{1}{2}]^2} \nabla\phi \cdot \nabla\Psi_{1,-1,\vec{0}} = \frac{\sqrt{2}}{32}(\hat{\imath} + \hat{\jmath}) \cdot [\vec{z}_1 - \vec{z}_4 + \vec{x}_5 - \vec{z}_8], \tag{4.5.14}$$

$$\int_{[0,\frac{1}{2}]^2} \nabla\phi \cdot \nabla\Psi_{2,-1,\vec{0}} = \frac{\sqrt{2}}{32}(\hat{\imath} - \hat{\jmath}) \cdot [-\vec{z}_2 + \vec{z}_3 - \vec{z}_6 + \vec{z}_7], \tag{4.5.15}$$

$$\int_{[0,\frac{1}{2}]^2} \nabla\phi \cdot \nabla\Psi_{3,-1,\vec{0}} = \frac{\sqrt{2}}{32}\hat{\jmath} \cdot (\vec{z}_1 - \vec{z}_2 - \vec{z}_5 + \vec{z}_6), \tag{4.5.16}$$

$$\int_{[0,\frac{1}{2}]^2} \nabla\phi \cdot \nabla\Psi_{4,-1,\vec{0}} = \frac{\sqrt{2}}{32}\hat{\imath} \cdot (-\vec{z}_3 + \vec{z}_4 + \vec{z}_7 - \vec{z}_8), \tag{4.5.17}$$

where we have combined vectors that are contracted with the same $\vec{z}_\mu$ and found the result to be $\pm$ the same vector for all $\vec{z}_\mu$ in a given case—namely, $-\hat{\imath} + \hat{\jmath}$ for $\iota = 1$,

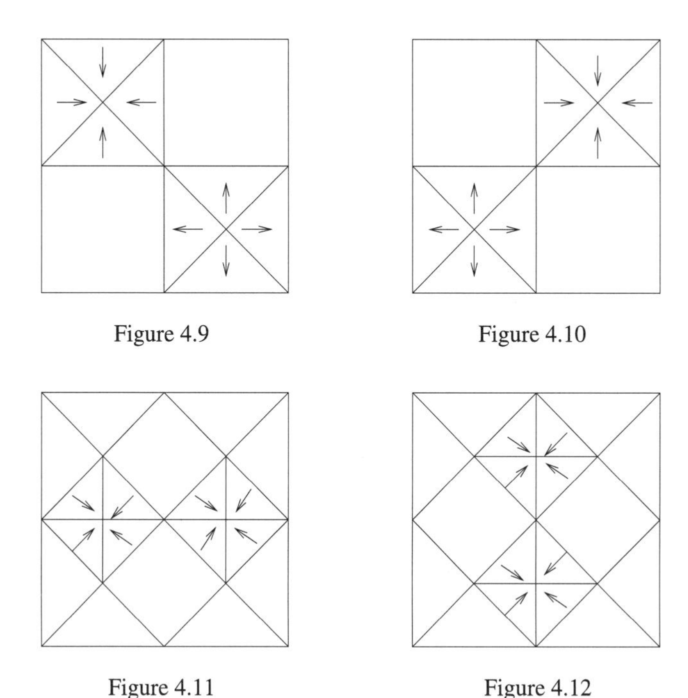

Figure 4.9

Figure 4.10

Figure 4.11

Figure 4.12

FIGURES 4.9–4.12. Diagrammatical description of $1/2$-scale wavelets.

$\hat{\imath} - \hat{\jmath}$ for $\iota = 2$, $\hat{\jmath}$ for $\iota = 3$, and $\hat{\imath}$ for $\iota = 4$. Obviously, we have accounted for the areas of the triangles as well. If we now insert (4.2.22)–(4.2.29), we obtain

$$\int_{[0,\frac{1}{2}]^2} \nabla\phi \cdot \nabla\Psi_{1,-1,\vec{0}} = \frac{\sqrt{2}}{16}(\sigma_1 - \sigma_3 + \sigma_4 + \sigma_5 - \sigma_7 + \sigma_8), \tag{4.5.18}$$

$$\int_{[0,\frac{1}{2}]^2} \nabla\phi \cdot \nabla\Phi_{2,-1,\vec{0}} = \frac{\sqrt{2}}{16}(-\sigma_1 + \sigma_2 - \sigma_3 - \sigma_5 + \sigma_6 - \sigma_7), \tag{4.5.19}$$

$$\int_{[0,\frac{1}{2}]^2} \nabla\phi \cdot \nabla\Psi_{3,-1,\vec{0}} = \frac{\sqrt{2}}{32}(2\sigma_1 - \sigma_2 - \sigma_4 - 2\sigma_5 + \sigma_6 + \sigma_8), \tag{4.5.20}$$

$$\int_{[0,\frac{1}{2}]^2} \nabla\phi \cdot \nabla\Psi_{4,-1,\vec{0}} = \frac{\sqrt{2}}{32}(-\sigma_2 + 2\sigma_3 - \sigma_4 + \sigma_6 - 2\sigma_7 + \sigma_8). \tag{4.5.21}$$

Combining this with (4.2.31), we reduce (4.5.13) to

$$\begin{aligned}(A^\infty\sigma, \sigma) = \frac{1}{32}\big[&2\sigma_1^2 + (\sigma_1 + \sigma_8)^2 + (\sigma_2 - \sigma_1)^2 + (\sigma_2 - \sigma_3)^2 + 2\sigma_3^2 \\ &+ (\sigma_3 - \sigma_4)^2 + 2\sigma_5^2 + (\sigma_4 + \sigma_5)^2 + (\sigma_6 - \sigma_5)^2 + 2\sigma_7^2 \\ &+ (\sigma_6 - \sigma_7)^2 + (\sigma_7 - \sigma_8)^2\big]\end{aligned}$$

$$
\begin{aligned}
&- \frac{1}{64}(\sigma_1 - \sigma_3 + \sigma_4 + \sigma_5 - \sigma_7 + \sigma_8)^2 \\
&- \frac{1}{64}(-\sigma_1 + \sigma_2 - \sigma_3 - \sigma_5 + \sigma_6 - \sigma_7)^2 \\
&- \frac{1}{256}(7 - \sqrt{33})(2\sigma_1 - \sigma_2 - \sigma_4 - 2\sigma_5 + \sigma_6 + \sigma_8)^2 \\
&- \frac{1}{256}(7 - \sqrt{33})(-\sigma_2 + 2\sigma_3 - \sigma_4 + \sigma_6 - 2\sigma_7 + \sigma_8)^2. \qquad (4.5.22)
\end{aligned}
$$

This is the quadratic form in the Isis variables for the Gaussian fixed point. The matrix elements of the symmetric 8×8 matrix are now easy to evaluate. We have

$$A_{11}^{\infty} = A_{33}^{\infty} = A_{55}^{\infty} = A_{77}^{\infty} = \frac{\sqrt{33} - 1}{64}, \qquad (4.5.23)$$

$$A_{22}^{\infty} = A_{44}^{\infty} = A_{66}^{\infty} = A_{88}^{\infty} = \frac{\sqrt{33} - 1}{128}, \qquad (4.5.24)$$

$$
\begin{aligned}
A_{12}^{\infty} &= A_{14}^{\infty} = A_{23}^{\infty} = A_{26}^{\infty} = A_{34}^{\infty} = A_{48}^{\infty} = A_{56}^{\infty} = A_{58}^{\infty} \\
&= A_{67}^{\infty} = A_{78}^{\infty} = \frac{5 - \sqrt{33}}{128}, \qquad (4.5.25)
\end{aligned}
$$

$$A_{16}^{\infty} = A_{18}^{\infty} = A_{25}^{\infty} = A_{27}^{\infty} = A_{36}^{\infty} = A_{38}^{\infty} = A_{45}^{\infty} = A_{47}^{\infty} = \frac{\sqrt{33} - 5}{128}, \qquad (4.5.26)$$

$$A_{15}^{\infty} = A_{37}^{\infty} = \frac{5 - \sqrt{33}}{64}, \qquad (4.5.27)$$

$$A_{24}^{\infty} = A_{68}^{\infty} = \frac{\sqrt{33} - 7}{128}, \qquad (4.5.28)$$

$$A_{28}^{\infty} = A_{46}^{\infty} = \frac{7 - \sqrt{33}}{128}, \qquad (4.5.29)$$

$$A_{13}^{\infty} = A_{17}^{\infty} = A_{35}^{\infty} = A_{57}^{\infty} = 0. \qquad (4.5.30)$$

4.6 Linearized RG transformation for the dipole perturbation

An important preliminary step in our RG analysis of the Osiris dipole gas is to calculate the Fréchet derivative of $\widehat{R}_{\infty}(\widehat{F})$ at $\widehat{F} \equiv 0$. Before we derive this formula, we check some of our numbers by explicitly deriving a couple of relationships that were implicit in previous calculations. First recall that

$$
\begin{aligned}
N(1, \vec{0}) &= (0, 0, 0, 0, 0, 0, 0, 0), && (4.6.1) \\
N(1, \hat{\imath}) &= \sqrt{2}(1, 1, 1, 1, -1, -1, -1, -1), && (4.6.2) \\
N(1, \hat{\jmath}) &= \sqrt{2}(-1, -1, -1, -1, 1, 1, 1, 1), && (4.6.3) \\
N(1, \hat{\imath} + \hat{\jmath}) &= (0, 0, 0, 0, 0, 0, 0, 0) && (4.6.4)
\end{aligned}
$$

$$N(2, \overrightarrow{0}) = \sqrt{2}(1, 1, 1, 1 - 1, -1, -1, -1), \tag{4.6.5}$$

$$N(2, \hat{i}) = (0, 0, 0, 0, 0, 0, 0, 0), \tag{4.6.6}$$

$$N(2, \hat{j}) = (0, 0, 0, 0, 0, 0, 0, 0), \tag{4.6.7}$$

$$N(2, \hat{i} + \hat{j}) = \sqrt{2}(-1, -1, -1, -1, 1, 1, 1, 1) \tag{4.6.8}$$

$$N(3, \overrightarrow{0}) = (0, 0, \sqrt{2}, 0, 0, 0, 0, 0), \tag{4.6.9}$$

$$N(3, \hat{i}) = (0, 0, \sqrt{2}, 0, 0, 0, 0, 0), \tag{4.6.10}$$

$$N(3, \hat{j}) = (0, 0, 0, 0, 0, 0, -\sqrt{2}, 0), \tag{4.6.11}$$

$$N(3, \hat{i} + \hat{j}) = (0, 0, 0, 0, 0, 0, -\sqrt{2}, 0) \tag{4.6.12}$$

$$N(4, \overrightarrow{0}) = (\sqrt{2}, 0, 0, 0, 0, 0, 0, 0,), \tag{4.6.13}$$

$$N(4, \hat{i}) = (0, 0, 0, 0, -\sqrt{2}, 0, 0, 0), \tag{4.6.14}$$

$$N(4, \hat{j}) = (\sqrt{2}, 0, 0, 0, 0, 0, 0, 0) \tag{4.6.15}$$

$$N(4, \hat{i} + \hat{j}) = (0, 0, 0, 0, -\sqrt{2}, 0, 0, 0). \tag{4.6.16}$$

Equivalently,

$$\sum_{\iota=1}^{4} \tau_\iota N(\iota, \overrightarrow{0}) = \sqrt{2}(\tau_2 + \tau_4, \tau_2, \tau_2 + \tau_3, \tau_2, -\tau_2, -\tau_2, -\tau_2, -\tau_2), \tag{4.6.17}$$

$$\sum_{\iota=1}^{4} \tau_\iota N(\iota, \hat{i}) = \sqrt{2}(\tau_1, \tau_1, \tau_1 + \tau_3, \tau_1, -\tau_1 - \tau_4, -\tau_1, -\tau_1, -\tau_1), \tag{4.6.18}$$

$$\sum_{\iota=1}^{4} \tau_\iota N(\iota, \hat{j}) = \sqrt{2}(-\tau_1 + \tau_4, -\tau_1, -\tau_1, -\tau_1, \tau_1, \tau_1, \tau_1 - \tau_3, \tau_1), \tag{4.6.19}$$

$$\sum_{\iota=1}^{4} \tau_\iota N(\iota, \hat{i} + \hat{j}) = \sqrt{2}(-\tau_2, -\tau_2, -\tau_2, -\tau_2, \tau_2 - \tau_4, \tau_2, \tau_2 - \tau_3, \tau_2). \tag{4.6.20}$$

Combining this with (4.2.31), we calculate

$$\begin{aligned} &\left(A^{(0)} \left(\sum_{\iota=1}^{4} \tau_\iota N(\iota, \overrightarrow{0}) \right), \sum_{\iota=1}^{4} \tau_\iota N(\iota, \overrightarrow{0}) \right) \\ &\quad = \frac{1}{8}[(\tau_2 + \tau_4)^2 + 2\tau_2^2 + (\tau_2 + \tau_3)^2 + \tau_3^2 + \tau_4^2], \end{aligned} \tag{4.6.21}$$

$$\begin{aligned} &\left(A^{(0)} \left(\sum_{\iota=1}^{4} \tau_\iota N(\iota, \hat{\imath}) \right), \sum_{\iota=1}^{4} \tau_\iota N(\iota, \hat{\imath}) \right) \\ &\quad = \frac{1}{8}[(\tau_1 + \tau_4)^2 + 2\tau_1^2 + (\tau_1 + \tau_3)^2 + \tau_3^2 + \tau_4^2], \end{aligned} \tag{4.6.22}$$

$$\left(A^{(0)} \left(\sum_{\iota=1}^{4} \tau_\iota N(\iota, \hat{j}) \right), \sum_{\iota=1}^{4} \tau_\iota N(\iota, \hat{j}) \right)$$

$$= \frac{1}{8}[(-\tau_1 + \tau_4)^2 + 2\tau_1^2 + (\tau_1 - \tau_3)^2 + \tau_3^2 + \tau_4^2], \tag{4.6.23}$$

$$\left(A^{(0)} \left(\sum_{\iota=1}^{4} \tau_\iota N(\iota, \hat{i} + \hat{j}) \right), \sum_{\iota=1}^{4} \tau_\iota N(\iota, \hat{i} + \hat{j}) \right)$$
$$= \frac{1}{8}[(\tau_2 - \tau_4)^2 + 2\tau_2^2 + (\tau_2 - \tau_3)^2 + \tau_3^2 + \tau_4^2]. \tag{4.6.24}$$

The sum of these equations reduces to

$$\sum_{\vec{\varepsilon} \in \{0,1\}^2} \left(A^{(0)} \left(\sum_{\iota=1}^{4} \tau_\iota N(\iota, \vec{\varepsilon}) \right), \sum_{\iota=1}^{4} \tau_\iota N(\iota, \vec{\varepsilon}) \right) = \sum_{\iota=1}^{4} \tau_\iota^2, \tag{4.6.25}$$

which is consistent with the basic observation that

$$\int_{[0,1]^2} \left(\sum_{\iota=1}^{4} \tau_\iota \nabla \Psi_\iota \right)^2 = \sum_{\iota=1}^{4} \tau_\iota^2. \tag{4.6.26}$$

This rules out possible mistakes with constants.

Now we derive the parallel formula for A^∞ by the same direct calculation. In Section 4.5, we found that

$$\begin{aligned}(A^\infty \sigma, \sigma) = (A^{(0)}\sigma, \sigma) &- \frac{1}{64}(\sigma_1 - \sigma_3 + \sigma_4 + \sigma_5 - \sigma_7 + \sigma_8)^2 \\ &- \frac{1}{64}(-\sigma_1 + \sigma_2 - \sigma_3 - \sigma_5 + \sigma_6 - \sigma_7)^2 \\ &- \frac{1}{256}(7 - \sqrt{33})(2\sigma_1 - \sigma_2 - \sigma_4 - 2\sigma_5 + \sigma_6 + \sigma_8)^2 \\ &- \frac{1}{256}(7 - \sqrt{33})(-\sigma_2 + 2\sigma_3 - \sigma_4 + \sigma_6 - 2\sigma_7 + \sigma_8),^2\end{aligned} \tag{4.6.27}$$

so if we insert $\sigma = \sum_{\iota=1}^4 \tau_\iota N(\iota, \vec{\varepsilon})$ for each $\vec{\varepsilon} \in \{0,1\}^2$, we obtain

$$\begin{aligned}&\left(A^\infty \left(\sum_{\iota=1}^{4} \tau_\iota N(\iota, \vec{\varepsilon}) \right), \sum_{\iota=1}^{4} \tau_\iota N(\iota, \vec{\varepsilon}) \right) \\ &= \left(A^{(0)} \left(\sum_{\iota=1}^{4} \tau_\iota N(\iota, \vec{\varepsilon}) \right), \sum_{\iota=1}^{4} \tau_\iota N(\iota, \vec{\varepsilon}) \right) \\ &\quad - \frac{1}{64}(\tau_3 + \tau_4)^2 - \frac{1}{64}(\tau_4 - \tau_3)^2 - \frac{1}{256}(7 - \sqrt{33})(2\tau_3)^2 \\ &\quad - \frac{1}{256}(7 - \sqrt{33})(2\tau_4)^2.\end{aligned} \tag{4.6.28}$$

Summing over $\vec{\varepsilon}$ and applying (4.6.25), we see that

$$\sum_{\vec{\varepsilon} \in \{0,1\}^2} \left(A^\infty \left(\sum_{\iota=1}^{4} \tau_\iota N(\iota, \vec{\varepsilon}) \right), \sum_{\iota=1}^{4} \tau_\iota N(\iota, \vec{\varepsilon}) \right)$$

$$= \sum_{\iota=1}^{4} \tau_\iota^2 - \frac{1}{16}(\tau_3+\tau_4)^2 - \frac{1}{16}(\tau_4-\tau_3)^2$$
$$- \frac{1}{16}(7-\sqrt{33})\tau_3^2 - \frac{1}{16}(7-\sqrt{33})\tau_4^2$$
$$= \tau_1^2 + \tau_2^2 + \left(\frac{7}{16}+\frac{1}{16}\sqrt{33}\right)\tau_3^2 + \left(\frac{7}{16}+\frac{1}{16}\sqrt{33}\right)\tau_4^2. \tag{4.6.29}$$

Thus, we have the expected formula

$$\sum_{\vec{\varepsilon}\in\{0,1\}^2} \left(A^\infty\left(\sum_{\iota=1}^{4}\tau_\iota N(\iota,\vec{\varepsilon})\right), \sum_{\iota=1}^{4}\tau_\iota N(\iota,\vec{\varepsilon})\right) = (G_\infty\tau,\tau), \tag{4.6.30}$$

and this derivation is yet another check on our calculations.

The formulas for the Gaussian integrals of monomials in the τ-variables are easily derived. For example, if we combine

$$\tau_\iota \exp\left(-\frac{1}{2}\sum_{\iota',\iota''=1}^{4}(G_\infty)_{\iota'\iota''}\tau_{\iota'}\tau_{\iota''}\right)$$
$$= -\sum_{\bar{\iota}=1}^{4}(G_\infty)^{-1}_{\iota\bar{\iota}}\frac{\partial}{\partial\tau_{\bar{\iota}}}\exp\left(-\frac{1}{2}\sum_{\iota',\iota''=1}^{4}(G_\infty)_{\iota'\iota''}\tau_{\iota'}\tau_{\iota''}\right) \tag{4.6.31}$$

with (4.3.10) and the identity (4.6.30), we obtain

$$\int \tau_\iota d\mu_\infty(\sigma')(\tau) = \frac{1}{2}\sum_{\iota=1}^{4}(G_\infty^{-1})_{\iota\bar{\iota}}\sum_{\vec{\varepsilon}\in\{0,1\}^2}(A^\infty L_{\vec{\varepsilon}}(\sigma'), N(\vec{\iota},\vec{\varepsilon})) \tag{4.6.32}$$

as a result of integrating by parts. For 2nd-degree monomials in the τ-variables, the formula

$$\tau_{\iota_1}\tau_{\iota_2} \exp\left(-\frac{1}{2}\sum_{\iota',\iota''=1}^{4}(G_\infty)_{\iota'\iota''}\tau_{\iota'}\tau_{\iota''}\right)$$
$$= \sum_{\bar{\iota}_1,\bar{\iota}_2=1}^{4}(G_\infty^{-1})_{\iota_1\bar{\iota}_1}(G_\infty^{-1})_{\iota_2\bar{\iota}_2}\left(\frac{\partial^2}{\partial\tau_{\bar{\iota}_1}\partial\tau_{\bar{\iota}_2}} + (G_\infty)_{\bar{\iota}_1\bar{\iota}_2}\right)$$
$$\times \exp\left(-\frac{1}{2}\sum_{\iota',\iota''=1}^{4}(G_\infty)_{\iota'\iota''}\tau_{\iota'}\tau_{\iota''}\right) \tag{4.6.33}$$

together with integration by parts yields

$$\int \tau_{\iota_1}\tau_{\iota_2}\, d\mu_\infty(\sigma')(\tau) = (G_\infty^{-1})_{\iota_1\iota_2} + a^\infty_{\iota_1}(\sigma')a^\infty_{\iota_2}(\sigma'), \tag{4.6.34}$$

$$a_\iota^\infty(\sigma') = \frac{1}{2}\sum_{\bar\iota=1}^{4}(G_\infty^{-1})_{\iota\bar\iota} \sum_{\vec\varepsilon\in\{0,1\}^2} (A^\infty L_{\vec\varepsilon}(\sigma'), N(\bar\iota, \vec\varepsilon)). \quad (4.6.35)$$

Similarly, the formula

$$\tau_{\iota_1}\tau_{\iota_2}\tau_{\iota_3} \exp\left(-\frac{1}{2}\sum_{\iota',\iota''=1}^{4}(G_\infty)_{\iota'\iota''}\tau_{\iota'}\tau_{\iota''}\right)$$
$$= \sum_{\bar\iota_1,\bar\iota_2,\bar\iota_3=1}^{4} (G_\infty^{-1})_{\iota_1\bar\iota_1}(G_\infty^{-1})_{\iota_2\bar\iota_2}(G_\infty^{-1})_{\iota_3\bar\iota_3}$$
$$\times\left(-\frac{\partial^3}{\partial\tau_{\bar\iota_1}\partial\tau_{\bar\iota_2}\partial\tau_{\bar\iota_3}} - (G_\infty)_{\bar\iota_1\bar\iota_2}\frac{\partial}{\partial\tau_{\bar\iota_3}} - (G_\infty)_{\bar\iota_3\bar\iota_1}\frac{\partial}{\partial\tau_{\bar\tau_2}} - (G_\infty)_{\bar\iota_2\bar\iota_3}\frac{\partial}{\partial\tau_{\bar\iota_1}}\right)$$
$$\times \exp\left(-\frac{1}{2}\sum_{\iota,\iota''=1}^{4}(G_\infty)_{\iota'\iota''}\tau_{\iota'}\tau_{\iota''}\right) \quad (4.6.36)$$

implies

$$\int \tau_{\iota_1}\tau_{\iota_2}\tau_{\iota_3}\, d\mu_\infty(\sigma')(\tau)$$
$$= -(G_\infty^{-1})_{\iota_1\iota_2}a_{\iota_3}^\infty(\sigma') - (G_\infty^{-1})_{\iota_3\iota_1}a_{\iota_2}^\infty(\sigma') - (G_\infty^{-1})_{\iota_2\iota_3}a_{\iota_1}^\infty(\sigma')$$
$$- a_{\iota_1}^\infty(\sigma')a_{\iota_2}^\infty(\sigma')a_{\iota_3}^\infty(\sigma'). \quad (4.6.37)$$

In general,

$$\int \tau_{\iota_1}\cdots\tau_{\iota_n}\, d\mu_\infty(\sigma')(\tau)$$
$$= (-1)^n \sum_{\substack{\Lambda\subset\{1,\ldots,n\}\\ \text{card }\Lambda\text{ even}}} \prod_{\kappa\notin\Lambda} a_{\iota_\kappa}^\infty(\sigma') \sum_{\substack{\text{Pairings}\\ P\text{ of }\Lambda}} \prod_{\{\kappa,\kappa'\}\in P} (G_\infty^{-1})_{\iota_\kappa\iota_{\kappa'}}. \quad (4.6.38)$$

We know that G_∞^{-1} is diagonal; indeed, G_∞^{-1} is given by (4.5.12). The linear functions $a_\iota^\infty(\sigma')$ are also quite explicit if we combine (4.6.1)–(4.6.16) with (4.5.23)–(4.5.30) and (4.2.32)–(4.2.35).

We are now in a position to derive the linearized RG transformation of the dipole perturbation of the quadratic form $(A^\infty\sigma,\sigma)$ in the neighborhood of that fixed point. The formula (4.6.38) obviously implies

$$\int\left(\sum_{\iota=1}^{4}\tau_\iota N(\iota,\vec\varepsilon)_{\mu_1}\right)\cdots\left(\sum_{\iota=1}^{4}\tau_\iota N(\iota,\vec\varepsilon)_{\mu_n}\right) d\mu_\infty(\sigma')(\tau)$$
$$= (-1)^n \sum_{\substack{\Lambda\subset\{1,\ldots,n\}\\ \text{card }\Lambda\text{ even}}} \left(\prod_{\kappa=1}^{n}\sum_{\iota_\kappa=1}^{4}\right)\prod_{\kappa=1}^{n} N(\iota_\kappa,\vec\varepsilon)_{\mu_\kappa} \prod_{\kappa\notin\Lambda} a_{\iota_\kappa}^\infty(\sigma')$$

$$\times \sum_{\substack{\text{Pairings} \\ P \text{ of } \Lambda}} \prod_{\{\kappa,\kappa'\}\in P} (G_\infty^{-1})_{\iota_\kappa,\iota_{\kappa'}}. \tag{4.6.39}$$

This particular formula is more useful for our purpose. The desired transformation is defined as the Fréchet derivative of $\widehat{R}$ at zero; i.e., it is given by

$$\widehat{D}_\infty(\widehat{F}) = \frac{d}{dt}\widehat{R}_\infty(t\widehat{F})|_{t=0}. \tag{4.6.40}$$

Since

$$\begin{aligned} \widehat{R}_\infty(t\widehat{F}) &= \ln \widehat{Z}_\infty(t\widehat{F}) \\ &\quad - \ln \int d\mu_\infty(\sigma')(\tau) \\ &\quad \times \exp\left(-t \sum_{\vec{\varepsilon}\in\{0,1\}^2} \widehat{F}\left(\frac{1}{2}L_{\vec{\varepsilon}}(\sigma') + \sum_{\iota=1}^{4}\tau_\iota N(\iota,\vec{\varepsilon})\right)\right) \end{aligned} \tag{4.6.41}$$

with $\widehat{Z}_\infty$ given by (4.3.21) and

$$\int d\mu_\infty(\sigma')(\tau) = 1 \tag{4.6.42}$$

independently of σ', we have

$$\begin{aligned} \widehat{D}_\infty(\widehat{F})(\sigma') = &\sum_{\vec{\varepsilon}\in\{0,1\}^2} \int d\mu_\infty(\sigma')(\tau)\widehat{F}\left(\frac{1}{2}L_{\vec{\varepsilon}}(\sigma') + \sum_{\iota=1}^{4}\tau_\iota N(\iota,\vec{\varepsilon})\right) \\ &- \sum_{\vec{\varepsilon}\in\{0,1\}^2} \int d\mu_\infty(0)(\tau)\widehat{F}\left(\sum_{\iota=1}^{4}\tau_\iota N(\iota,\vec{\varepsilon})\right). \end{aligned} \tag{4.6.43}$$

Naturally, we insert the Taylor expanson of $\widehat{F}$, but instead of the multi-index formula

$$\widehat{F}(\sigma) = \sum_{m=0}^{\infty}\sum_{|\omega|=m} \frac{1}{\omega!}\sigma_1^{\omega_1}\ldots\sigma_8^{\omega_8} \frac{\partial^{|\omega|}\widehat{F}(\sigma)}{\partial\sigma_1^{\omega_1}\ldots\partial\sigma_8^{\omega_8}}\bigg|_{\sigma_1=\cdots=\sigma_8=0}, \tag{4.6.44}$$

we prefer the equivalent formula

$$\widehat{F}(\sigma) = \sum_{n=0}^{\infty}\frac{1}{n!}\left(\prod_{\kappa=1}^{n}\sum_{\mu_\kappa=1}^{8}\right)\prod_{\kappa=1}^{n}\sigma_{\mu_\kappa}\left(\prod_{\kappa=1}^{n}\frac{\partial}{\partial\sigma_{\mu_\kappa}}\right)\widehat{F}(\sigma)\bigg|_{\sigma_1=\cdots=\sigma_8=0}. \tag{4.6.45}$$

Thus,

$$\widehat{D}_\infty(\widehat{F})(\sigma') = \sum_{\vec{\varepsilon}\in\{0,1\}^2}\sum_{n=0}^{\infty}\frac{1}{n!}\left(\prod_{\kappa=1}^{n}\sum_{\mu_\kappa=1}^{8}\right)\left(\prod_{\kappa=1}^{n}\frac{\partial}{\partial\sigma_{\mu_\kappa}}\right)\widehat{F}(\sigma)\bigg|_{\sigma_1=\cdots=\sigma_8=0}$$

$$\times \left[\int d\mu_\infty(\sigma')(\tau) \prod_{\kappa=1}^{n} \left(\frac{1}{2} L_{\vec{\varepsilon}}(\sigma')_{\mu_\kappa} + \sum_{\iota=1}^{4} \tau_\iota N(\iota, \vec{\varepsilon})_{\mu_\kappa} \right) - \int d\mu_\infty(0)(\tau) \prod_{\kappa=1}^{n} \left(\sum_{\iota=1}^{4} \tau_\iota N(\iota, \vec{\varepsilon})_{\mu_\kappa} \right) \right]$$

$$= \sum_{\vec{\varepsilon} \in \{0,1\}^2} \sum_{n=0}^{\infty} \frac{1}{n!} \left(\prod_{\kappa=1}^{n} \sum_{\mu_\kappa=1}^{8} \right) \left(\prod_{\kappa=1}^{n} \frac{\partial}{\partial \sigma_{\mu_\kappa}} \right) \widehat{F}(\sigma) \Bigg|_{\sigma_1=\cdots=\sigma_8=0}$$

$$\times \left[\sum_{\Gamma \subset \{1,\dots,n\}} \frac{1}{2^{n-\mathrm{card}\,\Gamma}} \prod_{\kappa \notin \Gamma} L_{\vec{\varepsilon}}(\sigma')_{\mu_\kappa} \int d\mu_\infty(\sigma')(\tau) \times \prod_{\kappa \in \Gamma} \left(\sum_{\iota=1}^{4} \tau_\iota N(\iota, \vec{\varepsilon})_{\mu_\kappa} \right) - \int d\mu_\infty(0)(\tau) \prod_{\kappa=1}^{n} \left(\sum_{\iota=1}^{4} \tau_\iota N(\iota, \vec{\varepsilon})_{\mu_\kappa} \right) \right], \tag{4.6.46}$$

while (4.6.39) implies

$$\int d\mu_\infty(0)(\tau) \prod_{\kappa=1}^{n} \left(\sum_{\iota=1}^{4} \tau_\iota N(\iota, \vec{\varepsilon})_{\mu_\kappa} \right) = \begin{cases} 0, & n \text{ odd}, \\ \left(\prod_{\kappa=1}^{n} \sum_{\iota_\kappa=1}^{4} \right) \prod_{\kappa=1}^{n} N(\iota_\kappa, \vec{\varepsilon})_{\mu_\kappa} \sum_{\substack{\text{Pairings } P \text{ of} \\ \{1,\dots,n\}}} \prod_{\{\kappa,\kappa'\} \in P} (G_\infty^{-1})_{\iota_\kappa, \iota_{\kappa'}}, & n \text{ even}, \end{cases} \tag{4.6.47}$$

$$\int d\mu_\infty(\sigma')(\tau) \prod_{\kappa \in \Gamma} \left(\sum_{\iota=1}^{4} \tau_\iota N(\iota, \vec{\varepsilon})_{\mu_\kappa} \right) = (-1)^{\mathrm{card}\,\Gamma} \sum_{\substack{\Lambda \subset \Gamma \\ \mathrm{card}\,\Lambda \text{ even}}} \left(\prod_{\kappa \in \Gamma} \sum_{\iota_\kappa=1}^{4} \right) \prod_{\kappa \in \Gamma} N(\iota_\kappa, \vec{\varepsilon})_{\mu_\kappa} \prod_{\kappa \in \Gamma \setminus \Lambda} a_{\iota_\kappa}^\infty(\sigma') \times \sum_{\substack{\text{Pairings} \\ P \text{ of } \Lambda}} \prod_{\{\kappa,\kappa'\} \in P} (G_\infty^{-1})_{\iota_\kappa \iota_{\kappa'}}. \tag{4.6.48}$$

The insertion of these formulas in (4.6.46) yields the formula for $\widehat{D}_\infty$.

Appendix

We need to verify that the three fixed points other than G_∞ give rise to undefined Gaussian states. For the moment, let $\widetilde{G}_\infty$ be one of them, and consider the variables $\alpha_{1\vec{0}}, \alpha_{2\vec{0}}, \alpha_{3\vec{0}}, \alpha_{4\vec{0}}$ associated with the unit square. If the Gaussian state exists, then

$$\langle \alpha_{\mu\vec{0}}\alpha_{\nu\vec{0}}\rangle_{\widetilde{G}_\infty} = \lim_{N\to\infty}\left[(Z^{(N)}_{\widetilde{G}_\infty})^{-1}\left(\prod_{\iota=1}^{4}\prod_{r=0}^{N}\prod_{\vec{m}\in\Gamma^{(N)}_r}\int_{-\infty}^{\infty} d\alpha^{(r)}_{\iota\vec{m}}\right)\alpha_{\mu\vec{0}}\alpha_{\nu\vec{0}} \times \exp\left(-\frac{1}{2}(S_{\widetilde{G}_\infty}\alpha,\alpha)\right)\right], \tag{4.A.1}$$

$$Z^{(N)}_{\widetilde{G}_\infty} = \left(\prod_{\iota=1}^{4}\prod_{r=0}^{N}\prod_{\vec{m}\in\Gamma^{(N)}_r}\int_{-\infty}^{\infty} d\alpha^{(r)}_{\iota\vec{m}}\right)\exp\left(-\frac{1}{2}(S_{\widetilde{G}_\infty}\alpha,\alpha)\right), \tag{4.A.2}$$

where

$$\Gamma^{(N)}_r = \{-2^{N-r}, -2^{N-r}+1,\dots,-1,0,1,\dots,2^{N-r}-1,2^{N-r}\}^2. \tag{4.A.3}$$

Since

$$(S_{\widetilde{G}_\infty}\alpha,\alpha) = \sum_{\iota,\kappa=1}^{4}\sum_{\vec{n}\in\mathbb{Z}^2}(\widetilde{G}_\infty)_{\iota\kappa}\alpha_{\iota\vec{n}}\alpha_{\kappa\vec{n}} + \sum_{r=1}^{\infty}\sum_{\iota=1}^{4}\sum_{\vec{n}\in\mathbb{Z}^2}\alpha^{(r)\,2}_{\iota\vec{n}} + 2\sum_{r=1}^{\infty}\sum_{\vec{m}\in\mathbb{Z}^2}\sum_{\vec{\varepsilon}\in\{0,1\}^2}\sum_{\iota,\iota'=1}^{4} Q^{\vec{\varepsilon}}_{\iota\iota'}\alpha^{(r-1)}_{\iota',2\vec{m}+\vec{\varepsilon}}\alpha^{(r)}_{\iota\vec{m}}, \tag{4.A.4}$$

we may combine the elementary formula

$$\alpha_{\mu\vec{0}}\alpha_{\nu\vec{0}}\exp\left(-\frac{1}{2}\sum_{\iota,\kappa=1}^{4}(\widetilde{G}_\infty)_{\iota\kappa}\alpha_{\iota\vec{0}}\alpha_{\kappa\vec{0}}\right) = \left((\widetilde{G}^{-1}_\infty)_{\mu\nu} + \sum_{\mu',\nu'=1}^{4}(\widetilde{G}^{-1}_\infty)_{\mu\mu'}(\widetilde{G}^{-1}_\infty)_{\nu\nu'}\frac{\partial^2}{\partial\alpha_{\mu'\vec{0}}\partial\alpha_{\nu'\vec{0}}}\right) \times \exp\left(-\frac{1}{2}\sum_{\iota,\kappa=1}^{4}(\widetilde{G}_\infty)_{\iota\kappa}\alpha_{\iota\vec{0}}\alpha_{\kappa\vec{0}}\right) \tag{4.A.5}$$

with integration by parts to obtain

$$\langle \alpha_{\mu\vec{0}}\alpha_{\nu\vec{0}}\rangle_{\widetilde{G}_\infty} = (\widetilde{G}^{-1}_\infty)_{\mu\nu} + \sum_{\mu',\nu'=1}^{4}(\widetilde{G}^{-1}_\infty)_{\mu\mu'}(\widetilde{G}^{-1}_\infty)_{\nu\nu'}\sum_{\iota,\bar{\iota}=1}^{4}Q^{\vec{0}}_{\iota\mu'}Q^{\vec{0}}_{\bar{\iota}\nu'}\langle\alpha^{(1)}_{\iota\vec{0}}\alpha^{(1)}_{\bar{\iota}\vec{0}}\rangle_{\widetilde{G}_\infty}. \tag{4.A.6}$$

On the other hand,

$$\langle \alpha^{(1)}_{\iota \vec{0}} \alpha^{(1)}_{\bar{\iota} \vec{0}} \rangle_{\widetilde{G}_\infty} = \langle \alpha_{\iota \vec{0}} \alpha_{\bar{\iota} \vec{0}} \rangle_{\widetilde{G}_\infty} \tag{4.A.7}$$

because this Gaussian is a fixed point with respect to the RG transformation. Therefore, the 4×4 matrix C defined by

$$C_{\mu\nu} = \langle \alpha_{\mu \vec{0}} \alpha_{\nu \vec{0}} \rangle_{\widetilde{G}_\infty} \tag{4.A.8}$$

satisfies the matrix equation

$$C = \widetilde{G}_\infty^{-1} + \widetilde{G}_\infty^{-1} Q^{\vec{0}*} C Q^{\vec{0}} \widetilde{G}_\infty^{-1}. \tag{4.A.9}$$

If we set

$$C' = Q^{\vec{0}*} C Q^{\vec{0}}, \tag{4.A.10}$$

it follows from (4.4.19) that

$$\begin{bmatrix} c'_{33} & c'_{34} \\ c'_{43} & c'_{44} \end{bmatrix} = \frac{1}{64} \begin{bmatrix} c_{44} & c_{43} \\ c_{34} & c_{33} \end{bmatrix} \tag{4.A.11}$$

with less simple expressions for the other matrix elements of C'.

Now for the $90°$ rotation in the counterclockwise direction we denote the effect on a function by $f \mapsto f^\perp$ and see that

$$\Psi_{1r\vec{n}}{}^\perp = -\Psi_{2,r,(-n_1,n_0)}, \tag{4.A.12}$$

$$\Psi_{2r\vec{n}}{}^\perp = \Psi_{1,r,(-n_1,n_0)}, \tag{4.A.13}$$

$$\Psi_{3r\vec{n}}{}^\perp = \Psi_{4,r,(-n_1,n_0)}, \tag{4.A.14}$$

$$\Psi_{4r\vec{n}}{}^\perp = \Psi_{3,r,(-n_1,n_0)}. \tag{4.A.15}$$

This means that the rotation preserves the space of Osiris constrained configurations

$$\phi(\vec{x}) = \sum_{\iota=1}^{4} \sum_{r=0}^{\infty} \sum_{\vec{n} \in \mathbb{Z}^2} \alpha^{(r)}_{\iota \vec{n}} \Psi_{\iota r \vec{n}}(\vec{x}) \tag{4.A.16}$$

and has the effect

$$\begin{aligned} \phi^\perp &= \sum_{\iota=1}^{4} \sum_{r=0}^{\infty} \sum_{\vec{n} \in \mathbb{Z}^2} \alpha^{(r)}_{\iota \vec{n}} \Psi_{\iota r \vec{n}}{}^\perp \\ &= \sum_{\iota=1}^{4} \sum_{r=0}^{\infty} \sum_{\vec{n} \in \mathbb{Z}^2} \alpha^{(r)}_{\iota \vec{n}}{}^\perp \Psi_{\iota r \vec{n}}, \end{aligned} \tag{4.A.17}$$

where we define

$$\alpha_{1r\vec{n}}{}^{\perp} = \alpha_{2,r,(n_1,-n_0)}, \tag{4.A.18}$$

$$\alpha_{2r\vec{n}}{}^{\perp} = -\alpha_{1,r,(n_1,-n_0)}, \tag{4.A.19}$$

$$\alpha_{3r\vec{n}}{}^{\perp} = \alpha_{4,r,(n_1,-n_0)}, \tag{4.A.20}$$

$$\alpha_{4r\vec{n}}{}^{\perp} = \alpha_{3,r,(n_1,-n_0)}. \tag{4.A.21}$$

On the other hand, the *canonical* quadratic form $(S\alpha, \alpha)$ is invariant with respect to this transformation; i.e.,

$$(S\alpha^{\perp}, \alpha^{\perp}) = \int (\nabla\phi^{\perp})^2 = \int (\nabla\phi)^2 = (S\alpha, \alpha). \tag{4.A.22}$$

Since

$$(S_{\widetilde{G}_\infty}\alpha, \alpha) = (S\alpha, \alpha) + \sum_{\iota,\kappa=1}^{4} \sum_{\vec{n}\in\mathbb{Z}^2} [(\widetilde{G}_\infty)_{\iota\kappa} - \delta_{\iota\kappa}]\alpha_{\iota\vec{n}}\alpha_{\kappa\vec{n}} \tag{4.A.23}$$

and $\widetilde{G}_\infty$ has the form (4.4.51), it follows that

$$(S_{\widetilde{G}_\infty}\alpha^{\perp}, \alpha^{\perp}) = (S_{\widetilde{G}_\infty}\alpha, \alpha). \tag{4.A.24}$$

Hence,

$$\langle \alpha_{4\vec{0}}{}^2 \rangle_{\widetilde{G}_\infty} = \langle (\alpha_{3\vec{0}}{}^{\perp})^2 \rangle_{\widetilde{G}_\infty} = \langle \alpha_{3\vec{0}}{}^2 \rangle_{\widetilde{G}_\infty}, \tag{4.A.25}$$

and so

$$c_{44} = c_{33} \equiv c_0. \tag{4.A.26}$$

With

$$c_{43} = c_{34} \equiv c_1 \tag{4.A.27}$$

as well, the lower right 2×2 submatrix of C is just $\left[\begin{smallmatrix} c_0 & c_1 \\ c_1 & c_0 \end{smallmatrix}\right]$. Now, by the form (4.4.51) for $\widetilde{G}_\infty$, we have

$$\widetilde{G}_\infty{}^{-1} = \frac{1}{b^2 - a^2} \begin{bmatrix} b^2 - a^2 & 0 & 0 & 0 \\ 0 & b^2 - a^2 & 0 & 0 \\ 0 & 0 & b & -a \\ 0 & 0 & -a & b \end{bmatrix}, \tag{4.A.28}$$

so if we set

$$C'' = \widetilde{G}_\infty^{-1} C' \widetilde{G}_\infty^{-1}, \tag{4.A.29}$$

we see that

$$\begin{bmatrix} c''_{33} & c''_{34} \\ c''_{43} & c''_{44} \end{bmatrix} = \frac{1}{(b^2 - a^2)^2} \begin{bmatrix} b & -a \\ -a & b \end{bmatrix} \begin{bmatrix} c'_{33} & c'_{34} \\ c'_{43} & c'_{44} \end{bmatrix} \begin{bmatrix} b & -a \\ -a & b \end{bmatrix}$$

$$= \frac{1}{64} \frac{1}{(b^2 - a^2)^2} \begin{bmatrix} b & -a \\ -a & b \end{bmatrix} \begin{bmatrix} c_0 & c_1 \\ c_1 & c_0 \end{bmatrix} \begin{bmatrix} b & -a \\ -a & b \end{bmatrix}$$
$$= \frac{1}{64} \frac{1}{(b^2 - a^2)^2} \begin{bmatrix} (a^2+b^2)c_0 - 2abc_1 & (a^2+b^2)c_1 - 2abc_0 \\ (a^2+b^2)c_1 - 2abc_0 & (a^2+b^2)c_0 - 2abc_1 \end{bmatrix}. \tag{4.A.30}$$

It follows from (4.A.9) that

$$\begin{bmatrix} c_0 & c_1 \\ c_1 & c_0 \end{bmatrix} = \frac{1}{b^2 - a^2} \begin{bmatrix} b & -a \\ -a & b \end{bmatrix} + \frac{1}{64(b^2 - a^2)^2} \begin{bmatrix} (a^2+b^2)c_0 - 2abc_1 & (a^2+b^2)c_1 - 2abc_0 \\ (a^2+b^2)c_1 - 2abc_0 & (a^2+b^2)c_0 - 2abc_1 \end{bmatrix}, \tag{4.A.31}$$

so we finally have the two equations

$$c_0 = \frac{b}{b^2 - a^2} + \frac{(a^2+b^2)c_0 - 2abc_1}{64(b^2 - a^2)^2}, \tag{4.A.32}$$

$$c_1 = -\frac{a}{b^2 - a^2} + \frac{(a^2+b^2)c_1 - 2abc_0}{64(b^2 - a^2)^2}. \tag{4.A.33}$$

We now consider the cases for $\widetilde{G}_\infty$. One case is given by $a = 0$ and $b = 7/16 - (1/16)\sqrt{33}$, which implies (since $b^{-1} = 7 + \sqrt{33}$ in this event)

$$c_0 = 7 + \sqrt{33} + \frac{41 + 7\sqrt{33}}{32} c_0, \tag{4.A.34}$$

$$c_1 = \frac{41 + 7\sqrt{33}}{32} c_1. \tag{4.A.35}$$

Clearly, $c_1 = 0$ and $c_0 < 0$. Since c_0 is a positive expectation, this is the desired contradiction. Next, consider the case where $a = (1/16)\sqrt{33}$ and $b = 7/16$. Then (since $b^2 - a^2 = 1/16$ in this event),

$$c_0 = 7 + \frac{41c_0 - 7\sqrt{33}c_1}{32}, \tag{4.A.36}$$

$$c_1 = -\sqrt{33} + \frac{41c_1 - 7\sqrt{33}c_0}{32}. \tag{4.A.37}$$

The solution is $(c_0, c_1) = (-7/2, (5/6)\sqrt{33})$. With $c_0 < 0$ in this case as well, we contradict the positivity of the same expectation. Finally, consider the case where $a = -(1/16)\sqrt{33}$ and $b = 7/16$. Then,

$$c_0 = 7 + \frac{41c_0 + 7\sqrt{33}c_1}{32}, \tag{4.A.38}$$

$$c_1 = \sqrt{33} + \frac{41c_1 + 7\sqrt{33}c_0}{32}. \tag{4.A.39}$$

The solution is $(c_0, c_1) = (-7/2, -(5/6)\sqrt{33})$, so we still have the desired contradiction.

This completes the proof that none of these three extraneous fixed points can exist as states.

References

[1] G. Battle, Osiris Wavelets and the Dipole Gas, in *Wavelet Transforms and Time-Frequency Signal Analysis*, L. Debnath, ed., Birkhäuser, Boston, 2001.
[2] G. Battle, Osiris Wavelets in three dimensions, *Ann. Phys.*, **286** (2000), 23–107.
[3] G. Battle, Structure of the RG Transformation for the Osiris Dipole Gas, Texas A&M University preprint.
[4] J. Kogut and K. Wilson, The renormalization group and the ε-expansion, *Phys. Rep.*, **12** (1974), 74–200.
[5] J. Glimm and A. Jaffe, *Quantum Physics: A Functional Integral Point of View*, Springer-Verlag, New York, 1987.
[6] K. Gawedzki and A. Kupiainen, A rigorous block spin approach to massless lattice theories, *Commun. Math. Phys.*, **77** (1980), 31–64.
[7] G. Battle, *Wavelets and Renormalization*, World Scientific, Singapore, 1999.
[8] P. Federbush, A phase cell approach to Yang Mills theory. I: Modes, lattice–continuum duality, *Commun. Math. Phys.*, **107** (1987), 319–329.
[9] T. Balaban, J. Imbrie, and A. Jaffe, Exact Renormalization Group for Gauge Theories, in *Progress in Gauge Field Theory*, G. t'Hooft et al., eds., Plenum Press, London, 1984.
[10] K. Gawedzki and A. Kupiainen, Block spin renormalization group for dipole gas and $(\nabla\phi)^4$, *Ann. Phys.*, **147** (1983), 198–243.
[11] K. Wilson, Renormalization group and critical phenomena, *Phys. Rev.*, **B4** (1971), 3174–3205.
[12] K. Gawedzki and A. Kupiainen, Massless lattice ϕ_4^4 theory: Rigorous control of a renormalizable asymptotically free model, *Commun. Math. Phys.*, **99** (1985), 197–252.
[13] J. Feldman, J. Magnen, V. Rivasseau, and R. Sénéor, Bounds on completely convergent Euclidean Feynman graphs, *Commun. Math. Phys.*, **98** (1985), 273–288.
[14] J. Feldman, J. Magnen, V. Rivasseau, and R. Sénéor, Bounds on a renormalized Feynman graph, *Commun. Math. Phys.*, **100** (1985), 23–55.
[15] J. Feldman, J. Magnen, V. Rivasseau, and R. Sénéor, Construction and Borel Summability of infrared ϕ_4^4 by a phase space expansion, *Commun. Math. Phys.* **109** (1987), 437–480.
[16] V. Rivasseau, *From Perturbative to Constructive Renormalization*, Princeton University Press, Princeton, 1991.
[17] H. Koch and P. Wittwer, A non-trivial renormalization group fixed point for the Dyson-Baker hierarchical model, *Commun. Math. Phys.*, **164** (1994), 627–647.
[18] G. Baker, Ising model with a scaling interaction, *Phys. Rev.*, **B5** (1972), 2622–2633.
[19] P. Collet and J. Eckmann, *A Renormalization Group Analysis of the Hierarchical Model in Statistical Mechanics*, Springer-Verlag, New York, 1978.
[20] K. Gawedzki and A. Kupiainen, Non-Gaussian fixed points of the block spin transformation. Hierarchical model approximation, *Commun. Math. Phys.*, **89** (1983), 191–220.
[21] P. Bleher and Y. Sinai, Critical indices for Dyson's asymptotically hierarchical models, *Commun. Math. Phys.*, **45** (1975), 247–278.

[22] G. Gallavotti, Some aspects of the renormalization problems in statistical mechanics and quantum field theory, *Atti. Acad. Naz. Lincei Rend. Cl. Sci. Fis. Mat. Natur.*, **(8)XV** (1978).
[23] G. Golner, Calculation of the critical exponent η via renormalization group recursion formulas, *Phys. Rev.*, **B8** (1973), 339–345.

5

Wavelets for Statistical Estimation and Detection

R. Chandramouli
K. M. Ramachandran

ABSTRACT In this chapter, we discuss two important statistical applications of wavelets: (a) signal estimation and (b) signal detection. Several popular wavelet-based signal estimation/de-noising schemes are reviewed, followed by the introduction of a new signal estimator that attempts to estimate and preserve some moments of the underlying original signal. A relationship between signal estimation and data compression is also discussed. Analytical and experimental results are presented to explain the performance of these signal estimation schemes.

The second part of this chapter deals with hypothesis testing-based signal detection. Wavelet decomposition level-dependent binary hypothesis tests are first presented, followed by global detection procedures that combine these decisions to obtain the final decision. Likelihood ratio-based statistical detectors are discussed toward this goal. The combinatorial explosion of the global detector design results in the investigation of two specific global detection fusion rules: OR and AND. Some mathematical approximations are exploited that aid in trading off complexity for the global detector performance. Theorems and their proofs relating to this issue are presented.

5.1 Introduction

In this chapter, we discuss the theory and applications of wavelet analysis in statistical estimation (filtering/de-noising) and hypothesis testing (detection). Signal estimation and detection are two important problems of both theoretical and practical importance. In signal estimation schemes, an approximate version of a clean signal (function) is estimated from its noisy version via filtering subject to a certain fidelity criterion. This problem is being widely studied by the statistical and engineering communities. Kernel smoothing, Wiener filtering, averaging, and thresholding-based de-noising are popular examples of signal estimation schemes. Filtering techniques are widely used in applications such as the reception of noisy signals in digital communication systems, radar image processing, and medical diagnostics. However, in many circumstances we are interested, not so much in retrieving the unknown signal from noise, but in detecting the presence or absence of it. Typical scenarios where such problems arise are electronic warfare, radar detection systems, and statistical quality control. Signal detection is usually studied as a statistical hypothesis test in such contexts. There are two basic approaches to filtering and detection of signals:

- Time domain processing.
- Frequency domain processing.

In time-domain methods, the noisy signal is filtered based on observations taken over a period of time. Averaging and smoothing based on weighting functions (kernels) are popular in this category. In general, these methods depend on the statistics of the noisy signal within a moving time window. The choice of the window size and the weighting function have a strong impact on the *quality* of the de-noised signal. Usually these parameters are chosen such that an optimal *bias-variance* trade-off is achieved. We know that oversmoothing increases the bias of the filtered signal but decreases its variance and undersmoothing produces the opposite effect. Proper kernel functions aid in achieving faster rates of convergence of the estimation error with increasing sample size.

Frequency-domain-based signal estimation relies heavily on the properties of the type of transforms being used. The Karhunen–Loeve transform (KLT) is known to be the optimal transform in the sense of transforming the signal into coefficients that have a diagonal covariance matrix. But the basis functions of the KLT are signal dependent. Therefore, it gives rise to practical implementational problems. This has led to research in de-noising based on transforms whose basis functions are independent of the signal. If the transform is such that the energy of the noisy signal is packed in only a few transform coefficients, then, by suitably modifying the coefficients whose energies are insignificant with respect to a fixed threshold, the original signal's transform coefficients can be estimated. Early methods of transform, based de-noising were based on the Fourier transform. Although the Fourier transform can be implemented very efficiently, it is not suitable for estimating nonstationary signals [1]. This has led to studies on the application of windowed Fourier transforms, and more recently, the wavelet transforms. Since this chapter deals with wavelet-based de-noising schemes, we briefly describe the differences between the Fourier and a wavelet transform for the sake of completeness. The Fourier transform of a function provides information about its frequency content over the entire support of the function. Some properties of the function, such as its jumps or discontinuities, are not localized in the Fourier transform domain. Also, changes in the statistics of a nonstationary signal are not well characterized by the Fourier coefficients. On the other hand, wavelet transform based multiresolution decomposition has proven to be a very effective tool in this regard [1]. It provides a convenient framework to study the time-frequency characteristics of the function. The block structure of the Fourier transform is replaced with a pyramidal or tree structure of the wavelet transform, and the function is decomposed into many resolutions. A discrete, or continuous-time function is decomposed into an *approximate function* and a *detail function*. This process is applied iteratively to the approximation function until a desired resolution is attained. The transform coefficients at the various resolutions tell us a great deal about the behavior of the function. By appropriately choosing the wavelet transform, we can obtain a sparse representation of the function. Hence, thresholding the wavelet coefficients leads to a smooth recovery of the function.

While signal estimation schemes attempt to recover an original signal from its noisy version, signal detection schemes detect the presence or absence of a signal of interest from noisy observations. For example, in applications such as radar target detection, the problem of interest is the detection of an object in a certain spatial location. Signal detection can be studied using techniques from statistical hypothesis tests. Note that the observations presented to a detector are noisy and also incomplete in many cases. The aim is to detect the signal within a desired probability of error using these noisy observations. A detection algorithm should be able to handle this problem recognizing the nature of the observations. In general, it is easier to design hypothesis tests for independent and identically distributed (i.i.d.) observations. Since the i.i.d. requirement need not be satisfied by the noisy observations studying the equivalent problem in the frequency domain is often helpful. For example, we know that the KLT decorrelates the data (though at an enormous computational cost). Therefore, if the observations are Gaussian, the KLT transform coefficients are i.i.d. However, a good approximation to the KLT can be implemented using the computationally simpler wavelet transform. Studies have shown that it is reasonable to assume that the wavelet transform coefficients of an observed noisy signal are approximately i.i.d. This assumption has been widely used by many researchers studying the detection of transient signals. Performance of the detector when there is mismatch in this assumption can be studied through asymptotic methods.

The first part of this chapter deals with signal estimation. Signal estimation for low to medium signal-to-noise ratios (SNRs) is the focus. Most of the existing wavelet thresholding-based de-noising algorithms work well for high SNRs. When the noise energy increases, a considerable degradation in the performance is usually observed. One approach to circumvent this problem is to exploit some empirical statistics of the original signal. We explain later how to choose these statistics and describe a de-noising methodology based on these statistics with the following properties:

- It is simple and uses the empirical statistical moments of the signal for de-noising.
- It is robust to the errors in the empirical estimates.
- It achieves de-noising and data compression jointly.

The second part of this chapter deals with wavelet-based hypothesis testing. We discuss a likelihood ratio-based wavelet level-dependent hypothesis test. First, hypothesis tests are performed at each level of the wavelet decomposition, and then a global fusion test is performed to combine these individual decisions into a global decision. Performances of these tests are based on the false alarm and the miss probabilities of detection.

This chapter is organized as follows. A brief introduction to wavelet analysis is given in Section 5.2, Section 5.3 deals with the wavelet-based statistical estimation method, and Section 5.4 deals with hypothesis testing. Concluding remarks are presented in Section 5.5.

5.2 Wavelet analysis

We refer to [2], [3], and [1] for an introduction to wavelet analysis. At the heart of wavelet analysis is the multiresolution decomposition, wherein approximations that are linear combinations of dilations and translations of a scaling function are constructed, resulting in approximations at different levels of resolution. This is an efficient way to extract the information content of a signal.

A multiresolution analysis in $L^2(\Re)$ ($\Re$ denotes the set of reals) consists of a sequence of embedded closed subspaces $\{V_j : j \in \mathcal{Z}\}$ ($\mathcal{Z}$ denotes the set of integers) satisfying the following:

(a) $V_j \subset V_{j-1};\ j \in \mathcal{Z}$ (embedding)
(b) $\overline{\bigcup_{j\in\mathcal{Z}} V_j} = L^2(\Re)$ (upward completeness)
(c) $\bigcap_{j\in\mathcal{Z}} V_j = \{0\}$ (downward completeness)
(d) $f(x) \in V_j \Longleftrightarrow f(2^j x) \in V_0$ (scale invariance)
(e) $f(x) \in V_0 \Longleftrightarrow f(x-n) \in V_0$ for all $n \in \mathcal{Z}$ (shift invariance)
(f) $\exists$ a $\phi \in V_0$ s.t. $\{\phi(x-n) | n \in \mathcal{Z}\}$ is an orthonormal basis for V_0 (existence of a basis).

The function $\phi(x)$ is called the *scaling function* and integrates to unity. We also note that $\phi_{j,n}(x) = \{2^{j/2}\phi(2^j x - n) | n \in \mathcal{Z}\}$ is a basis for V_{-j}. The orthogonality of $\phi(x)$ is not necessary, since a nonorthogonal basis with the shift property can be orthogonalized. Due to the embedding and the scaling property, it can be verified that the scaling function $\phi(t)$ satisfies a two-scale equation. Since $\phi(t) \in V_0 \subset V_{-1}$, there exists a unique l^2 sequence $\{h_n : n \in \mathcal{Z}\}$, called the two-scale sequence of ϕ, that satisfies the two-scale equation

$$\phi(x) = \sqrt{2} \sum_{n\in\mathcal{Z}} h_n \phi(2x - n). \tag{5.2.1}$$

Denote the orthogonal complement of V_{j+1} in V_j as W_{j+1} for each j. That is,

$$V_j = V_{j+1} \oplus W_{j+1}, \tag{5.2.2}$$

where $\oplus$ represents a direct sum with $V_{j+1} \perp W_{j+1}$. Each subscript j usually corresponds to a resolution 2^j (it could be any integer raised to the power of j). Hence, the maximum resolution, say, 2^J is denoted by J. Then, we see that

$$V_0 = W_1 \oplus W_2 \oplus \cdots \oplus W_J \oplus V_J. \tag{5.2.3}$$

The spaces $\{W_i\}_{i=1}^J$ are called the *detail* or *wavelet* spaces, and the space V_J is called the *approximation* space at resolution 2^J. Higher values of the index j correspond to coarser levels of the wavelet decomposition. Given a multiresolution analysis, there exists an orthonormal basis for $L^2(\Re)$

$$\psi_{j,n}(x) = 2^{-j/2}\psi(2^{-j}x - n), \qquad j, n \in \mathcal{Z} \tag{5.2.4}$$

called the *wavelet basis* such that $\psi_{j,n}(x)$, $n \in \mathcal{Z}$ is an orthonormal basis for W_j. In fact, the wavelet ψ can be constructed in terms of the scaling function ϕ by

$$\psi(x) = \sqrt{2}\sum_{n\in\mathcal{Z}}(-1)^n h_{-n+1}\phi(2x - n). \tag{5.2.5}$$

Hence, we have the following discrete wavelet transform representation for $f \in L^2(\Re)$:

$$f = \sum_{n\in\mathcal{Z}} c_{J,n}\phi_{J,n} + \sum_{j\leq J}\sum_{n\in\mathcal{Z}} d_{j,n}\psi_{j,n}, \tag{5.2.6}$$

where the scaling coefficients are $c_{J,n} = \int_{-\infty}^{\infty} f(x)\phi_{J,n}(x)dx$, and the wavelet coefficients $d_{j,n} = \int_{-\infty}^{\infty} f(x)\psi_{j,n}(x)$; $j \leq J$. Here $\sum_{n\in\mathcal{Z}} c_{J,n}\phi_{J,n}$ is the approximation of the function f at resolution J. $\sum_{j\leq J}\sum_{n\in\mathcal{Z}} d_{j,n}\psi_{j,n}$ represents the fine details in terms of the dilated and translated wavelets. As the wavelet bases are spatially adapted, the wavelet coefficients at spatial indices n with $n2^j = t$, for $j = 1, 2, \ldots, J$, allow us to discuss the behavior of the function f near t. Large wavelet coefficients at the various levels occur exclusively in the region of major spatial activity. It is fairly simple to show that the following relations hold:

$$c_{j,n} = \sum_{l\in\mathcal{Z}} h_{l-2n}c_{j-1,l}; \qquad j, n \in \mathcal{Z} \tag{5.2.7}$$

$$d_{j,n} = \sum_{l\in\mathcal{Z}} g_{l-2n}c_{j-1,l}; \qquad j, n \in \mathcal{Z} \tag{5.2.8}$$

$$c_{j-1,n} = \sum_{l\in\mathcal{Z}} h_{n-2l}c_{j,l} + \sum_{l\in\mathcal{Z}} g_{n-2l}d_{j,l}; \qquad j, n \in \mathcal{Z}, \tag{5.2.9}$$

where $g_n = (-1)^n h_{-n+1}$. Therefore, the scaling and the wavelet coefficients can be computed recursively. The fast discrete wavelet transform algorithm can be used to compute these coefficients with $O(N)$ complexity [2].

5.3 Statistical estimation

Filtering a noisy signal by thresholding the coefficients of its multiresolution wavelet decomposition has recently gained prominence [4]– [7]. Consider a discrete-time noisy signal that is assumed to be of the form

$$y_i = f(t_i) + \sigma z_i, \qquad i = 0, \ldots, N - 1, \tag{5.3.1}$$

where $t_i = i/N$, N is a power of 2, and $f \in L^2(\Re)$ is the unknown signal of interest to be separated from the zero-mean, unit variance white noise process, $\{z_0, z_1, \dots, z_{N-1}\}$. We assume that the noise is independent of the signal. The parameter σ, in conjunction with f, determines the SNR. The estimate of f proposed here is based on the wavelet decomposition of the observations $\{y_i\}$ and is hence a nonparametric estimate. Wavelets are used extensively in data compression, communications, and biomedical engineering, and are preferred due to their good time-frequency localization property. The multiresolution analysis implemented by wavelets carries out the estimation process at a succession of levels—lower levels capture the overall structure, and successive levels add details. Thus, one would expect to see local features captured by higher level coefficients. At higher levels, only a few *significant* wavelet coefficients may suffice to capture the essential structure of the function.

Wavelet shrinkage [4]– [7] refers to a class of algorithms that works on the principle of de-noising by "thresholding" the wavelet coefficients of the noisy signal. The signal samples $\{y_i\}$ are first transformed using a discrete wavelet transform. The empirical wavelet coefficients are then *shrunk* toward zero based on a threshold value—coefficients that are greater than the threshold are retained and the others are set to zero. Finally, the de-noised signal is obtained by the inverse wavelet transform applied to the thresholded coefficients. Different shrinkage estimators can be derived based on the choice of the threshold. For example, the universal threshold estimator [5] uses a threshold $T = \sigma\sqrt{2 \log N}$. Data enters this thresholding scheme only through the estimate of σ. Only the coefficients with magnitudes above a threshold are retained. When the noise variance is sufficiently high compared to the signal strength, i.e., in cases with low SNRs, ignoring the coefficients below the threshold may lead to significant loss of information. Low SNRs occur in practice [8]; however, not many studies deal with this scenario.

In this chapter, we propose and discuss a framework based on wavelet decomposition for jointly de-noising and compressing the data for the low SNR situation by using moments of the wavelet coefficients in addition to thresholds. Moment-based filters have been previously studied in applications such as image processing, pattern recognition, and compression [9]– [11].

Quantization of the noisy empirical wavelet coefficients leads to de-noising. This is the principle on which the shrinkage estimators are based. By appropriately choosing the quantizer, it is possible to achieve data compression along with de-noising. We know that the Lloyd–Max quantizer is optimal in the L^2 sense for a desired compression ratio when the probability distribution of the noisy wavelet coefficients is completely known [8]. However, in practice, the probability distribution is not known exactly. Weak convergence of the distributions of the de-noised samples to the distribution of the signal f can be proved by establishing convergence of the moments. By quantizing these coefficients so that all the moments of the quantized coefficients are equal to those of the original signal f, the convergence in distribution of the de-noised signal to f can be established. Other approaches to de-noising via quantization can be found in [12]–[14].

We note that the probability distribution of the wavelet coefficients has been approximated with a Laplace or a normal probability distribution function [2], [12]. This assumption has also been successfully used in applications such as image de-noising and compression using wavelets [12]. Sometimes a generalized Gaussian distribution is also used. As the Laplace and normal distribution can be described completely via their mean and variance, it seems reasonable to base the quantization on just the first two moments of the distribution. Closed-form expressions for the quantizer can be easily derived; this is, in general, not possible for the Lloyd–Max quantizer. The empirical moments of the noisy wavelet coefficients are used to achieve filtering via quantization in such a manner that the first two moments of the original signal f are preserved by the quantized (de-noised) signal samples. The method can also be generalized so that the first m moments are preserved, where m is finite. Hard and soft quantizers that preserve the first two moments are derived explicitly.

The performance of the proposed moment preserving estimator (MPE) is compared to the universal threshold estimator (UTE) [5], Stein's unbiased risk estimator (SURE) [6], the minimax threshold estimator (ME) [7], and the Wiener filter (WF). We find that the desirable properties of the other shrinkage estimators can be incorporated into the MPE by using their threshold selection procedures. Through the quantization used by the MPE, high compression ratios are also achieved.

5.3.1 Popular wavelet-based signal estimation schemes

In this section, we outline some popular estimators that will be used in performance comparisons with the moment preserving threshold estimator proposed here.

5.3.1.1 Wavelet shrinkage estimators

Define the following vectors:

$$\mathbf{y} = (y_0, y_1, y_2, \dots, y_{N-1}) \tag{5.3.2}$$

$$\mathbf{f} = (f(t_0), f(t_1), \dots, f(t_{N-1})) \tag{5.3.3}$$

$$\mathbf{z} = (z_0, z_1, z_2, \dots, z_{N-1}), \tag{5.3.4}$$

where

$$y_i = f(t_i) + \sigma z_i, \qquad i = 0, \dots, N-1, \tag{5.3.5}$$

and where $t_i = i/N$, σ is the standard deviation of the noise, and the $\{z_i\}$ are i.i.d. random variables distributed as $N(0,1)$. Performing a discrete wavelet transform (denoted by $\mathcal{W}$) on $\mathbf{y}$, for the jth resolution we get

$$x_{j,i} = w_{j,i} + \sigma \eta_{j,i}, \qquad j = 1, 2, \dots, J;\ i = 1, \dots, N/2^j, \tag{5.3.6}$$

where $x_{j,i}$, $w_{j,i}$, and $\eta_{j,i}$ are the ith coordinates at the jth resolution. For orthonormal wavelet basis functions, the wavelet transform of the white noise, z_i, is also white. In most situations, the value of σ is unknown, but it can be estimated using the median of the absolute deviation of the finest scale wavelet coefficients [6], namely,

$$\hat{\sigma} = \frac{\text{median}_{k=1,2,\ldots,N/2}(|x_{1,k} - \text{median}_k(x_{1,k})|)}{0.6745}. \tag{5.3.7}$$

It can be seen from Eq. (5.3.6) that each empirical wavelet coefficient $x_{j,i}$ is noisy with the variance of the noise equal to σ^2. For a spatially inhomogeneous function f, at the higher resolution levels, only few of these noisy wavelet coefficients correspond to the signal. Thus, it is natural to reconstruct the function using only the largest empirical coefficients through some kind of thresholding. This is the idea underlying wavelet thresholding—coefficients at the detail levels are thresholded; the coarse level approximations are left as they are.

Thresholding can be either global, i.e., the same threshold is used for every resolution level, or level dependent, i.e., a different threshold is used for each resolution level. Also, depending on the kind of thresholding function used, we obtain *hard thresholding* or *soft thresholding*. Hard thresholding functions either keep or kill the wavelet coefficients, depending on the results of a comparison with a threshold. Soft thresholding uses multiple levels to shrink the coefficient towards zero according to some fixed rule. For example, a universal hard thresholding estimator, $Q_h(\cdot)$, is described by

$$Q_h(x_{j,i}) = \begin{cases} x_{j,i} & \text{if } |x_{j,i}| > T \\ 0 & \text{if } |x_{j,i}| \leq T \end{cases} \tag{5.3.8}$$

and a soft thresholding estimator is

$$Q(x_{j,i}) = \begin{cases} x_{j,i} - T & \text{if } x_{j,i} \geq T \\ x_{j,i} + T & \text{if } x_{j,i} \leq -T \\ 0 & \text{if } |x_{j,i}| < T \end{cases}. \tag{5.3.9}$$

The choice of thresholds makes a big difference to the reconstruction, hence choice of the threshold is a fundamental issue. Thresholds that produce relatively small differences quantitatively, in terms of the estimated risk, could differ significantly qualitatively. Three of the methods described below are wavelet threshold estimators, while the fourth is the WF.

5.3.1.1.1 Minimax thresholding

In this thresholding scheme, a minimax threshold that depends on the sample size minimizes an upper bound of the L^2 risk in estimating $f(x)$. The threshold has to be computed numerically, as there is no closed-form solution. In fact, asymptotically as $n \to \infty$ the minimax threshold is equivalent to the universal threshold [5] to be discussed in the next subsection. We refer to [5] for the details.

5.3.1.1.2 UTE

A UTE [5] produces an estimate of the signal $f(x)$ by thresholding the wavelet coefficients of its noisy version using a universal threshold, $T = \sigma\sqrt{2\log N}$. The choice of this threshold is justified by using the following result [15]:

$$\lim_{N\to\infty} P\left(\max_{1\le i\le N} |v_i| > \sigma\sqrt{2\log_e N}\right) = 0, \tag{5.3.10}$$

where $\{v_i\}$ is a Gaussian stochastic process with zero mean and variance σ^2. Therefore, for large N the maximum amplitude of the zero-mean white noise with variance σ^2 has a high probability of being smaller than $T = \sigma\sqrt{2\log_e N}$ while remaining close to T.

5.3.1.1.3 SURE thresholding

The universal threshold is conservative, i.e., it rarely allows a truly zero signal to be estimated with a nonzero reconstruction. However, it may set a low signal value to zero. Better performance (in the L^2 norm) could be obtained by using smaller values for the threshold. An unbiased risk estimate is obtained by a method due to [16] called *Stein's unbiased risk estimate* (SURE) in [6]. For i.i.d. Gaussian noise, this method gives an unbiased estimate for the mean-squared soft-thresholding error for each choice of threshold T_j corresponding to the wavelet coefficients at level j. The threshold is computed based on the noisy wavelet data by minimizing the estimate with respect to the thresholds over $[0, \sigma\sqrt{2\log N}]$. This can be done because it is possible to minimize the risk in the wavelet domain which, in turn, minimizes the risk in the original domain. However, a drawback of this method is that it does not perform well when the wavelet representation at any level is sparse.

5.3.1.1.4 Wiener filtering

The WF gives a linear estimate of a signal embedded in additive noise by minimizing the mean-squared error (MSE). For zero-mean, white Gaussian noise where the signal is also zero-mean Gaussian with variance σ_f^2,

$$\tilde{x} = \frac{\sigma_f^2 \psi}{\sigma_f^2 + \sigma^2} \tag{5.3.11}$$

represents the estimate of the WF, and ψ is a generic Fourier coefficient of the signal. When applied to the subbands of the wavelet transform, the estimate is an approximation to the WF. Unlike the shrinkage estimators, no explicit thresholding is performed. Instead, the noisy wavelet coefficients are rescaled depending on the variance of the signal and noise. This requires a knowledge of both the signal and the noise statistics. The variance of the signal is computed as the difference between the empirical variance of the wavelet coefficients and $\hat{\sigma}^2$.

5.3.2 MPE

As described in Sections 5.2 and 5.3.1, the motivation for thresholding is that only the largest wavelet coefficients are needed to reconstruct the signal, and all coefficients smaller than the threshold are discarded. However, especially in the case of low SNRs, some of the coefficients that are below the threshold may make valuable contributions to the signal. We propose a moment-based thresholding scheme that is particularly useful in situations of this kind. As moments represent features of a signal [8], this scheme preserves features that may be discarded by other thresholding schemes described earlier. The wavelet coefficients at level j describe the signal localized to spatial positions near $t = n2^j$ and frequencies near 2^{-j}. Due to this localization property, any significant change in the function near t is reflected in the moments of these coefficients. For example, if the wavelet coefficients are large, then this means that there is a significant change in the function. In this case, we can expect the first moment to be nonzero and the second moment to be large. The following result further emphasizes the importance of moments.

Theorem 5.1 ([19]). *Let P be a probability measure on the line having finite moments $\alpha_k = \int_{-\infty}^{\infty} x^k P(dx)$ of all orders. If the power series $\sum_k \alpha_k r^k/k!$ has a positive radius of convergence, then P is the only probability measure with the moments $\alpha_1, \alpha_2, \ldots$*

Such a probability measure is said to be *determined by its moments*. For the standard normal distribution $|\alpha_k| \leq k!$, and so it is determined by its moments. If the additive noise is normal, then the noisy wavelet coefficients also possess a normal distribution. Since the normal distribution is described completely by its first two moments, it may be sufficient to preserve only these two moments. Hence, if the moments of the underlying signal are preserved during spatially adaptive de-noising, it could result in an improvement in the performance. For example, in applications like image de-noising and compression, it is known that the MSE is not perfectly correlated to the visual quality [8], whereas the moments are frequently used as key image features [17]. Moreover, it is known that the second moment of the output of a minimum mean-squared quantizer in general is less than that of the original signal [18]. The empirical moments, which are easy to compute, can be used as approximations to the actual moments. By preserving the empirical moments, the estimator becomes entirely data driven. Moreover, the convergence of the estimate in distribution can be studied by Parseval's relation and the following theorem [19].

Theorem 5.2. *Suppose that the distribution of a random variable X is determined by its moments, that the sequence of random variables $\{X_N\}$ have moments of all orders, and that* $\lim_N E[X_N^r] = E[X^r]$ *for* $r = 1, 2, \ldots$*. Then* $X_N \Longrightarrow X$.

Due to Parseval's relation, we know that if the estimated wavelet coefficients are fairly *close* to the original wavelet coefficients, then the estimated function should be close to the original function. Therefore, if the conditions of Theorem 5.2 are satisfied, an estimator that preserves all the moments of the underlying signal will

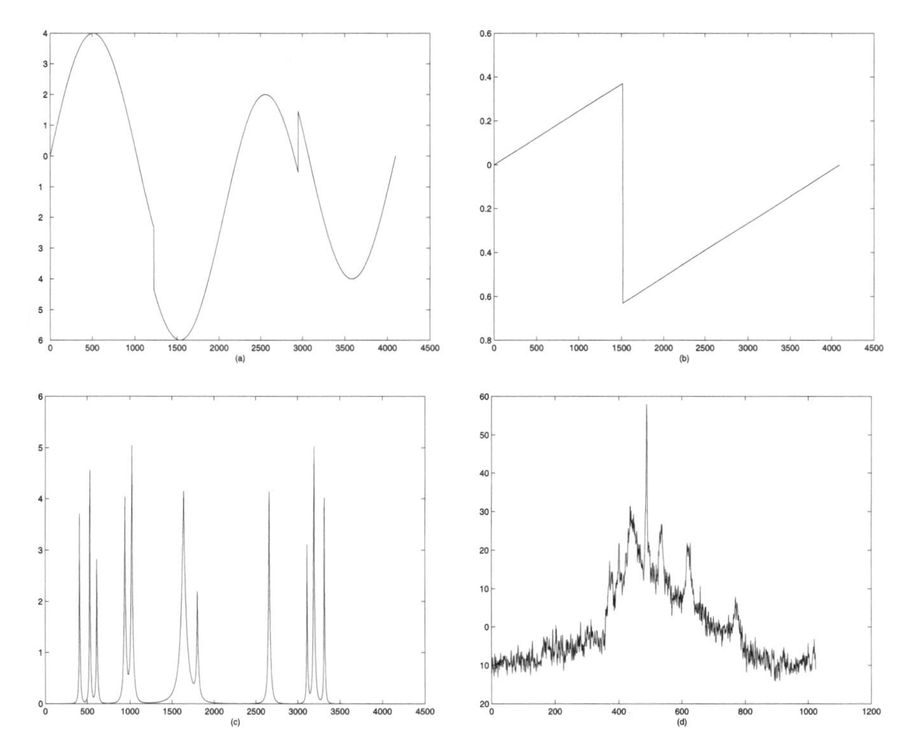

FIGURE 5.1. Original signals, (a) Heavisine, (b) Ramp, (c) Bumps, (d) nuclear magnetic resonance (NMR).

converge in distribution when the number of signal samples increases. A sufficient condition for a probability distribution of a continuous random variable X to be characterized by its moments (Hamburger moment problem) for an infinite interval is given by

$$\sum_{n=1}^{\infty} \left(\inf_{i \leq n} (E[X^{2i}])^{\frac{1}{2i}} \right)^{-1} = \infty. \tag{5.3.12}$$

Since the wavelet coefficients usually have a Laplace or a normal distribution [20] for which the preceding condition holds, their distributions are characterized by the moments.

Our experiments show that the coefficients exhibit a distribution close to Laplacian or normal even when the original signal has a skewed distribution. As an example, consider the Bumps signal in Figure 5.1 whose histogram is shown in Figure 5.2. Figure 5.3 gives the histogram of the wavelet coefficients at the 6th level of the decomposition for this signal. We see that it is close to a Laplace-like probability distribution. Figure 5.4 in Subsection 5.3.3.1 gives the MPE for this case. As the normal and Laplacian distributions are completely determined by their first two

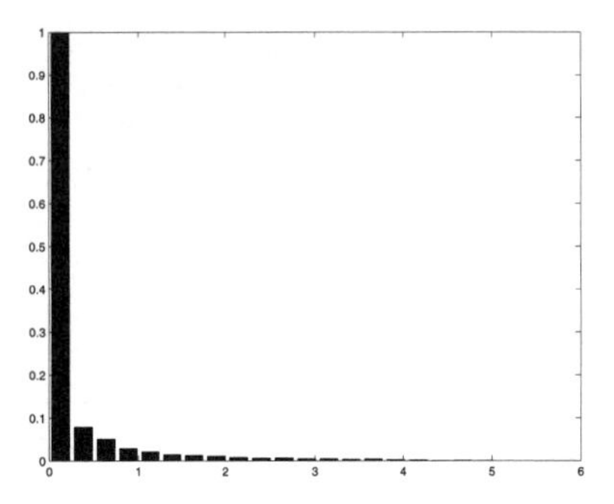

FIGURE 5.2. Histogram of Bumps.

moments, it appears reasonable to base the quantization in these cases to preserve just the first two moments. Our simulations for the MPE confirm this. However, for cases where the probability distribution of the wavelet coefficients are not Laplacian or normal, it may be necessary to preserve the higher order moments. We now derive the quantization parameters for a general m-moment preserving quantizer.

Applying Mallat's wavelet decomposition algorithm on the set of noisy observations, the wavelet coefficients at various levels are obtained. Let $S_j = \{x_{j,i}\}$, $i = 1, \dots, N_j$, where $N_j = N/2^j$, denote the set of wavelet coefficients at the jth level of the decomposition. The quantizer is designed such that the first m moments of the underlying (noiseless) signal are preserved in the de-noised version. For fixed thresholds $\{T_i\}_{i=0}^{m}$, where $T_0 = -\infty$ and $T_m = \infty$, the quantization function $Q(x) : \Re \to \Re$ is defined as

$$Q(x) = a_{j,k} \text{ if } x \in (T_{k-1}, T_k], \qquad k = 1, \dots, m, \tag{5.3.13}$$

where x denotes a generic wavelet coefficient and $\{a_{j,k}\}$ are the representation levels of the quantizer. If $q_{j,k} = \#\{x_{j,i} \in S_j \text{ s.t. } x_{j,i} \in (T_{k-1}, T_k]\}$, $k = 1, \dots, m$, then for the first m moments of the signal to be preserved in the output of the quantizer at the jth level, we solve the following system of equations written in matrix form:

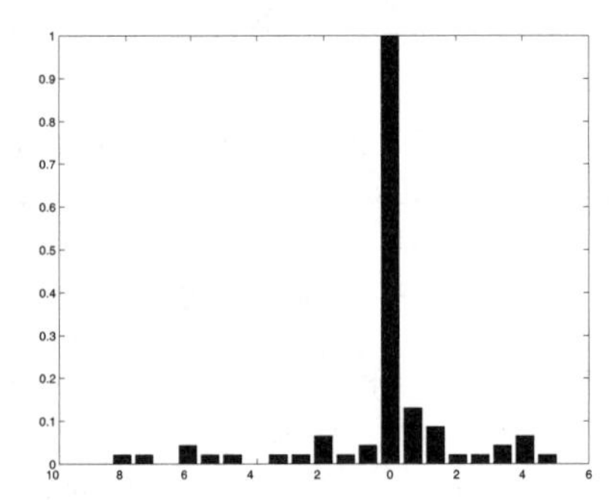

FIGURE 5.3. Histogram of the wavelet coefficients at level 6 for the Bumps signal.

$$(\hat{\mu}^j)^t = \frac{1}{N_j}\mathbf{q}_j{}^t\mathbf{A} \tag{5.3.14}$$

where

$$\hat{\mu}^j = \begin{bmatrix} \hat{\mu}_1^j \\ \hat{\mu}_2^j \\ \vdots \\ \hat{\mu}_m^j \end{bmatrix}, \quad \mathbf{q}_j = \begin{bmatrix} q_{j,1} \\ q_{j,2} \\ \vdots \\ q_{j,m} \end{bmatrix}, \quad \text{and} \quad \mathbf{A} = \begin{bmatrix} a_{j,1} & a_{j,1}^2 & \cdots & a_{j,1}^m \\ a_{j,2} & a_{j,2}^2 & \cdots & a_{j,2}^m \\ \vdots & \vdots & & \vdots \\ a_{j,m} & a_{j,m}^2 & \cdots & a_{j,m}^m \end{bmatrix} \tag{5.3.15}$$

to obtain the values of $\{a_{j,i}\}_{i=1}^m$. The superscript t denotes the transpose of the matrix. $\hat{\mu}_i^j$, $i = 1, 2, \ldots, m$, $j = 1, 2, \ldots, J$ are estimated empirically from the data. We obtain these equations because the right side of this system of equations consists of the empirical moments of the quantized coefficients, and the left side corresponds the empirical estimates of the moments of the underlying signal. To show that $T_0 < a_{.,1} \leq T_1 \leq a_{.,2} \leq T_2 \ldots T_{m-1} \leq a_{.,m} < T_m$, we follow an approach similar to that of [21]. Let w_j denote a generic wavelet coefficient of the signal at level j. Then, at level j, $j = 1, 2, \ldots, J$, the kth moment, $k = 1, 2, \ldots, m$ of the signal is given by

$$\mu_k^j = \int_{-\infty}^{\infty} w_j^k dF(w) = \sum_{i=1}^{m} a_{j,i}^k P_i, \tag{5.3.16}$$

where $F(w)$ is the probability distribution of wavelet coefficients corresponding to the signal. The integral in Equation (5.3.16) is a Lebesgue–Stieltjes integral and $P_i = F(T_i) - F(T_{i-1})$. Eq. (5.3.16) is a form of the Gauss–Jacobi mechanical quadrature [22]. The representation levels $a_{j,i}$ of an m-level moment preserving quantizer are the zeros of the mth degree polynomial associated with $F(w)$. The P_i are Christoffel numbers and the T_k and $a_{j,k}$ alternate by the separation theorem of Chebyshev–Markov–Stieltjes [22].

5.3.2.1 Moment-based hard quantization

In this subsection we derive the moment preserving estimate of the noisy signal by the hard quantization of its wavelet coefficients. When the first two moments of the underlying signal are preserved, i.e., for $m = 2$, the hard quantization function $Q_h(x) : \Re \rightarrow \Re$ is given by

$$Q_h(x) = \begin{cases} a_{j,1} & \text{if } x \leq T_1 \\ a_{j,2} & \text{if } x > T_1, \end{cases} \tag{5.3.17}$$

where x denotes a generic coefficient. For the first two moments to be preserved, we have from Eq. (5.3.14)

$$N_j \hat{\mu}_1^j = q_{j,1}a_{j,1} + (N_j - q_{j,1})a_{j,2} \tag{5.3.18}$$

$$N_j\hat{\mu}_2^j = q_{j,1}a_{j,1}^2 + (N_j - q_{j,1})a_{j,2}^2, \tag{5.3.19}$$

where

$$\hat{\mu}_1^j = \frac{1}{N_j}\sum_{i=1}^{N_j} x_{j,i},$$
$$\hat{\mu}_2^j = \frac{1}{N_j}\sum_{i=1}^{N_j} x_{j,i}^2 - (\hat{\sigma})^2. \tag{5.3.20}$$

$\hat{\sigma}^2$ is the estimated variance of the noise computed using Eq. (5.3.7). The solution to these equations gives

$$a_{j,1} = \hat{\mu}_1^j - \tilde{\sigma}_j\sqrt{\frac{N - q_{j,1}}{q_{j,1}}} \tag{5.3.21}$$

$$a_{j,2} = \hat{\mu}_1^j + \tilde{\sigma}_j\sqrt{\frac{q_{j,1}}{N - q_{j,1}}}, \tag{5.3.22}$$

where $\tilde{\sigma}_j = \sqrt{\hat{\mu}_2^j - (\hat{\mu}_1^j)^2}$ denotes the estimated variance of the noiseless signal. Note that the threshold T_1 is a parameter of our choice. It can be chosen in such a way that the estimate is compatible with the existing threshold-based estimators.

5.3.2.2 Moment-based soft quantization

We now derive the parameters of the soft threshold estimator that preserves the first two moments. When we do not make a hard decision on the wavelet coefficients (quantized to one of two levels), we call the method *soft* quantization. In order to mimic the nonlinear soft thresholding operation that has been widely used in wavelet based de-noising [23], [24], we choose $a_{j,2} = 0, j = 1, 2, \ldots, J$ *a priori*. Let $a_{j,1} = -b$, $a_{j,3} = a$, $T_1 = -\tilde{T}$, and $T_3 = \tilde{T}$, where $\tilde{T} > 0$. Then the soft quantization function $Q_s : \Re \rightarrow \Re$ quantizes the jth-level wavelet coefficients using the following function:

$$Q_s(x) = \begin{cases} a & \text{if } x > \tilde{T} \\ 0 & \text{if } |x| \leq \tilde{T} \\ -b & \text{if } x < -\tilde{T}, \end{cases} \tag{5.3.23}$$

where x is a general wavelet coefficient. If $q_{j,1} = \#\{x_{j,i} \in S_j \text{ s.t. } x < -T_3\}$ and $q_{j,3} = \#\{x_{j,i} \in S_j \text{ s.t. } x > T_3\}$, then for the two moments to be preserved at the output of the quantizer we must have

$$N_j\hat{\mu}_1^j = q_{j,3}a - q_{j,1}b \tag{5.3.24}$$
$$N_j\hat{\mu}_2^j = q_{j,3}a^2 + q_{j,1}b^2. \tag{5.3.25}$$

This implies

$$b = \frac{-q_{j,1} N_j \hat{\mu}_1^j + \sqrt{q_{j,1} q_{j,3} N_j \hat{\mu}_2^j (q_{j,1} + q_{j,3}) - q_{j,1} q_{j,3} N_j^2 (\hat{\mu}_1^j)^2}}{q_{j,1}(q_{j,1} + q_{j,3})}$$

$$a = \frac{N_j \hat{\mu}_1^j}{q_{j,3}} - \frac{q_{j,1} b}{q_{j,3}}. \quad (5.3.26)$$

As in the previous case, the threshold $\tilde{T}$ can be chosen to be compatible with other estimation schemes.

Moment preserving wavelet de-noising is obtained by quantizing the noisy wavelet coefficients using the hard and soft quantizers. The quantized wavelet coefficients preserve the first two moments of the noiseless wavelet coefficients.

5.3.3 *Performance analysis*

We now compare the performance of the various estimators described in the previous sections in terms of SNR of the reconstructed (de-noised) signal (RSNR), robustness to the inaccuracies in the parameter estimates, and compression ratio. We use standard simulated data and one real-life data set for the experiments. Figure 5.1 shows the original signals generated using WaveLab [26] in which Figure 5.1(d) is real-life NMR data [25]. The first three signals are noiseless whereas the real-life NMR data is noisy. For the experiments we synthetically added noise to the noiseless, signals. The input SNR is defined as

$$SNR = \frac{1}{N\sigma^2} \sum_{i=1}^{N} f^2(t_i) \quad (5.3.27)$$

and, if $\hat{f}(t)$ denotes the estimate of $f(t)$, then the SNR after de-noising the signal is given by

$$RSNR = \frac{\sum_{i=1}^{N} f^2(t_i)}{\sum_{i=1}^{N} (\hat{f}(t_i) - f(t_i))^2}. \quad (5.3.28)$$

For all the experiments we use the compactly supported mother wavelet D_8 from the Daubechies family followed by decimation after each level of decomposition. Similar performance was also obtained with various other filters. The original signal is corrupted with white Gaussian noise with variance σ^2. In all cases the universal threshold was used in the MPE. SNRs that range from low to very low (ranging from 1 to 5) are studied because in many practical situations the signal is *buried* in noise [8]. All the results presented are for a sample size $N = 4096$, averaged over 20 runs of the simulation, and using eight levels of wavelet decomposition. The coarse level approximation coefficients were not de-noised. For other standard signals generated by WaveLab [26] we find that the estimators behave similarly.

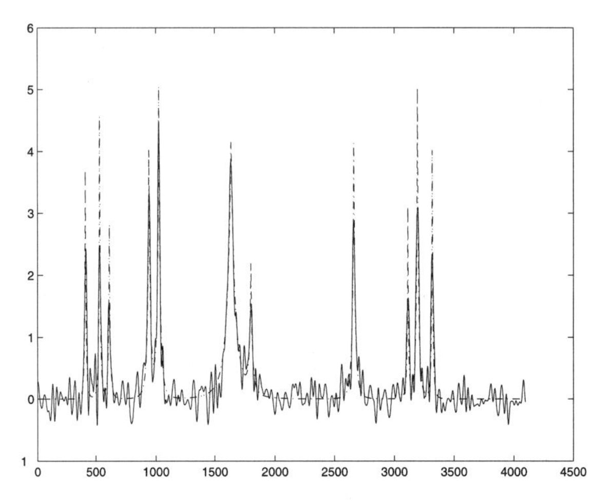

FIGURE 5.4. De-noised Bumps signal using MPE with hard quantization when SNR = 1.

5.3.3.1 SNR of the de-noised signal

Before we present graphically the results from the various estimation schemes under comparison, we illustrate that coefficient distributions that are not exactly Laplacian do not appear to present a problem. Consider the Bumps signal in Figure 5.1. Figure 5.4 shows the original Bumps signal (represented by a solid line) and its de-noised version (dotted line). We observe from this figure that the proposed moment-based estimator produces an acceptable estimate using the noisy Bumps signal, even though its coefficients do not have an exact Laplace distribution.

Figure 5.5 and Figure 5.6 show the de-noised signals for hard and soft thresholding, respectively, when the SNR is 1. Clearly, the MPE and UTE produce visually pleasing estimates of the original signal compared to the other methods. The UTE smoothes the data considerably. Visually, however, both the estimates are similar and are quite close to the original. The SURE and ME do not perform well under this low SNR condition. Large spikes are seen in the estimates. The Wiener filtered signal does not have as large spikes, but it is not smooth. Figure 5.7 is the result of de-noising real-life biomedical data using the MPE. Visual inspection shows that the MPE produces acceptable performance for this real-life data set also.

The numerical values of the RSNR for Ramp and Heavisine are shown in Figures 5.8 and 5.9. It is clear that the MPE and UTE outperform the other estimators. They produce a higher RSNR and hence a smaller MSE. Also, the difference in performance between MPE and UTE is only marginal.

5.3.3.2 Robustness

Since, the MPE and UTE give better de-noised signals in terms of RSNR and visual quality, we compare the robustness of these methods. The threshold estimates

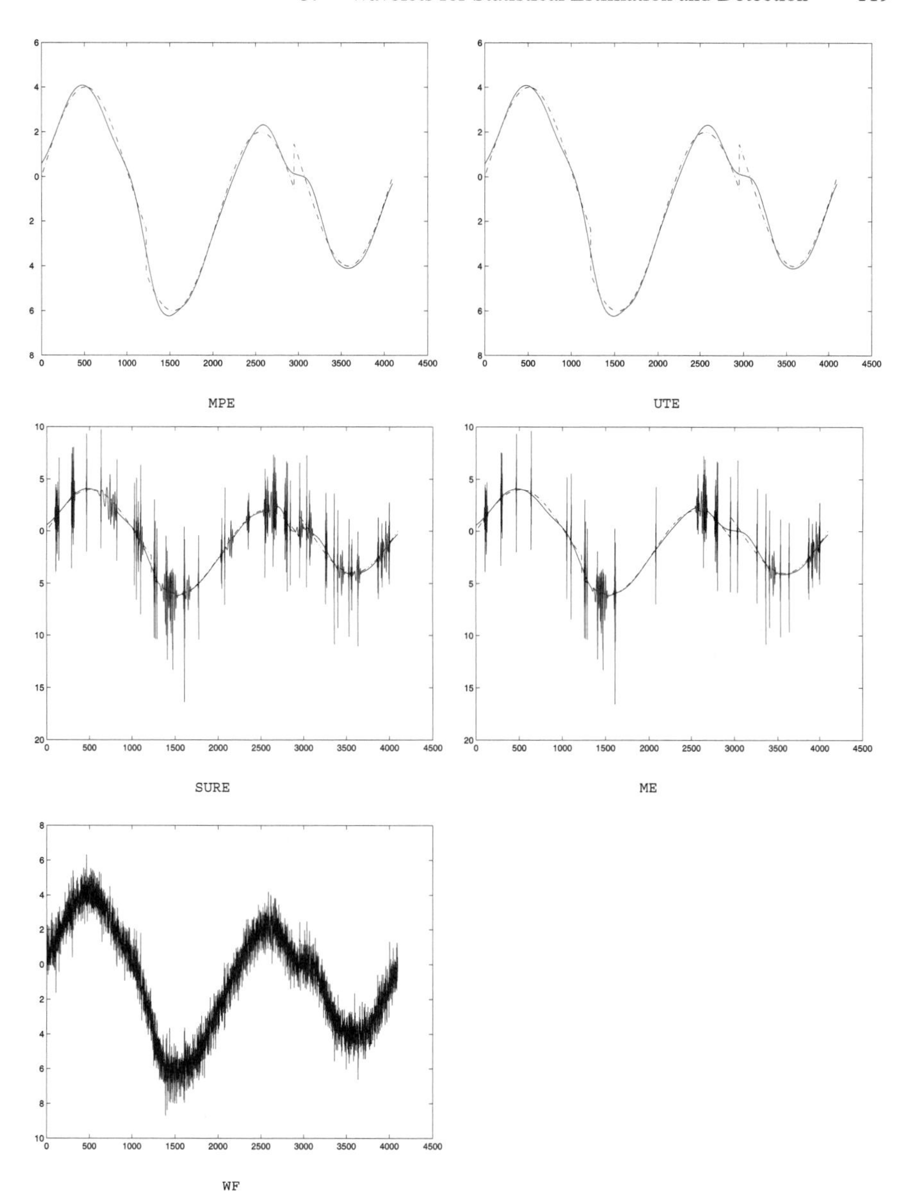

FIGURE 5.5. De-noised Heavisine signal using MPE, UTE, SURE, and ME with hard quantization and the WF when SNR = 1. Solid line represents the original signal; dotted line represents the de-noised version.

depend on the accuracy with which the various parameters of the signal and noise are estimated. For example, the estimated noise variance $\hat{\sigma}^2$ may fail to give a good estimate under certain circumstances. This could lead to considerable loss in qual-

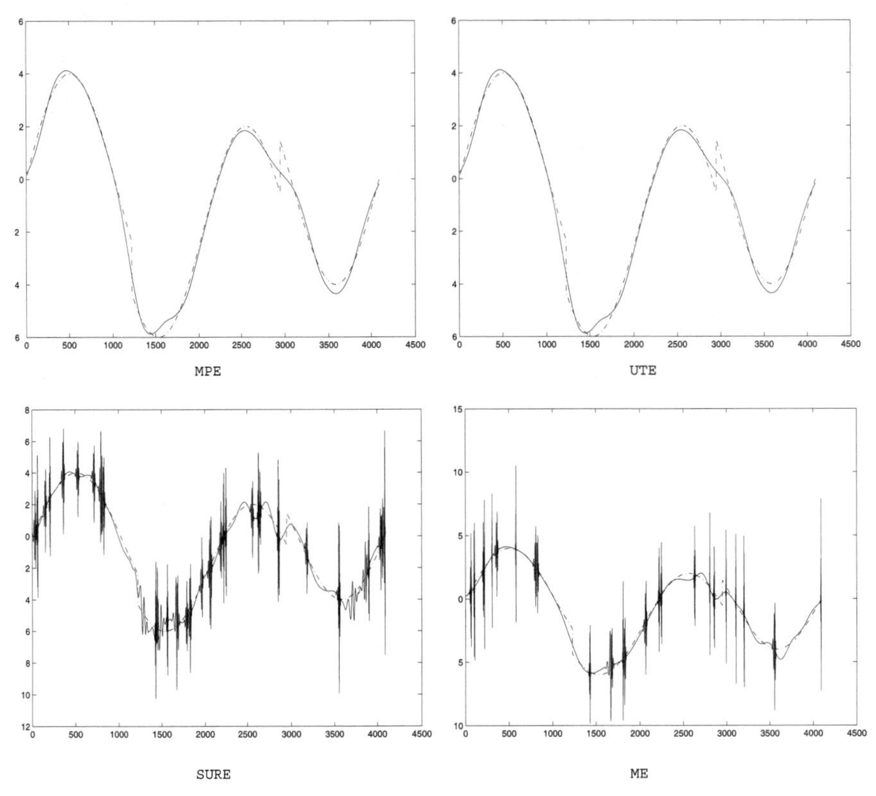

FIGURE 5.6. De-noised Heavisine signal using the estimators with soft quantization when SNR = 1. Solid line represents the original signal; dotted line represents the de-noised version.

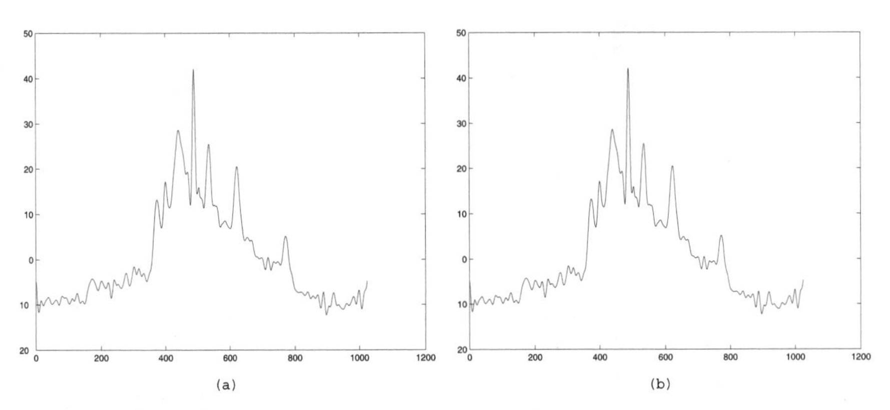

FIGURE 5.7. De-noised NMR signal using MPE: (a) hard quantization, (b) soft quantization.

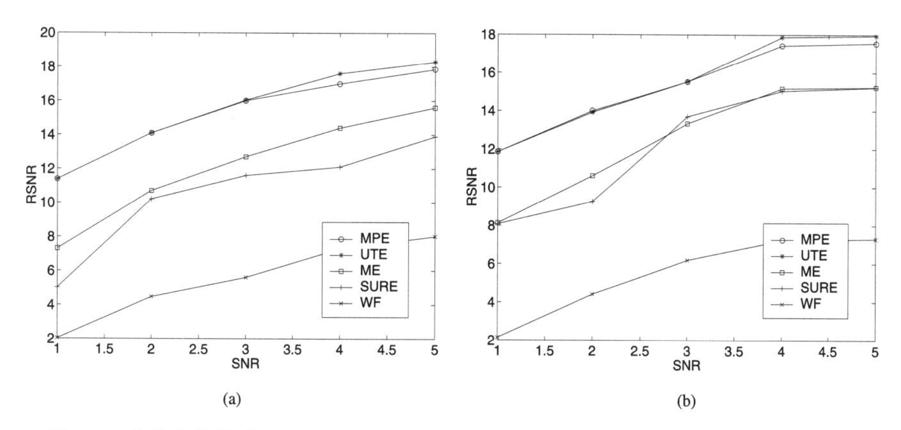

FIGURE 5.8. RSNR for Ramp signal using (a) hard quantization, (b) soft quantization.

ity of the estimated signal. Therefore, it is desirable that the estimators be robust to such errors. In order to study this, the following estimates for the noise variance were used: $\hat{\sigma} = k\sigma$, $k = 0.2, 0.4, \ldots, 2$. This enables a study of the effect of both under- and overestimates of the variance. From Figures 5.10 and 5.11 it is seen that the MPE is highly robust to such changes, unlike the UTE. Almost a constant RSNR is maintained by the MPE. This is due to the fact that in the UTE the wavelet coefficients are thresholded based only on $\hat{\sigma}$. Any error in the estimation of this parameter leads to a degradation in the de-noising. However, the MPE uses not only the estimated variance but also the moments of the underlying signal. This adds robustness to its estimates.

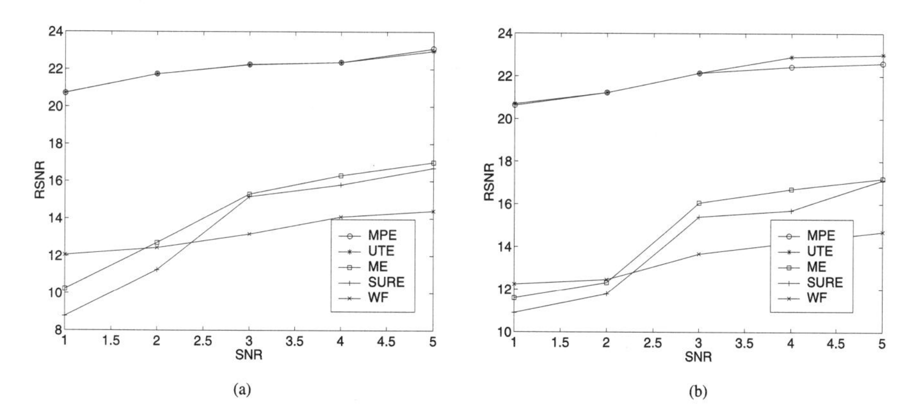

FIGURE 5.9. RSNR for Heavisine using (a) hard quantization, (b) soft quantization.

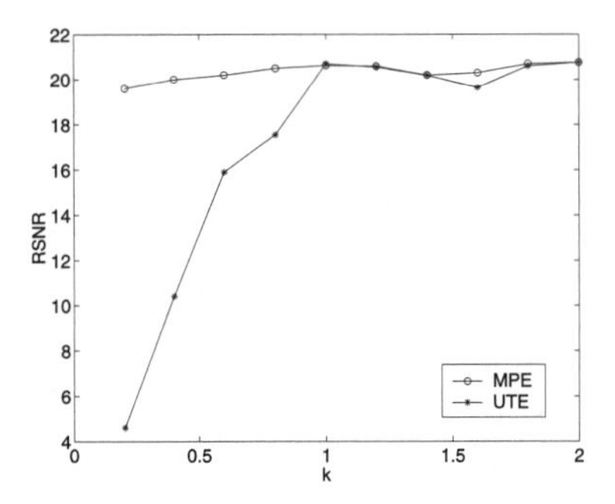

FIGURE 5.10. Comparison of the robustness of the MPE and UTE for hard quantization.

5.3.3.3 Compression ratio

Compression is critical in memory intensive applications such as medical images. We define the compression as

$$\text{Compression ratio} = \frac{\#\ \text{bits to represent original signal}}{\#\ \text{bits to represent the de-noised signal}}. \tag{5.3.29}$$

The higher the compression ratio, the smaller the number of bits required to store the de-noised signal. We used a 32-bit representation for the unquantized wavelet coefficients in our implementation. The low-pass coefficients were not quantized. Therefore, the total number of bits required to store the wavelet coefficients is equal to the sum of the total number of bits to store the coarse level approximation coefficients and the detail coefficients.

Tables 5.1 and 5.2 give the compression ratios achieved using the MPE and UTE for hard and soft quantization, respectively. For the MPE with hard quantization we use 1 bit/coefficient for the detail coefficients and for the soft quantization we use 2 bits/coefficient. This is evident from the fact that we have two representation levels for the hard quantizer and three for the soft quantizer. The compression ratio

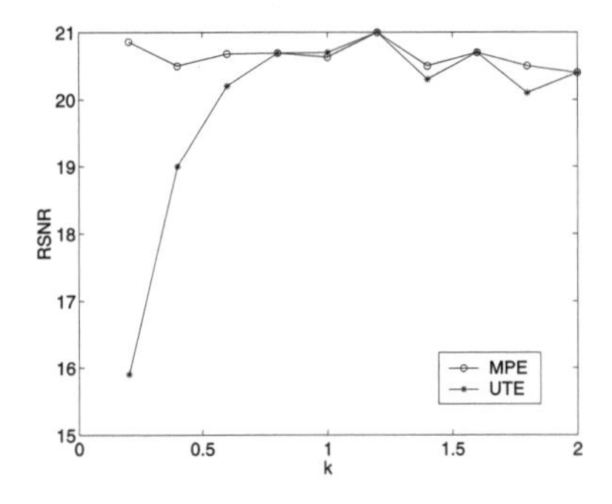

FIGURE 5.11. Comparison of the robustness of the MPE and UTE for soft quantization.

SNR	MPE	UTE
1	28.54	4.39
2	"	3.90
3	"	3.42
4	"	3.01
5	"	2.56

TABLE 5.1. Compression ratio for Heavisine using hard quantization.

SNR	MPE	UTE
1	15.11	1.7
2	"	1.57
3	"	1.52
4	"	1.48
5	"	1.41

TABLE 5.2. Compression ratio for Heavisine using soft quantization.

while using UTE is determined by the number of detail coefficients that survive the thresholding. A similar idea called the *dead-zone limiter* [8] is used in data compression. Here, coefficients that do not have significantly high magnitude are zeroed out to achieve higher compression ratios. We observe from the tables that, as the SNR increases the noise level decreases. Therefore, many coefficients pass the universal threshold test, leading to a reduction in the compression ratio. We see from the tables that the MPE nearly achieves a compression ratio 7 times higher than that of the UTE for hard quantization and 8 times higher for soft quantization for low SNR. Note that the MPE uses explicit quantization, while the UTE does not. This comparison points out that through this explicit quantization the MPE produces almost the same RSNR with fewer bits to store the de-noised signal.

5.4 Hypothesis testing

This section deals with the problem of detecting a known, deterministic signal buried in noise using a hypothesis testing procedure. Assume that there are two possible hypotheses H_0 and H_1 corresponding to two possible probability distributions P_0 and P_1, respectively, on the measurable space $(\Re^N, \mathcal{B}^N)$; where $\mathcal{B}^N$ is the Borel σ-algebra over $\Re^N$. The hypothesis test is defined as the statistical test between

$$
\begin{aligned}
H_1 &: r_i = \sqrt{E_0} s_i + \sigma z_i \\
H_0 &: r_i = \sigma z_i, \quad i = 1, \dots, N,
\end{aligned}
\tag{5.4.1}
$$

where the random vector $\mathbf{r} = (r_1, r_2, \ldots, r_N)$ denotes the noisy observations used by the statistical test, $\mathbf{s} = (s_1, s_2, \ldots, s_N)$ stands for the vector of samples of the deterministic signal, and $\mathbf{z} = (z_1, z_2, \ldots, z_N)$ is the random vector consisting of samples of a stationary, zero-mean, unit variance white Gaussian noise. In Eq. (5.4.1), $\sqrt{E_0}$ stands for the SNR parameter. Further, we assume that the *normalization condition* holds, namely, $\sum_{i=1}^{N} s_i^2 = 1$. Clearly, the noise distribution is determined by the marginal density $p \sim \mathcal{N}(0, \sigma^2)$ on $\Re^N$. Suppose that the signal is not known; then the procedure discussed previously can be used to first estimate the signal, followed by detection. We also assume N to be a power of 2, say, $N = 2^J$ for an integer J, for the sake of simplicity. Taking the wavelet transform of the observation vector, we then get

$$
\begin{aligned}
H_1 &: \bar{r}_{j,i} = \sqrt{E_0}\bar{s}_{j,i} + \sigma\eta_{j,i} \\
H_0 &: \bar{r}_{j,i} = \sigma\eta_{j,i}, \quad i = 1, 2, \ldots, J; \; i = 1, 2, \ldots, N_j = N/2^j
\end{aligned}
\tag{5.4.2}
$$

Then at level j of the wavelet decomposition the log-likelihood ratio is given by

$$
\begin{aligned}
l_j(\bar{r}_{j,1}, \bar{r}_{j,2}, \ldots, \bar{r}_{j,N_j}) &= \log \prod_{i=1}^{N_j} \frac{p(\bar{r}_{j,i} - \sqrt{E_0}\bar{s}_{j,i})}{p(\bar{r}_{j,i})} \\
&= \log \prod_{i=1}^{N_j} \frac{\frac{1}{2\sigma^2} e^{-\left(\bar{r}_{j,i} - \sqrt{E_0}\bar{s}_{j,i}\right)^2/2\sigma^2}}{\frac{1}{2\sigma^2} e^{-\bar{r}_{j,i}^2/2\sigma^2}} \\
&= \frac{\sqrt{E_0}}{\sigma^2} \sum_{i=1}^{N_j} \bar{r}_{j,i}\bar{s}_{j,i} - \frac{E_0}{2\sigma^2} \sum_{i=1}^{N_j} \bar{s}_{j,i}^2.
\end{aligned}
\tag{5.4.3}
$$

The *decision function* $\phi_j : \Re \to \{0, 1\}$ measurable with respect to the σ-algebra generated by $\{\bar{r}_{j,1}, \bar{r}_{j,2}, \ldots, \bar{r}_{j,N_j}\}$ gives the final decision of the hypothesis test at level j where H_0 and H_1 are denoted by $\{0, 1\}$ in that order. We define the false alarm probability as $\alpha = P_0(\text{decide } H_1)$ and the miss probability to be $\beta = P_1(\text{decide } H_0)$. The false alarm probability constraint is assumed to be the same for all j. Performing a likelihood ratio test at each level j for the local decision, we get

$$
L_j = \frac{1}{\sigma^2} \sum_{i=1}^{N_j} \bar{r}_{j,i}\bar{s}_{j,i}
\begin{cases}
\geq & T_j \text{ decide } H_1 \\
< & T_j \text{ decide } H_0
\end{cases},
\tag{5.4.4}
$$

where T_j is a decision threshold. This test statistic can also be rewritten as $\phi_j = 1(L_j \geq T_j)$, where $1(\cdot)$ is the indicator function. The conditional means of the sufficient statistic L_j are given by

$$
\begin{aligned}
\tilde{\mu}_{j,1} &= E[L_j | H_1] \\
&= \frac{1}{\sigma^2} E\left[\sum_{i=1}^{N_j} \bar{r}_{j,i}\bar{s}_{j,i} | H_1\right]
\end{aligned}
$$

$$
\begin{aligned}
&= \frac{1}{\sigma^2}\sum_{i=1}^{N_j} E\left[\sqrt{E_0}\bar{s}_{j,i}^2 + \bar{s}_{j,i}\eta_{j,i}\right] \\
&= \frac{\sqrt{E_0}}{\sigma^2}\sum_{i=1}^{N_j} \bar{s}_{j,i}^2 \qquad (5.4.5)\\
\tilde{\mu}_{j,0} &= E[L_j|H_0] \\
&= \frac{1}{\sigma^2}\sum_{i=1}^{N_j} E\left[\eta_{j,i}\bar{s}_{j,i}\right] \\
&= 0, \qquad (5.4.6)
\end{aligned}
$$

and the conditional variances are

$$
\begin{aligned}
\tilde{\sigma}_{j,1}^2 &= \mathrm{Var}[L_j|H_1] \\
&= \frac{1}{\sigma^4}\sum_{i=1}^{N_j} \mathrm{Var}\left[\sqrt{E_0}\bar{s}_{j,i}^2 + \bar{s}_{j,i}\right] \qquad (5.4.7)\\
&= \frac{1}{\sigma^2}\sum_{i=1}^{N_j} \bar{s}_{j,i}^2 \\
\tilde{\sigma}_{j,0}^2 &= \mathrm{Var}[L_j|H_0] \\
&= \frac{1}{\sigma^2}\sum_{i=1}^{N_j} \bar{s}_{j,i}^2. \qquad (5.4.8)
\end{aligned}
$$

Hence, let $\tilde{\sigma}_{j,1}^2 = \tilde{\sigma}_{j,0}^2 = \tilde{\sigma}_j$. We also note that $L_j \sim N(\tilde{\mu}_j, \tilde{\sigma}_j^2)$ under both H_1 and H_0. Further, we assume that $\{L_j\}_{j=1}^J$ are independent due to the wavelet decomposition. If the noise standard deviation is estimated using Eq. (5.3.7) and used for all the decomposition levels, then it is easy to see that $T_j = T\ \forall j$ when α is the same for all levels.

Once the detector outputs are known at every level of the wavelet decomposition, the next task is to combine all these decisions for one global decision about the true hypothesis. It is immediately clear that this is a combinatorial problem because for J levels of decomposition there are 2^J possible ways of combining the local decision at each of the J levels to obtain a global decision. Therefore, we resort to two (intuitive) heuristic decision fusion rules that can combine the decisions $\{\phi_j\}_{j=1}^J$ into a global decision. The first is the OR rule, and the second one is the AND rule. We discuss these two fusion rules next.

5.4.1 *OR decision fusion*

In the OR rule, the global decision is H_1 if it is accepted by at least one ϕ_j, $j = 1, 2, \ldots, J$, i.e., the global decision, $\phi_{\mathrm{OR}}^* = 1(\bigcup_j [\phi_j = 1])$. The global false alarm probability is therefore given by

$$\alpha_{\mathrm{OR}} = 1 - P_0(\phi_1 = 0, \phi_2 = 0, \ldots, \phi_J = 0)$$
$$= 1 - \prod_{j=1}^{J} P_0(\phi_j = 0) \ \text{(due to independence)}. \tag{5.4.9}$$

Let $f_{j,k}$ denote the probability density function (pdf) of L_j conditioned on hypothesis H_k, $k = 0, 1$. Then,

$$P_0(\phi_j = 0) = \int_{-\infty}^{T} f_{j,0}(l)dl$$
$$= \int_{-\infty}^{T} \frac{1}{\sqrt{2\pi\tilde{\sigma}_j^2}} e^{-l^2/2\tilde{\sigma}_j^2} dl$$
$$= \frac{1}{2} + \frac{1}{2} erf\left[\frac{T}{\sqrt{2}\tilde{\sigma}_j}\right], \tag{5.4.10}$$

where $erf(x) = (2/\sqrt{\pi}) \int_0^x e^{-t^2} dt$. Therefore, from Eq. (5.4.9) we get

$$\alpha_{\mathrm{OR}} = 1 - \prod_{j=1}^{J} \left[\frac{1}{2} + \frac{1}{2} erf\left[\frac{T}{\sqrt{2}\tilde{\sigma}_j}\right]\right]. \tag{5.4.11}$$

5.4.1.1 Choice of decision threshold (T)

Solving Eq. (5.4.11) to get a closed-form solution for the optimal value of T is difficult. Therefore, we use approximations to solve for T. The trade-off as a result of the approximations is also addressed. Due to the normalization condition we see that $\sum_{j=1}^{J} \sum_{i=1}^{N_j} \bar{s}_{j,i}^2 = 1$. Hence, from Eq. (5.4.8) we observe that

$$\tilde{\sigma}_j^2 = \frac{1}{\sigma^2} \sum_{i=1}^{N_j} \bar{s}_{j,i}^2$$
$$\leq \frac{1}{\sigma^2}. \tag{5.4.12}$$

We then approximate $\tilde{\sigma}_j^2$ by σ^2 ($\leq 1/\tilde{\sigma}_j^2$) in Eq. (5.4.11). Using the fact that the error function $erf(\cdot)$ is an increasing function, we see that the new false alarm probability is, say, $\tilde{\alpha}_{\mathrm{OR}} \geq \alpha_{\mathrm{OR}}$. So, it may seem that this approximation resulting in a higher false alarm probability is not useful. However, we will show below that there is an increase in the power (detection probability) of the test corresponding to the loss in the false alarm. Since the power of the test is more important in many applications, it is worthwhile to explore this way of computing the decision threshold. Now,

$$\alpha_{\mathrm{OR}} \leq 1 - \prod_{j=1}^{J} \left[\frac{1}{2} + \frac{1}{2} erf\left[\frac{\sigma T}{\sqrt{2}}\right]\right] \Rightarrow \left[\frac{1}{2} + \frac{1}{2} erf\left[\frac{\sigma T}{\sqrt{2}}\right]\right]^J \leq 1 - \alpha_{\mathrm{OR}}$$

$$\Rightarrow T \leq \frac{\sqrt{2}}{\sigma} erf^{-1}[2(1-\alpha_{\mathrm{OR}})^{1/J} - 1]. \tag{5.4.13}$$

This inequality can be taken as an equality to choose the decision threshold, T. It is natural to study the effect of this approximation on the power of the hypothesis test. Since the power of the test is equal to $1 - \beta_{\mathrm{OR}}$, we obtain

$$\begin{aligned} 1 - \beta_{\mathrm{OR}} &= 1 - P_1\left(\bigcap_j [\phi_j = 0]\right) \\ &= 1 - \prod_{j=1}^{J} P_1(\phi_j = 0) \qquad \text{(due to independence)} \\ &= 1 - \prod_{j=1}^{J}\left[\frac{1}{2} + \frac{1}{2} erf\left[\frac{T - \tilde{\mu}_{j,1}}{\sqrt{2}\tilde{\sigma}_j}\right]\right]. \end{aligned} \tag{5.4.14}$$

Using the approximation $\sigma^2 \leq 1/\tilde{\sigma}_j^2$ in this equation and calling the resulting power $1 - \tilde{\beta}_{\mathrm{OR}}$, we see that $1 - \beta_{\mathrm{OR}} \leq 1 - \tilde{\beta}_{\mathrm{OR}}$ due to the increasing nature of the error function. Therefore, the power of the test increases by the choice of T given in Eq. (5.4.13).

5.4.2 AND decision fusion

The AND fusion is the second fusion rule that we investigate to make a global decision regarding the true hypothesis. Here, we accept H_1 only if it is accepted at all the levels, $j = 1, 2, \ldots, J$. This can be written as $\phi^*_{\mathrm{AND}} = 1(\bigcap_j [\phi_j = 1])$. The global false alarm probability can then be computed as

$$\begin{aligned} \alpha_{\mathrm{AND}} &= P_0(\phi_1 = 1, \phi_2 = 1, \ldots, \phi_J = 1) \\ &= \prod_{j=1}^{J} P_0(\phi_j = 1) \\ &= \prod_{j=1}^{J}\left[\frac{1}{2} - \frac{1}{2} erf\left[\frac{T}{\sqrt{2}\tilde{\sigma}_j}\right]\right]. \end{aligned} \tag{5.4.15}$$

Again, substituting $\sigma^2 \leq 1/\tilde{\sigma}_j$ in this equation, we see that the resulting false alarm satisfies $\tilde{\alpha}_{\mathrm{AND}} \leq \alpha_{\mathrm{AND}}$. Therefore,

$$\alpha_{\mathrm{AND}} \leq \left[\frac{1}{2} - \frac{1}{2} erf\left[\frac{\sigma T}{\sqrt{2}}\right]\right]^J \Rightarrow T \leq \frac{\sqrt{2}}{\sigma} erf^{-1}\left[1 - 2\alpha_{\mathrm{AND}}^{1/J}\right]. \tag{5.4.16}$$

Thus, T can be computed by considering the preceding inequality to be an equality. Now, the power of the test is given by

$$\begin{aligned}
1-\beta_{\text{AND}} &= P_1(\phi_1=1,\phi_2=1,\ldots,\phi_J=1) \\
&= \prod_{j=1}^{J} P_1(\phi_j=1) \\
&= \prod_{j=1}^{J} [1-P_1(\phi_j=0)],
\end{aligned} \tag{5.4.17}$$

where

$$\begin{aligned}
P_1(\phi_j=0) &= \int_{-\infty}^{T} f_{j,1}(l)dl \\
&= \frac{1}{2}+\frac{1}{2}erf\left[\frac{T-\tilde{\mu}_{j,1}}{\sqrt{2}\tilde{\sigma}_j}\right].
\end{aligned} \tag{5.4.18}$$

By using Eq. (5.4.17), Eq. (5.4.18), and $\sigma \leq 1/\tilde{\sigma}_j$, we obtain

$$\begin{aligned}
1-\beta_{\text{AND}} &= \prod_{j=1}^{J}\left[\frac{1}{2}-\frac{1}{2}erf\left[\frac{T-\tilde{\mu}_{j,1}}{\sqrt{2}\tilde{\sigma}_j}\right]\right] \\
&\leq \prod_{j=1}^{J}\left[\frac{1}{2}-\frac{1}{2}erf\left[\frac{\sigma(T-\tilde{\mu}_{j,1})}{\sqrt{2}}\right]\right] = 1-\tilde{\beta}_{\text{AND}}.
\end{aligned} \tag{5.4.19}$$

We observe again that using the approximation to compute the value of the decision threshold results in a loss in the false alarm but a gain in the power of the test.

5.4.3 *Asymptotic analysis*

We now study the asymptotic performance of the two decision fusion rules as the number of observations becomes infinitely large. The following two theorems sum up the asymptotic behavior of the multilevel signal detection procedure.

Theorem 5.3. $\tilde{\alpha}_{\text{OR}} \to 0$ *and* $\tilde{\beta}_{\text{OR}} \to 1$ *as* $N \to \infty$.

Proof: When Eq. (5.4.13) is used to compute the value of T, we get

$$\begin{aligned}
T &= \frac{\sqrt{2}}{\sigma}erf^{-1}\left[2(1-\alpha_{\text{OR}})^{1/J}-1\right] \\
&= \frac{\sqrt{2}}{\sigma}erf^{-1}\left[2(1-\alpha_{\text{OR}})^{1/\log N}-1\right].
\end{aligned} \tag{5.4.20}$$

Since $N \to \infty$, $2(1-\alpha_{\text{OR}})^{1/\log N}-1 \to 1$, we see that $T \to \infty$. Therefore,

$$\tilde{\alpha}_{\text{OR}} = 1-\left[\frac{1}{2}+\frac{1}{2}erf\left[\frac{\sigma T}{\sqrt{2}}\right]\right]$$

$$= 0 \qquad \text{as } N \to \infty. \tag{5.4.21}$$

Similarly,

$$\begin{aligned}\tilde{\beta}_{\text{OR}} &= 1 - \prod_{j=1}^{J}\left[\frac{1}{2} + \frac{1}{2}erf\left[\frac{(T - \tilde{\mu}_{j,1})\sigma}{\sqrt{2}}\right]\right] \\ &= 1 \qquad \text{as } N \to \infty \text{ since } erf(\infty) = 1. \qquad \square\end{aligned} \tag{5.4.22}$$

An analogous result for the AND decision fusion is given by the following theorem.

Theorem 5.4. $\tilde{\alpha}_{\text{AND}} \to 1$ *and* $\tilde{\beta}_{\text{AND}} \to 0$ *as* $N \to \infty$.

Proof: From Eq. (5.4.16),

$$\begin{aligned}T &= \frac{\sqrt{2}}{\sigma}erf^{-1}\left[1 - 2\alpha_{\text{AND}}^{1/\log N}\right] \\ &= \frac{\sqrt{2}}{\sigma}erf^{-1}[-1] \text{ as } N \to \infty \\ &= -\infty.\end{aligned} \tag{5.4.23}$$

Therefore,

$$\begin{aligned}\tilde{\alpha}_{\text{AND}} &= \left[\frac{1}{2} - \frac{1}{2}erf\left[\frac{\sigma T}{\sqrt{2}}\right]\right]^{\log N} \\ &= 1 \qquad \text{as } N \to \infty\end{aligned} \tag{5.4.24}$$

and

$$\begin{aligned}\tilde{\beta}_{\text{AND}} &= \prod_{j=1}^{J}\left[\frac{1}{2} - \frac{1}{2}erf\left[\frac{(T - \tilde{\mu}_{j,1})\sigma}{\sqrt{2}}\right]\right] \\ &= 0 \qquad \text{since } erf(-\infty) = -1. \qquad \square\end{aligned} \tag{5.4.25}$$

These results show that there is a nice way to trade off complexity for performance by using approximations and exploiting the properties of some well-known functions.

5.5 Conclusion

This chapter consists of two parts: wavelet-based signal estimation followed by signal detection. A brief survey of some popular wavelet-based de-noising techniques is first presented. Then a moment preserving de-noising method is discussed. Analytical and experimental techniques are used to present the pros and cons of these methods.

A wavelet decomposition level-dependent binary hypothesis test is investigated for signal detection. Outputs of likelihood ratio test-based local detectors are combined by a fusion rule to obtain a global decision about the true hypothesis. Due to the combinatorial nature of this detection fusion problem, two specific fusion rules, the OR and AND decision fusion rules, are chosen and discussed in detail. A method to compute the global decision thresholds and the asymptotic properties of the hypothesis test is also derived.

References

[1] I. Daubechies, Ten Lectures on Wavelets, *SIAM CBMS-NSF Regional Conference Series in Applied Mathematics*, 1992.

[2] S. G. Mallat, A Theory for Multiresolution Signal Decomposition: The Wavelet Representation, *IEEE Trans. on PAMI*, vol. 11, no. 7, July 1989, pp. 674–693.

[3] C. S. Burrus, R. A. Gopinath, and H. Guo, *Introduction to wavelets and wavelet transforms: a primer*, Englewood Cliffs, NJ: Prentice-Hall, 1998.

[4] D. L. Donoho, Nonlinear wavelet methods for recovery of signals, densities, and spectra from indirect and noisy data, *Proc. Symposia Appl. Math.*, vol. 47, 1993, pp. 173–205.

[5] D. L. Donoho and I. M. Johnstone, Ideal spatial adaptation via wavelet shrinkage, *Biometrika*, vol. 81, 1994, pp. 425–455.

[6] D. L. Donoho and I. M. Johnstone, Adapting to unknown smoothness via wavelet shrinkage, *J. American Stat. Assoc.*, vol. 90, 1995, pp. 1200–1224.

[7] D. L. Donoho and I. M. Johnstone, Minimax estimation via wavelet shrinkage, Available on-line at http://www-stat.stanford.edu/~donoho/Reports/index.html

[8] N. S. Jayant and P. Noll, *Digital Coding of Waveforms, Principles and Applications to Speech and Video*, Englewood Cliffs, NJ: Prentice-Hall, 1984.

[9] Y. Wu and S. Tai, Efficient BTC image compression technique, *IEEE Trans. on Consumer Electronics*, vol. 44, May 1998, pp. 317–324.

[10] E. J. Delp and O. R. Mitchell, Image compression using block truncation coding, *IEEE Trans. on Communications*, vol. 27, Sept. 1979, pp. 1335–1342.

[11] T. B. Nguyen and B. J. Oommen, Moment-preserving piecewise linear approximations of signals and systems, *IEEE Trans. on Pattern Analysis and Machine Intelligence*, vol. 19, Jan. 1997, pp. 84–91.

[12] S. G. Chang, B. Yu, and M. Vetterli, Image denoising via lossy compression and wavelet thresholding, *Proc. of International Conf. on Image Processing*, vol. 1, 1997, pp. 604–607.

[13] M. Hansen and B. Yu, Assessing MDL in wavelet denoising and compression, *Information Theory and Networking Workshop*, 1999, p. 34.

[14] G. Chang, B. Yu, and M. Vetterli, Bridging compression to wavelet thresholding as a denoising method," *Proceedings of 1997 Conference on Information Sciences and Systems*, Baltimore, MD, 1997.

[15] S. M. Berman, *Sojourns and Extremes of Stochastic Processes*, Wadsworth, Reading, MA, 1989.

[16] C. Stein, Estimation of the mean of a multivariate normal distribution, *Ann. Statistics*, vol. 9, 1981, pp. 1135–1151.

[17] A. K. Jain, *Fundamentals of digital image processing*, Englewood Cliffs, NJ: Prentice-Hall, 1989.

[18] J. A. Bucklew and N. C. Gallagher, A note on optimum quantization, *IEEE Trans. on Information Theory*, vol. 24, May 1979, pp. 365–366.

[19] P. Billingsley, *Probability and Measure*, New York: John Wiley, 1979.

[20] S. Mallat and S. Zhong, Characterization of signals from multiscale edges, *IEEE Trans. Patt. Recog. and Mach. Intell.*, vol. 14, July 1992, pp. 710–732.
[21] E. J. Delp and O. R. Mitchell, Moment preserving quantization, *IEEE Trans. on Communications*, vol. 39, no. 11, Nov. 1991, pp. 1549–1558.
[22] G. Szego, *Orthogonal Polynomials*, vol. 23, American Mathematical Society, Providence, RI, 1975.
[23] D. L. Donoho, De-noising by soft-thresholding, *IEEE Trans. on Information Theory*, vol. 41, no. 3, May 1995, pp. 613–627.
[24] R. A. DeVore and B. J. Lucier, Fast wavelet techniques for near optimal processing, *IEEE Military Communications Conf.*, 1992, pp. 48.3.1–48.3.7.
[25] Source: Dr. Adrain Maudsley, MRS Unit, VA Medical Center, San Francisco, CA.
[26] MATLAB software package: http://www-stat.stanford.edu/~wavelab/

Part II

Signal Processing and Time-Frequency Signal Analysis

6

High Performance Time-Frequency Distributions for Practical Applications

Boualem Boashash
Victor Sucic

ABSTRACT This chapter presents in three interrelated sections the key concepts and techniques needed to design and use high performance time-frequency distributions (TFDs) in real-world practical applications.

Section 6.1 first presents, in a heuristic approach, the *core concepts* forming the field of time-frequency signal processing, incorporating recent developments, such as the design of high resolution quadratic TFDs for multicomponent signal analysis.

Section 6.2 outlines *methods of assessment* of the performance of time-frequency techniques, in terms of the resolution performance of TFDs in separating closely spaced components in the time-frequency domain. A performance measure is defined using key attributes of TFDs, such as the components' mainlobes and sidelobes, and cross-terms. This method of assessment of TFDs performance has led to improvements in designing high resolution quadratic TFD for time-frequency analysis of multicomponent signals.

Section 6.3 presents a *methodology* for selecting the optimal TFD for a given real-life signal under application-specific constraints. The methodology, based on the performance measure, allows for emphasis of signal features in specific regions of interest in the time-frequency domain.

6.1 The core concepts of time-frequency signal processing

6.1.1 Rationale for using time-frequency distributions

In order to provide more insight into the nature of nonstationary signals, a new field of science and engineering has emerged: the field of joint time-frequency signal processing (TFSP).

The introduction of TFSP has led to new tools to represent and characterize the time-varying contents of nonstationary signals using time-frequency distributions (TFDs), the most popular ones belonging to the class of quadratic distributions (see

Eq. (6.1.20)). By distributing the signal energy over the time-frequency plane, TFDs provide the analyst with information unavailable from the signal time-domain representation or its frequency-domain representation. This includes the number of components present in the signal, the time durations and frequency bands over which these components are defined, the components' relative amplitudes, phase information, and the instantaneous frequency (IF) laws that components follow in the time-frequency plane.

The essential characteristic of TFSP is that it comprises a set of signal processing methods, techniques, and algorithms in which the two natural variables time, t, and frequency, f, are used *concurrently*. This contrasts with the traditional signal processing methods in which time and frequency variables are used *exclusively* and *independently*. Nature shows us in our daily experiences that the two variables, t and f, are usually simultaneously present in signals (e.g., the speech of a person, the song of a bird, or music played on the radio). Such signals are called "nonstationary" signals because their spectral characteristics vary with time. This chapter considers signals that have finite duration T and nearly finite bandwidth B; these signals are often referred to as asymptotic, with the degree of asymptoticity expressed by the BT product [3]. TFSP is designed to deal effectively with such signals by allowing a more detailed analysis and processing.

In addition to existing methods of TFSP, special-purpose TFDs could be defined to suit the particular class of signals under investigation. This chapter presents a simple approach for TFD design. For this purpose, we first revisit the core concepts of TFSP from a practical point of view (Section 6.1), and investigate the performance criteria for TFDs from a user's point of view. The performance of TFDs [1,2] is assessed in terms of *concentration* and *resolution*, and an optimization procedure is used to select the optimal parameter values of TFDs (Section 6.2). Finally, we define a methodology for selecting the most suitable TFD in a given practical situation (Section 6.3).

6.1.2 Limitations of traditional signal representations

The spectrum of a signal (deterministic or random) gives no indication as to how the frequency content of a signal changes with time—information that is important when one deals with frequency modulated (FM) signals or other kinds of nonstationary signals. This frequency variation often contains crucial information about the signal and process studied in applications.

The limitation of "classical" spectral representations is better illustrated by the fact that we can find totally different signals (related to different physical phenomena), $s_a(t)$ and $s_b(t)$, which yet have the same "spectra" (that is, magnitude spectra). The following example shows the inherent limitation of conventional spectral analysis.

Example 6.1. Consider the two signals, $s_a(t)$ (finite duration linear FM) and $s_b(t)$ (infinite duration sinc function), defined as

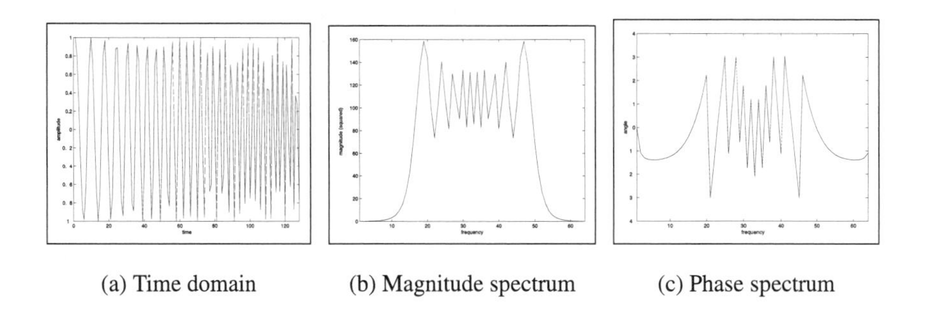

(a) Time domain (b) Magnitude spectrum (c) Phase spectrum

FIGURE 6.1. Time-domain and frequency-domain representations of the finite duration linear FM signal $s_a(t)$ defined by Eq. (6.1.1).

$$s_a(t) = \begin{cases} \cos\left[2\pi\left(f_0 t + \alpha\frac{t^2}{2}\right)\right], & 0 < t < T \\ 0, & \text{elsewhere} \end{cases} \tag{6.1.1}$$

$$s_b(t) = \frac{\sin(\pi B t)}{\pi t}\cos(2\pi f_c t), \tag{6.1.2}$$

where f_0 is the start frequency, $\alpha = B/T$ represents the rate of the frequency change, and $f_c = f_0 + B/2$.

The signal $s_a(t)$ is nonstationary: it is a linear FM signal commonly used in radar and seismic applications (it is analogous to a musical note with a steadily rising pitch). Equation (6.1.1) indicates that the signal $s_a(t)$ is a cosine function in the interval $(0, T)$, and zero outside this interval. Figures 6.1 and 6.2 show the time representation and the frequency representation of the signals $s_a(t)$ and $s_b(t)$, respectively.

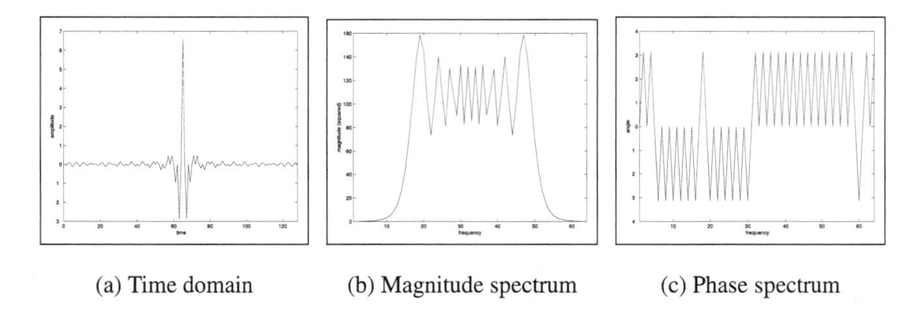

(a) Time domain (b) Magnitude spectrum (c) Phase spectrum

FIGURE 6.2. Time-domain and frequency-domain representations of the infinite duration sinc signal $s_b(t)$ defined by Eq. (6.1.2).

Although the signals $s_a(t)$ and $s_b(t)$ are fundamentally different, they have the same magnitude squared spectrum (Figures 6.1(b) and 6.2(b)). The information that allows one to discriminate between them is contained in their phase spectra (Figures 6.1(c) and 6.2(c)), which is lost when we form the square modulus of the Fourier transform (FT). The frequency in the signal $s_a(t)$ is steadily rising with time, a fact not revealed by the signal spectrum displayed in Figure 6.1(b), which only shows a constant amplitude broadband spectrum.

It follows then that the magnitude spectrum by itself is insufficient (and hence inadequate) for representing signals in a way useful for precise characterization and identification. This serves as a part of the motivation for devising a more "sophisticated" and "practical" nonstationary signal analysis tool, which preserves "all" the information of the signal and discriminates signals in a better way, using a single complete representation instead of attempting to interpret signals' magnitude and phase spectra separately.

6.1.3 Positive frequencies and the analytic signal

An important property of the FT is that for $s(t)$ real, its FT, $S(f)$, is complex and has Hermitian symmetry ($S(f) = S^*(-f)$); i.e., its real part and magnitude are even symmetric, and its imaginary part and phase are odd symmetric. The information contained in the negative frequencies is therefore a duplication of the information contained in the positive frequencies. As a frequency represents the number of oscillations of a signal observed in a second, a frequency is expected to be positive. The negative frequencies appear in the spectrum of a real signal as a consequence of the mathematical model of the FT. In practice, a signal analyst would prefer to deal only with positive frequencies. This is achieved by introducing a complex signal called the analytic signal.

Definition 6.1. An analytic signal $z(t)$ is a complex signal that contains only positive frequencies, such that its FT, $Z(f)$, satisfies

$$Z(f) = 0, \qquad \text{for } f < 0. \tag{6.1.3}$$

A real signal $x(t)$ cannot have only positive frequencies, as the FT of $x(t)$ has a Hermitian symmetry, which contradicts the hypothesis of $X(f) = 0$ for $f < 0$.

Theorem 6.1. *A complex signal $z(t) = x(t) + j\,y(t)$ is analytic if and only if its real and imaginary parts are related as follows:*

$$y(t) = x(t) \star \frac{1}{\pi t} \equiv \mathcal{H}\{x(t)\}, \tag{6.1.4}$$

where $\star$ denotes the convolution in time t and $\mathcal{H}\{\cdot\}$ is the Hilbert transform:

$$\mathcal{H}\{x(t)\} = p.v. \left\{ \int_{-\infty}^{\infty} \frac{x(t-\tau)}{\pi\tau}\, d\tau \right\} \tag{6.1.5}$$

with p.v. being the principal value of Cauchy's integral.

Theorem 6.2. *If $s(t)$ is a real signal expressed as $s(t) = a(t)\cos\phi(t)$, then under the conditions outlined in [4], and using the fact that $\mathcal{H}\{a(t)\cos\phi(t)\} = a(t)\,\mathcal{H}\{\cos\phi(t)\}$, we can construct an analytic signal $z(t)$ which corresponds to $s(t)$ by replacing $\cos\phi(t)$ by $e^{j\phi(t)}$; that is, $z(t) = a(t)\,e^{j\phi(t)}$ and $s(t)$ is the real part of $z(t)$.*

This theorem is important for practical applications because it shows a relation of equivalence between the real signal $s(t)$ and its analytic associate $z(t)$.

6.1.4 Joint time-frequency representations

6.1.4.1 Uncovering hidden information using time-frequency representations

Revealing the time and frequency dependence of a signal, such as the signal $s_a(t)$ in Example 6.1, is achieved by using a joint time-frequency representation. In Figure 6.3, the Wigner–Ville distribution (Eq. (6.1.16)) is used to represent $s_a(t)$, as it provides the optimal joint time-frequency representation for a chirp (linear FM) signal [3].

In Figure 6.3, the signal start and stop times are clearly identifiable from its time-frequency representation, as is the time variation of the frequency content of the

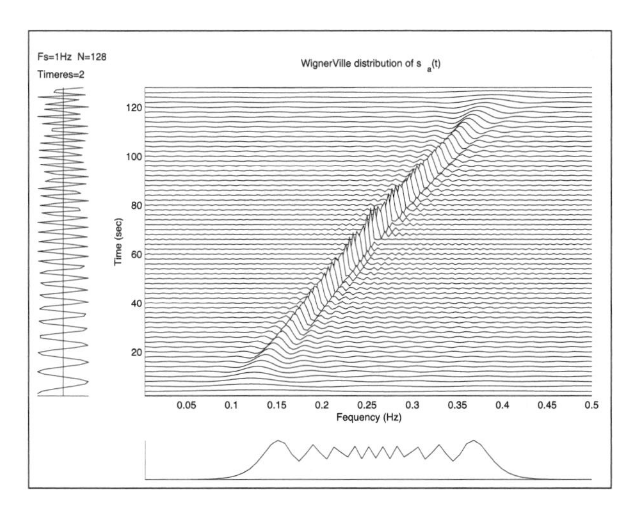

FIGURE 6.3. Time-frequency representation (Wigner–Ville distribution) of the linear FM signal $s_a(t)$ defined by Eq. (6.1.1). The signal's time domain representation is given on the left, and its magnitude spectrum on the bottom.

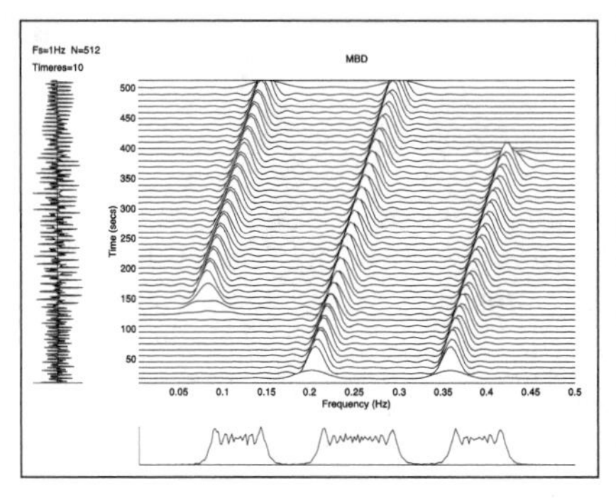

(a) Modified B distribution of the first signal

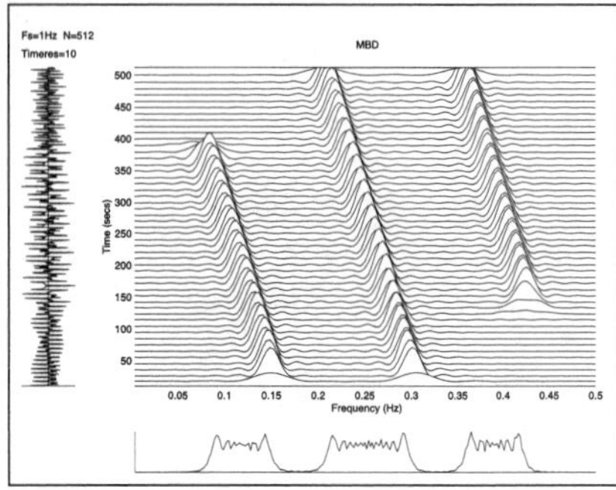

(b) Modified B distribution of the second signal

FIGURE 6.4. Time-frequency representation (Modified B distribution) of two different three-component signals.

signal. This information cannot be retrieved solely from either the signal instantaneous power $|s(t)|^2$ or its spectrum $|S(f)|^2$ representation. The ease of interpreting plots such as the plot in Figure 6.3 makes the concept of a joint time-frequency signal representation attractive.

Example 6.2. Figure 6.4 illustrates another case of two different signals having similar magnitude spectra. Both signals contain three linear FM components. The differences in the time intervals, frequency bands, and the FM laws that characterise the signals' components are not shown clearly in the t domain and the f domain, but do appear clearly in the signals' (t, f) representation (here the Modified B distribution (MBD) [5]).

Example 6.2 shows that another advantage of using a joint time-frequency representation of signals is that it reveals whether the signal is monocomponent or multicomponent [3, 6], a fact that cannot be easily obtained from the signal time-domain or frequency-domain representation.

6.1.4.2 What is a time-frequency representation?

TFSP is a "natural" extension of both time-domain and frequency-domain processing, that involves representing signals in a complete space that can display "all" the signal information in a more accessible way [6]. Such a representation is intended to provide a *distribution* of signal energy versus both time and frequency simultaneously. For this reason, the representation is commonly called a TFD, and is denoted $\rho_z(t, f)$.

A concept intimately related to joint time-frequency representation is that of IF and time delay. The IF corresponds to the frequency of a sine wave that locally (at a given time) fits the signal under analysis. The time delay is a measure of the "order of arrival" of the frequencies.

The TFD is expected to visually exhibit in the (t, f) domain the time-frequency law of each signal component, thereby making the estimation of their IFs and time delays easier, and should also provide additional information about components' relative amplitudes, and the spectral spread of the components around their IFs (the spread is known as the instantaneous bandwidth [3, 6, 7]). For example, the ridges in Figure 6.4 indicate that the three signal components have equal amplitudes, and the peaks of these ridges reveal the linear FM laws of the components. All components have the same instantaneous bandwidth around their IFs, as indicated by the same concentration the ridges attain about their peaks.

6.1.4.2.1 Instantaneous frequency and time delay

The instantaneous frequency of a monocomponent signal is a measure of the localization in time of the individual frequency components of the signal [4].

The IF, $f_i(t)$, of a monocomponent *analytic* signal $z(t) = a(t)e^{j\phi(t)}$ is given by

$$f_i(t) = \frac{1}{2\pi}\frac{d\phi(t)}{dt}. \tag{6.1.6}$$

The IF of a monocomponent *real* signal $s(t) = a(t)\cos\phi(t)$ is defined as the IF of the analytic signal $z(t)$ corresponding to $s(t)$.

The dual of the IF concept in the frequency domain is called the "time delay." The time delay, $\tau_d(f)$, of a monocomponent *analytic* signal $z(t)$ is defined as

$$\tau_d(f) = -\frac{1}{2\pi}\frac{d\theta(f)}{df}, \tag{6.1.7}$$

where

$$\mathcal{F}\{z(t)\} = Z(f) = A(f)e^{j\theta(f)} \tag{6.1.8}$$

is the FT of $z(t)$.

The time delay of a monocomponent *real* signal $s(t)$ is defined as the time delay of the analytic signal $z(t)$ corresponding to $s(t)$.

The order of appearance of each time-varying frequency component is the time delay. The global order of appearance of the frequencies is called the *group delay* (a mean value of individual time delays). The group delay equivalent for IFs is the *mean* IF.

The IF and the time delay describe the "internal organization" of the signal. The use of time-domain representations *or* frequency-domain representations leads one to effectively neglect this information, resulting in a scrambling of the information contained in the signal, as illustrated in Examples 6.1 and 6.2.

6.1.4.2.2 *Physical interpretation of TFDs*

Most TFDs used for a practical time-frequency signal analysis are not necessarily positive, so they do not represent an instantaneous energy spectral density at time t and frequency f. For example, the Page distribution is defined as the time derivative of the "running spectrum" (spectrum of the signal from $-\infty$ to time t), and hence can take both positive and negative values [8]:

$$\rho_z(t,f) = \frac{\partial}{\partial t}\left|\int_{-\infty}^{t} z(u)e^{-j2\pi f u}du\right|^2 .$$

To relate $\rho_z(t,f)$ to the physical quantities used in practical experimentation, we interpret $\rho_z(t,f)$ as a measure of energy flow through the spectral window $(f - \Delta f/2, f + \Delta f/2)$ during the time interval $(t - \Delta t/2, t + \Delta t/2)$. The signal energy localized in the time-frequency region $(\Delta t, \Delta f)$ is then given by [3]

$$E_{\Delta t,\Delta f} = \int_{t-\Delta t/2}^{t+\Delta t/2}\int_{f-\Delta f/2}^{f+\Delta f/2} \rho_z(t,f)\, df\, dt. \tag{6.1.9}$$

The spectral window should be chosen large enough so that the product $\Delta t\Delta f$ satisfies the Heisenberg uncertainty relation [9]:

$$\Delta t\, \Delta f \geq 1/(4\pi). \tag{6.1.10}$$

With this interpretation, negative values of $\rho_z(t,f)$ are then accounted for, and positivity can be removed as a requirement. This is especially important since positivity is also incompatible with the requirement that the signal IF is the first moment of its TFD with respect to frequency [3].

6.1.4.2.3 *Desirable properties and criteria for a TFD*

The TFD is expected to satisfy a certain number of properties that are intuitively desirable for a practical analysis. It was reported in [3] that a TFD should satisfy the marginals; i.e., a TFD should reduce to the spectrum and instantaneous power by integrating over t, respectively f. Furthermore, a TFD is expected to have the IF as its first moment with respect to frequency. These strict constraints on the TFD design led to the terminology of Cohen's class [10].

However, our approach is different and more in line with actual *usage and practice*. We first note that many popular TFDs (e.g., the spectrogram) do not satisfy the marginals and the IF moment condition. Yet the spectrogram has been a valuable tool in many practical applications, suggesting that the time and the frequency marginal and the IF moment constraints may not be strictly needed in practice. What is more important in most practical applications is to maximize the energy concentration about the IF for monocomponent signals and improve the resolution for multicomponent signals.

We have found from practical experience that a time-frequency representation of signal $z(t)$, $\rho_z(t,f)$, needs to satisfy the following properties, so that it can be useful for practical purposes and not just for theoretical interest [2, Part I].

• P1 (*Reality and energy*): The TFD should be real, and its integration over the whole time-frequency domain should be equal to the energy of the signal $z(t)$, E_z:

$$\int_{-\infty}^{\infty}\int_{-\infty}^{\infty} \rho_z(t,f)\,dt\,df \equiv E_z. \tag{6.1.11}$$

• P2 (*Distribution of energy*): The signal energy in a certain region R in the (t,f) plane, E_{z_R}, should be obtained by integrating $\rho_z(t,f)$ over the boundaries $(\Delta t, \Delta f)$ of the region R:

$$\int_{\Delta t}\int_{\Delta f} \rho_z(t,f)\,df\,dt = E_{z_R}. \tag{6.1.12}$$

E_{z_R} is a portion of signal energy in the frequency band Δf and time interval Δt. Note that Δf and Δt need to be selected in such a way that the uncertainty principle (Eq. (6.1.10)) is satisfied. Property P2 is equivalent to Eq. (6.1.9).

• P3 (*IF peak property*): The peak of the time-frequency representation of a monocomponent FM signal with respect to frequency should reflect the IF of the signal:

$$\left.\frac{\partial \rho_z(t,f)}{\partial f}\right|_{f=f_i(t)} = 0. \tag{6.1.13}$$

For multicomponent signals, the same property should apply to the individual components.

• P4 (*Concentration and resolution*): The TFD of a monocomponent FM signal is expected to have a good energy *concentration* around the signal IF law. For a multicomponent FM signal, a TFD is expected to provide a good (t,f) *resolution* of the signal components. This requires a good energy concentration for each of the components and the suppression of any undesirable artifacts.

• P5 (*Robustness to noise*): When a signal embedded in additive white noise is analyzed in the joint time-frequency plane, the noise becomes evenly spread in the plane [3]. For moderate to high signal-to-noise ratios, the signal components dominate over noise in the time-frequency domain, so their IF laws can be estimated from the TFD's dominant peaks (Property P3), provided the TFD satisfies Property P4. It is desirable that such obtained IF estimates are as close as possible to the signal components' true IF laws, i.e., that they are minimum bias and minimum variance IF estimates.

Naturally, the following questions arise: Is there a TFD that meets specifications P1–P5? If yes, how do we construct it? What are the significant signal characteristics and parameters that impact on the design of such a TFD? How do these relate to the TFD itself? How do we obtain them from the actual designed TFD? These questions are answered in the following sections of this chapter.

6.1.5 Heuristic formulation of the class of quadratic TFDs

6.1.5.1 Time-varying spectrum and the Wigner–Ville distribution

Considering the basic definition of the power spectrum density (PSD), let us determine why time information seems to disappear when we take the PSD, and how it can be restored.

Let us consider a complex random process $Z(t)$ with realizations $z(t, \epsilon)$, where ϵ represents the ensemble index identifying each realization. Using context, rather than explicit rigorous mathematical formulation, to improve clarity we simply replace $z(t, \epsilon)$ by $z(t)$. Thus, ϵ is implicit when we say that $z(t)$ is random.

The Wiener–Khintchine theorem states that for a stationary signal, the signal PSD equals the FT of its autocorrelation function [11]. By extension to nonstationary random signals, the time-varying PSD, $S_z(t, f)$, is then defined as the FT of the time-varying autocorrelation function, $R_z(t, \tau)$.

The time-varying autocorrelation function of $z(t)$ is defined in symmetric form as [3]

$$\begin{aligned} R_z(t,\tau) &= \mathcal{E}\{z(t+\tau/2)z^*(t-\tau/2)\} \\ &= \mathcal{E}\{K_z(t,\tau)\}, \end{aligned} \tag{6.1.14}$$

where $K_z(t, \tau) = z(t + \tau/2)z^*(t - \tau/2)$ is the *signal kernel* and $\mathcal{E}\{\cdot\}$ defines the expected value operator.

Therefore, the signal time-varying PSD $S_z(t, f)$ is given as

$$\begin{aligned} S_z(t,f) &= \underset{\tau\to f}{\mathcal{F}} \{R_z(t,\tau)\} \\ &= \underset{\tau\to f}{\mathcal{F}} \{\mathcal{E}\{K_z(t,\tau)\}\} \\ &= \mathcal{E}\left\{\underset{\tau\to f}{\mathcal{F}} \{K_z(t,\tau)\}\right\} \\ &= \mathcal{E}\left\{W_z(t,f)\right\}, \end{aligned} \tag{6.1.15}$$

where the interchange of $\mathcal{E}\{\cdot\}$ and FT is made under assumptions verified by most practical signals and asymptotic signals [3]; and $W_z(t, f)$ denotes the Wigner–Ville distribution (WVD) of $z(t)$, expressed as

$$\begin{aligned} W_z(t,f) &= \underset{\tau\to f}{\mathcal{F}}\{K_z(t,\tau)\} \\ &= \underset{\tau\to f}{\mathcal{F}}\{z(t+\tau/2)z^*(t-\tau/2)\} \\ &= \int_{-\infty}^{\infty} z(t+\tau/2)z^*(t-\tau/2)\, e^{-j2\pi f\tau}\, d\tau. \end{aligned} \tag{6.1.16}$$

The problem of estimating the time-varying PSD of a random process $z(t)$ thus reduces to averaging the WVD of the process over ϵ. If only one realization of the process is available, assuming the process is locally ergodic [11] over a time window, an estimate of the process time-varying PSD is obtained by smoothing the WVD over the window of local ergodicity [3].

6.1.5.2 Time-varying spectrum estimates and quadratic TFDs

If $z(t)$ is deterministic, from Eq. (6.1.15), $S_z(t,f)$ reduces to $W_z(t,f)$:

$$S_z(t,f) = \mathcal{E}\{W_z(t,f)\} = W_z(t,f). \tag{6.1.17}$$

The signals we consider in this chapter are asymptotic (most real-life signals), having a finite duration T, and a finite bandwidth B [3]. The finite duration of a signal is obtained by windowing the signal with a finite duration time function $g(t)$, hence convolving $S_z(t,f)$ in the frequency domain with $G(f) = \underset{t\to f}{\mathcal{F}}\{g(t)\}$.

The finite bandwidth restriction is met by windowing the signal in the frequency domain with a band-limited function $H(f)$, hence convolving $S_z(t,f)$ in the time domain with $h(t) = \underset{f\to t}{\mathcal{F}}^{-1}\{H(f)\}$.

By combining the separate windowing effects of $g(t)$ in time and $H(f)$ in frequency, the estimate of the signal time-varying PSD can be defined as

$$\hat{S}_z(t,f) = G(f) \star_f W_z(t,f) \star_t h(t), \tag{6.1.18}$$

where $G(f)$ and $h(t)$ are even functions (such as those traditionally used in spectral analysis and digital filter design), and $\star_t$ and $\star_f$, respectively, denote the convolution in time and convolution in frequency.

The preceding formulation for $\hat{S}_z(t,f)$ was introduced step by step to illustrate the two-dimensional (2D) convolution that is inherent to most real-life signals. This formulation, however, corresponds to a special case where the 2D windowing in t and f is separable.

In the general case, we need to introduce an even function $\gamma(t,f)$, which may or may not be separable, that reflects the signal overall duration-bandwidth limitations in both time and frequency. This leads to the following general formulation of time-frequency representations $\rho_z(t,f)$ of signal $z(t)$ [2, Part I]:

$$\hat{S}_z(t,f) \equiv \rho_z(t,f) = W_z(t,f) \star_t \star_f \gamma(t,f), \tag{6.1.19}$$

where $\gamma(t,f)$ is an even function that defines the TFD $\rho_z(t,f)$ and its properties, and $\star_t\star_f$ denotes the convolution in both time and frequency.

By expanding $W_z(t,f)$ and the double convolution in Eq. (6.1.19), we obtain the following *general quadratic form* of TFDs [2, Part I]:

$$\rho_z(t,f) = \int_{-\infty}^{\infty}\int_{-\infty}^{\infty}\int_{-\infty}^{\infty} e^{j2\pi\nu(u-t)} g(\tau,\nu) z\Big(u+\frac{\tau}{2}\Big) z^*\Big(u-\frac{\tau}{2}\Big) e^{-j2\pi f\tau} d\nu\, du\, d\tau, \tag{6.1.20}$$

where $g(\tau,\nu)$, known as the TFD *kernel filter*, is the 2D FT of $\gamma(t,f)$.

The function $g(\tau,\nu)$ is analogous to the windows used in classical spectral analysis, and it determines how the signal energy is distributed in time and frequency.

By appropriately choosing $g(\tau,\nu)$ in Eq. (6.1.20), we can obtain most of the popular time-frequency representations of $z(t)$, such as those defined in [2, Part I].

Eq. (6.1.20) differs from Cohen's class [7,10] by the sign in the first exponential. This allows for fewer restrictions on the choice of kernel filter, as discussed in Subsection 6.2.3, and the interpretation of $\rho_z(t,f)$ as a filtered version of the WVD, as explained in Subsection 6.1.5.3.

6.1.5.3 Time, lag, frequency, Doppler domains, and the ambiguity function

We have defined in Eq. (6.1.16) the WVD of signal $z(t)$ as the FT of the signal kernel $K_z(t,\tau)$:

$$W_z(t,f) = \underset{\tau\to f}{\mathcal{F}}\{K_z(t,\tau)\}. \quad (6.1.21)$$

The representation $W_z(t,f)$ is a function of two variables, t and f.

The time-averaged autocorrelation function of signal $z(t)$ may be defined as

$$R_z(\tau) = \int_{-\infty}^{\infty} z(t+\frac{\tau}{2})z^*(t-\frac{\tau}{2})\,dt. \quad (6.1.22)$$

$R_z(\tau)$ describes the similarity between the signal and its time-delayed copies. It is obtained by taking the integral over time of the signal kernel $K_z(t,\tau)$.

Another quantity that relates to $K_z(t,\tau)$ is the symmetric ambiguity function [2,3]:

$$\begin{aligned} A_z(\tau,\nu) &= \int_{-\infty}^{\infty} K_z(t,\tau)\,e^{-j2\pi\nu t}dt \\ &= \int_{-\infty}^{\infty} z(t+\frac{\tau}{2})z^*(t-\frac{\tau}{2})\,e^{-j2\pi\nu t}dt. \end{aligned} \quad (6.1.23)$$

Equation (6.1.23) represents the basic radar equation obtained by correlating a signal with the same signal delayed in time by lag τ and shifted in frequency by Doppler ν.

Thus, $K_z(t,\tau)$ is the key element in the formulation of many important TFSP concepts. Time-frequency (t,f), time-lag (t,τ), lag-Doppler (τ,ν), and Doppler-frequency (ν,f) representations can be obtained from the signal kernel $K_z(t,\tau)$ by means of the FT and the inverse FT [2, Part I]. This is illustrated in Figure 6.5.

As shown in Figure 6.5, the 2D FT (indicated by two vertical arrows) of the WVD of signal $z(t)$ equals the symmetric ambiguity function $A_z(\tau,\nu)$ of $z(t)$ [2, Part I]:

$$W_z(t,f) \;\underset{f\;\;\tau}{\overset{t\;\;\nu}{\rightleftharpoons}}\; A_z(\tau,\nu). \quad (6.1.24)$$

This is an important relationship in TFSP because it indicates that the dual of the time-frequency domain is the lag-Doppler domain (also called the ambiguity domain). The (t,f) domain represents the signal as a function of actual time and actual frequency, and the (τ,ν) domain represents the signal as a function of time shifts and frequency shifts.

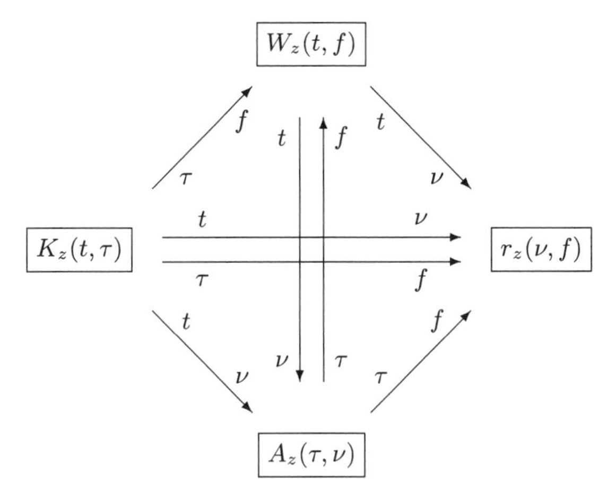

FIGURE 6.5. Different dual domain representations of signal $z(t)$, obtained from the signal kernel $K_z(t,\tau)$.

Given that a double convolution in (t, f) results, by 2D FT, in a multiplication in the ambiguity domain, $\mathcal{A}_z(\tau,\nu)$ becomes a windowed version of $A_z(\tau,\nu)$. This leads to the interpretation of TFD design, defined by Eq. (6.1.19), as a 2D filtering procedure in the ambiguity domain.

Equation (6.1.20) can therefore be rewritten as the 2D FT of the symmetric ambiguity function $A_z(\tau,\nu)$ filtered by the TFD kernel filter $g(\tau,\nu)$ [2]:

$$A_z(\tau,\nu)\, g(\tau,\nu) = \mathcal{A}_z(\tau,\nu) \underset{\nu \rightleftharpoons t}{\overset{\tau \rightleftharpoons f}{}} \rho_z(t,f) = \int_{-\infty}^{\infty}\int_{-\infty}^{\infty} g(\tau,\nu) A_z(\tau,\nu) e^{j2\pi(\nu t - f\tau)}\, d\nu\, d\tau. \tag{6.1.25}$$

Choosing the kernel filter $g(\tau,\nu)$ in Eq. (6.1.25) most relevant to an application results in a specific TFD. For an all-pass filter, $g(\tau,\nu) = 1$, $\rho_z(t,f)$ reduces to the WVD. For $g(\tau,\nu)$ chosen to be the ambiguity function of a time analysis window $w(t)$, $\rho_z(t,f)$ corresponds to the spectrogram with window $w(t)$ [2, 3].

6.1.5.4 Quadratic TFDs, multicomponent signals, and cross-terms reduction

Equation (6.1.20) defines TFDs that are quadratic (or bilinear) in the signal $z(t)$. This implies that if $z(t)$ includes two components $z_1(t)$ and $z_2(t)$, then its quadratic formulation will not only include these two components but also additional components corresponding to their cross product $z_1(t)\, z_2(t)$. These additional components are often called cross-terms and are considered as "artifacts" or "ghosts" appearing unexpectedly in the (t, f) representation. In Figure 6.6 (the WVD of two linear FMs), the "ghost" component is located halfway between the two signal components, and has an (oscillating) amplitude that exhibits large positive and negative values [3]. An effect similar to cross-terms appearance in (t, f) representations occurs when we

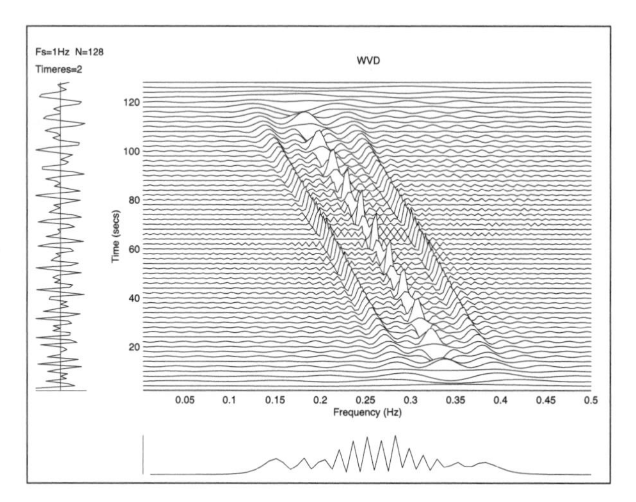

FIGURE 6.6. WVD of a signal composed of two linear FMs. The middle "component," exhibiting large positive and negative amplitudes, is known as the cross-term.

take the spectrum of $z_1(t) + z_2(t)$; we obtain cross-spectral components that are zero only when $z_1(t)$ and $z_2(t)$ do not overlap in frequency.

Thus, the introduction of either noise or some other deterministic components in $z(t)$ introduces cross-terms in its representations. In some applications the cross-terms may be useful, as they provide additional features that can be used for signal identification and recognition [6, Chapter 18]. However, in most cases, they are considered as undesirable interference terms that distort the reading of the representation. So it is generally desired to design TFDs that suppress them "best."

6.1.5.4.1 Location of cross-terms

It was shown that a signal mapped by $A_z(\tau, \nu)$ into the lag-Doppler domain always traverses the origin of that plane, while the cross-terms, having oscillating amplitude in the time-frequency domain, are located away from the origin in the lag-Doppler plane, the distance being directly proportional to the time and frequency separation of the signal components [2, Articles 4.2 and 5.1], [12]. This is illustrated in Figure 6.7 for a two-component signal whose (Gaussian-like) components are centered in the time-frequency domain at (t_1, f_1) and (t_2, f_2), respectively.

Since the WVD is related to the ambiguity function by a 2D FT (Eq. (6.1.25)), the simplest way to reduce the cross-terms of the WVD would be by filtering them out in the ambiguity domain, before taking a two-dimensional FT to return to the (t, f) domain. The two-dimensional filter $g(\tau, \nu)$ needs to be chosen such that it "passes" the region of the ambiguity plane close to the origin (where the signal components are located), and at the same time attenuates the rest of the plane. Note that any resulting truncation of the signal terms by the kernel filter $g(\tau, \nu)$ would result in the signal components spreading in the (t, f) domain, and so in the loss of their time-frequency resolution. So, a compromise is needed when defining the kernel filter in

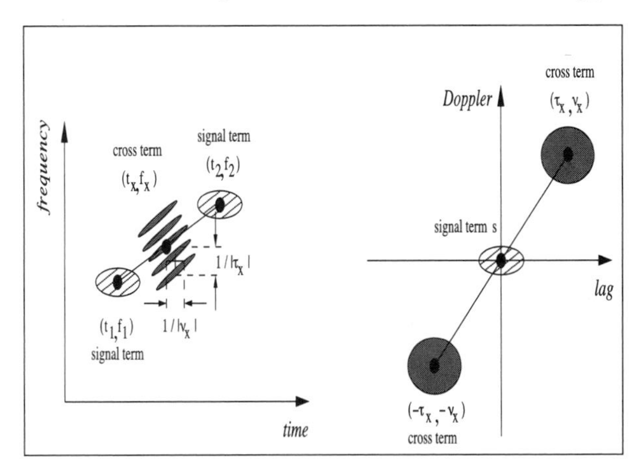

FIGURE 6.7. Location of signal terms and cross-terms in the (t, f), left, and (τ, ν), right, domain.

the ambiguity domain (and setting its parameters) such that the best possible trade-off is achieved between time-frequency resolution and cross-terms suppression.

6.1.5.4.2 *Reduced interference TFDs*

Many TFDs have been designed for the purpose of cross-terms suppression, the most popular among them belonging to the class of reduced interference distributions (RIDs) [6, 13]. One of the first RIDs was the Choi–Williams distribution (CWD), whose $g(\tau, \nu)$ is a 2D Gaussian function centered around the origin in the ambiguity plane, with its spread controlled by a parameter σ [13].

Another, more recent, reduced interference TFD is the MBD [5], whose time-lag kernel filter is defined as

$$G(t, \tau) = G(t) = \frac{k}{\cosh^{2\beta}(t)}, \tag{6.1.26}$$

where $k = \Gamma(2\beta)/(2^{2\beta-1}\Gamma^2(\beta))$ is the normalizing factor ($\Gamma(\cdot)$ stands for the gamma function), and β is a real, positive number that controls the sharpness of the cutoff of the kernel filter in the ambiguity domain, and so also controls the trade-off between time-frequency resolution and cross-terms suppression.

The MBD was found to be the closest to the ideal compromise; it is almost cross-terms free and has high components' resolution in the time-frequency plane [5]. In addition, the MBD allows an efficient IF estimation for multicomponent signals [5]. The original B distribution (BD) [14] also performs well, but it does not meet the traditional RIDs requirements. Both the BD and MBD are practically "equivalent" to the WVD in the analysis and estimation of a monocomponent linear FM signal [5, 14].

6.1.5.4.3 Reduced interference TFDs with separable kernel filters

A simple way to achieve a good cross-terms suppression, and at the same time preserve the signal components' time-frequency resolution, is to design a TFD using a *separable* kernel filter expressed as

$$g(\tau, \nu) = g_2(\tau)\, G_1(\nu). \tag{6.1.27}$$

This product form allows us to formulate separate filter constraints for each variable.

A TFD corresponding to the separable kernel filter $g(\tau, \nu)$ is then defined, using Eq. (6.1.20), as

$$\rho_z(t, f) = g_1(t) \star_t W_z(t, f) \star_f G_2(f), \tag{6.1.28}$$

where $g_1(t) = \mathop{\mathcal{F}}_{\nu \to t}^{-1}\{G_1(\nu)\}$ and $G_2(f) = \mathop{\mathcal{F}}_{\tau \to f}\{g_2(\tau)\}$.

Separable kernel filters, therefore, allow for separate convolutions of the WVD in both time and frequency directions. This makes the design of such TFDs easier to implement. Note that the (t, f) domain form of the separable kernel filter $g(\tau, \nu)$ is also a separable kernel filter $\gamma(t, f) = g_1(t) G_2(f)$ [2].

Two special subclasses of separable kernel filters are:

- the Doppler-independent $(G_1(\nu) = 1)$ kernel filters, which allow for smoothing of the WVD in the frequency direction only, and
- the lag-independent $(g_2(\tau) = 1)$ kernel filters, which allow for the WVD smoothing in the time direction only.

The properties of the Doppler-independent and the lag-independent kernel filters are studied in detail in [2, Article 5.7].

For the WVD of a multicomponent signal, the inner interference terms [3] resulting from a nonlinear FM law of the signal components alternate in the frequency direction, so they can be successfully suppressed by a Doppler-independent kernel filter [2, Article 5.7]. The cross-terms of the WVD oscillate in the time direction [3], and so they can be significantly suppressed by filtering the WVD with a time window, which is the inverse FT of a lag-independent kernel filter [2, Article 5.7]. To successfully suppress both types of these interfering terms in the WVD of a multicomponent signal, a full *separable* kernel filter can be used.

The BD kernel filter [14], $g(\tau, \nu) = (|\tau| / \cosh^2(t))^\beta$, is an example of separable kernels. It was shown in [14] that, for small values (close to zero) of its parameter β, this TFD in general performs well for different types of signals. For small values of β, the BD kernel can be approximated by the lag-independent kernel filter $g_1(t) = 1/\cosh^{2\beta}(t)$ [2, Article 5.7], which, when normalized, defines the MBD kernel filter (see Eq. (6.1.26)). The MBD achieves in particular good cross-terms suppression and good resolution of closely spaced components of multicomponent signals when the components have slowly varying IF laws, as shown in [2, Article 5.7], and illustrated by several examples in this chapter. The definitions and properties of other classical and popular quadratic TFDs are provided in [2].

For a more conclusive assessment of the relative worth of each TFD, a quantitative comparison of their performance is needed. This requires the introduction of specific criteria that take into account usual key attributes of TFDs (such as the amplitudes of signal components and cross-terms, components' instantaneous bandwidths, sidelobes' amplitudes, etc.) [1], as detailed in the next section.

6.2 Performance assessment for quadratic TFDs

6.2.1 Visual comparison of TFDs

Stationary signals are usually analyzed and compared in either the time or the frequency domain. The autocorrelation function in time is examined by looking at the position and relative amplitudes of each lobe, as well as the "correlation width" of the mainlobe. These characteristics are used for examination and comparison. The PSD in frequency is examined by looking at the position and relative amplitude of spectral peaks. It is desired to resolve closely located spectral peaks in the PSD.

For nonstationary signals with a time-varying spectrum, TFSP techniques are needed to represent the signals in the joint time-frequency domain using an appropriate choice of TFD.

Just as some spectral estimates are better than others, some TFDs outperform others when used to analyze certain classes of signals [3, 15–17]. For example, the WVD is known to be optimal for monocomponent signals with the quadratic phase law (linear frequency modulation (LFM)), since it achieves the best energy concentration around the signal IF law [2,3] (see Figure 6.3). The spectrogram, on the other hand, although still regarded as one of the most popular quadratic TFDs, results in an undesirable smoothing of the signal energy around its IF [3].

This example is just a simple illustration of the fact that choosing the right TFD to analyze the given signal is not straightforward, even for monocomponent signals. The task, then, appears to be more complex when one deals with multicomponent signals.

For illustration, let us consider a multicomponent bird song signal [18], represented in the (t, f) domain using the Born–Jordan distribution [2], the CWD, the MBD, the Rihaczek distribution [10], the smoothed WVD [3, 6], the spectrogram, the WVD, and the Zhao–Atlas–Marks distribution [19] (Figure 6.8).

How do we determine which of the TFDs in Figure 6.8 best represents the given signal? To answer this question, according to the common practice in TFSP, one would *visually* compare the eight plots and choose the one that is most appealing. From Figure 6.8, we can see that the MBD, the smoothed WVD, and the spectrogram have "cleaner" plots (less interference, better components' concentration) than the other considered TFDs. However, selecting the best time-frequency distribution among those three TFDs, based *only* on the visual comparison of their plots, is difficult.

The need to *objectively* compare the plots in Figure 6.8 requires the definition of a quantitative performance measure for TFDs. Some theoretical measures that deal

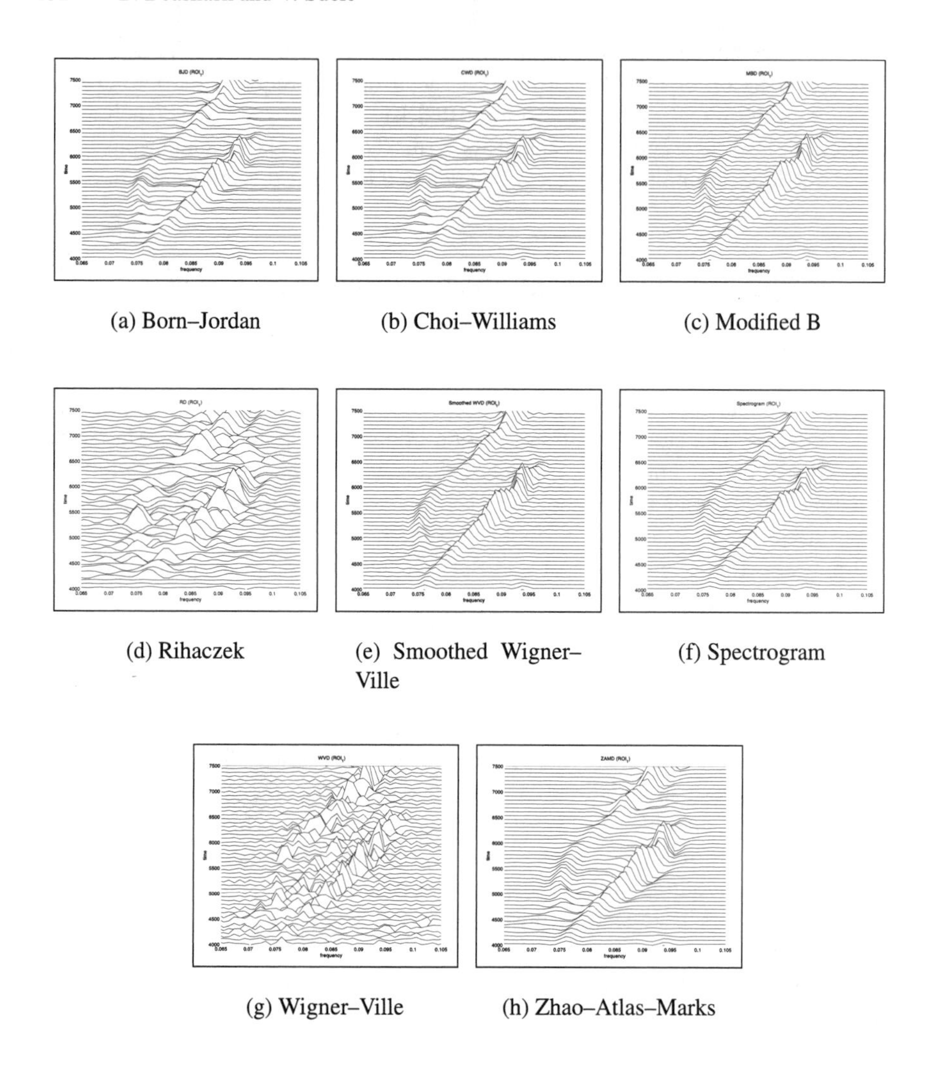

(a) Born–Jordan (b) Choi–Williams (c) Modified B

(d) Rihaczek (e) Smoothed Wigner–Ville (f) Spectrogram

(g) Wigner–Ville (h) Zhao–Atlas–Marks

FIGURE 6.8. TFDs of a multicomponent bird song signal.

essentially with signal concentration have been proposed in the literature [20–23]. This chapter uses objective quantitative measure criteria that take into account not only concentration but also resolution aspects for a practical analysis in the case of closely spaced components. The characteristics of TFDs that influence their resolution, such as energy concentration, mainlobes separation, and sidelobes and cross-terms minimization, need to be combined to define these quantitative measure criteria, as explained in the following section.

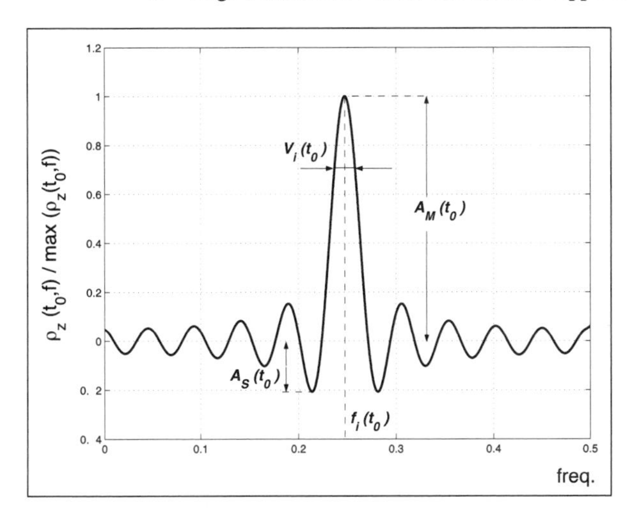

FIGURE 6.9. Instantaneous ($t = t_0$) TFD resolution performance for a monocomponent FM signal.

6.2.2 *Performance criteria for TFDs*

6.2.2.1 Monocomponent signal

For a *monocomponent* FM signal, performance of its TFD is usually defined in terms of the energy concentration that the TFD achieves about the signal IF [4]. One desires to minimize sidelobe amplitude $A_S(t)$ relative to mainlobe amplitude $A_M(t)$, and to minimize instantaneous (mainlobe) bandwidth $V_i(t)$ about the signal IF $f_i(t)$.

For a given time slice of a TFD of a monocomponent signal, as illustrated in Figure 6.9, we may then quantify the signal TFD performance by the measure p expressed as

$$p(t) = \left|\frac{A_S(t)}{A_M(t)}\right| \frac{V_i(t)}{f_i(t)}. \tag{6.2.1}$$

For clarity of presentation, we limit ourselves to measuring the instantaneous bandwidth at 0.7071 of the component normalized amplitude. Note also that in Eq. (6.2.1) we have normalized the bandwidth $V_i(t)$ with the IF $f_i(t)$. However, other possible normalization factors are currently being studied, such as the signal sampling frequency, or the signal half-power bandwidth [11].

From Eq. (6.2.1), good performance of a TFD is characterized by a small (close to zero) value of its measure p. The WVD of an LFM signal with infinite duration, for example, has $p = 0$ [2, Article 7.4].

6.2.2.2 Multicomponent signal

For a *multicomponent* FM signal, performance of its TFD involves not only the energy concentration the TFD attains about the respective IFs of each component,

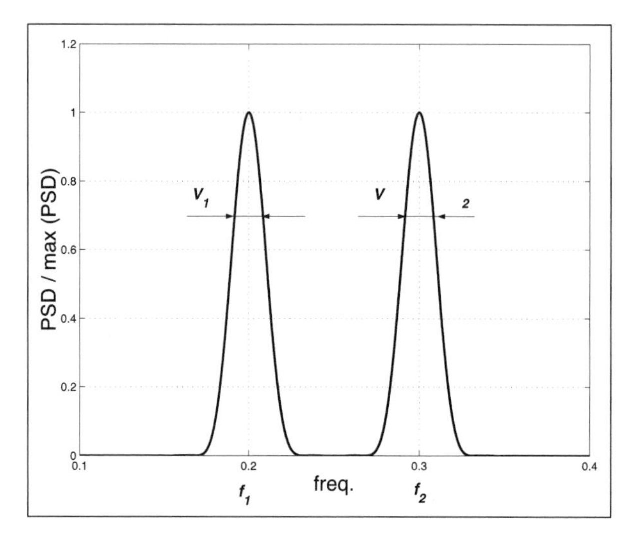

FIGURE 6.10. PSD estimate resolution performance for two sinusoids.

but also the resolution, as measured by the minimum frequency separation between the components' mainlobes for which their amplitudes and bandwidths are just preserved. Let us define these two notions.

6.2.2.2.1 Energy concentration

By extending the concept introduced in Subsection 6.2.2.1 for a monocomponent signal, we say that a TFD has best energy concentration for a given multicomponent signal if *for each* signal component it yields the smallest:

- instantaneous bandwidth relative to the component IF $(V_i(t)/f_i(t))$,
- sidelobe magnitude relative to mainlobe magnitude $(|A_S(t)/A_M(t)|)$.

6.2.2.2.2 Resolution

The frequency resolution in a PSD estimate of a signal composed of two single tones, f_1 and f_2, is defined as the minimum difference $f_2 - f_1$ for which the following inequality holds:

$$f_1 + \frac{V_1}{2} < f_2 - \frac{V_2}{2}, \qquad f_1 < f_2, \tag{6.2.2}$$

where V_1 and V_2 are the bandwidths of the first and the second sinusoid, respectively, as illustrated in Figure 6.10.

For most quadratic TFDs of a two-component signal, however, we also need to account for the effect of cross-terms on resolution, as illustrated by Figure 6.11. From Figure 6.11, we can see that in order to make Eq. (6.2.2) applicable to TFD slices, TFDs should first minimize the cross-terms relative to signal components. Therefore,

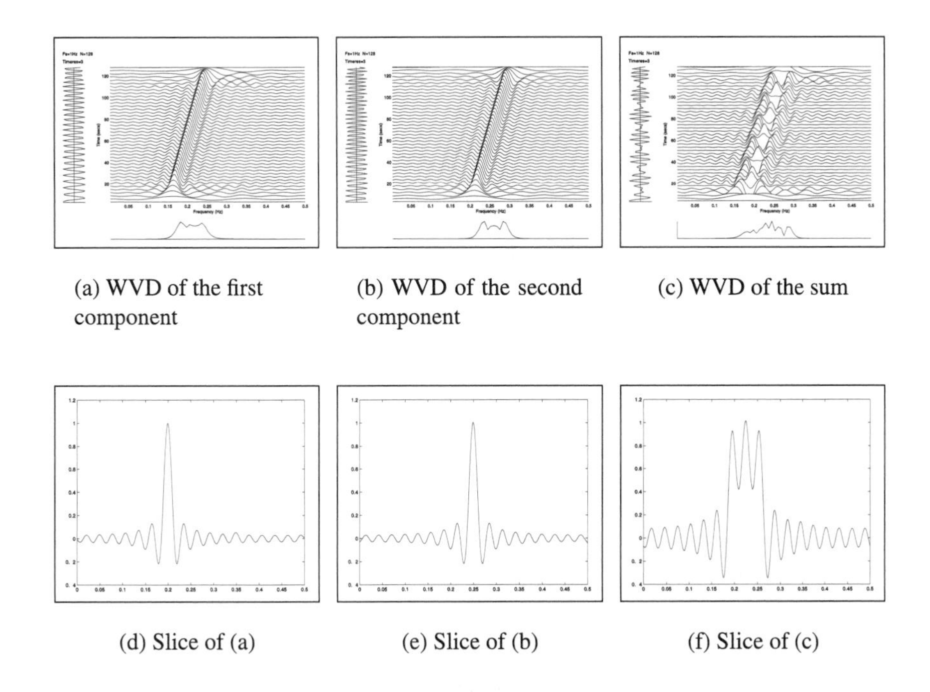

(a) WVD of the first component

(b) WVD of the second component

(c) WVD of the sum

(d) Slice of (a)

(e) Slice of (b)

(f) Slice of (c)

FIGURE 6.11. WVD (a–c) and its slice at the middle of the signal duration interval (d–f) for a two-component signal, considered individually for each contributing term and combined.

the concentration requirement for each signal component needs to be complemented by the cross-term suppression requirement when evaluating the resolution performance of TFDs. All important signal parameters that are needed for this purpose are shown in Figure 6.12, which represents a time slice ($t = t_0$) of a typical quadratic TFD.

In Figure 6.12, $V_{i_1}(t_0)$, $f_{i_1}(t_0)$, $A_{S_1}(t_0)$ and $A_{M_1}(t_0)$ denote, respectively, the instantaneous bandwidth, the IF, the sidelobe amplitude, and the mainlobe amplitude of the first component at time $t = t_0$. Similarly, $V_{i_2}(t_0)$, $f_{i_2}(t_0)$, $A_{S_2}(t_0)$, and $A_{M_2}(t_0)$ represent the instantaneous bandwidth, the IF, the sidelobe amplitude, and the mainlobe amplitude of the second component at the same time t_0. $A_X(t_0)$ is the cross-terms amplitude.

An example of a TFD with nonresolved components is shown in Figure 6.13, where the signal components and the cross-term have all merged in a single lobe.

6.2.2.3 Resolution performance measure for quadratic TFDs

From Eq. (6.2.2) and Figure 6.12, the frequency resolution of a TFD for a pair of components in a multicomponent signal may be quantified by the minimum difference $f_{i_2}(t) - f_{i_1}(t)$ ($f_{i_1}(t) < f_{i_2}(t)$) for which a separation measure D between the

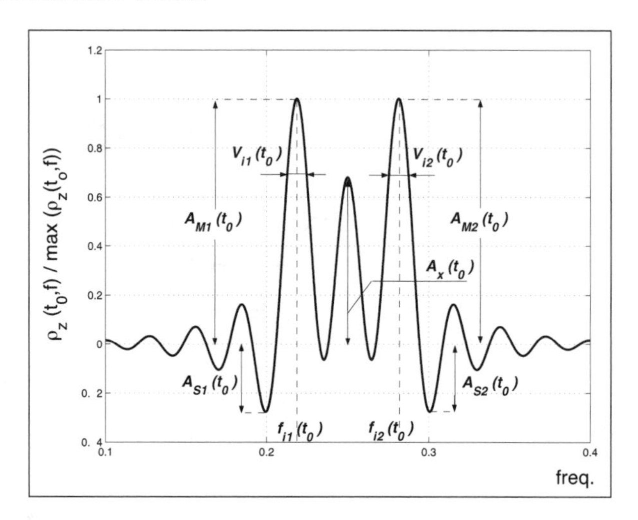

FIGURE 6.12. Instantaneous ($t = t_0$) TFD resolution performance for a two-component FM signal with *resolved* components.

components' mainlobes, centered about their respective IFs f_{i_2} and f_{i_1}, is positive. We define the components' separation measure D as

$$D(t) = \frac{(f_{i_2}(t) - V_{i_2}(t)/2) - (f_{i_1}(t) + V_{i_1}(t)/2)}{f_{i_2}(t) - f_{i_1}(t)}$$
$$= 1 - \frac{1}{2}\left(\frac{V_{i_1}(t) + V_{i_2}(t)}{f_{i_2}(t) - f_{i_1}(t)}\right). \tag{6.2.3}$$

For a better resolution performance of quadratic TFDs:

- the separation measure D should be maximized (close to 1), and *concurrently*,
- the interfering terms (cross-terms and components' sidelobes) should be minimized.

To maximize D, we need to maximize the energy concentration for each component in a pair of signal components under analysis by minimizing their instantaneous bandwidths about the components IFs.

The interfering terms, on the other hand, can be minimized by minimizing both the sidelobe–mainlobe amplitude ratio $|A_S(t)/A_M(t)|$ (so also improving the energy concentration) for each component, and the cross-term—components' mainlobe amplitudes ratios $|A_X(t)/A_M(t)|$.

Therefore, an overall measure P of the resolution performance of a TFD for a pair of components in a multicomponent signal can then be expressed as [1]

$$P(t) = \left|\frac{A_S(t)}{A_M(t)}\right| \left|\frac{A_X(t)}{A_M(t)}\right| \frac{1}{D(t)}, \tag{6.2.4}$$

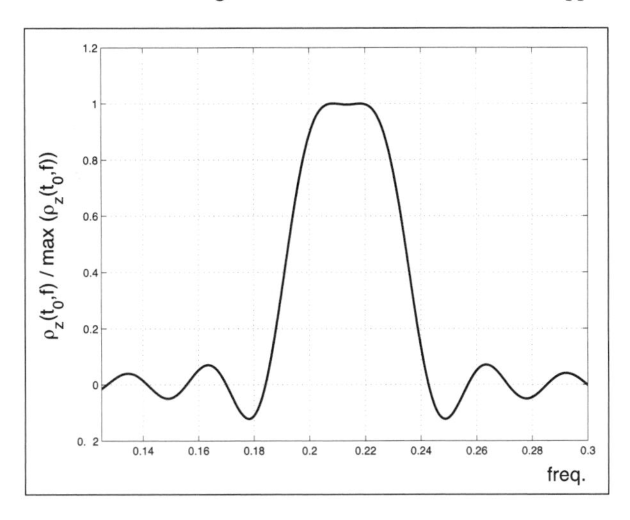

FIGURE 6.13. Instantaneous ($t = t_0$) TFD resolution performance for a two-component FM signal with *unresolved* components.

where $A_M(t)$, $A_S(t)$, and $A_X(t)$ are, respectively, the average amplitude of the components' mainlobes, the average amplitude of the components' sidelobes, and the cross-term amplitude, and $D(t)$, defined by Eq. (6.2.3), is a measure of the components' separation in frequency.

From Eq. (6.2.4) we can see that a good resolution performance of a TFD for a given pair of components in a multicomponent signal is characterized by a small (close to zero) positive value of the measure P.

6.2.2.4 Assessment of the resolution performance of TFDs for a two-LFM-component signal

As an illustration of the use of the performance measure P in Eq. (6.2.4), let us consider a multicomponent signal $s_1(t)$ of duration $T = 128$ given by

$$s_1(t) = \cos(2\pi(0.15\,t + 0.0004\,t^2)) + \cos(2\pi(0.2\,t + 0.0004\,t^2)). \qquad (6.2.5)$$

The signal $s_1(t)$ is represented in the time-frequency domain using a selection of TFDs (see Figure 6.14). In this example, we compare the TFDs' resolution performance at the middle of the signal duration interval. So, for each TFD we take a slice at $t = 64$ and measure the parameters $A_M(64)$, $A_S(64)$, $A_X(64)$ and $V_i(64)$ (the average of the components' instantaneous bandwidths). Using these, we then calculate the frequency separation measure of the components $D(64)$, defined by Eq. (6.2.3), and the resolution performance measure $P(64)$, defined by Eq. (6.2.4). The measurements results are recorded in Table 6.1, and the slices of the TFDs at $t = 64$ are shown in Figure 6.15.

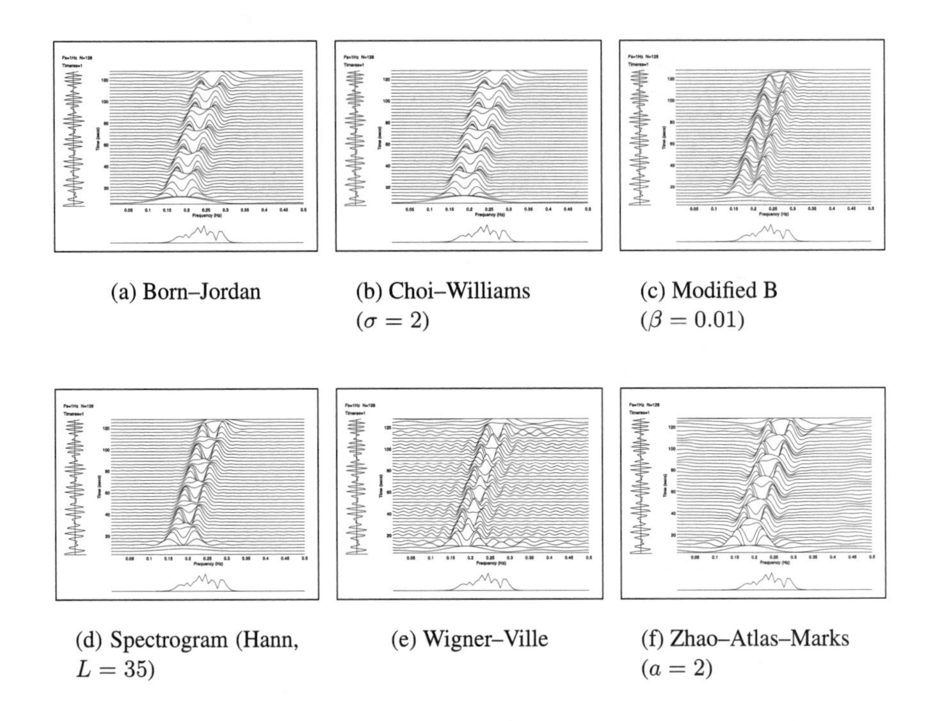

(a) Born–Jordan (b) Choi–Williams ($\sigma = 2$) (c) Modified B ($\beta = 0.01$)

(d) Spectrogram (Hann, $L = 35$) (e) Wigner–Ville (f) Zhao–Atlas–Marks ($a = 2$)

FIGURE 6.14. TFDs of signal $s_1(t)$ defined by Eq. (6.2.5).

As indicated by Eq. (6.2.4), a TFD that, at a given time instant, has the smallest positive value of the measure P is the TFD with the best resolution performance at that time instant for the signal under analysis. From Table 6.1, the MBD ($\beta = 0.01$) of signal $s_1(t)$ gives the smallest value for P at time $t = 64$, and hence should be selected as the best-performing TFD of $s_1(t)$ at $t = 64$.

TFD	$A_M(64)$	$A_S(64)$	$A_X(64)$	$V_i(64)$	$D(64)$	$P(64)$
Born–Jordan	0.9320	0.1222	0.3798	0.0219	0.5512	0.0969
Choi–Williams	0.9355	0.0178	0.4415	0.0238	0.5174	0.0174
Modified B	0.9676	0.0099	0.0983	0.0185	0.6483	0.0016
Spectrogram	0.9119	0.0087	0.5527	0.0266	0.5309	0.0109
Wigner–Ville	0.9153	0.3365	1	0.0130	0.7735	0.5193
Zhao–Atlas–Marks	0.9146	0.4847	0.4796	0.0214	0.4905	0.5666

TABLE 6.1. Parameters and the resolution performance measure P of the signal $s_1(t)$ (Eq. (6.2.5)) TFDs' slices, shown in Figure 6.15.

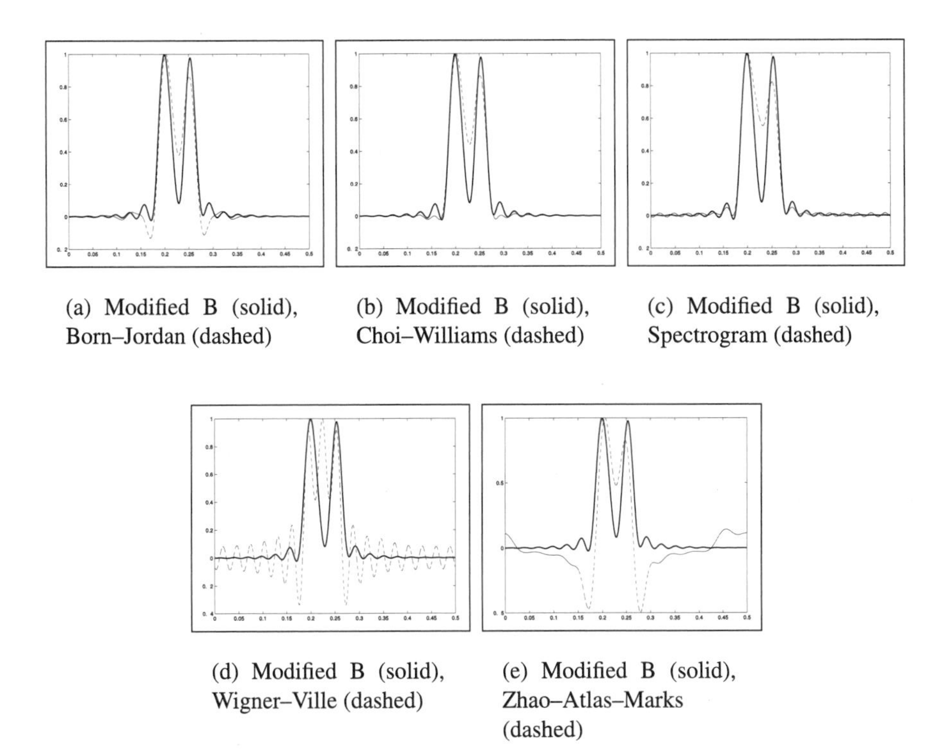

(a) Modified B (solid), Born–Jordan (dashed)

(b) Modified B (solid), Choi–Williams (dashed)

(c) Modified B (solid), Spectrogram (dashed)

(d) Modified B (solid), Wigner–Ville (dashed)

(e) Modified B (solid), Zhao–Atlas–Marks (dashed)

FIGURE 6.15. Normalized slices of TFDs plotted in Figure 6.14 taken at a half of the time interval ($t = 64$) of signal $s_1(t)$ (Eq. (6.2.5)).

6.2.2.5 Alternative (normalized) resolution performance measure

An alternative to the measure P, defined by Eq. (6.2.4), is to combine $|A_S(t)/A_M(t)|$, $|A_X(t)/A_M(t)|$, and $D(t)$ into a sum, rather than a product, and so account for their effects more independently. This results in the following definition for the resolution performance measure [24]:

$$P_1(t) = \left|\frac{A_S(t)}{A_M(t)}\right| + \left|\frac{A_X(t)}{A_M(t)}\right| + \frac{1}{D(t)}. \tag{6.2.6}$$

We would like to have a performance measure that is close to one for TFDs that perform well, and close to zero for poor-performing ones. This requires P_1 to be normalized, for which we need to normalize each of its three contributing terms.

Since, from Eq. (6.2.3), $0 < D(t) < 1$ for resolved components, and therefore $1 < 1/D(t) < \infty$ cannot be normalized, we have chosen to use $(1 - D(t))$ instead of $1/D(t)$ in Eq. (6.2.6). Note that we want to minimize $0 < (1 - D(t)) < 1$ in order to achieve good frequency separation of signal components.

On the other hand, the sidelobe magnitude $|A_S(t)|$ will always be smaller than the mainlobe magnitude $|A_M(t)|$, while the cross-term magnitude $|A_X(t)|$ can be as large as twice the magnitude of the signal components [3] (we assume here normalized amplitudes). Therefore,

$$0 < \left| \frac{A_S(t)}{A_M(t)} \right| < 1, \quad \text{and} \quad 0 < \frac{1}{2} \left| \frac{A_X(t)}{A_M(t)} \right| < 1. \tag{6.2.7}$$

By combining the three normalized quantities into a sum, and normalizing this sum, another resolution performance measure P_2 is obtained [24]:

$$P_2(t) = \frac{1}{3} \left\{ \left| \frac{A_S(t)}{A_M(t)} \right| + \frac{1}{2} \left| \frac{A_X(t)}{A_M(t)} \right| + (1 - D(t)) \right\}, \tag{6.2.8}$$

where the scaling factor $1/3$ normalizes the summation.

Smaller values of P_2 mean better resolution performance of TFDs. To make this measure be close to 1 for well-performing TFDs and 0 for poor-performing ones (TFDs with large interfering terms and poorly resolved components), we have defined the *normalized instantaneous resolution performance measure* P_i as [2, Article 7.4]

$$P_i(t) = 1 - \frac{1}{3} \left\{ \left| \frac{A_S(t)}{A_M(t)} \right| + \frac{1}{2} \left| \frac{A_X(t)}{A_M(t)} \right| + (1 - D(t)) \right\}, \quad 0 < P_i(t) < 1. \tag{6.2.9}$$

6.2.3 Performance specifications for the design of high resolution TFDs

In order to design TFDs that behave optimally in the difficult case of closely spaced multicomponent signals, we need to revisit the constraints for TFD design, discussed in Subsection 6.1.4.2.3, and relate them to the performance measure P_i defined by Eq. (6.2.9).

Subsection 6.1.4.2.3 lists five properties. The first two ensure a physical interpretation of TFDs (i.e., energy conservation), while the last two directly relate to the performance measure P_i. The third property ensures that the IF peak requirement is verified. So, these five properties could be further reduced to essentially three combined ones.

Following this logic, to be a suitable tool for a *practical* high resolution time-frequency analysis, a TFD must verify the following properties:

1. Preserve and concentrate signal energy (Properties P1 and P2 of Subsection 6.1.4.2.3)
2. Reveal the IF law(s) of a signal by its peak(s) (Property P3 of Subsection 6.1.4.2.3)
3. Maximize the measure P_i, defined by Eq. (6.2.9), in order to reduce the cross-terms while preserving time-frequency resolution and improving noise performance (Properties P4 and P5 of Subsection 6.1.4.2.3).

In order to define a quadratic TFD that best meets these constraints, we could use the separable kernel TFD design procedure of Subsection 6.1.5.4.3, and then vary the windows $G_1(\nu)$ and $g_2(\tau)$ until the value of the performance measure P_i is maximum for a given class of signals.

One outcome of such a procedure is the MBD [5], whose kernel filter is defined by Eq. (6.1.26). The kernel filter of the MBD was chosen in the ambiguity domain to be a 2D low-pass function centered around the origin, with sharp cutoff edges. In this way, the kernel retains as much of the auto-terms as possible, while attenuating cross-terms. The amounts of auto-terms and cross-terms kept and filtered out are functions of the volume underneath the kernel filter $g(\tau, \nu)$. This volume can be changed by varying the parameter β.

The MBD and its original version the BD, meet the properties listed above, although, like the spectrogram, they do not satisfy the marginals [5, 14, 25].

6.3 Selecting the best TFD for a given practical application

6.3.1 Assessment of performance for selection of the best TFD

6.3.1.1 Assessment and selection procedure

The procedure for selecting the optimal time-frequency representation for a given multicomponent signal consists of the following steps:

1. *Define a set of criteria for comparison of TFDs*: The criteria must be related to the information we seek from a TFD (e.g., the number of signal components, their relative amplitudes, components' frequency modulation laws, etc.). A set of such criteria is defined in Subsection 6.2.2 for both mono- and multicomponent signals.
2. *Define a quantitative measure for evaluating TFDs performance based on these criteria*: The measure P_i (Eq. (6.2.9)), for example, has been defined for evaluating the resolution performance of quadratic TFDs based on the comparison criteria for FM signals described in Subsection 6.2.2.
3. *Optimize TFDs to match the comparison criteria as closely as possible*: The optimization procedure for a TFD with parameter α (e.g., MBD with parameter β, or CWD with parameter σ) can be done as follows. First, we choose the initial value for the parameter α (a value closest to its lower bound) and calculate the TFD of a given signal. For each time instant in the time interval of interest (time instants over which we want to compare the performance of different TFDs), we take a slice of the TFD and measure its instantaneous performance (e.g., use P_i defined by Eq. (6.2.9)). The average of all instantaneous measures defines the interval performance measure of the TFD for the given value of α.
   The procedure is repeated for the next value of the parameter α. The increment in α should be neither too small (long computation time) nor too large (too "coarse"

optimization results). The optimal value of the parameter α is that value of α that maximizes the TFD interval performance measure. This maximum value of the TFD interval performance measure is called the TFD optimal performance measure.

4. *Select the best TFD for the given signal*: When all TFDs used to represent the given signal in the joint (t, f) domain are optimized, the TFD with the largest value of its optimal performance measure is selected as the *best* TFD for the analyzed signal.

6.3.1.2 Resolution performance comparison of TFDs for a two-LFM-component signal in additive white Gaussian noise

The following example illustrates how to use the above procedure for selecting the optimal TFD for the two-component signal $s_1(t)$ (Eq. (6.2.5)) embedded in noise, i.e.,

$$s_2(t) = \cos(2\pi(0.15\,t + 0.0004\,t^2)) + \cos(2\pi(0.2\,t + 0.0004\,t^2)) + n(t), \tag{6.3.1}$$

where $n(t)$ is additive white Gaussian noise, with a signal-to-noise ratio (SNR) of 10 dB.

The signal $s_2(t)$, of duration $T = 128$, is analysed in the (t, f) domain using the Born–Jordan distribution, the CWD, the MBD, the spectrogram, the WVD, and the Zhao–Atlas–Marks distribution.

To find the TFD that best resolves the two LFM components of the signal $s_2(t)$, we first optimize each of the considered TFDs, as defined in step 3 of the procedure in Subsection 6.3.1.1 (note that the WVD and the Born–Jordan distribution (α [2] equals $1/2$, by convention) have no kernel filter parameters, and hence need no optimizing). The resolution performance of TFDs is compared based on the criteria defined in Subsection 6.2.2, using the comparison measure P_i (Eq. (6.2.9)). The optimal resolution performance measure, P_{opt}, is found for each of the TFDs. Table 6.2 contains the results of the optimization process. It shows that the optimal TFD for signal $s_2(t)$ is the MBD with parameter $\beta = 0.002$, as it has the largest value of P_{opt}. The time-frequency plots of the optimized TFDs are shown in Figure 6.16.

TFD	*Optimal kernel filter parameter*	P_{opt}
Born–Jordan	N/A	0.7542
Choi–Williams	$\sigma = 0.5$	0.7976
Modified B	$\beta = 0.002$	0.8605
Spectrogram	Bartlett window, length 47	0.8448
Wigner–Ville	N/A	0.6694
Zhao–Atlas–Marks	$a = 8$	0.6550

TABLE 6.2. Optimization results for TFDs of signal $s_2(t)$ defined by Eq. (6.3.1).

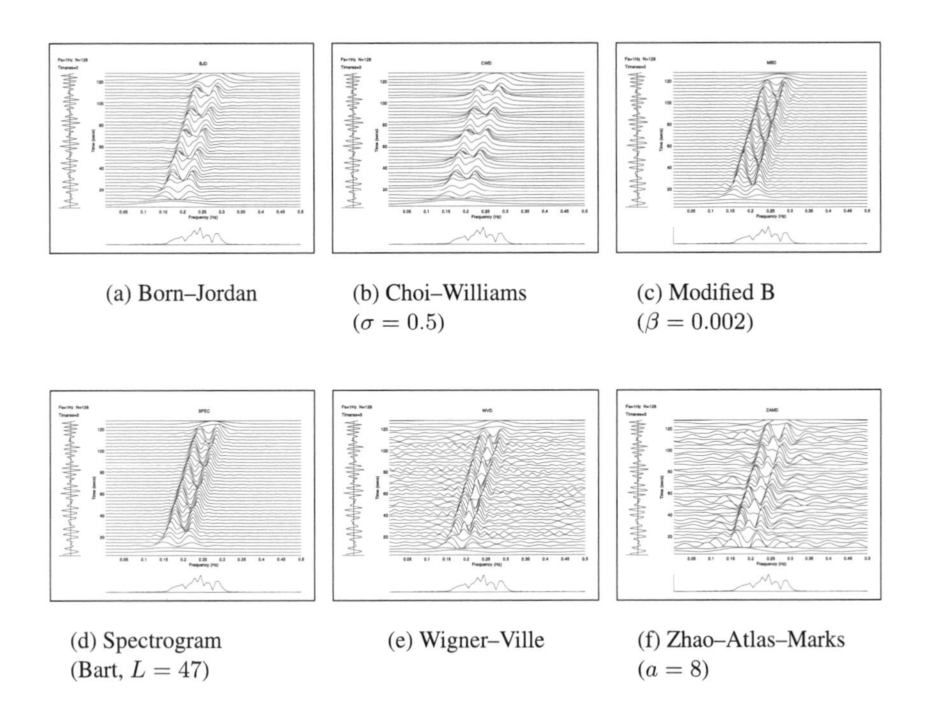

(a) Born–Jordan

(b) Choi–Williams ($\sigma = 0.5$)

(c) Modified B ($\beta = 0.002$)

(d) Spectrogram (Bart, $L = 47$)

(e) Wigner–Ville

(f) Zhao–Atlas–Marks ($a = 8$)

FIGURE 6.16. Optimised TFDs of signal $s_2(t)$ defined by Eq. (6.3.1).

As explained in Subsections 6.1.4.2.3 and 6.2.3, for a practical time-frequency analysis, an important property that a TFD should satisfy is to accurately reveal the IF laws of signal components by its peaks. In Figure 6.17 we compare the true IF laws of the two signal components with those measured from the peaks of the optimized MBD (the best-performing TFD for $s_2(t)$), and the optimized spectrogram (second best TFD for $s_2(t)$).

The quality of the components' IF estimates is measured using the mean-squared error (MSE). As indicated by the MSE values, recorded in Figure 6.17 next to each of the two components, the MBD provides more accurate estimates (smaller MSEs) than the spectrogram for *both* components of the noisy two-component signal $s_2(t)$.

6.3.1.3 Resolution performance comparison of TFDs for a signal with a linear and a nonlinear FM component embedded in additive white Gaussian noise

The optimal TFD methodology is now illustrated for a two-component signal $s_3(t)$ comprising a quadratic FM component whose frequency varies from 0.1 Hz to 0.3 Hz (the sampling frequency is $f_s = 1$ Hz) over the time interval $t \in [1, 128]$, and a component that linearly decreases in frequency from 0.4 Hz to 0.275 Hz over the

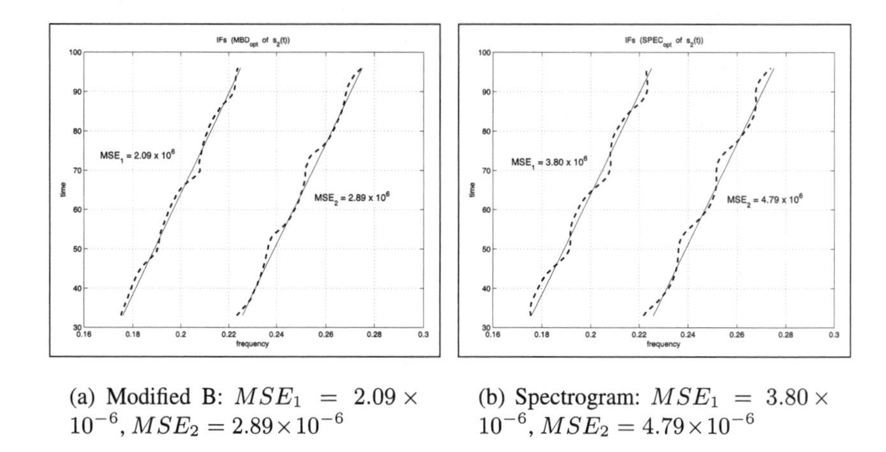

(a) Modified B: $MSE_1 = 2.09 \times 10^{-6}$, $MSE_2 = 2.89 \times 10^{-6}$

(b) Spectrogram: $MSE_1 = 3.80 \times 10^{-6}$, $MSE_2 = 4.79 \times 10^{-6}$

FIGURE 6.17. Comparison of the measured (dashed) and true (solid) IF laws of the two linear FM components of signal $s_2(t)$, defined by Eq. (6.3.1), for the optimized MBD (a) and the optimized spectrogram (b) of $s_2(t)$.

same time interval but has nonzero rectangular amplitude values only for $t \in [33, 96]$. A diagram illustrating the IF laws of the signal $s_3(t)$ is shown in Figure 6.18.

We measure the optimized performance P_{opt} of the Born–Jordan, Choi–Williams, Modified B, spectrogram, Wigner–Ville, and Zhao–Atlas–Marks distributions, when these TFDs represent $s_3(t)$ in the joint time-frequency domain. Table 6.3 lists the TFDs' optimal performance measures and the corresponding optimal values of their kernel filter parameters. It shows that the signal best-performing TFD is the spectrogram ($P_{opt} = 0.84$), closely followed by the MBD ($P_{opt} = 0.8361$).

If we repeat this experiment, but embed the signal $s_3(t)$ in additive white Gaussian noise for two different values of SNR, then:

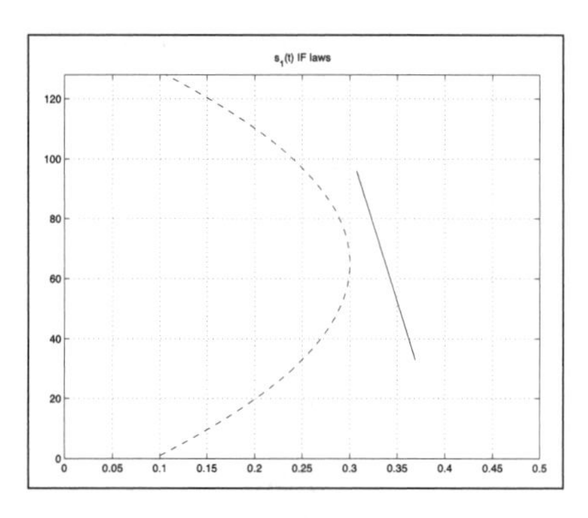

FIGURE 6.18. Diagram illustrating the IF laws of signal $s_3(t)$.

TFD	*Optimal kernel filter parameter*	P_{opt}
Born–Jordan	N/A	0.7114
Choi–Williams	$\sigma = 0.7$	0.7660
Modified B	$\beta = 2.6 \times 10^{-3}$	0.8361
Spectrogram	Hamming window, length 53	0.8400
Wigner–Ville	N/A	0.6232
Zhao–Atlas–Marks	$a = 7$	0.6359

TABLE 6.3. Optimization results for TFDs of signal $s_3(t)$.

TFD	*Optimal kernel filter parameter*	P_{opt}
Born–Jordan	N/A	0.7073
Choi–Williams	$\sigma = 0.5$	0.7414
Modified B	$\beta = 7 \times 10^{-3}$	0.8316
Spectrogram	Bartlett window, length 57	0.8401
Wigner–Ville	N/A	0.5791
Zhao–Atlas–Marks	$a = 7$	0.6471

TABLE 6.4. Optimization results for TFDs of signal $s_3(t)$ in 5 dB additive white Gaussian noise.

TFD	*Optimal kernel filter parameter*	P_{opt}
Born–Jordan	N/A	0.6548
Choi–Williams	$\sigma = 45$	0.5811
Modified B	$\beta = 2.6 \times 10^{-3}$	0.8219
Spectrogram	Hanning window, length 47	0.7955
Wigner–Ville	N/A	0.5647
Zhao–Atlas–Marks	$a = 2$	0.5703

TABLE 6.5. Optimization results for TFDs of signal $s_3(t)$ in 0 dB additive white Gaussian noise.

1. For SNR = 0 dB, the optimal TFD is found to be the MBD, whose optimal resolution performance measure is $P_{opt} = 0.8219$. The spectrogram is the second best TFD with $P_{opt} = 0.7955$. Figure 6.19 shows the plots of the TFDs of signal $s_3(t)$ in 0 dB additive white Gaussian noise.
2. For SNR = 5 dB, the optimal TFD is the spectrogram ($P_{opt} = 0.8401$), while the MBD is the second best TFD ($P_{opt} = 0.8316$).

The performance of other TFDs for $s_3(t)$ in 5 and 0 dB noise is shown in Tables 6.4 and 6.5, respectively.

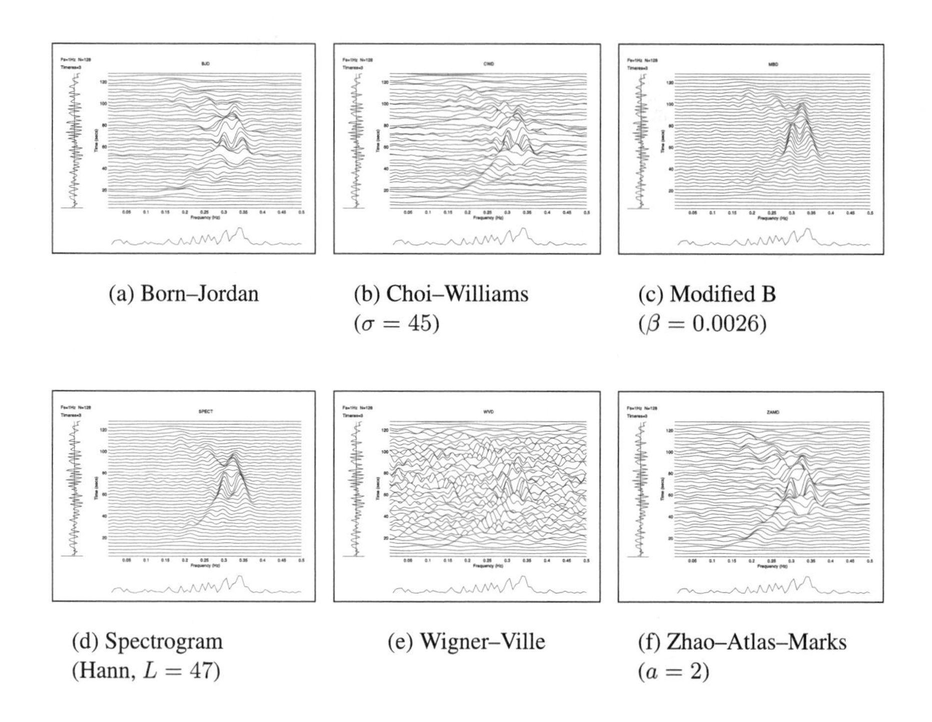

(a) Born–Jordan

(b) Choi–Williams ($\sigma = 45$)

(c) Modified B ($\beta = 0.0026$)

(d) Spectrogram (Hann, $L = 47$)

(e) Wigner–Ville

(f) Zhao–Atlas–Marks ($a = 2$)

FIGURE 6.19. Optimized TFDs of signal $s_3(t)$, with the components' IF laws defined in Figure 6.18, embedded in 0 dB additive white Gaussian noise.

The optimized spectrogram and the optimized MBD of signal $s_3(t)$ perform (out and in noise) similarly, with both TFDs clearly outperforming all other considered TFDs. In situations when two or more TFDs have similar performance for a given signal, the final choice of the signal's optimal TFD should be based on application-specific constraints, so that the TFD that more closely meets these constraints is selected as best over the other "optimal" TFD(s). For example, we have seen that the MBD and the spectrogram have similar resolution performance for the signal $s_2(t)$. However, if, in addition to the resolution requirement, the peak IF criterion (defined in Subsection 6.2.3) is considered, the MBD outperforms the spectrogram, and hence is selected as the optimal TFD for $s_2(t)$. The issue of selecting the optimal TFD under given constraints is further discussed, and illustrated on a real-life signal example, in Subsection 6.3.2.

6.3.1.4 Testing the resolution performance of TFDs for a three-component signal with *different frequency separation* between the components

We now compare the resolution performance of TFDs of a signal with *three* frequency components, with different separations between its consecutive components in the time-frequency plane. The measure P_i, defined by Eq. (6.2.9), evaluates the TFD performance for a given pair of the signal components.

For this test, it suffices to consider a signal $s_4(t)$, which is a sum of three sinusoids: $f_1 = 0.1$ Hz, $f_2 = 0.2$ Hz, and $f_3 = 0.4$ Hz, all three defined over $t \in [1, 128]$. We define P_{i_1} to be the resolution performance measure for the first pair of components $\{f_1, f_2\}$ (closer components), and P_{i_2} the resolution performance measure of the second pair $\{f_2, f_3\}$ (more separated components). Table 6.6 shows the values of $P_{i_1}(64)$ and $P_{i_2}(64)$ for the Born–Jordan, Choi–Williams, Modified B, spectrogram, and Zhao–Atlas–Marks distributions. We do not consider the WVD in this example since the cross-terms significantly degrade its performance whenever there are more than two components in the signal.

Table 6.6 indicates that the MBD is the best-performing TFD for the pair of closer components, followed by the spectrogram, Choi–Williams, Born–Jordan, and Zhao–Atlas–Marks distributions. The *same* ranking of the considered TFDs is obtained for the pair of more separated components.

Thus, for signals with more than two components, we only need to determine a TFD's resolution performance measure for the pair of closest components. This value of the performance measure provides an indication of the TFD's resolution performance, with respect to other TFDs, for all other pairs of consecutive (in frequency) signal components.

6.3.1.5 Testing the resolution performance of TFDs for a signal with components of *different amplitudes*

We have shown that the spectrogram and the MBD are, based on the resolution performance measure P_i, two best-performing TFDs for representing the signal $s_4(t)$ consisting of three sinusoids. We now assess how the performance of these TFDs varies when the components' amplitudes of a two-sinusoid signal $s_5(t)$: (a) are equal, and (b) the second component amplitude is half of the first component amplitude (in time); both cases are studied for the following three separations between the two sinusoids: $\Delta f = 0.2$ Hz, $\Delta f = 0.1$ Hz, and $\Delta f = 0.05$ Hz.

TFD	$P_{i_1}(64)$	$P_{i_2}(64)$
Born–Jordan	0.8887	0.9029
Choi–Williams	0.9387	0.9580
Modified B	0.9566	0.9675
Spectrogram	0.9520	0.9631
Zhao–Atlas–Marks	0.8054	0.8128

TABLE 6.6. Instantaneous ($t = 64$) resolution performance assessment measure for TFDs of a three-component signal $s_4(t)$ for the pair of closer ($P_{i_1}(64)$) and the pair of more separated ($P_{i_2}(64)$) signal components.

Window	$\Delta f = f_2 - f_1$	A_2/A_1	$L = 127$	$L = 95$	$L = 63$	$L = 31$
Rectangular	0.2	1	0.8623	0.8782	0.8797	0.8686
		1/2	0.8235	0.8533	0.8458	0.8253
	0.1	1	0.8952	0.8643	0.8741	0.8281
		1/2	0.8746	0.8328	0.8324	0.8097
	0.05	1	0.8279	0.8369	0.8663	0.8010
		1/2	0.7941	0.8010	0.8087	0.5116
Hamming	0.2	1	0.9682	0.9646	0.9570	0.9316
		1/2	0.9644	0.9589	0.9519	0.9282
	0.1	1	0.9576	0.9545	0.9381	0.9151
		1/2	0.9534	0.9495	0.9334	0.9129
	0.05	1	0.9403	0.9163	0.8682	0
		1/2	0.9351	0.9175	0.8782	0.8704

TABLE 6.7. Assessment of the resolution performance (at $t = 64$) for the spectrogram of a two-sinusoid signal $s_5(t)$ for different ratios of the components' mainlobe amplitudes A_2/A_1 and different components' frequency separations Δf. The spectrogram is calculated using the rectangular and Hamming windows of length L.

Table 6.7 summarizes the performance results (at the time instant $t = 64$) obtained for the signal spectrogram with the rectangular and Hamming windows (of four different lengths). We observe that the resolution performance of the spectrogram with the Hamming window continuously decreases as the window length decreases from $L = 127$ to $L = 31$ for both equal and nonequal amplitudes of the components for *all* three frequency separations between these components. The difference in the P_i values becomes larger as the components get closer in frequency, and so their resolution becomes more critical. The spectrogram with the rectangular window does not show this trend of the continuous decrease in its resolution performance, since, unlike the Hamming window case, its optimal window length ($L_{opt} = 107$) is smaller than the signal duration $T = 128$.

Table 6.8 gives the performance results (at the time instant $t = 64$) for the MBD of signal $s_5(t)$ for a range of its kernel filter parameter values. It indicates that for both the equal and nonequal amplitude components the resolution performance of the MBD continuously decreases as the parameter β increases, regardless of the components' frequency separation. This is due to the fact that, for large β, cross-terms are less attenuated by the more-spread kernel filter in the ambiguity domain, and the large values of the cross-term amplitude A_X in the (t, f) domain reduce the resolution performance measure P_i.

This test illustrates that the measure P_i provides a good assessment of TFDs' resolution performance for signals with different components' amplitudes and different time-varying components' frequency separations.

$\Delta f = f_2 - f_1$	A_2/A_1	$\beta = 10^{-4}$	$\beta = 10^{-3}$	$\beta = 10^{-2}$	$\beta = 10^{-1}$
0.2	1	0.9700	0.9211	0.8757	0.8683
	1/2	0.9664	0.9054	0.8365	0.8263
0.1	1	0.9611	0.9236	0.8933	0.8813
	1/2	0.9568	0.9065	0.8622	0.8590
0.05	1	0.9436	0.8938	0.8520	0.7011
	1/2	0.9412	0.8822	0.8199	0.4610

TABLE 6.8. Assessment of the resolution performance (at $t = 64$) for the MBD (for four values of its kernel filter parameter β) of a two-sinusoid signal $s_5(t)$ for different ratios of the components' mainlobe amplitudes A_2/A_1 and different components' frequency separations Δf.

6.3.2 *Selecting the optimal TFD for real-life signals under given constraints*

As mentioned earlier, different TFDs display information in the time-frequency plane with a varying amount of detail and "accuracy." In an application, selecting a TFD that does this in the optimal way is a critical factor when applying time-frequency analysis to real-life signals.

Based on the concept of multiple view TFDs [26], we define in this subsection the methodology for the selection of the optimal TFD, under application-specific constraints, for real-life signals in time-frequency regions where signal features of interest are located.

6.3.2.1 Methodology

The methodology consists of the following steps.

1. Represent the signal in the (t, f) domain using the WVD, the spectrogram, and the MBD. These three TFDs will provide us with indications of the main signal features in the time-frequency plane: the number of components, their relative amplitudes, the components' durations and bandwidths (features obtained from the MBD and the spectrogram), as well as the cross-terms locations (obtained from the signal WVD).
   The WVD is essentially a parameter-free TFD, while the key parameters of the spectrogram and MBD (the window type and length, and the kernel filter parameter β, respectively) need to be initialized. For the spectrogram, we use the Hanning window as the initial window type, and from several window lengths (short, medium, and long—whose values depend on the signal time duration) select as the initial window length the one that results in the "cleanest" time-frequency plot. This window length is also used as the initial effective window length [3] for all other TFDs considered in the application. In the case of MBD, the initial value for β is set to 0.001. The choice of these initial parameter values

is based on the fact that the spectrogram with the Hanning window and MBD with $\beta \leq 0.001$ have been found to perform well for the majority of signals whose TFDs' resolution performance we have analyzed in [1,2,27,28].
2. Using these three TFDs, identify regions in the (t, f) domain where application-specific signal features (e.g., closely-spaced components, crossing components, sudden changes in component IF laws) are located. We call such regions the regions of interest (ROIs). The ROIs are rectangles in the time-frequency plane whose dimensions are the "time of interest" and the "frequency of interest." The time (resp. frequency) of interest is the time interval (resp. frequency band) of the TFD where the specific signal features are located in time (resp. frequency).
3. For each selected ROI, optimize different TFDs under application-given constraints.
4. Select as the signal optimal TFD, the TFD which among other considered optimised TFDs best satisfies *each* of the given constraints in *all* observed ROIs.

6.3.2.2 Optimal TFD selection for the unbiased, efficient, multicomponent IF estimation of the Noisy Minor song signal

In this section we illustrate the use of our methodology for the "Noisy Minor" (*Manorina melanocephala*) Australian bird song signal [18]. The constraint that needs to be satisfied is the unbiased, efficient, multicomponent IF estimation in (t, f) regions where the signal components are closely-spaced.

Figure 6.20(a) shows the signal time-domain plot, which indicates how its amplitude varies with time, and Figure 6.20(b) (the signal PSD) shows which frequencies (with which magnitudes) are present in the signal. However, neither of the two can provide us with information on the signal internal structure (the number of signal

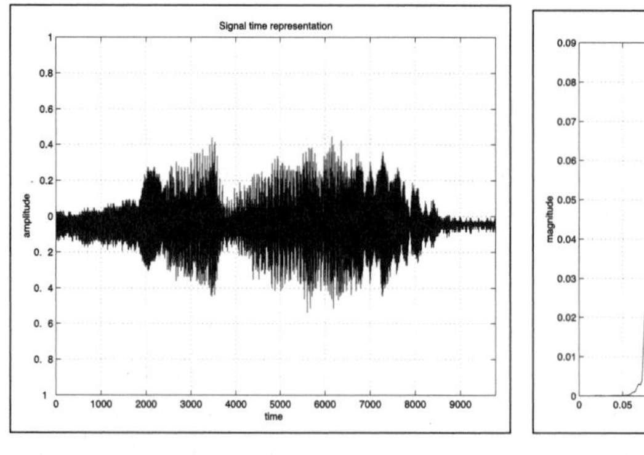

(a) Time-domain

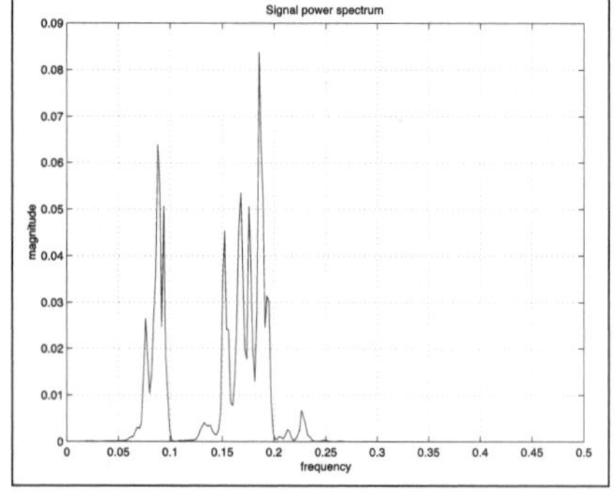

(b) Frequency-domain

FIGURE 6.20. Time-domain and frequency-domain representations of the Noisy Minor song signal.

components, their relative amplitudes, IF laws, time intervals and frequency bands these components occupy, etc.).

To have a more complete "picture" of this signal, we analyze it in the (t, f) domain. This requires identifying a TFD that is optimal for the signal. By employing the steps of the above-defined methodology, we find in this example the regionally optimal TFD of the Noisy Minor song signal for efficient and unbiased estimations of the signal components' IFs from the TFD's peaks [2, 3, 6] in the (t, f) regions where these components form closely spaced pairs.

We start by representing the signal in the (t, f) plane using the WVD, MBD ($\beta = 0.001$), and the spectrogram with Hanning window of length 511. In selecting the optimal window length for the spectrogram, other lengths (127, 255, 1023) were also tested, but length 511 resulted in the most clear (t, f) plot of the signal spectrogram. The effective window length for the WVD and the MBD is also initially set to 511. By comparing the plots of these TFDs, several components (dominant ridges) are identified in the signal, and four ROI are defined where these components are closely spaced to each other in the time-frequency plane (see Figure 6.21).

In order to meet the accurate IFs estimation constraint, we first need to optimize the resolution performance of TFDs in the selected ROIs by applying the procedure defined in Subsection 6.3.1.1. Table 6.9 lists the TFDs' resolution optimization results. It shows that the MBD with $\beta = 5 \times 10^{-5}$, having the largest optimal performance measure P_{opt} (the regional performance measure corresponding to the optimal value of the TFD kernel filter parameter) among the eight considered TFDs, achieves the best resolution of the signal components in ROI_1. It slightly outperforms the smoothed WVD (with the Hamming window of length 415 chosen as

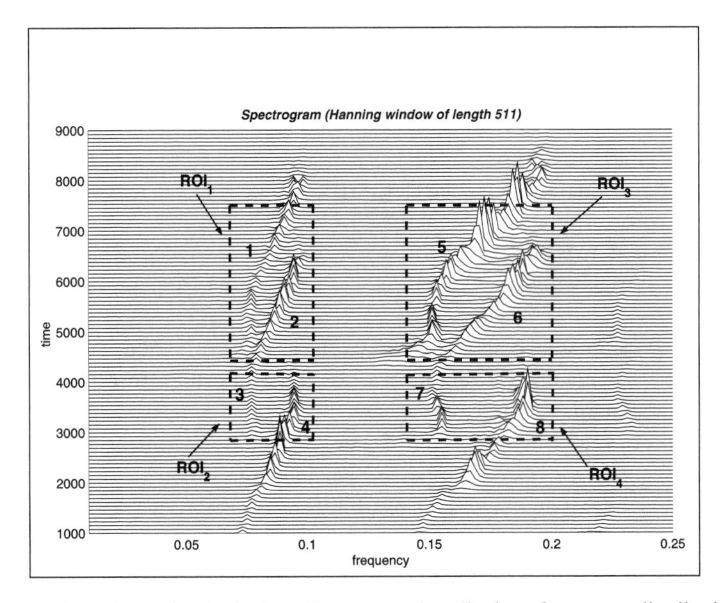

FIGURE 6.21. ROIs for the Noisy Minor song signal's time-frequency distributions.

the smoothing window) and the spectrogram (with the Hanning window of length 447). Other TFDs, even after being optimized in ROI_1, still do not achieve as good resolution performance as the MBD does. The Rihaczek distribution and WVD in particular perform poorly, as indicated by their P_{opt} values of 0.5825 and 0.4984, respectively.

Figure 6.8 shows the ROI_1 of the optimized TFDs of the Noisy Minor song signal. Note that P_{opt} ranks the TFDs in accordance with our visual impressions of their (t, f) plots; i.e., the clearer the plot is (good components' concentration and less interference present in the TFD), the larger P_{opt} is obtained.

Similar analyses can be done for ROI_2, ROI_3 and ROI_4. What is interesting to observe is that the spectrogram has the best resolution performance in each of ROI_2, ROI_3, and ROI_4. The spectrogram's performance closely approaches that of the MBD in ROI_1 too. The three best-performing TFDs for all ROIs considered in this example, with very similar resolution performances, are the MBD, the smoothed WVD, and the spectrogram. The fact that more than one TFD perform well for a certain signal can be of benefit to the signal analyst, giving him/her more freedom in selecting the signal optimal TFD under application-specific constraints.

Having dealt with the requirement for a good (t, f) resolution of closely spaced components of the Noisy Minor song signal, we now select the signal TFD that allows for the most accurate components' IFs estimation (from the peaks of the TFD's dominant ridges) in each of the observed ROIs. Among the three best-performing TFDs, only the MBD provides unbiased and efficient multicomponent FM signals' IF estimates [5]. Therefore, under the given accurate multicomponent IF estimation constraint, we select the Modified B distribution as the *optimal TFD* of the Noisy Minor song signal, *for all four ROIs*. The optimal value of the MBD kernel filter parameter β slightly varies across the ROIs, so that in each of these regions the best resolution of the signal components is achieved; as required by the robust multicomponent IF estimation techniques [2].

6.4 Discussion and conclusions

This chapter brings together, in a heuristic way, three key recent developments that are fundamental to a better understanding and use of time-frequency signal analysis tools.

The first development consists of a concise, updated presentation of the core concepts of TFSP, including a simple way of designing high resolution TFDs for real-life practical applications.

The second development consists of using a quantitative measure of goodness for TFDs to compare the resolution performance of time-frequency distributions for multicomponent signals analysis. This result fills an obvious need for the practitioner in that, until recently, the comparison of the resolution performance of TFDs was primarily based on a visual impression of the TFDs' plots. The introduction of the resolution performance measure for quadratic TFDs has led to an improvement in the design of tools for high resolution time-frequency analysis of multicomponent

TFD	ROI_1		ROI_2		ROI_3		ROI_4	
	P_{opt}	*Parameter*	P_{opt}	*Parameter*	P_{opt}	*Parameter*	P_{opt}	*Parameter*
Born–Jordan	0.7574	N/A	0.9005	N/A	0.8647	N/A	0.8837	N/A
Choi–Williams	0.8755	3	0.9323	0.09	0.8723	0.3	0.9077	3
Modified B	0.9198	5×10^{-5}	0.9433	6×10^{-5}	0.8907	4×10^{-4}	0.9602	3×10^{-5}
Rihaczek	0.5825	N/A	0.7883	N/A	0.6260	N/A	0.6916	N/A
Smoothed Wigner–Ville	0.9195	Hamm, 415	0.9420	Bart, 383	0.8879	Bart, 159	0.9534	Rect, 287
Spectrogram	0.9145	Hann, 447	0.9574	Bart, 383	0.9261	Hann, 223	0.9710	Bart, 447
Wigner–Ville	0.4984	N/A	0.7858	N/A	0.6077	N/A	0.5813	N/A
Zhao–Atlas–Marks	0.7900	1	0.8409	1	0.7922	3	0.8191	2

TABLE 6.9. The optimal resolution performance measures P_{opt} and the corresponding optimal kernel filter parameter values of TFDs of the Noisy Minor song signal for the four time-frequency ROIs defined in Figure 6.21.

signals, by removing unnecessary limitations in the way desirable properties of TFDs were previously chosen. The MBD, which outperforms existing TFDs in terms of (t, f) resolution for signals with closely spaced components, was introduced in this way.

Thirdly, a methodology for selecting an optimal TFD for a given real-life signal under application-specific constraints has been defined. The use of the methodology in practice was illustrated on an Australian bird song signal, for which the optimal TFD for an accurate multicomponent IF estimation was found in several time-frequency ROI.

The results presented in this chapter are important for the field of TFSP. They not only allow for the optimal selection of a specific application-dependent TFD, but also open the way for further research in developing high resolution digital signal processing tools for nonstationary signals, by providing a measure of quality of TFDs and removing often unnecessary TFD design limitations.

References

[1] B. Boashash and V. Sucic, A resolution performance measure for quadratic time-frequency distributions. In *Proc. 10th IEEE Workshop on Statistical Signal and Array Processing, SSAP 2000*, pages 584–588, Pocono Manor, PA, USA, August 2000.

[2] B. Boashash, editor, *Time-Frequency Signal Analysis and Processing*. Prentice-Hall, Englewood Cliffs, NJ, 2003.

[3] B. Boashash, Time–frequency signal analysis. In S. Haykin, editor, *Advances in Spectrum Analysis and Array Processing*, volume 1, chapter 9, pages 418–517. Prentice-Hall, Englewood Cliffs, NJ, 1991.

[4] B. Boashash, Estimating and interpreting the instantaneous frequency of a signal—part 1: Fundamentals, part 2: Algorithms and applications. *Proceedings of the IEEE*, 80(4):519–568, April 1992.

[5] Z. M. Hussain and B. Boashash, Multi-component IF estimation. In *Proc.* 10th *IEEE Workshop on Statistical Signal and Array Processing, SSAP 2000*, pages 559–563, Pocono Manor, PA, USA, August 2000.

[6] B. Boashash, editor, *Time-Frequency Signal Analysis. Methods and Applications*. Longman–Cheshire/Wiley, Melbourne/New York, 1992.

[7] L. Cohen, *Time-Frequency Analysis*. Prentice-Hall, Englewood Cliffs, NJ, 1995.

[8] C. H. Page, Instantaneous power spectra. *Journal of Applied Physics*, 23(1):103–106, January 1952.

[9] D. Gabor, Theory of communication. *Journal of IEE*, 93:429–457, 1946.

[10] L. Cohen, Time-frequency distributions—a review. *Proceedings of the IEEE*, 77(7):941–981, July 1989.

[11] P. Z. Peebles, *Probability, Random Variables and Random Signal Principles*. McGraw-Hill, New York, NY, 4th edition, 2001.

[12] P. Flandrin, Some features of time-frequency representation of multicomponent signals. In *Proc. IEEE Int. Conf. on Acoustics, Speech and Signal Processing, ICASSP 1984*, pages 41.B.4.1–4, San Diego, CA, USA, March 1984.

[13] H. Choi and W. Williams, Improved time-frequency representation of multicomponent signals using exponential kernels. *IEEE Transactions on Acoustics, Speech and Signal Processing*, 37(6):862–871, June 1989.

[14] B. Barkat and B. Boashash, A high–resolution quadratic time–frequency distribution for multicomponent signals analysis. *IEEE Transactions on Signal Processing*, 49(10):2232–2239, October 2001.

[15] Lj. Stankovic, An analysis of some time-frequency and time-scale distributions. *Annales–des–Telecommunications*, 49(9–10):505–517, September/October 1994.

[16] Lj. Stankovic, Auto-term representation by the reduced interference distributions: A procedure for kernel design. *IEEE Transactions on Signal Processing*, 44(6):1557–1563, June 1996.

[17] M. G. Amin and W. Williams, High spectral resolution time-frequency distribution kernels. *IEEE Transactions on Signal Processing*, 46(10):2796–2803, October 1998.

[18] John Gould's Birds of Australia, RMB 4375, Seymour 3660, Victoria, Australia, On CD-ROM.

[19] Y. Zhao, R. J. Marks, and L. E. Atlas, The use of cone-shaped kernels for generalised time-frequency representations of nonstationary signals. *IEEE Transactions on Acoustics, Speech and Signal Processing*, 38(7):1082–1091, July 1990.

[20] D. Jones and T. Parks, A high resolution data-adaptive time-frequency representation. *IEEE Transactions on Acoustics, Speech and Signal Processing*, 38(12):2127–2135, December 1990.

[21] W. J. Williams and T. Sang, Adaptive RID kernels which minimize time-frequency uncertainty. In *Proc. IEEE–SP Int. Symposium on Time-Frequency and Time-Scale Analysis*, pages 96–99, Philadelphia, PA, USA, October 1994.

[22] T. H. Sang and W. J. Williams, Renyi information and signal-dependent optimal kernel design. In *Proc. IEEE Int. Conf. on Acoustics, Speech and Signal Processing, ICASSP 1995*, volume 2, pages 997–1000, Detroit, MI, USA, May 1995.

[23] P. Oliviera and V. Barosso, Uncertainty in the time-frequency plane. In *Proc. 10th IEEE Workshop on Statistical Signal and Array Processing, SSAP 2000*, pages 607–611, Pocono Manor, PA, USA, August 2000.

[24] V. Sucic and B. Boashash, On the selection of quadratic time-frequency distributions and optimisation of their parameters. In *Proc. 3rd Australasian Workshop on Signal Processing Applications, WoSPA 2000*, Brisbane, Australia, December 2000. On CD-ROM.

[25] V. Sucic, B. Barkat, and B. Boashash, Performance evaluation of the B-distribution. In *Proc. 5th Int. Symposium on Signal Processing and Its Applications, ISSPA 1999*, volume 1, pages 267–270, Brisbane, Australia, August 1999.

[26] G. Frazer and B. Boashash, Multiple view time-frequency distributions. In *Proc. 27th Asilomar Conference on Signals, Systems and Computers*, volume 1, pages 513–517, Los Alamitos, CA, USA, 1993.

[27] V. Sucic and B. Boashash, Parameter selection for optimising time-frequency distributions and measurements of time-frequency characteristics of non-stationary signals. In *Proc. IEEE Int. Conf. on Acoustics, Speech and Signal Processing, ICASSP 2001*, volume 6, pages 3557–3560, Salt Lake City, UT, USA, May 2001.

[28] V. Sucic and B. Boashash, Optimisation algorithm for selecting quadratic time-frequency distributions: Performance results and calibration. In *Proc. 6th Int. Symposium on Signal Processing and Its Applications, ISSPA 2001*, volume 1, pages 331–334, Kuala-Lumpur, Malaysia, August 2001.

7

Covariant Time-Frequency Analysis

Franz Hlawatsch
Georg Tauböck
Teresa Twaroch

ABSTRACT We present a theory of linear and bilinear/quadratic time-frequency (TF) representations that satisfy a covariance property with respect to "TF displacement operators." These operators cause TF displacements such as (possibly dispersive) TF shifts and dilations/compressions. Our covariance theory establishes a unified framework for important classes of linear TF representations (e.g., the short-time Fourier transform and continuous wavelet transform) as well as bilinear TF representations (e.g., Cohen's class and the affine class). It yields a theoretical basis for TF analysis and allows the systematic construction of covariant TF representations.

The covariance principle is developed both in the group domain and in the TF domain. Fundamental properties of the *displacement function* connecting these two domains and their far-reaching consequences are studied, and a method for constructing the displacement function is presented.

We also introduce important classes of operator families (modulation and warping operators, dual and affine operators), and we apply the results of the covariance theory to these operator classes. It is shown that for dual operator pairs the characteristic function method for constructing bilinear TF representations is equivalent to the covariance method.

7.1 Introduction

Linear and bilinear/quadratic time-frequency (TF) representations are powerful tools for signal analysis [1–4]. Many important classes of TF representations can be defined by a *covariance property*. Covariance means that a TF representation reacts to certain unitary signal transformations (such as a TF shift, for example) by an associated TF coordinate transform.

In this chapter, we develop a general *covariance theory of TF analysis*, i.e., a theory of linear and bilinear TF representations based on the covariance principle. The covariance theory provides a unified framework for important linear TF representations (e.g., the short-time Fourier transform and continuous wavelet transform) as well as bilinear TF representations (e.g., Cohen's class and the affine class). It is a *constructive* theory since it allows the systematic construction of covariant TF representations.

We will illustrate the covariance principle by considering some covariant TF representations of fundamental importance. Let us first define some elementary unitary signal transformations, namely, the time-shift operator $\mathbf{T}_\tau$, frequency-shift operator $\mathbf{F}_\nu$, and TF scaling operator $\mathbf{C}_\sigma$:

$$(\mathbf{T}_\tau x)(t) = x(t-\tau), \qquad \tau \in \mathbb{R} \tag{7.1.1}$$

$$(\mathbf{F}_\nu x)(t) = e^{j2\pi\nu t} x(t), \qquad \nu \in \mathbb{R} \tag{7.1.2}$$

$$(\mathbf{C}_\sigma x)(t) = \frac{1}{\sqrt{|\sigma|}} x\Big(\frac{t}{\sigma}\Big), \qquad \sigma \in \mathbb{R} \setminus \{0\} \quad \text{or} \quad \sigma \in \mathbb{R}^+. \tag{7.1.3}$$

Using these operators, we define the TF shift operator $\mathbf{S}_{\tau,\nu}$ and the time-shift/TF scaling operator $\mathbf{R}_{\sigma,\tau}$ as

$$\mathbf{S}_{\tau,\nu} = \mathbf{F}_\nu \mathbf{T}_\tau, \qquad \mathbf{R}_{\sigma,\tau} = \mathbf{T}_\tau \mathbf{C}_\sigma.$$

We first consider some elementary families of *linear* TF representations (LTFRs), i.e., TF representations that depend on a signal under analysis, $x(t)$, in a linear manner. The classical LTFR family is the *short-time Fourier transform* [1,3,4]

$$\mathrm{STFT}_x(t,f) = \int_{t'} x(t') h^*(t'-t) e^{-j2\pi f t'} dt', \tag{7.1.4}$$

where t and f denote time and frequency, respectively, $h(t)$ is a function (known as the "window") that does not depend on $x(t)$, and integration is over the entire support of the function integrated. The short-time Fourier transform family can be shown to consist of all LTFRs L that are "covariant" (up to a phase factor) to TF shifts, in the sense that

$$L_{\mathbf{S}_{\tau,\nu}x}(t,f) = e^{-j2\pi\tau(f-\nu)}\, L_x(t-\tau, f-\nu). \tag{7.1.5}$$

We can say that within the entire class of LTFRs, the short-time Fourier transform is axiomatically defined by the TF shift covariance property (7.1.5), i.e., by covariance with respect to the operator $\mathbf{S}_{\tau,\nu}$.

A second important LTFR family is the *continuous wavelet transform* [3–7]

$$\mathrm{WT}_x(t,f) = \sqrt{|f|} \int_{t'} x(t') h^*\big(f(t'-t)\big) dt', \qquad f \neq 0,$$

with $h(t)$ a signal-independent function (the "mother wavelet"). The wavelet transform family consists of all LTFRs L covariant to time shifts and TF scalings:

$$L_{\mathbf{R}_{\sigma,\tau}x}(t,f) = L_x\left(\frac{t-\tau}{\sigma}, \sigma f\right).$$

Other LTFR families are the *hyperbolic wavelet transform* [8], the *power wavelet transform* [9], and all other LTFR families that can be derived from the short-time

Fourier transform or the wavelet transform using the principle of *unitary equivalence* [10–12]. All these LTFR families can be defined by covariance properties with respect to specific operators of the generic form $\mathbf{D}_{\alpha,\beta} = \mathbf{B}_\beta \mathbf{A}_\alpha$.

Let us now turn to important families of *bilinear* TF representations (BTFRs), which depend on two signals $x(t), y(t)$ in a bilinear (strictly speaking, sesquilinear) manner [13]. The classical BTFR family is *Cohen's class* (with signal-independent kernel) [1–3]

$$C_{x,y}(t,f) = \int_{t_1}\int_{t_2} x(t_1)y^*(t_2)h^*(t_1-t, t_2-t)e^{-j2\pi f(t_1-t_2)}dt_1dt_2, \quad (7.1.6)$$

where $h(t_1,t_2)$ is a signal-independent function (the "kernel" function). Cohen's class is well known to consist of all BTFRs B that are covariant to TF shifts [1,3,14],

$$B_{\mathbf{S}_{\tau,\nu}x,\mathbf{S}_{\tau,\nu}y}(t,f) = B_{x,y}(t-\tau, f-\nu). \quad (7.1.7)$$

Hence, Cohen's class is axiomatically defined by the TF shift covariance property (7.1.7).

A second important BTFR class is the *affine class* [1, 3, 15–19]

$$A_{x,y}(t,f) = |f| \int_{t_1}\int_{t_2} x(t_1)y^*(t_2)\; h^*\big(f(t_1-t), f(t_2-t)\big)dt_1dt_2.$$

The affine class consists of all BTFRs B covariant to time shifts and TF scalings [15–17],

$$B_{\mathbf{R}_{\sigma,\tau}x,\mathbf{R}_{\sigma,\tau}y}(t,f) = B_{x,y}\left(\frac{t-\tau}{\sigma}, \sigma f\right).$$

Other BTFR classes are the *hyperbolic class* [8, 20, 21], the *power classes* [9, 21, 22], the *exponential class* [23, 24], and all other BTFR classes that can be derived from Cohen's class or the affine class using the unitary equivalence principle [8–12, 21, 24]. All these BTFR classes can be defined by covariance properties with respect to specific operators of the generic form $\mathbf{D}_{\alpha,\beta} = \mathbf{B}_\beta \mathbf{A}_\alpha$.

The fact that important LTFR and BTFR classes can be defined by a covariance property has led to the development of a unified *covariance theory of TF analysis* [25–28]. The present chapter provides a discussion of this theory that takes into account later contributions [12, 28–35] and adds important new aspects. The covariance theory is based on "displacement operators" of the form (possibly up to a phase factor) $\mathbf{D}_{\alpha,\beta} = \mathbf{B}_\beta \mathbf{A}_\alpha$ which generalize the operators $\mathbf{S}_{\tau,\nu} = \mathbf{F}_\nu \mathbf{T}_\tau$ and $\mathbf{R}_{\sigma,\tau} = \mathbf{T}_\tau \mathbf{C}_\sigma$. An important attribute of a displacement operator is its *displacement function*, which describes its action in the TF plane.

This chapter is organized as follows. Section 7.2 discusses unitary and projective group representations that provide mathematical models for displacement operators. Two important types of unitary group representations—termed *modulation operators* and *warping operators*—are considered in some detail.

In the next two sections, two typical relations between modulation and warping operators are extended in a canonical way. Section 7.3 defines pairs of *dual operators* (generalizing the operator pair $(\mathbf{T}_\tau, \mathbf{F}_\nu)$), and Section 7.4 defines pairs of *affine operators* (generalizing the operator pair $(\mathbf{C}_\sigma, \mathbf{T}_\tau)$).

Section 7.5 introduces the concept of *displacement operators* $\mathbf{D}_{\alpha,\beta}$ and derives the classes of all linear and bilinear "(α, β)-representations" covariant to a given displacement operator. These covariant (α, β)-representations are an intermediate step in the construction of covariant TF representations.

Section 7.6 defines the *displacement function* of a displacement operator as a mapping between the (α, β)-domain and the TF domain. A systematic method for constructing the displacement function is presented in Section 7.7. The displacement function is then used in Section 7.8 to convert the covariant (α, β)-representations of Section 7.5 into covariant TF representations.

Section 7.9 considers the *characteristic function method* for the construction of BTFRs. It is shown that for *dual* operators the characteristic function method is equivalent to the covariance method. Finally, Section 7.10 outlines the extension of the covariance theory to general locally compact abelian (LCA) groups.

7.2 Groups and group representations

The covariance theory is based on families of unitary operators $\mathbf{A}_\alpha$, $\mathbf{B}_\beta$, and $\mathbf{D}_{\alpha,\beta}$ whose parameters α, β, or $\theta = (\alpha, \beta)$ belong to a *group*. These operator families have the mathematical structure of *unitary or projective group representations*. This section reviews some fundamentals and then introduces two special types of unitary group representations termed *modulation operators* and *warping operators*. Modulation and warping operators frequently occur in signal theory and will be important in later sections.

7.2.1 Groups

We shall first review some group theory fundamentals [36–39]. Let $\mathcal{G}$ be a set of elements with a binary operation $\star$ that assigns to every ordered pair of elements (g_1, g_2) with $g_1 \in \mathcal{G}$ and $g_2 \in \mathcal{G}$ a unique element $g_1 \star g_2 \in \mathcal{G}$. $(\mathcal{G}, \star)$ is called a *group* if it satisfies the following properties [36]:

1. There exists an *identity element* $g_0 \in \mathcal{G}$ such that $g \star g_0 = g_0 \star g = g$ for all $g \in \mathcal{G}$.
2. To every $g \in \mathcal{G}$ there exists an *inverse element* $g^{-1} \in \mathcal{G}$ such that $g \star g^{-1} = g^{-1} \star g = g_0$. We note that $(g_1 \star g_2)^{-1} = g_2^{-1} \star g_1^{-1}$.
3. The operation $\star$ is *associative*, i.e., $g_1 \star (g_2 \star g_3) = (g_1 \star g_2) \star g_3$, for all $g_1, g_2, g_3 \in \mathcal{G}$.

If, in addition, $g_1 \star g_2 = g_2 \star g_1$ for all $g_1, g_2 \in \mathcal{G}$, the group is called *commutative* or *abelian*.

Some elementary examples of groups are the following:

- $(\mathbb{R}, +)$, the set of all real numbers with addition, i.e., $g_1 \star g_2 = g_1 + g_2$, is a commutative group with identity element $g_0 = 0$ and inverse elements $g^{-1} = -g$.
- $(\mathbb{R}^+, \cdot)$, the set of all positive real numbers with multiplication, i.e., $g_1 \star g_2 = g_1 g_2$, is a commutative group with identity element $g_0 = 1$ and inverse elements $g^{-1} = 1/g$.
- $(\mathbb{R}^2, +)$, the set of all ordered pairs (g, h), where $g \in \mathbb{R}$ and $h \in \mathbb{R}$, together with elementwise addition, i.e., $(g_1, h_1) \star (g_2, h_2) = (g_1 + g_2, h_1 + h_2)$, is a commutative group. The identity element is $(0, 0)$, and the inverse elements are $(g, h)^{-1} = (-g, -h)$.
- An example of a noncommutative group is the *affine group* of ordered pairs (g, h), where $g \in \mathbb{R}^+$, $h \in \mathbb{R}$, with group operation $(g_1, h_1) \star (g_2, h_2) = (g_1 g_2, h_1 g_2 + h_2)$. The identity element is $(1, 0)$, and the inverse elements are $(g, h)^{-1} = (1/g, -h/g)$.

These are examples of *topological* groups [36, 37, 40, 41] for which $(g_1, g_2) \mapsto g_1 \star g_2^{-1}$ is a continuous map of the product space $\mathcal{G} \times \mathcal{G}$ onto $\mathcal{G}$; here, continuity is defined with respect to some topology on $\mathcal{G}$. For instance, in our first example, the mapping $(g_1, g_2) \mapsto g_1 - g_2$ from $\mathbb{R} \times \mathbb{R}$ to $\mathbb{R}$ is continuous.

Let us now consider two groups $(\mathcal{G}, \star)$ and $(\mathcal{H}, \diamond)$. $(\mathcal{G}, \star)$ is said to be *isomorphic* to $(\mathcal{H}, \diamond)$ with *isomorphism* ψ if the mapping $\psi : \mathcal{G} \to \mathcal{H}$ is invertible (i.e., one-to-one and onto) and if

$$\psi(g_1 \star g_2) = \psi(g_1) \diamond \psi(g_2) \qquad \text{for all } g_1, g_2 \in \mathcal{G}.$$

If $(\mathcal{G}, \star)$ is isomorphic to $(\mathcal{H}, \diamond)$ with isomorphism ψ, then conversely $(\mathcal{H}, \diamond)$ is isomorphic to $(\mathcal{G}, \star)$ with isomorphism ψ^{-1}, i.e., $\psi^{-1}(h_1 \diamond h_2) = \psi^{-1}(h_1) \star \psi^{-1}(h_2)$ for all $h_1, h_2 \in \mathcal{H}$. For $\mathcal{G}$ and $\mathcal{H}$ topological groups, $\mathcal{G}$ is said to be *topologically isomorphic* to $\mathcal{H}$ if $(\mathcal{G}, \star)$ is isomorphic to $(\mathcal{H}, \diamond)$ and the isomorphism ψ and the inverse function ψ^{-1} are continuous. For example, $(\mathbb{R}, +)$ is topologically isomorphic to $(\mathbb{R}^+, \cdot)$ with isomorphism $\psi(g) = e^g$, $g \in \mathbb{R}$. Of special importance for our discussion are groups $(\mathcal{G}, \star)$ that are topologically isomorphic to $(\mathbb{R}, +)$. Such groups are commutative, and the isomorphism $\psi_{\mathcal{G}} : \mathcal{G} \to \mathbb{R}$ satisfies $\psi_{\mathcal{G}}(g_1 \star g_2) = \psi_{\mathcal{G}}(g_1) + \psi_{\mathcal{G}}(g_2)$ for all $g_1, g_2 \in \mathcal{G}$. *Hereafter, "isomorphic" will always mean "topologically isomorphic."*

7.2.2 Unitary and projective group representations

Let $(\mathcal{G}, \star)$ be a topological group with identity element g_0 and let $\mathcal{X}$ be an infinite-dimensional Hilbert space of functions (signals) $x(t)$, i.e., the space is equipped with an inner product $\langle \cdot, \cdot \rangle$ and complete. Usually, $\mathcal{X}$ will be $L^2(\Omega)$ or a subspace of $L^2(\Omega)$, where $\Omega \subseteq \mathbb{R}$, and the inner product is of the form $\langle x, y \rangle = \int_\Omega x(t) y^*(t) dt$. A family of unitary linear operators $\{\mathbf{G}_g\}_{g \in (\mathcal{G}, \star)}$ where each operator $\mathbf{G}_g$ maps $\mathcal{X}$ onto itself is called a *unitary representation of the group $(\mathcal{G}, \star)$ on $\mathcal{X}$* if the following conditions are satisfied [38, 39]:

1a. The map $g \mapsto \mathbf{G}_g$ is such that

$$\mathbf{G}_{g_0} = \mathbf{I} \quad \text{and} \quad \mathbf{G}_{g_2}\mathbf{G}_{g_1} = \mathbf{G}_{g_1 \star g_2} \quad \text{for all } g_1, g_2 \in \mathcal{G}, \tag{7.2.8}$$

where $\mathbf{I}$ is the identity operator on $\mathcal{X}$. Note that if $(\mathcal{G}, \star)$ is a commutative group, then $g_1 \star g_2 = g_2 \star g_1$, which implies $\mathbf{G}_{g_2}\mathbf{G}_{g_1} = \mathbf{G}_{g_1}\mathbf{G}_{g_2}$.

2a. For every fixed $x \in \mathcal{X}$, the map $\mathcal{G} \to \mathcal{X}, g \mapsto \mathbf{G}_g x$ is continuous.

Elementary examples of unitary group representations are the time-shift operator $\mathbf{T}_\tau$, frequency-shift operator $\mathbf{F}_\nu$, TF scaling operator $\mathbf{C}_\sigma$, and the operators $\mathbf{R}_{\sigma,\tau} = \mathbf{T}_\tau \mathbf{C}_\sigma$ and $\mathbf{R}'_{\sigma,\tau} = \mathbf{C}_\sigma \mathbf{T}_\tau$.

Similarly, a family of unitary operators $\{\mathbf{G}_g\}_{g \in (\mathcal{G},\star)}$ defined on a Hilbert space $\mathcal{X}$ is called a *projective (or ray) representation of the group* $(\mathcal{G}, \star)$ *on* $\mathcal{X}$ if the following conditions are satisfied [42, 43]:

1b. The map $g \mapsto \mathbf{G}_g$ is such that

$$\mathbf{G}_{g_0} = \mathbf{I} \quad \text{and} \quad \mathbf{G}_{g_2}\mathbf{G}_{g_1} = c(g_1, g_2)\mathbf{G}_{g_1 \star g_2} \quad \text{for all } g_1, g_2 \in \mathcal{G}, \tag{7.2.9}$$

with a continuous function $c : \mathcal{G} \times \mathcal{G} \to \mathbb{C}$.

2b. For every fixed $x \in \mathcal{X}$, the map $\mathcal{G} \to \mathcal{X}, g \mapsto \mathbf{G}_g x$ is continuous.

Note that if $c(g_1, g_2) \equiv 1$, then $\mathbf{G}_g$ is a unitary representation of $\mathcal{G}$. Hence, projective representations can be viewed as a generalization of unitary representations. The function $c(g_1, g_2)$ is called the *cocycle* of $\mathbf{G}_g$. From (7.2.9), it follows that the cocycle c has the following properties:

$$|c(g_1, g_2)| = 1 \quad \text{or} \quad \frac{1}{c(g_1, g_2)} = c^*(g_1, g_2) \tag{7.2.10}$$

$$c(g, g_0) = c(g_0, g) = 1$$

$$c(g_1, g_2)c(g_1 \star g_2, g_3) = c(g_1, g_2 \star g_3)c(g_2, g_3)$$

$$c(g_1, g_2^{-1})c(g_1 \star g_2^{-1}, g_2) = c(g_2^{-1}, g_2) \tag{7.2.11}$$

$$c(g, g^{-1}) = c(g^{-1}, g), \tag{7.2.12}$$

for all $g_1, g_2, g_3 \in \mathcal{G}$. Equation (7.2.10) shows that the cocycle is a phase factor, i.e., $c(g_1, g_2) = e^{j\phi(g_1, g_2)}$, with some phase function $\phi(g_1, g_2)$ that satisfies $\phi(g, g_0) = \phi(g_0, g) = 0 \bmod 2\pi$ and $\phi(g_1, g_2) + \phi(g_1 \star g_2, g_3) = \phi(g_1, g_2 \star g_3) + \phi(g_2, g_3) \bmod 2\pi$. Furthermore, it follows from (7.2.9) and (7.2.10) with $g_1 = g$ and $g_2 = g^{-1}$ that

$$\mathbf{G}_g^{-1} = c^*(g, g^{-1})\mathbf{G}_{g^{-1}}. \tag{7.2.13}$$

Elementary examples of projective group representations are the TF shift operators $\mathbf{S}_{\tau,\nu} = \mathbf{F}_\nu \mathbf{T}_\tau$ and $\mathbf{S}'_{\tau,\nu} = \mathbf{T}_\tau \mathbf{F}_\nu$ with cocycles $c\big((\tau_1, \nu_1), (\tau_2, \nu_2)\big) = e^{-j2\pi\nu_1\tau_2}$ and $c'\big((\tau_1, \nu_1), (\tau_2, \nu_2)\big) = e^{j2\pi\nu_2\tau_1}$, respectively.

A unitary or projective group representation $\mathbf{G}_g$ on a Hilbert space $\mathcal{X}$ is called *irreducible* if $\mathcal{X}$ is *minimal invariant* under $\mathbf{G}_g$, i.e., if there are no (nonzero) closed subspaces of $\mathcal{X}$ that are preserved by the action of $\mathbf{G}_g$. It is called *faithful* if $\mathbf{G}_{g_1} = \mathbf{G}_{g_2}$ implies $g_1 = g_2$ [38].

7.2.3 Modulation operators

We shall now discuss an important type of unitary group representation, previously considered in [21, 33], that generalizes the frequency-shift operator $\mathbf{F}_\nu$ in (7.1.2).

Definition 7.1. ([33]). Consider a set $\Omega \subseteq \mathbb{R}$, two groups $(\mathcal{G}, \star)$ and $(\mathcal{H}, \diamond)$ isomorphic to $(\mathbb{R}, +)$ with isomorphisms $\psi_\mathcal{G}$ and $\psi_\mathcal{H}$, respectively, and an invertible function $m(t)$ with domain Ω and range $m(\Omega) \subseteq \mathcal{G}$. Let $\tilde{m}(t) \triangleq \psi_\mathcal{G}(m(t))$. A family of linear operators $\{\mathbf{M}_h\}_{h \in (\mathcal{H}, \diamond)}$ defined on $L^2(\Omega)$ as

$$(\mathbf{M}_h x)(t) = e^{j2\pi\psi_\mathcal{H}(h)\tilde{m}(t)} x(t), \qquad h \in (\mathcal{H}, \diamond), \quad t \in \Omega \tag{7.2.14}$$

will be called a *modulation operator (family)*. The function $m(t)$ will be called the *modulation function*.

We note that $\tilde{m}(t) = \psi_\mathcal{G}(m(t))$ has domain $\Omega \subseteq \mathbb{R}$ and its range is $\subseteq \mathbb{R}$. It is easily checked that the modulation operator $\{\mathbf{M}_h\}_{h \in \mathcal{H}}$ is a unitary representation of the group $(\mathcal{H}, \diamond)$; in particular, $\mathbf{M}_h$ is unitary and satisfies $\mathbf{M}_{h_2}\mathbf{M}_{h_1} = \mathbf{M}_{h_1}\mathbf{M}_{h_2} = \mathbf{M}_{h_1 \diamond h_2}$ for all $h_1, h_2 \in \mathcal{H}$. Two examples of modulation operators are the frequency-shift operator $\mathbf{F}_\nu$, with $\nu \in (\mathbb{R}, +)$ and $x \in L^2(\mathbb{R})$, and the "hyperbolic frequency-shift operator" defined as $(\mathbf{M}_h x)(t) = e^{j2\pi h \ln t} x(t)$ with $h \in (\mathbb{R}, +)$, $t \in \mathbb{R}^+$, and $x \in L^2(\mathbb{R}^+)$. Furthermore, a dual (nonequivalent) definition of modulation operators can be given in the frequency domain; examples of such "frequency-domain modulation operators" are the time-shift operator $\mathbf{T}_\tau$ in (7.1.1) and the hyperbolic time-shift operator defined in [8].

It can be shown that the (generalized [44–46]) eigenfunctions and eigenvalues of $\mathbf{M}_h$ are given by

$$u_g^\mathbf{M}(t) = r(t)\delta\big(\tilde{m}(t) - \psi_\mathcal{G}(g)\big), \qquad t \in \Omega, \;\; g \in (\mathcal{G}, \star) \tag{7.2.15}$$
$$\lambda_{h,g}^\mathbf{M} = e^{j2\pi\psi_\mathcal{H}(h)\psi_\mathcal{G}(g)}, \qquad h \in (\mathcal{H}, \diamond), \;\; g \in (\mathcal{G}, \star),$$

respectively, where $\delta(\cdot)$ is the Dirac delta function. Assuming $\tilde{m}(t)$ to be differentiable and $m(\Omega) = \mathcal{G}$, the choice $r(t) = \sqrt{|\tilde{m}'(t)|}$ (where $'$ denotes differentiation) guarantees that the eigenfunctions $\{u_g^\mathbf{M}(t)\}_{g \in \mathcal{G}}$ are complete and orthonormal (in a generalized functions sense [44–46]) in $L^2(\Omega)$. Any modulation operator $\mathbf{M}_h$ is unitarily equivalent (up to the parameter transformation $\nu = \psi_\mathcal{H}(h)$) to the frequency-shift operator $\mathbf{F}_\nu$,

$$\mathbf{M}_h = \mathbf{U}\mathbf{F}_{\psi_\mathcal{H}(h)}\mathbf{U}^{-1}, \tag{7.2.16}$$

with $\mathbf{U} : L^2(\mathbb{R}) \to L^2(\Omega)$, $(\mathbf{U}x)(t) = \sqrt{|\tilde{m}'(t)|}\, x\big(\tilde{m}(t)\big)$.

7.2.4 Warping operators

A second important type of unitary group representation, previously introduced in [47], generalizes the time-shift operator $\mathbf{T}_\tau$ in (7.1.1) and the TF scaling operator $\mathbf{C}_\sigma$ in (7.1.3):

Definition 7.2. ([33]). Consider a set $\Theta \subseteq \mathbb{R}$ and a group $(\mathcal{G}, \star)$ isomorphic to $(\mathbb{R}, +)$ with isomorphism $\psi_{\mathcal{G}}$. Let $w_g(t)$ with $g \in (\mathcal{G}, \star)$ and $t \in \Theta$ be an indexed function with the following properties:

- For fixed $g \in (\mathcal{G}, \star)$, $w_g(t)$ is an invertible, continuously differentiable function that maps Θ onto Θ with nonzero derivative (i.e., a *diffeomorphism*) and that satisfies the composition property

$$w_{g_1}\big(w_{g_2}(t)\big) = w_{g_1 \star g_2}(t). \qquad (7.2.17)$$

  (Note that this implies $w_{g_0}(t) = t$.)
- For fixed t, the map $g \rightarrow w_g(t)$ is continuous.

Then, a family of linear operators $\{\mathbf{W}_g\}_{g \in (\mathcal{G}, \star)}$ defined on $L^2(\Theta)$ as

$$(\mathbf{W}_g x)(t) = \sqrt{|w_g'(t)|}\, x\big(w_g(t)\big), \qquad g \in (\mathcal{G}, \star), \quad t \in \Theta \qquad (7.2.18)$$

will be called a *warping operator (family)*. The function family $w_g(t)$ will be called the *warping function.*

The warping operator $\{\mathbf{W}_g\}_{g \in \mathcal{G}}$ is a unitary representation of the group $(\mathcal{G}, \star)$. Elementary examples are the operators $\mathbf{T}_\tau$, with $\tau \in (\mathbb{R}, +)$ and $x \in L^2(\mathbb{R})$, and $\mathbf{C}_\sigma$, with $\sigma \in (\mathbb{R}^+, \cdot)$ and $x \in L^2(\mathbb{R}^+)$. Furthermore, a dual (nonequivalent) definition of warping operators can be given in the frequency domain.

The following theorem states a fundamental relation between warping and modulation functions. For the special case $(\mathcal{G}, \star) = (\mathbb{R}, +)$, a related result has been proved in [48] and (using a different argument) in [47].

Theorem 7.3. ([33]). *Let $\Omega \subseteq \mathbb{R}$ and let $(\mathcal{G}, \star)$ be a group isomorphic to $(\mathbb{R}, +)$ with isomorphism $\psi_{\mathcal{G}}$. If $m(t)$ is a modulation function on Ω with range $m(\Omega) = \mathcal{G}$, i.e., $m(t)$ is an invertible mapping of Ω onto $\mathcal{G}$, then*

$$w_g(t) \triangleq m^{-1}\big(m(t) \star g^{-1}\big) = \tilde{m}^{-1}\big(\tilde{m}(t) - \psi_{\mathcal{G}}(g)\big), \quad t \in \Omega, \quad g \in (\mathcal{G}, \star),$$

with $\tilde{m}(t) = \psi_{\mathcal{G}}(m(t))$, is invertible with range $w_g(\Omega) = \Omega$, and it satisfies the composition property (7.2.17). (Note that $m^{-1}(\cdot)$ denotes the inverse function associated to $m(t)$, whereas g^{-1} denotes the group inverse element associated to g. Furthermore, note that $m(\Omega) = \mathcal{G}$ is equivalent to $\tilde{m}(t) = \mathbb{R}$.) If $\tilde{m}(t) = \psi_{\mathcal{G}}(m(t))$ is a diffeomorphism, then so is $w_g(t)$ (with g fixed), and the map $g \mapsto w_g(t)$ is continuous for fixed $t \in \Omega$, i.e., $w_g(t)$ is a warping function on Ω.

Conversely, consider a group $(\mathcal{G}, \star)$ isomorphic to $(\mathbb{R}, +)$ with isomorphism $\psi_{\mathcal{G}}$, and a function $w_g(t)$ with $g \in (\mathcal{G}, \star)$ and $t \in \Theta \subseteq \mathbb{R}$ that satisfies (7.2.17). If $f(g) \triangleq w_{g^{-1}}(t_0)$ ($g \in \mathcal{G}$; $t_0 \in \Theta$ arbitrary but fixed) is invertible with range $f(\mathcal{G}) = \Omega \subseteq \Theta$, then $w_g(t) = f\big(f^{-1}(t) \star g^{-1}\big)$ for $t \in \Omega$, $g \in (\mathcal{G}, \star)$, or equivalently, setting $m(t) \triangleq f^{-1}(t)$,

$$w_g(t) = m^{-1}\big(m(t) \star g^{-1}\big), \qquad t \in \Omega, \quad g \in (\mathcal{G}, \star). \qquad (7.2.19)$$

Here, $m : \Omega \to \mathcal{G}$ is invertible and unique up to group translations, i.e., $m_1^{-1}\big(m_1(t) \star g^{-1}\big) \equiv m_2^{-1}\big(m_2(t) \star g^{-1}\big)$ implies $m_2(t) \equiv m_1(t) \star \kappa$ with some $\kappa \in (\mathcal{G}, \star)$. Finally, there is $w_g(\Omega) = \Omega$.

Proof: Since $m(t)$ is invertible so is $m^{-1}(g)$, and hence $w_g(t) = m^{-1}\big(m(t) \star g^{-1}\big)$ is invertible for fixed g. From $m(\Omega) = \mathcal{G}$ and $m^{-1}(\mathcal{G}) = \Omega$, it follows that $w_g(\Omega) = \Omega$. It is easily verified that $w_g(t) = m^{-1}\big(m(t) \star g^{-1}\big)$ satisfies the composition property (7.2.17). Finally, if $\tilde{m}$ is a diffeomorphism, the inverse function $\tilde{m}^{-1}$ is also a diffeomorphism, from which it follows that $w_g(t) = \tilde{m}^{-1}\big(\tilde{m}(t) - \psi_{\mathcal{G}}(g)\big)$ is a diffeomorphism for fixed g. Since both $\tilde{m}^{-1}$ and $\psi_{\mathcal{G}}$ are continuous, $w_g(t)$ is continuous with respect to g.

Our proof of the converse statement generalizes a proof in [48]. Let $f(g) \triangleq w_{g^{-1}}(t_0)$ with fixed $t_0 \in \Theta$. Then $w_g\big(f(g_1)\big) = w_g\big(w_{g_1^{-1}}(t_0)\big) = w_{g \star g_1^{-1}}(t_0) = f(g_1 \star g^{-1})$. Setting $f(g_1) = t \in \Omega \subseteq \Theta$, we obtain

$$w_g(t) = f\big(f^{-1}(t) \star g^{-1}\big). \tag{7.2.20}$$

Since, by assumption, $f(g)$ is invertible on $f(\mathcal{G}) = \Omega \subseteq \Theta$, Eq. (7.2.20) holds for all $t \in \Omega \subseteq \Theta$ and $g \in \mathcal{G}$. Furthermore, (7.2.20) implies that the range of $w_g(t)$ is $w_g(\Omega) = f(\mathcal{G}) = \Omega$. To show that f is unique up to group translations, assume there is a function $\tilde{f}$ such that $w_g(t) = \tilde{f}\big(\tilde{f}^{-1}(t) \star g^{-1}\big)$. Then $w_{g_1}\big(\tilde{f}(g_2)\big) = \tilde{f}(g_2 \star g_1^{-1})$, but at the same time $w_{g_1}\big(\tilde{f}(g_2)\big) = f\big(f^{-1}\big(\tilde{f}(g_2)\big) \star g_1^{-1}\big)$. Hence, setting $g_1^{-1} = g$ and $g_2 = g_0$, we obtain $\tilde{f}(g \star g_0) = f\big(f^{-1}\big(\tilde{f}(g_0)\big) \star g\big)$, i.e., $\tilde{f}(g) = f(\kappa \star g)$ with $\kappa = f^{-1}\big(\tilde{f}(g_0)\big) \in \mathcal{G}$ fixed. □

Theorem 7.3 states that, under appropriate assumptions (in particular, $m(\Omega) = \mathcal{G}$ and not just $m(\Omega) \subseteq \mathcal{G}$ as required in Definition 7.1), every modulation function $m(t)$ induces an associated warping function $w_g(t) = m^{-1}\big(m(t) \star g^{-1}\big)$, and conversely, again under appropriate assumptions, every warping function $w_g(t)$ on a set Θ is generated—at least on a suitable subset $\Omega \subseteq \Theta$—by an associated modulation function $m(t)$. If $\Omega \subset \Theta$, i.e., $\Omega \neq \Theta$, then Θ can be partitioned into disjoint subsets Ω_i with associated modulation functions $m_i(t)$, $t \in \Omega_i$, such that $w_g(t) = m_i^{-1}\big(m_i(t) \star g^{-1}\big)$ for $t \in \Omega_i$ [47,49].

This relation between modulation and warping functions immediately translates into a relation between modulation and warping operators as stated by the next corollary.

Corollary 7.4. *Let $(\mathcal{G}, \star)$ and $(\mathcal{H}, \diamond)$ be groups isomorphic to $(\mathbb{R}, +)$ with isomorphisms $\psi_{\mathcal{G}}$ and $\psi_{\mathcal{H}}$. Let $\{\mathbf{M}_h\}_{h \in \mathcal{H}}$ be a modulation operator defined on the Hilbert space $\mathcal{X} = L^2(\Omega)$, based on an invertible modulation function $m : \Omega \to \mathcal{G}$ with $m(\Omega) = \mathcal{G}$. Then, if $\tilde{m}(t) = \psi_{\mathcal{G}}(m(t))$ is a diffeomorphism, there exists a warping operator $\{\mathbf{W}_g\}_{g \in (\mathcal{G}, \star)}$ on $\mathcal{X}$ with warping function $w_g(t) = m^{-1}(m(t) \star g^{-1})$.*

Conversely, let $\{\mathbf{W}_g\}_{g \in (\mathcal{G}, \star)}$ be a warping operator on $\mathcal{X} = L^2(\Theta)$ whose warping function $w_g(t)$ satisfies the condition of the converse part of Theorem 7.3. Then there exist a set $\Omega \subseteq \Theta$ and a modulation operator $\mathbf{M}_h$ on $\mathcal{X}_\Omega = L^2(\Omega) \subseteq \mathcal{X}$

whose modulation function $m(t)$ is associated to $w_g(t)$ according to $w_g(t) = m^{-1}(m(t) \star g^{-1})$. This modulation operator is a unitary representation of any group $(\mathcal{H}, \diamond)$ that is isomorphic to $(\mathbb{R}, +)$ via some isomorphism $\psi_{\mathcal{H}}$; it is uniquely defined up to a phase factor of the form $e^{j2\pi\psi_{\mathcal{H}}(\kappa)\psi_{\mathcal{G}}(\gamma)}$, where $\gamma \in \mathcal{G}, \kappa \in \mathcal{H}$.

We shall illustrate the relation between modulation and warping operators by discussing two examples. First, consider the modulation function $m(t) = t$ on $\Omega = \mathbb{R}$. Since $m(\Omega) = \mathbb{R}$, we must choose $(\mathcal{G}, \star) = (\mathbb{R}, +)$, whence $\psi_{\mathcal{G}}(g) = g$ and $\tilde{m}(t) \equiv \psi_{\mathcal{G}}(m(t)) = t$. Setting $(\mathcal{H}, \diamond) = (\mathbb{R}, +)$, whence $\psi_{\mathcal{H}}(h) = h$, the modulation operator is obtained as the frequency-shift operator, $(\mathbf{M}_h x)(t) = e^{j2\pi ht} x(t)$ with $t \in \Omega = \mathbb{R}$, $h \in (\mathcal{H}, \diamond) = (\mathbb{R}, +)$. The associated warping function is $w_g(t) \equiv m^{-1}(m(t) \star g^{-1}) = t - g$ with $t \in \Omega = \mathbb{R}$, $g \in (\mathcal{G}, \star) = (\mathbb{R}, +)$, so that the associated warping operator is the time-shift operator, $(\mathbf{W}_g x)(t) = x(t-g)$.

Next, consider $m(t) = t$ on $\Omega = \mathbb{R}^+$. Here, $m(\Omega) = \mathbb{R}^+$ so that we must choose $(\mathcal{G}, \star) = (\mathbb{R}^+, \cdot)$, isomorphic to $(\mathbb{R}, +)$ by $\psi_{\mathcal{G}}(g) = \ln g$. Hence, $\tilde{m}(t) \equiv \psi_{\mathcal{G}}(m(t)) = \ln t$. Setting $(\mathcal{H}, \diamond) = (\mathbb{R}, +)$, whence $\psi_{\mathcal{H}}(h) = h$, the modulation operator is obtained as $(\mathbf{M}_h x)(t) = e^{j2\pi h \ln t} x(t)$, $t \in \mathbb{R}^+$, $h \in (\mathbb{R}, +)$. The associated warping function and warping operator are $w_g(t) \equiv m^{-1}(m(t) \star g^{-1}) = t/g$ and $(\mathbf{W}_g x)(t) = \sqrt{1/g}\, x(t/g)$, respectively, with $t \in \mathbb{R}^+$, $g \in (\mathbb{R}^+, \cdot)$. These and other examples are listed in Table 7.1.

Using the modulation function $m(t)$ (or, equivalently, $\tilde{m}(t) = \psi_{\mathcal{G}}(m(t))$ associated to $w_g(t)$ according to (7.2.19), we can give the following expressions for the generalized eigenfunctions and eigenvalues of $\mathbf{W}_g$:

$$
\begin{aligned}
u_h^{\mathbf{W}}(t) &= \sqrt{|\tilde{m}'(t)|}\, e^{j2\pi\psi_{\mathcal{H}}(h)\tilde{m}(t)}, \qquad t \in \Omega,\ \ h \in (\mathcal{H}, \diamond) \\
\lambda_{g,h}^{\mathbf{W}} &= e^{-j2\pi\psi_{\mathcal{G}}(g)\psi_{\mathcal{H}}(h)}, \qquad g \in (\mathcal{G}, \star),\ \ h \in (\mathcal{H}, \diamond),
\end{aligned}
\tag{7.2.21}
$$

where $(\mathcal{H}, \diamond)$ is any group isomorphic to $(\mathbb{R}, +)$ with some isomorphism $\psi_{\mathcal{H}}$. The eigenfunctions $u_h^{\mathbf{W}}(t)$ are complete and orthonormal in $L^2(\Omega)$ in a generalized functions sense. Furthermore, it can be shown that on $L^2(\Omega)$, $\mathbf{W}_g$ is unitarily equivalent

$m(t)$	Ω or Ω_i	$(\mathcal{G}, \star)$	$w_g(t)$	Θ
t	$\mathbb{R}$	$(\mathbb{R}, +)$	$t - g$	$\mathbb{R}$
$\lvert t\rvert$	$\mathbb{R}^+, \mathbb{R}^-$	$(\mathbb{R}^+, \cdot)$	t/g	$\mathbb{R}$
$\lvert t\rvert^\kappa$ $(\kappa > 0)$	$\mathbb{R}^+, \mathbb{R}^-$	$(\mathbb{R}^+, \cdot)$	$g^{-1/\kappa} t$	$\mathbb{R}$
$\operatorname{sgn}(t)\lvert t\rvert^\kappa$ $(\kappa > 0)$	$\mathbb{R}$	$(\mathbb{R}, +)$	$\operatorname{sgn}(\operatorname{sgn}(t)\lvert t\rvert^\kappa - g) \cdot \lvert\operatorname{sgn}(t)\lvert t\rvert^\kappa - g\rvert^{1/\kappa}$	$\mathbb{R}$
$\ln\lvert t\rvert$	$\mathbb{R}^+, \mathbb{R}^-$	$(\mathbb{R}, +)$	$e^{-g} t$	$\mathbb{R}$
$\lvert\ln t\rvert$	$(0,1), (1,\infty)$	$(\mathbb{R}^+, \cdot)$	$t^{1/g}$	$\mathbb{R}^+$
e^t	$\mathbb{R}$	$(\mathbb{R}^+, \cdot)$	$t - \ln g$	$\mathbb{R}$

TABLE 7.1. Some modulation functions and associated warping functions. Note that $m(\Omega) = \mathcal{G}$ and $g \in (\mathcal{G}, \star)$.

to $\mathbf{T}_\tau$ via the same operator as in (7.2.16), i.e.,

$$\mathbf{W}_g = \mathbf{U}\mathbf{T}_{\psi_{\mathcal{G}}(g)}\mathbf{U}^{-1}, \tag{7.2.22}$$

with $\mathbf{U} : L^2(\mathbb{R}) \to L^2(\Omega)$, $(\mathbf{U}x)(t) = \sqrt{|\tilde{m}'(t)|}\,x\big(\tilde{m}(t)\big)$.

7.3 Dual operators

The time-shift operator $\mathbf{T}_\tau$ and the frequency-shift operator $\mathbf{F}_\nu$ are unitary representations of the group $(\mathbb{R}, +)$ on the Hilbert space $L^2(\mathbb{R})$. They satisfy the commutation relation

$$\mathbf{F}_\nu \mathbf{T}_\tau = e^{j2\pi\tau\nu}\mathbf{T}_\tau \mathbf{F}_\nu$$

and several other interesting relations. The strong relationship existing between $\mathbf{T}_\tau$ and $\mathbf{F}_\nu$ is generalized by the concept of *dual operators* (originally termed *conjugate operators* in [26, 30]). In this section, we define dual operators, study some of their properties, and show that associated modulation and warping operators are dual.

Definition 7.5. ([29, 30]). Let $(\mathcal{G}, \star)$ and $(\mathcal{H}, \diamond)$ be two groups isomorphic to $(\mathbb{R}, +)$ with isomorphisms $\psi_{\mathcal{G}}$ and $\psi_{\mathcal{H}}$, respectively, and consider two unitary representations $\{\mathbf{G}_g\}_{g\in\mathcal{G}}$ and $\{\mathbf{H}_h\}_{h\in\mathcal{H}}$ of $(\mathcal{G}, \star)$ and $(\mathcal{H}, \diamond)$, respectively, on a minimal invariant Hilbert space $\mathcal{X}$ (i.e., $\mathcal{X}$ is the smallest nonzero space that is invariant under both $\mathbf{G}_g$ and $\mathbf{H}_h$). Then, $\{\mathbf{H}_h\}_{h\in\mathcal{H}}$ is called *dual to* $\{\mathbf{G}_g\}_{g\in\mathcal{G}}$ if

$$\mathbf{H}_h\mathbf{G}_g = e^{j2\pi\psi_{\mathcal{G}}(g)\psi_{\mathcal{H}}(h)}\mathbf{G}_g\mathbf{H}_h, \quad \text{for all } \ g \in (\mathcal{G}, \star),\ h \in (\mathcal{H}, \diamond). \tag{7.3.23}$$

Note that when $\mathbf{H}_h$ is dual to $\mathbf{G}_g$, then at the same time $\mathbf{G}_{g^{-1}}$ is dual to $\mathbf{H}_h$ and $\mathbf{G}_g$ is dual to $\mathbf{H}_{h^{-1}}$. The time-shift operator $\mathbf{T}_\tau$ and the frequency-shift operator $\mathbf{F}_\nu$ constitute the most elementary example of dual operators ($\mathbf{F}_\nu$ is dual to $\mathbf{T}_\tau$). Further examples will be considered in Subsection 7.3.2.

7.3.1 Properties of dual operators

The following theorem states a relation of dual operators with the two-parameter group $(\mathbb{R}^2, +)$. As we will see later, this result implies that we can use the composition $\mathbf{D}_{g,h} = \mathbf{H}_h\mathbf{G}_g$ or $\mathbf{D}'_{g,h} = \mathbf{G}_g\mathbf{H}_h$ of dual operators $\mathbf{G}_g$, $\mathbf{H}_h$ as "displacement operators" for constructing covariant TF signal representations.

Theorem 7.6. *Let $(\mathcal{G}, \star)$ and $(\mathcal{H}, \diamond)$ be isomorphic to $(\mathbb{R}, +)$ with isomorphisms $\psi_{\mathcal{G}}$ and $\psi_{\mathcal{H}}$, respectively. Let $\{\mathbf{H}_h\}_{h\in\mathcal{H}}$ be dual to $\{\mathbf{G}_g\}_{g\in\mathcal{G}}$ on a minimal invariant Hilbert space $\mathcal{X}$. Then the unitary operator $\mathbf{D}_{g,h} \triangleq \mathbf{H}_h\mathbf{G}_g$ is an irreducible and faithful projective representation of the group $(\mathcal{G}, \star) \times (\mathcal{H}, \diamond)$ that is isomorphic to $(\mathbb{R}^2, +)$ with isomorphism $\psi_{\mathcal{G}\times\mathcal{H}}(g, h) = \big(\psi_{\mathcal{G}}(g), \psi_{\mathcal{H}}(h)\big)$, on $\mathcal{X}$, i.e.,*

$$\mathbf{D}_{g_0,h_0} = \mathbf{I}, \qquad \mathbf{D}_{g_2,h_2}\mathbf{D}_{g_1,h_1} = e^{-j2\pi\psi_{\mathcal{G}}(g_2)\psi_{\mathcal{H}}(h_1)}\mathbf{D}_{g_1\star g_2, h_1\diamond h_2}. \tag{7.3.24}$$

Similarly, also $\mathbf{D}'_{g,h} \triangleq \mathbf{G}_g \mathbf{H}_h$ *is an irreducible and faithful projective representation of* $(\mathcal{G}, \star) \times (\mathcal{H}, \diamond)$ *on* $\mathcal{X}$,

$$\mathbf{D}'_{g_0,h_0} = \mathbf{I}, \qquad \mathbf{D}'_{g_2,h_2}\mathbf{D}'_{g_1,h_1} = e^{j2\pi\psi_{\mathcal{G}}(g_1)\psi_{\mathcal{H}}(h_2)}\mathbf{D}'_{g_1 \star g_2, h_1 \diamond h_2}.$$

Proof: Clearly, $\mathbf{D}_{g_0,h_0} = \mathbf{H}_{h_0}\mathbf{G}_{g_0} = \mathbf{I}$. Furthermore, due to (7.3.23) and (7.2.8) we have

$$\begin{aligned}\mathbf{D}_{g_2,h_2}\mathbf{D}_{g_1,h_1} &= \mathbf{H}_{h_2}\mathbf{G}_{g_2}\mathbf{H}_{h_1}\mathbf{G}_{g_1} = e^{-j2\pi\psi_{\mathcal{G}}(g_2)\psi_{\mathcal{H}}(h_1)}\mathbf{H}_{h_2}\mathbf{H}_{h_1}\mathbf{G}_{g_2}\mathbf{G}_{g_1} \\ &= e^{-j2\pi\psi_{\mathcal{G}}(g_2)\psi_{\mathcal{H}}(h_1)}\mathbf{H}_{h_1 \diamond h_2}\mathbf{G}_{g_1 \star g_2} = e^{-j2\pi\psi_{\mathcal{G}}(g_2)\psi_{\mathcal{H}}(h_1)}\mathbf{D}_{g_1 \star g_2, h_1 \diamond h_2},\end{aligned}$$

which is the second equation in (7.3.24). Hence, $\mathbf{D}_{g,h}$ is a projective representation of the group $(\mathcal{G} \times \mathcal{H}, \circ) = (\mathcal{G}, \star) \times (\mathcal{H}, \diamond)$ with operation $(g_1, h_1) \circ (g_2, h_2) = (g_1 \star g_2, h_1 \diamond h_2)$. This group is easily seen to be isomorphic to $(\mathbb{R}^2, +)$ with isomorphism $\psi_{\mathcal{G}\times\mathcal{H}}(g, h) = (\psi_{\mathcal{G}}(g), \psi_{\mathcal{H}}(h))$. Finally, since $\mathcal{X}$ is minimal invariant under $\mathbf{G}_g$ and $\mathbf{H}_h$, it is also minimal invariant under the composition $\mathbf{D}_{g,h} = \mathbf{H}_h\mathbf{G}_g$; hence, $\mathbf{D}_{g,h}$ is irreducible. The faithfulness of $\mathbf{D}_{g,h}$ follows from Theorem 7.7. The proof for $\mathbf{D}'_{g,h}$ is analogous. □

The next theorem, proved in [50–52] (see also [29]), states a relation of dual operators to $\mathbf{T}_\tau$ and $\mathbf{F}_\nu$.

Theorem 7.7. *Let* $(\mathcal{G}, \star)$ *and* $(\mathcal{H}, \diamond)$ *be isomorphic to* $(\mathbb{R}, +)$ *with isomorphisms* $\psi_{\mathcal{G}}$ *and* $\psi_{\mathcal{H}}$, *respectively, and let* $\{\mathbf{G}_g\}_{g\in\mathcal{G}}$ *and* $\{\mathbf{H}_h\}_{h\in\mathcal{H}}$ *be two unitary representations on a minimal invariant Hilbert space* $\mathcal{X}$. *Then,* $\{\mathbf{H}_h\}_{h\in\mathcal{H}}$ *is dual to* $\{\mathbf{G}_g\}_{g\in\mathcal{G}}$ *if and only if* $\mathbf{G}_g$ *and* $\mathbf{H}_h$ *are unitarily equivalent, up to parameter transformations by* $\psi_{\mathcal{G}}$ *and* $\psi_{\mathcal{H}}$, *to (respectively)* $\mathbf{T}_\tau$ *and* $\mathbf{F}_\nu$ *on* $L^2(\mathbb{R})$, *i.e.,*

$$\mathbf{G}_g = \mathbf{U}\mathbf{T}_{\psi_{\mathcal{G}}(g)}\mathbf{U}^{-1} \quad \text{and} \quad \mathbf{H}_h = \mathbf{U}\mathbf{F}_{\psi_{\mathcal{H}}(h)}\mathbf{U}^{-1}, \tag{7.3.25}$$

where $\mathbf{U} : L^2(\mathbb{R}) \to \mathcal{X}$ *is an invertible and isometric (i.e., inner product preserving) operator.*

For a Hilbert space $\mathcal{X} = L^2(\Omega)$, the operators $\mathbf{U}$ and $\mathbf{U}^{-1}$ are explicitly given in terms of the (generalized) eigenfunctions of $\mathbf{G}_g$ [29,49]. The eigenfunctions $u_h^{\mathbf{G}}(t)$ of $\mathbf{G}_g$ do not depend on g, and they are parameterized by a parameter h that belongs to some group isomorphic to $(\mathbb{R}, +)$. This parameterization can always be chosen such that this group is $\mathcal{H}$. The system of eigenfunctions $\{u_h^{\mathbf{G}}(t)\}_{h\in\mathcal{H}}$ is complete and orthogonal in $\mathcal{X}$ in a generalized functions sense. For $\mathcal{X} = L^2(\Omega)$, we then have for $\mathbf{U}$ and $\mathbf{U}^{-1}$,

$$\mathbf{U} = \widetilde{\mathbf{U}}\mathbb{F}^{-1} \quad \text{with} \quad (\widetilde{\mathbf{U}}y)(t) = \int_{\mathbb{R}} y(r)u^{\mathbf{G}}_{\psi_{\mathcal{H}}^{-1}(r)}(t)dr, \qquad t \in \Omega \tag{7.3.26}$$

$$\begin{aligned}\mathbf{U}^{-1} = \mathbb{F}\widetilde{\mathbf{U}}^{-1} \quad \text{with} \quad (\widetilde{\mathbf{U}}^{-1}x)(r) &= \left\langle x, u^{\mathbf{G}}_{\psi_{\mathcal{H}}^{-1}(r)} \right\rangle \\ &= \int_{\Omega} x(t)u^{\mathbf{G}*}_{\psi_{\mathcal{H}}^{-1}(r)}(t)dt, \qquad r \in \mathbb{R},\end{aligned}$$

where $\mathbb{F}$ denotes the Fourier transform. We note that the transformation $\widetilde{\mathbf{U}}^{-1}$ is a reparameterized version of the $\mathbf{G}_g$-*Fourier transform* [29,45].

These results allow the construction of the dual operator $\mathbf{H}_h$ for a given operator $\mathbf{G}_g$: Let $\{\mathbf{G}_g\}_{g\in\mathcal{G}}$ be a unitary representation of $(\mathcal{G},\star)$ on $\mathcal{X}$ that is unitarily equivalent to $\mathbf{T}_\tau$ as in (7.3.25). Then the eigenfunctions of $\mathbf{G}_g$ define the operator $\mathbf{U}$ according to (7.3.26), and the operator $\mathbf{H}_h$ dual to $\mathbf{G}_g$ is obtained from the second equation in (7.3.25).

For later use, we finally note two eigenfunction relations of dual operators. Let $\{\mathbf{H}_h\}_{h\in(\mathcal{H},\diamond)}$ be dual to $\{\mathbf{G}_g\}_{g\in(\mathcal{G},\star)}$, and let $\{u_h^{\mathbf{G}}(t)\}_{h\in(\mathcal{H},\diamond)}$ and $\{u_g^{\mathbf{H}}(t)\}_{g\in(\mathcal{G},\star)}$ denote the eigenfunctions of $\mathbf{G}_g$ and $\mathbf{H}_h$, respectively. Then it can be shown [26,30] that, assuming suitable parameterization of the eigenfunctions,

$$\big(\mathbf{G}_{g_1} u_g^{\mathbf{H}}\big)(t) = u_{g\star g_1}^{\mathbf{H}}(t), \qquad \big(\mathbf{H}_{h_1} u_h^{\mathbf{G}}\big)(t) = u_{h\diamond h_1}^{\mathbf{G}}(t). \tag{7.3.27}$$

That is, $\mathbf{G}_g$ maps an eigenfunction $u_g^{\mathbf{H}}(t)$ of the dual operator $\mathbf{H}_h$ again onto an eigenfunction of $\mathbf{H}_h$ with the eigenfunction parameter g translated by g_1, and similarly for $\mathbf{H}_h$.

7.3.2 Modulation and warping operators as dual operators

Let $(\mathcal{G},\star)$ and $(\mathcal{H},\diamond)$ be groups isomorphic to $(\mathbb{R},+)$ with isomorphisms $\psi_{\mathcal{G}}$ and $\psi_{\mathcal{H}}$, respectively. Consider the modulation operator $\{\mathbf{M}_h\}_{h\in(\mathcal{H},\diamond)}$ generated by some modulation function $m(t)$ defined on some set Ω, with $m(\Omega)=\mathcal{G}$, and the warping operator $\{\mathbf{W}_g\}_{g\in(\mathcal{G},\star)}$ that is *associated* to $\{\mathbf{M}_h\}_{h\in(\mathcal{H},\diamond)}$ in the sense of Corollary 7.4, i.e., the underlying warping function $w_g(t)$ is induced by $m(t)$ according to $w_g(t) = m^{-1}\big(m(t)\star g^{-1}\big)$. Note that $\mathbf{M}_h$ and $\mathbf{W}_g$ are defined on the same Hilbert space $\mathcal{X}=L^2(\Omega)$.

According to (7.2.16) and (7.2.22), $\mathbf{M}_h$ and $\mathbf{W}_g$ are unitarily equivalent to $\mathbf{F}_\nu$ and $\mathbf{T}_\tau$, respectively, i.e., $\mathbf{M}_h = \mathbf{U}\mathbf{F}_{\psi_{\mathcal{H}}(h)}\mathbf{U}^{-1}$ and $\mathbf{W}_g = \mathbf{U}\mathbf{T}_{\psi_{\mathcal{G}}(g)}\mathbf{U}^{-1}$, with $(\mathbf{U}x)(t) = \sqrt{|\tilde{m}'(t)|}\,x\big(\tilde{m}(t)\big)$. With Theorem 7.7, we then obtain the important result that *associated modulation and warping operators are dual* ($\mathbf{M}_h$ is dual to $\mathbf{W}_g$). Some examples of dual modulation and warping operators are listed in Table 7.2 (see also Table 7.1).

$m(t)$	Ω	$(\mathcal{H},\diamond)$	$(\mathcal{G},\star)$	$w_g(t)$	$(\mathbf{M}_h x)(t)$	$(\mathbf{W}_g x)(t)$
t	$\mathbb{R}$	$(\mathbb{R},+)$	$(\mathbb{R},+)$	$t-g$	$e^{j2\pi ht}x(t)$	$x(t-g)$
t	$\mathbb{R}^+$	$(\mathbb{R},+)$	$(\mathbb{R}^+,\cdot)$	t/g	$e^{j2\pi h\ln t}x(t)$	$(1/\sqrt{g})x(t/g)$
$\ln t$	$(1,\infty)$	$(\mathbb{R},+)$	$(\mathbb{R}^+,\cdot)$	$t^{1/g}$	$e^{j2\pi h\ln(\ln t)}x(t)$	$[(1/g)t^{1/g-1}]^{1/2}x(t^{1/g})$
e^t	$\mathbb{R}$	$(\mathbb{R},+)$	$(\mathbb{R}^+,\cdot)$	$t-\ln g$	$e^{j2\pi ht}x(t)$	$x(t-\ln g)$
$1/t$	$\mathbb{R}^+$	$(\mathbb{R},+)$	$(\mathbb{R}^+,\cdot)$	gt	$e^{-j2\pi h\ln t}x(t)$	$\sqrt{g}x(gt)$

TABLE 7.2. Some dual modulation and warping operators along with the underlying modulation and warping functions. We note that $m(\Omega)=\mathcal{G}$, $g\in(\mathcal{G},\star)$, and $h\in(\mathcal{H},\diamond)$.

7.4 Affine operators

The TF scaling operator $\mathbf{C}_\sigma$ and the time-shift operator $\mathbf{T}_\tau$ are unitary representations of the groups $(\mathbb{R}^+, \cdot)$ and $(\mathbb{R}, +)$, respectively, on the Hardy space $H^2(\mathbb{R})$, and they satisfy the commutation relation

$$\mathbf{T}_{\sigma\tau}\mathbf{C}_\sigma = \mathbf{C}_\sigma\mathbf{T}_\tau. \tag{7.4.28}$$

(We note that the Hardy space $H^2(\mathbb{R})$ is a subspace of $L^2(\mathbb{R})$ that contains all functions whose Fourier transforms are zero for negative frequencies.) This relation between $\mathbf{T}_\tau$ and $\mathbf{C}_\sigma$ is generalized by the concept of *affine operators*. In this section, we introduce affine operators, discuss some of their properties, and show that suitably defined pairs of modulation and warping operators are affine operators.

Definition 7.8. Let $(\mathcal{G}, \star)$ and $(\mathcal{H}, \diamond)$ be two groups isomorphic to $(\mathbb{R}, +)$ with isomorphisms $\psi_\mathcal{G}$ and $\psi_\mathcal{H}$, respectively, and consider two unitary representations $\{\mathbf{G}_g\}_{g\in\mathcal{G}}$ and $\{\mathbf{H}_h\}_{h\in\mathcal{H}}$ of $(\mathcal{G}, \star)$ and $(\mathcal{H}, \diamond)$, respectively, on a minimal invariant Hilbert space $\mathcal{X}$. Then, $\{\mathbf{H}_h\}_{h\in\mathcal{H}}$ is called *affine to* $\{\mathbf{G}_g\}_{g\in\mathcal{G}}$ if

$$\mathbf{H}_{\mu(g,h)}\mathbf{G}_g = \mathbf{G}_g\mathbf{H}_h, \qquad \text{for all } g \in (\mathcal{G}, \star),\ h \in (\mathcal{H}, \diamond),$$
$$\text{with } \mu(g,h) = \psi_\mathcal{H}^{-1}\big(\psi_\mathcal{H}(h)\exp\big(\psi_\mathcal{G}(g)\big)\big). \tag{7.4.29}$$

An elementary example of affine operators is given by the time-shift operator $\mathbf{T}_\tau$ and the scaling operator $\mathbf{C}_\sigma$ ($\mathbf{T}_\tau$ is affine to $\mathbf{C}_\sigma$). Further examples will be considered in Subsection 7.4.2.

7.4.1 *Properties of affine operators*

The next theorem states a relation of affine operators with the affine group (see Subsection 7.2.1), thus explaining the term "affine operators." Later, we will see that this result implies that we can use the composition $\mathbf{D}_{g,h} = \mathbf{H}_h\mathbf{G}_g$ or $\mathbf{D}'_{g,h} = \mathbf{G}_g\mathbf{H}_h$ of affine operators as "displacement operators."

Theorem 7.9. *Let* $(\mathcal{G}, \star)$ *and* $(\mathcal{H}, \diamond)$ *be isomorphic to* $(\mathbb{R}, +)$ *with isomorphisms* $\psi_\mathcal{G}$ *and* $\psi_\mathcal{H}$*, respectively. Let* $\{\mathbf{H}_h\}_{h\in\mathcal{H}}$ *be affine to* $\{\mathbf{G}_g\}_{g\in\mathcal{G}}$ *on a minimal invariant Hilbert space* $\mathcal{X}$*. Then* $\mathbf{D}_{g,h} \triangleq \mathbf{H}_h\mathbf{G}_g$ *is an irreducible and faithful unitary representation of the group* $(\mathcal{G} \times \mathcal{H}, \circ)$ *that is isomorphic to the affine group with isomorphism*

$$\psi_{\mathcal{G}\times\mathcal{H}}(g,h) = \big(\exp\big(\psi_\mathcal{G}(g)\big), \psi_\mathcal{H}(h)\big),$$

on $\mathcal{X}$*, i.e.,*

$$\mathbf{D}_{g_0,h_0} = \mathbf{I}, \qquad \mathbf{D}_{g_2,h_2}\mathbf{D}_{g_1,h_1} = \mathbf{D}_{g_1\star g_2,\mu(g_2,h_1)\,\diamond\, h_2}, \tag{7.4.30}$$

with $\mu(g,h) = \psi_\mathcal{H}^{-1}\big(\psi_\mathcal{H}(h)\exp\big(\psi_\mathcal{G}(g)\big)\big)$*. Similarly,* $\mathbf{D}'_{g,h} \triangleq \mathbf{G}_g\mathbf{H}_h$ *is an irreducible and faithful unitary representation of the group* $(\mathcal{G}\times\mathcal{H}, \circ')$ *that is isomorphic to the affine group with isomorphism*

$$\psi_{\mathcal{G}\times\mathcal{H}}(g,h) = \big(\exp\big(\psi_{\mathcal{G}}(g)\big), \psi_{\mathcal{H}}(h)\exp\big(\psi_{\mathcal{G}}(g)\big)\big),$$

on $\mathcal{X}$, i.e.,

$$\mathbf{D}'_{g_0,h_0} = \mathbf{I}, \qquad \mathbf{D}'_{g_2,h_2}\mathbf{D}'_{g_1,h_1} = \mathbf{D}'_{g_1\star g_2, h_1\diamond\mu(g_1^{-1},h_2)}.$$

Proof: Clearly, $\mathbf{D}_{g_0,h_0} = \mathbf{H}_{h_0}\mathbf{G}_{g_0} = \mathbf{I}$. Furthermore, due to (7.4.29) and (7.2.8) we have

$$\begin{aligned}\mathbf{D}_{g_2,h_2}\mathbf{D}_{g_1,h_1} &= \mathbf{H}_{h_2}\mathbf{G}_{g_2}\mathbf{H}_{h_1}\mathbf{G}_{g_1} = \mathbf{H}_{h_2}\mathbf{H}_{\mu(g_2,h_1)}\mathbf{G}_{g_2}\mathbf{G}_{g_1}\\ &= \mathbf{H}_{\mu(g_2,h_1)\diamond h_2}\mathbf{G}_{g_1\star g_2} = \mathbf{D}_{g_1\star g_2,\mu(g_2,h_1)\diamond h_2},\end{aligned}$$

which is the second equation in (7.4.30). Hence, $\mathbf{D}_{g,h}$ is a unitary representation of $(\mathcal{G}\times\mathcal{H},\circ)$ with operation $(g_1,h_1)\circ(g_2,h_2) = \big(g_1\star g_2, \mu(g_2,h_1)\diamond h_2\big)$. To see that this group is isomorphic to the affine group, we apply the isomorphism $(\tilde{g},\tilde{h}) = \psi_{\mathcal{G}\times\mathcal{H}}(g,h) = \big(\exp\big(\psi_{\mathcal{G}}(g)\big), \psi_{\mathcal{H}}(h)\big)$ to this group operation and obtain, after some intermediate steps,

$$\begin{aligned}&\psi_{\mathcal{G}\times\mathcal{H}}\big((g_1,h_1)\circ(g_2,h_2)\big)\\ &\quad= \Big(\exp\big(\psi_{\mathcal{G}}(g_1)\big)\exp\big(\psi_{\mathcal{G}}(g_2)\big), \psi_{\mathcal{H}}(h_1)\exp\big(\psi_{\mathcal{G}}(g_2)\big) + \psi_{\mathcal{H}}(h_2)\Big),\end{aligned}$$

or equivalently, $(\tilde{g}_1,\tilde{h}_1)\tilde{\circ}(\tilde{g}_2,\tilde{h}_2) = (\tilde{g}_1\tilde{g}_2, \tilde{h}_1\tilde{g}_2 + \tilde{h}_2)$, which is the group operation of the affine group. Finally, since $\mathcal{X}$ is minimal invariant under $\mathbf{G}_g$ and $\mathbf{H}_h$, it is also minimal invariant under the composition $\mathbf{D}_{g,h} = \mathbf{H}_h\mathbf{G}_g$; hence, $\mathbf{D}_{g,h}$ is irreducible. The faithfulness of $\mathbf{D}_{g,h}$ follows from Theorem 7.10. The proof for $\mathbf{D}'_{g,h}$ is analogous. □

The next theorem states a relation of affine operators to the operators $\mathbf{C}_\sigma$ and $\mathbf{T}_\tau$.

Theorem 7.10. *Let $(\mathcal{G},\star)$ and $(\mathcal{H},\diamond)$ be isomorphic to $(\mathbb{R},+)$ with isomorphisms $\psi_{\mathcal{G}}$ and $\psi_{\mathcal{H}}$, respectively, and let $\{\mathbf{G}_g\}_{g\in\mathcal{G}}$ and $\{\mathbf{H}_h\}_{h\in\mathcal{H}}$ be two unitary representations on a minimal invariant Hilbert space $\mathcal{X}$. Then, $\{\mathbf{H}_h\}_{h\in\mathcal{H}}$ is affine to $\{\mathbf{G}_g\}_{g\in\mathcal{G}}$ if and only if $\mathbf{G}_g$ and $\mathbf{H}_h$ are unitarily equivalent, up to parameter transformations specified below, to (respectively) $\mathbf{C}_\sigma$ and $\mathbf{T}_\tau$ on $H^2(\mathbb{R})$, i.e.,*

$$\mathbf{G}_g = \mathbf{V}\mathbf{C}_{\exp(\psi_{\mathcal{G}}(g))}\mathbf{V}^{-1} \quad \text{and} \quad \mathbf{H}_h = \mathbf{V}\mathbf{T}_{\pm\psi_{\mathcal{H}}(h)}\mathbf{V}^{-1}, \tag{7.4.31}$$

where $\mathbf{V}: H^2(\mathbb{R}) \to \mathcal{X}$ is an invertible and isometric operator.

Proof: Using (7.4.28) and $\mathbf{V}^{-1}\mathbf{V} = \mathbf{I}$, Eq. (7.4.31) implies

$$\begin{aligned}\mathbf{G}_g\mathbf{H}_h &= \mathbf{V}\mathbf{C}_{\exp(\psi_{\mathcal{G}}(g))}\mathbf{V}^{-1}\mathbf{V}\mathbf{T}_{\pm\psi_{\mathcal{H}}(h)}\mathbf{V}^{-1} = \mathbf{V}\mathbf{C}_{\exp(\psi_{\mathcal{G}}(g))}\mathbf{T}_{\pm\psi_{\mathcal{H}}(h)}\mathbf{V}^{-1}\\ &= \mathbf{V}\mathbf{T}_{\pm\psi_{\mathcal{H}}(h)\exp(\psi_{\mathcal{G}}(g))}\mathbf{C}_{\exp(\psi_{\mathcal{G}}(g))}\mathbf{V}^{-1} = \mathbf{H}_{\mu(g,h)}\mathbf{G}_g,\end{aligned}$$

which is equal to (7.4.29). Hence, $\mathbf{H}_h$ is affine to $\mathbf{G}_g$.

Conversely, let $\{\mathbf{H}_h\}_{h\in\mathcal{H}}$ be affine to $\{\mathbf{G}_g\}_{g\in\mathcal{G}}$. Due to Theorem 7.9, $\mathbf{D}_{g,h} = \mathbf{H}_h\mathbf{G}_g$ is an irreducible unitary representation of a group isomorphic to the affine group. According to [53, 54], there then exists an invertible and isometric operator $\mathbf{V} : H^2(\mathbb{R}) \to \mathcal{X}$ such that $\mathbf{D}_{g,h} = \mathbf{V}\mathbf{T}_{\pm\psi_\mathcal{H}(h)}\mathbf{C}_{\exp(\psi_\mathcal{G}(g))}\mathbf{V}^{-1}$ or, equivalently, $\mathbf{H}_h\mathbf{G}_g = \mathbf{V}\mathbf{T}_{\pm\psi_\mathcal{H}(h)}\mathbf{V}^{-1}\mathbf{V}\,\mathbf{C}_{\exp(\psi_\mathcal{G}(g))}\mathbf{V}^{-1}$. With $\mathbf{T}_{\pm\psi_\mathcal{H}(h_0)} = \mathbf{C}_{\exp(\psi_\mathcal{G}(g_0))} = \mathbf{I}$, this implies

$$\mathbf{G}_g = \mathbf{H}_{h_0}\mathbf{G}_g = \mathbf{V}\mathbf{I}\mathbf{V}^{-1}\mathbf{V}\mathbf{C}_{\exp(\psi_\mathcal{G}(g))}\mathbf{V}^{-1} = \mathbf{V}\mathbf{C}_{\exp(\psi_\mathcal{G}(g))}\mathbf{V}^{-1}$$
$$\mathbf{H}_h = \mathbf{H}_h\mathbf{G}_{g_0} = \mathbf{V}\mathbf{T}_{\pm\psi_\mathcal{H}(h)}\mathbf{V}^{-1}\mathbf{V}\mathbf{I}\mathbf{V}^{-1} = \mathbf{V}\mathbf{T}_{\pm\psi_\mathcal{H}(h)}\mathbf{V}^{-1}. \qquad \square$$

We note that for $\mathcal{X} = L^2(\Omega)$, the operator $\mathbf{V}$ is given by $\mathbf{V} = \widetilde{\mathbf{U}}\mathbf{L}$, where $\widetilde{\mathbf{U}}$ was defined in (7.3.26) and $\mathbf{L}$ is given in the frequency domain (the frequency-domain version of $\mathbf{L}$ will be denoted as $\hat{\mathbf{L}}$) by $\mathbf{L} : H^2(\mathbb{R}) \to L^2(\mathbb{R})$, $(\hat{\mathbf{L}}X)(f) = \sqrt{e^{-f}}X(e^{-f})$. This can be used to construct an "affine operator" $\mathbf{H}_h$ to a given operator $\mathbf{G}_g$: Let $\{\mathbf{G}_g\}_{g\in\mathcal{G}}$ be a unitary representation of $\mathcal{G}$ on $\mathcal{X}$ that is unitarily equivalent to $\mathbf{C}_\sigma$ as in (7.4.31). Then the eigenfunctions of $\mathbf{G}_g$ define the operator $\widetilde{\mathbf{U}}$ by (7.3.26), and $\mathbf{H}_h$ is obtained from the second equation in (7.4.31) with $\mathbf{V} = \widetilde{\mathbf{U}}\mathbf{L}$.

7.4.2 *Modulation and warping operators as affine operators*

Let $(\mathcal{G}, \star)$ and $(\mathcal{H}, \diamond)$ be groups isomorphic to $(\mathbb{R}, +)$ with isomorphisms $\psi_\mathcal{G}$ and $\psi_\mathcal{H}$, respectively. Consider the modulation operator $\{\mathbf{M}_h\}_{h\in(\mathcal{H},\diamond)}$ generated by a modulation function $m : \Omega \to \mathcal{G}$ with some set Ω, such that $\tilde{m}(\Omega) = \mathbb{R}^+$ or $\tilde{m}(\Omega) = \mathbb{R}^-$ or $\tilde{m}(\Omega) = \mathbb{R}$, where $\tilde{m}(t) \triangleq \psi_\mathcal{G}(m(t))$. Furthermore, consider the warping operator $\{\mathbf{W}_g\}_{g\in(\mathcal{G},\star)}$ whose warping function is related to $m(t)$ according to

$$w_g(t) = \tilde{m}^{-1}\big(\tilde{m}(t)\exp\big(\psi_\mathcal{G}(g)\big)\big). \tag{7.4.32}$$

Both $\mathbf{M}_h$ and $\mathbf{W}_g$ are defined on the Hilbert space $\mathcal{X} = L^2(\Omega)$. Note that $\mathbf{M}_h$ and $\mathbf{W}_g$ are *not* dual operators—indeed, the warping operator to which $\mathbf{M}_h$ is dual would have the warping function $w_g(t) = \tilde{m}^{-1}\big(\tilde{m}(t) - \psi_\mathcal{G}(g)\big)$. Instead, $\mathbf{M}_h$ is *affine* to $\mathbf{W}_g$, as is easily verified: we have

$$\begin{aligned}\big(\mathbf{M}_{\mu(g,h)}\mathbf{W}_g x\big)(t) &= e^{j2\pi\psi_\mathcal{H}(h)\exp(\psi_\mathcal{G}(g))\tilde{m}(t)}\sqrt{|w_g'(t)|}\, x\big(w_g(t)\big)\\ &= \sqrt{|w_g'(t)|}\, e^{j2\pi\psi_\mathcal{H}(h)\tilde{m}(w_g(t))}x\big(w_g(t)\big) = \big(\mathbf{W}_g\mathbf{M}_h x\big)(t).\end{aligned}$$

Affine modulation and warping operators as formulated above are unitarily equivalent to $\mathbf{T}_\tau$ and $\mathbf{C}_\sigma$ (defined on $H^2(\mathbb{R})$) according to

$$\mathbf{M}_h = \mathbf{V}\mathbf{T}_{\pm\psi_\mathcal{H}(h)}\mathbf{V}^{-1} \quad \text{and} \quad \mathbf{W}_g = \mathbf{V}\mathbf{C}_{\exp(\psi_\mathcal{G}(g))}\mathbf{V}^{-1}, \tag{7.4.33}$$

where $\mathbf{V} = \mathbf{U}\mathbb{F}$ with $(\mathbf{U}x)(t) = \sqrt{|\tilde{m}'(t)|}x\big(|\tilde{m}(t)|\big)$. In (7.4.33), the $+$ $(-)$ sign applies for $\tilde{m}(\Omega) = \mathbb{R}^-$ $(\tilde{m}(\Omega) = \mathbb{R}^+)$. Some examples of affine modulation and warping operators are listed in Table 7.3 (cf. Table 7.1).

$m(t)$	Ω	$(\mathcal{H},\diamond)$ $(\mathcal{G},\star)$	$w_g(t)$	$(\mathbf{M}_h x)(t)$ $(\mathbf{W}_g x)(t)$
t	$\mathbb{R}$	$(\mathbb{R},+)$ $(\mathbb{R},+)$	te^g	$e^{j2\pi ht}x(t)$ $e^{g/2}x(te^g)$
t	$\mathbb{R}^+$	$(\mathbb{R},+)$ $(\mathbb{R}^+,\cdot)$	t^g	$e^{j2\pi h\ln t}x(t)$ $\sqrt{gt^{g-1}}x(t^g)$
$\ln t$	$(1,\infty)$	$(\mathbb{R},+)$ $(\mathbb{R}^+,\cdot)$	$\exp\left((\ln t)^g\right)$	$e^{j2\pi h\ln(\ln t)}x(t)$ $\sqrt{w_g(t)\frac{g}{t}(\ln t)^{g-1}}x\left(w_g(t)\right)$
e^t	$\mathbb{R}$	$(\mathbb{R},+)$ $(\mathbb{R}^+,\cdot)$	gt	$e^{j2\pi ht}x(t)$ $\sqrt{g}x(gt)$
$\exp\left(\mathrm{sgn}(t)\lvert t\rvert^{1/\kappa}\right)$, $\kappa>0$	$\mathbb{R}$	$(\mathbb{R},+)$ $(\mathbb{R}^+,\cdot)$	$g^\kappa t$	$e^{j2\pi h\mathrm{sgn}(t)\lvert t\rvert^{1/\kappa}}x(t)$ $g^{\kappa/2}x(g^\kappa t)$

TABLE 7.3. Some affine modulation and warping operators along with the underlying modulation and warping functions. We note that $m(\Omega) \subseteq \mathcal{G}$ such that $\tilde{m}(\Omega) = \mathbb{R}^+, \mathbb{R}^-$, or $\mathbb{R}$ (where $\tilde{m}(t) \triangleq \psi_{\mathcal{G}}(m(t))$), $g \in (\mathcal{G},\star)$, and $h \in (\mathcal{H},\diamond)$.

7.5 Covariant signal representations: group domain

So far, we have discussed modulation and warping operators as well as dual and affine pairs of operators. In this section, we begin our development of the covariance theory. This theory is based on unitary "displacement operators" (DOs) $\mathbf{D}_{\alpha,\beta}$, where the "displacement parameter" (α,β) belongs to a group. After a formal definition and discussion of DOs, we shall consider the systematic construction of linear and bilinear signal representations that are covariant to a given DO $\mathbf{D}_{\alpha,\beta}$. These covariant signal representations are functions of the displacement parameter (α,β). In later sections, we will consider the conversion of covariant (α,β)-representations into covariant TF representations.

7.5.1 Displacement operators

The notion underlying DOs $\mathbf{D}_{\alpha,\beta}$ is that they displace signals in the TF plane without changing their energy. More specifically, if a signal x is concentrated about a TF point (t,f), then the transformed signal $\mathbf{D}_{\alpha,\beta}x$ is concentrated about a TF point (t',f') that depends on (t,f) and on (α,β). Elementary examples of DOs are the operators $\mathbf{S}_{\tau,\nu} = \mathbf{F}_\nu\mathbf{T}_\tau$ and $\mathbf{R}_{\sigma,\tau} = \mathbf{T}_\tau\mathbf{C}_\sigma$ for which $(t',f') = (t+\tau, f+\nu)$ and $(t',f') = (\sigma t+\tau, f/\sigma)$, respectively. In what follows, we will often use the shorthand notation $\theta \triangleq (\alpha,\beta)$, so that DOs $\mathbf{D}_{\alpha,\beta}$ will be briefly written as $\mathbf{D}_\theta$.

We first discuss some basic assumptions about TF displacements and DOs as well as the mathematical structure of DOs that results from these assumptions.

1. We assume that, for each θ, the operator $\mathbf{D}_\theta$ maps a Hilbert space $\mathcal{X}$ onto itself. We also assume that $\mathbf{D}_\theta$ does not change a signal's energy and that all displacements can be reversed; hence, it is natural to model $\mathbf{D}_\theta$ as a *unitary* (i.e., invertible and isometric) operator on $\mathcal{X}$.
2. We further assume that displacing a signal first by θ_1 and then by θ_2 is essentially equivalent to a single displacement by $\theta_1 \circ \theta_2$, where $\circ$ is some group law; hence, $\theta = (\alpha, \beta)$ must belong to some group $(\mathcal{D}, \circ)$. In terms of $\mathbf{D}_\theta$, this assumption suggests the composition law $\mathbf{D}_{\theta_2}\mathbf{D}_{\theta_1} = \mathbf{D}_{\theta_1 \circ \theta_2}$, i.e., that $\mathbf{D}_\theta$ is a unitary representation of the group $(\mathcal{D}, \circ)$. However, for the TF shift operator we have $\mathbf{S}_{\tau_2,\nu_2}\mathbf{S}_{\tau_1,\nu_1} = e^{-j2\pi\nu_1\tau_2}\mathbf{S}_{\tau_1+\tau_2,\nu_1+\nu_2}$. Allowing for a phase factor in our composition law implies that $\mathbf{D}_\theta$ is a *projective* representation of the group $(\mathcal{D}, \circ)$ (see Subsection 7.2.2).
3. We also assume that the projective group representation $\mathbf{D}_\theta$ is *irreducible* on $\mathcal{X}$, i.e., $\mathcal{X}$ is minimal invariant under $\mathbf{D}_\theta$. Translated into the TF domain, this corresponds to the requirement that all TF points (of our underlying TF point set) can be reached via displacements from other TF points.
4. Finally, we assume $\mathbf{D}_\theta$ to be a *faithful* group representation, i.e., $\mathbf{D}_\theta = \mathbf{D}_{\theta'}$ implies $\theta = \theta'$. This means that each TF displacement corresponds to a uniquely defined TF displacement parameter θ.

These assumptions will now be summarized in the following formal definition of a DO.

Definition 7.11. ([25, 28]). Let $(\mathcal{D}, \circ)$ be a topological group with group identity element θ_0. We call a unitary operator family $\{\mathbf{D}_\theta\}_{\theta \in \mathcal{D}}$ defined on an infinite-dimensional Hilbert space $\mathcal{X}$ a *displacement operator (DO)* if $\mathbf{D}_\theta$ is an irreducible and faithful projective representation of $(\mathcal{D}, \circ)$ on $\mathcal{X}$. In particular, this implies that $\mathcal{X}$ is minimal invariant under $\mathbf{D}_\theta$ and that

$$\mathbf{D}_{\theta_0} = \mathbf{I} \quad \text{and} \quad \mathbf{D}_{\theta_2}\mathbf{D}_{\theta_1} = c(\theta_1, \theta_2)\mathbf{D}_{\theta_1 \circ \theta_2} \qquad \text{for all } \theta_1, \theta_2 \in \mathcal{D}, \tag{7.5.34}$$

with a continuous function $c : \mathcal{D} \times \mathcal{D} \to \mathbb{C}$.

An important example of DOs is provided by the composition of dual or affine warping and modulation operators based on groups $(\mathcal{A}, \bullet)$ and $(\mathcal{B}, *)$ that are isomorphic to $(\mathbb{R}, +)$:

$$\begin{aligned}(\mathbf{D}_{\alpha,\beta}x)(t) &\\ &= (\mathbf{M}_\beta \mathbf{W}_\alpha x)(t) \\ &= e^{j2\pi\psi_{\mathcal{B}}(\beta)\tilde{m}(t)} \sqrt{|w'_\alpha(t)|}\, x\big(w_\alpha(t)\big), \qquad \alpha \in (\mathcal{A}, \bullet),\ \beta \in (\mathcal{B}, *),\ t \in \Omega,\end{aligned} \tag{7.5.35}$$

with $\tilde{m}(t) = \psi_{\mathcal{A}}(m(t))$. Here, $\mathbf{M}_\beta$ is either dual to $\mathbf{W}_\alpha$, i.e.,

$$w_\alpha(t) = m^{-1}\big(m(t) \bullet \alpha^{-1}\big) = \tilde{m}^{-1}\big(\tilde{m}(t) - \psi_{\mathcal{A}}(\alpha)\big),$$

or $\mathbf{M}_\beta$ is affine to $\mathbf{W}_\alpha$, i.e.,

$$w_\alpha(t) = \tilde{m}^{-1}\big(\tilde{m}(t) \exp\big(\psi_{\mathcal{A}}(\alpha)\big)\big).$$

The cocycle of $\mathbf{D}_{\alpha,\beta}$ is $c((\alpha_1,\beta_1),(\alpha_2,\beta_2)) = \exp(-j2\pi\psi_{\mathcal{A}}(\alpha_2)\psi_{\mathcal{B}}(\beta_1))$ in the dual case and $c((\alpha_1,\beta_1),(\alpha_2,\beta_2)) \equiv 1$ in the affine case.

The DO property is preserved by unitary equivalence transformations and invertible parameter transformations. Let $\{\mathbf{D}_\theta\}_{\theta\in(\mathcal{D},\circ)}$ be a DO for a group $(\mathcal{D},\circ)$ on a Hilbert space $\mathcal{X}$. Let $\mathbf{U}$ be an invertible and isometric operator mapping $\mathcal{X}$ onto a Hilbert space $\mathcal{X}'$ (generally different from $\mathcal{X}$), and let $(\mathcal{D}',\circ')$ be a group isomorphic to $(\mathcal{D},\circ)$ via some isomorphism ψ. Then, the operator family

$$\mathbf{D}'_{\theta'} = \mathbf{U}\mathbf{D}_{\psi(\theta')}\mathbf{U}^{-1}, \qquad \theta' \in (\mathcal{D}',\circ') \tag{7.5.36}$$

can be shown to be a DO for the group $(\mathcal{D}',\circ')$, on the Hilbert space $\mathcal{X}'$.

Further important properties of DOs will be considered in Subsections 7.6.2 and 7.6.3.

7.5.2 *Covariant linear θ-representations*

Let us now consider linear θ-representations $L_x(\theta) = L_x(\alpha,\beta)$, i.e., signal representations that are functions of the displacement parameter $\theta \in (\mathcal{D},\circ)$ and that depend on the signal $x(t)$ in a linear manner. We note that any linear θ-representation can be written in the form

$$L_x(\theta) = \langle x, h_\theta\rangle = \int_t x(t)h_\theta^*(t)dt, \tag{7.5.37}$$

where $h_\theta(t)$ is a function that depends on the displacement parameter θ but not on the signal x.

Definition 7.12. ([25]). Let $\{\mathbf{D}_\theta\}_{\theta\in(\mathcal{D},\circ)}$ be a DO with cocycle c, defined on a Hilbert space $\mathcal{X}$. A linear θ-representation $L_x(\theta)$ is called *covariant to the DO* $\mathbf{D}_\theta$ if for all $x \in \mathcal{X}$

$$L_{\mathbf{D}_{\theta_1}x}(\theta) = c(\theta\circ\theta_1^{-1},\theta_1)L_x(\theta\circ\theta_1^{-1}), \qquad \text{for all } \theta,\theta_1 \in \mathcal{D}. \tag{7.5.38}$$

Hence, the representation of the displaced signal $(\mathbf{D}_{\theta_1}x)(t)$ equals (possibly up to a phase factor $c(\theta\circ\theta_1^{-1},\theta_1)$) the displaced representation of the original signal $x(t)$, $L_x(\theta\circ\theta_1^{-1})$. We note that there is some arbitrariness regarding the exact definition of the phase factor in (7.5.38). Using $c(\theta\circ\theta_1^{-1},\theta_1)$ as in (7.5.38) has the advantage that the expression for $L_x(\theta)$ to be obtained in (7.5.40) does not contain a phase factor.

For example, in the case of the TF shift operator $\mathbf{S}_{\tau,\nu} = \mathbf{F}_\nu \mathbf{T}_\tau$, whose cocycle is $c\big((\tau_1,\nu_1),(\tau_2,\nu_2)\big) = e^{-j2\pi\nu_1\tau_2}$, the covariance property (7.5.38) becomes

$$L_{\mathbf{S}_{\tau_1,\nu_1}x}(\tau,\nu) = e^{-j2\pi\tau_1(\nu-\nu_1)} L_x(\tau-\tau_1,\nu-\nu_1). \tag{7.5.39}$$

Further examples of covariance properties are listed in Table 7.4.

The following fundamental theorem characterizes all covariant linear θ-representations. (We note that in [25], the theorems of this section were first formulated and proved directly for TF representations rather than θ-representations. In the following sections, we will see that this is in fact equivalent.)

Theorem 7.13 ([25]). *All linear θ-representations covariant to a DO $\{\mathbf{D}_\theta\}_{\theta\in(\mathcal{D},\circ)}$ on a Hilbert space $\mathcal{X}$ are given by the expression*

$$L_x(\theta) = \langle x, \mathbf{D}_\theta h\rangle = \int_t x(t)(\mathbf{D}_\theta h)^*(t)dt, \qquad \theta\in\mathcal{D}, \tag{7.5.40}$$

where h is an arbitrary function in $\mathcal{X}$.

Proof: First we show that the linear θ-representation in (7.5.40) satisfies the covariance property (7.5.38):

$$\begin{aligned}
L_{\mathbf{D}_{\theta_1}x}(\theta) &= \langle \mathbf{D}_{\theta_1}x, \mathbf{D}_\theta h\rangle = \big\langle x, \mathbf{D}_{\theta_1}^{-1}\mathbf{D}_\theta h\big\rangle = \big\langle x, c^*(\theta_1^{-1},\theta_1)\mathbf{D}_{\theta_1^{-1}}\mathbf{D}_\theta h\big\rangle \\
&= \big\langle x, c^*(\theta_1^{-1},\theta_1)c(\theta,\theta_1^{-1})\mathbf{D}_{\theta\circ\theta_1^{-1}}h\big\rangle \\
&= c(\theta_1^{-1},\theta_1)c^*(\theta,\theta_1^{-1})L_x(\theta\circ\theta_1^{-1}) = c(\theta\circ\theta_1^{-1},\theta_1)L_x(\theta\circ\theta_1^{-1}),
\end{aligned}$$

where we have used the unitarity of $\mathbf{D}_\theta$ as well as (7.2.13), (7.2.12), (7.5.34), (7.2.10), and (7.2.11).

Conversely, let $L_x(\theta)$ satisfy the covariance property (7.5.38). With (7.5.37), Eq. (7.5.38) becomes $\langle \mathbf{D}_{\theta_1}x, h_\theta\rangle = c(\theta\circ\theta_1^{-1},\theta_1)\big\langle x, h_{\theta\circ\theta_1^{-1}}\big\rangle$. Substituting $\theta=\theta_0$ and $\theta_1^{-1}=\theta$ yields $\langle \mathbf{D}_{\theta^{-1}}x, h_{\theta_0}\rangle = c(\theta_0\circ\theta,\theta^{-1})\langle x, h_{\theta_0\circ\theta}\rangle = c(\theta,\theta^{-1})\langle x, h_\theta\rangle$ and hence

$$\begin{aligned}
L_x(\theta) &= \langle x, h_\theta\rangle = c^*(\theta,\theta^{-1})\big\langle \mathbf{D}_{\theta^{-1}}x, h_{\theta_0}\big\rangle = |c(\theta,\theta^{-1})|^2\big\langle \mathbf{D}_\theta^{-1}x, h_{\theta_0}\big\rangle \\
&= \big\langle \mathbf{D}_\theta^{-1}x, h_{\theta_0}\big\rangle = \langle x, \mathbf{D}_\theta h\rangle,
\end{aligned}$$

with $h(t) \triangleq h_{\theta_0}(t)$. □

For example, the expression (7.5.40) of all linear θ-representations covariant to $\mathbf{S}_{\tau,\nu} = \mathbf{F}_\nu\mathbf{T}_\tau$ becomes

$$L_x(\tau,\nu) = \langle x, \mathbf{S}_{\tau,\nu}h\rangle = \int_t x(t)h^*(t-\tau)e^{-j2\pi\nu t}dt, \tag{7.5.41}$$

which is the short-time Fourier transform in (7.1.4). Further examples are listed in Table 7.4.

Short-time Fourier transform	
$\mathbf{A}_\alpha$	$(\mathbf{A}_\alpha x)(t) = x(t-\alpha), \quad \alpha \in (\mathbb{R},+)$
$\mathbf{B}_\beta$	$(\mathbf{B}_\beta x)(t) = e^{j2\pi\beta t}x(t), \quad \beta \in (\mathbb{R},+)$
Cocycle	$c(\theta_1,\theta_2) = e^{-j2\pi\beta_1\alpha_2}$
Composition property	$\mathbf{D}_{\alpha_2,\beta_2}\mathbf{D}_{\alpha_1,\beta_1} = e^{-j2\pi\beta_1\alpha_2}\mathbf{D}_{\alpha_1+\alpha_2,\beta_1+\beta_2}$
Covariance property	$L_{\mathbf{D}_{\alpha_1,\beta_1}x}(\alpha,\beta) = e^{-j2\pi\alpha_1(\beta-\beta_1)}L_x(\alpha-\alpha_1,\beta-\beta_1)$
Covariant lin. θ-repres.	$L_x(\alpha,\beta) = \int_{-\infty}^{\infty} x(t)h^*(t-\alpha)e^{-j2\pi\beta t}dt$

Wavelet transform	
$\mathbf{A}_\alpha$	$(\mathbf{A}_\alpha x)(t) = \frac{1}{\sqrt{\alpha}}x\left(\frac{t}{\alpha}\right), \quad \alpha \in (\mathbb{R}^+,\cdot)$
$\mathbf{B}_\beta$	$(\mathbf{B}_\beta x)(t) = x(t-\beta), \quad \beta \in (\mathbb{R},+)$
Cocycle	$c(\theta_1,\theta_2) \equiv 1$
Composition property	$\mathbf{D}_{\alpha_2,\beta_2}\mathbf{D}_{\alpha_1,\beta_1} = \mathbf{D}_{\alpha_1\alpha_2,\beta_1\alpha_2+\beta_2}$
Covariance property	$L_{\mathbf{D}_{\alpha_1,\beta_1}x}(\alpha,\beta) = L_x\left(\frac{\alpha}{\alpha_1},\frac{\beta-\beta_1}{\alpha_1}\right)$
Covariant lin. θ-repres.	$L_x(\alpha,\beta) = \frac{1}{\sqrt{\alpha}}\int_{-\infty}^{\infty} x(t)h^*\left(\frac{t-\beta}{\alpha}\right)dt$

Hyperbolic wavelet transform	
$\mathbf{A}_\alpha$	$(\mathbf{A}_\alpha x)(t) = \frac{1}{\sqrt{\alpha}}x\left(\frac{t}{\alpha}\right), \quad \alpha \in (\mathbb{R}^+,\cdot)$
$\mathbf{B}_\beta$	$(\hat{\mathbf{B}}_\beta X)(f) = e^{-j2\pi\beta\ln f}X(f), \quad \beta \in (\mathbb{R},+),\ f>0$
Cocycle	$c(\theta_1,\theta_2) = e^{-j2\pi\beta_1\ln\alpha_2}$
Composition property	$\mathbf{D}_{\alpha_2,\beta_2}\mathbf{D}_{\alpha_1,\beta_1} = e^{-j2\pi\beta_1\ln\alpha_2}\mathbf{D}_{\alpha_1\alpha_2,\beta_1+\beta_2}$
Covariance property	$L_{\mathbf{D}_{\alpha_1,\beta_1}x}(\alpha,\beta) = e^{-j2\pi(\beta-\beta_1)\ln\alpha_1}L_x\left(\frac{\alpha}{\alpha_1},\beta-\beta_1\right)$
Covariant lin. θ-repres.	$L_x(\alpha,\beta) = \sqrt{\alpha}\int_0^{\infty} X(f)H^*(\alpha f)\,e^{j2\pi\beta\ln f}df$

Power wavelet transform	
$\mathbf{A}_\alpha$	$(\mathbf{A}_\alpha x)(t) = \frac{1}{\sqrt{\alpha}}x\left(\frac{t}{\alpha}\right), \quad \alpha \in (\mathbb{R}^+,\cdot)$
$\mathbf{B}_\beta$	$(\hat{\mathbf{B}}_\beta X)(f) = e^{-j2\pi\beta\xi_\kappa(f)}X(f), \quad \beta \in (\mathbb{R},+),$ with $\xi_\kappa(f) = \text{sgn}(f)\lvert f\rvert^\kappa$
Cocycle	$c(\theta_1,\theta_2) \equiv 1$
Composition property	$\mathbf{D}_{\alpha_2,\beta_2}\mathbf{D}_{\alpha_1,\beta_1} = \mathbf{D}_{\alpha_1\alpha_2,\beta_1\xi_\kappa(\alpha_2)+\beta_2}$
Covariance property	$L_{\mathbf{D}_{\alpha_1,\beta_1}x}(\alpha,\beta) = L_x\left(\frac{\alpha}{\alpha_1},\frac{\beta-\beta_1}{\xi_\kappa(\alpha_1)}\right)$
Covariant lin. θ-repres. (extended to $X(f)$ on $f \in \mathbb{R}$)	$L_x(\alpha,\beta) = \sqrt{\alpha}\int_{-\infty}^{\infty} X(f)H^*(\alpha f)\,e^{j2\pi\beta\xi_\kappa(f)}df$

TABLE 7.4. Some covariant linear θ-representations. Note that $\theta = (\alpha,\beta)$, $\mathbf{D}_{\alpha,\beta} = \mathbf{B}_\beta\mathbf{A}_\alpha$, $X(f)$ denotes the Fourier transform of $x(t)$, and $\hat{\mathbf{B}}_\beta$ denotes the frequency-domain version of $\mathbf{B}_\beta$.

Let us now consider unitarily equivalent DOs $\mathbf{D}_\theta$ and $\mathbf{D}'_{\theta'} = \mathbf{U}\mathbf{D}_{\psi(\theta')}\mathbf{U}^{-1}$ on $\mathcal{X}$ and $\mathcal{X}'$, respectively (see (7.5.36)). By Theorem 7.13, the class of all linear θ'-representations covariant to $\mathbf{D}'_{\theta'}$ is given by

$$\begin{aligned} L'_x(\theta') &= \langle x, \mathbf{D}'_{\theta'} h' \rangle = \langle x, \mathbf{U}\mathbf{D}_{\psi(\theta')}\mathbf{U}^{-1} h' \rangle = \langle \mathbf{U}^{-1} x, \mathbf{D}_{\psi(\theta')}\mathbf{U}^{-1} h' \rangle \\ &= \langle \mathbf{U}^{-1} x, \mathbf{D}_{\psi(\theta')} h \rangle = L_{\mathbf{U}^{-1} x}(\psi(\theta')), \end{aligned}$$

where $h = \mathbf{U}^{-1} h'$ is an arbitrary function in $\mathcal{X}$ and L is a linear θ-representation covariant to $\mathbf{D}_\theta$. Hence, there exists a simple relation between the covariant classes $\{L_x(\theta)\}$ and $\{L'_x(\theta')\}$.

7.5.3 *Covariant bilinear θ-representations*

Next, we consider bilinear θ-representations $B_{x,y}(\theta) = B_{x,y}(\alpha, \beta)$ with $\theta \in (\mathcal{D}, \circ)$. Any bilinear θ-representation can be written in the form [13]

$$B_{x,y}(\theta) = \langle x, \mathbf{K}_\theta y \rangle = \int_{t_1} \int_{t_2} x(t_1) y^*(t_2) k_\theta^*(t_1, t_2) dt_1 dt_2, \tag{7.5.42}$$

where $\mathbf{K}_\theta$ is a linear operator that depends on the displacement parameter θ but not on x, y, and $k_\theta(t_1, t_2)$ is its kernel. The following definition and theorem are analogous to the linear case previously discussed.

Definition 7.14. ([25,26,28,31,32]). Let $\{\mathbf{D}_\theta\}_{\theta \in (\mathcal{D}, \circ)}$ be a DO defined on a Hilbert space $\mathcal{X}$. A bilinear θ-representation $B_{x,y}(\theta)$ is called *covariant to the DO* $\mathbf{D}_\theta$ if for all $x, y \in \mathcal{X}$

$$B_{\mathbf{D}_{\theta_1} x, \mathbf{D}_{\theta_1} y}(\theta) = B_{x,y}(\theta \circ \theta_1^{-1}), \quad \text{for all } \theta, \theta_1 \in \mathcal{D}. \tag{7.5.43}$$

We note that this definition of covariance differs from the covariance definition in the linear case, (7.5.38), by the absence of the DO's cocycle c. This difference is caused by the bilinear structure of B.

Theorem 7.15 ([25,26,28,31,32])**.** *All bilinear θ-representations covariant to a DO $\{\mathbf{D}_\theta\}_{\theta \in (\mathcal{D}, \circ)}$ on a Hilbert space $\mathcal{X}$ are given by the expression*

$$\begin{aligned} B_{x,y}(\theta) &= \langle x, \mathbf{D}_\theta \mathbf{K} \mathbf{D}_\theta^{-1} y \rangle \\ &= \int_{t_1} \int_{t_2} x(t_1) y^*(t_2) \big[\mathbf{D}_\theta \mathbf{K} \mathbf{D}_\theta^{-1}\big]^*(t_1, t_2) dt_1 dt_2, \qquad \theta \in \mathcal{D}, \end{aligned} \tag{7.5.44}$$

where $\mathbf{K}$ is an arbitrary linear operator on $\mathcal{X}$ and $\big[\mathbf{D}_\theta \mathbf{K} \mathbf{D}_\theta^{-1}\big](t_1, t_2)$ denotes the kernel of $\mathbf{D}_\theta \mathbf{K} \mathbf{D}_\theta^{-1}$.

Proof: We shall use the proof given in [31,32] since it is simpler than the approach taken in [25]. First we show that the bilinear θ-representation in (7.5.44) satisfies the covariance property (7.5.43):

$$\begin{aligned}
B_{\mathbf{D}_{\theta_1}x,\mathbf{D}_{\theta_1}y}(\theta) &= \big\langle \mathbf{D}_{\theta_1}x, \mathbf{D}_\theta \mathbf{K}\mathbf{D}_\theta^{-1}\mathbf{D}_{\theta_1}y\big\rangle = \big\langle x, \mathbf{D}_{\theta_1}^{-1}\mathbf{D}_\theta \mathbf{K}(\mathbf{D}_{\theta_1}^{-1}\mathbf{D}_\theta)^{-1}y\big\rangle \\
&= \big\langle x, c^*(\theta_1,\theta_1^{-1})\mathbf{D}_{\theta_1^{-1}}\mathbf{D}_\theta \mathbf{K}\big[c^*(\theta_1,\theta_1^{-1})\mathbf{D}_{\theta_1^{-1}}\mathbf{D}_\theta\big]^{-1}y\big\rangle \\
&= \big\langle x, \mathbf{D}_{\theta_1^{-1}}\mathbf{D}_\theta \mathbf{K}(\mathbf{D}_{\theta_1^{-1}}\mathbf{D}_\theta)^{-1}y\big\rangle \\
&= \big\langle x, c(\theta,\theta_1^{-1})\mathbf{D}_{\theta\circ\theta_1^{-1}}\mathbf{K}\big[c(\theta,\theta_1^{-1})\mathbf{D}_{\theta\circ\theta_1^{-1}}\big]^{-1}y\big\rangle \\
&= B_{x,y}(\theta\circ\theta_1^{-1}),
\end{aligned}$$

where we have used the unitarity of $\mathbf{D}_\theta$ as well as (7.2.13) and (7.5.34).

Conversely, let $B_{x,y}(\theta)$ satisfy the covariance property (7.5.43). With (7.5.42), Eq. (7.5.43) becomes $\langle \mathbf{D}_{\theta_1}x, \mathbf{K}_\theta \mathbf{D}_{\theta_1}y\rangle = \langle x, \mathbf{K}_{\theta\circ\theta_1^{-1}}y\rangle$. The substitution $\theta=\theta_0$, $\theta_1^{-1}=\theta$ yields $\langle \mathbf{D}_{\theta^{-1}}x, \mathbf{K}_{\theta_0}\mathbf{D}_{\theta^{-1}}y\rangle = \langle x, \mathbf{K}_{\theta_0\circ\theta}y\rangle = \langle x, \mathbf{K}_\theta y\rangle$, and hence

$$\begin{aligned}
B_{x,y}(\theta) &= \langle x, \mathbf{K}_\theta y\rangle = \langle \mathbf{D}_{\theta^{-1}}x, \mathbf{K}_{\theta_0}\mathbf{D}_{\theta^{-1}}y\rangle \\
&= |c(\theta,\theta^{-1})|^2\langle \mathbf{D}_\theta^{-1}x, \mathbf{K}_{\theta_0}\mathbf{D}_\theta^{-1}y\rangle = \langle x, \mathbf{D}_\theta \mathbf{K}\mathbf{D}_\theta^{-1}y\rangle,
\end{aligned}$$

with $\mathbf{K} \triangleq \mathbf{K}_{\theta_0}$. □

For example, in the case of the TF shift operator $\mathbf{S}_{\tau,\nu}=\mathbf{F}_\nu\mathbf{T}_\tau$, the covariance property (7.5.43) reads

$$B_{\mathbf{S}_{\tau_1,\nu_1}x,\mathbf{S}_{\tau_1,\nu_1}y}(\tau,\nu) = B_{x,y}(\tau-\tau_1,\nu-\nu_1),$$

and the canonical expression (7.5.44) of all bilinear θ-representations covariant to $\mathbf{S}_{\tau,\nu}$ becomes

$$\begin{aligned}
B_{x,y}(\tau,\nu) &= \langle x, \mathbf{S}_{\tau,\nu}\mathbf{K}\mathbf{S}_{\tau,\nu}^{-1}y\rangle \\
&= \int_{t_1}\int_{t_2} x(t_1)y^*(t_2)\, k^*(t_1-\tau,t_2-\tau)\, e^{-j2\pi\nu(t_1-t_2)}dt_1dt_2,
\end{aligned}$$

which is recognized as Cohen's class (7.1.6). Further examples of covariance properties and the resulting canonical expressions of covariant BTFRs are listed in Table 7.5.

Using Theorem 7.15, one easily shows [12] that the bilinear θ-representations covariant to a DO $\mathbf{D}_\theta$ are related to the bilinear θ'-representations covariant to the DO $\mathbf{D}'_{\theta'}=\mathbf{U}\mathbf{D}_{\psi(\theta')}\mathbf{U}^{-1}$ (see (7.5.36)) according to

$$B'_{x,y}(\theta') = B_{\mathbf{U}^{-1}x,\mathbf{U}^{-1}y}(\psi(\theta')),$$

where the operators $\mathbf{K}$ and $\mathbf{K}'$ used in $B_{x,y}(\theta)=\langle x, \mathbf{D}_\theta\mathbf{K}\mathbf{D}_\theta^{-1}y\rangle$ and in $B'_{x,y}(\theta') = \langle x, \mathbf{D}'_{\theta'}\mathbf{K}'\mathbf{D}'^{-1}_{\theta'}y\rangle$, respectively, are related as $\mathbf{K}=\mathbf{U}^{-1}\mathbf{K}'\mathbf{U}$.

Cohen's class	
$\mathbf{A}_\alpha$	$(\mathbf{A}_\alpha x)(t) = x(t-\alpha), \quad \alpha \in (\mathbb{R},+)$
$\mathbf{B}_\beta$	$(\mathbf{B}_\beta x)(t) = e^{j2\pi\beta t} x(t), \quad \beta \in (\mathbb{R},+)$
Composition property	$\mathbf{D}_{\alpha_2,\beta_2}\mathbf{D}_{\alpha_1,\beta_1} = e^{-j2\pi\beta_1\alpha_2}\mathbf{D}_{\alpha_1+\alpha_2,\beta_1+\beta_2}$
Covariance property	$B_{\mathbf{D}_{\alpha_1,\beta_1}x,\mathbf{D}_{\alpha_1,\beta_1}y}(\alpha,\beta) = B_{x,y}(\alpha-\alpha_1,\beta-\beta_1)$
Covar. bilin. θ-repres.	$B_{x,y}(\alpha,\beta) = \int_{-\infty}^{\infty}\int_{-\infty}^{\infty} x(t_1)y^*(t_2)\, k^*(t_1-\alpha,t_2-\alpha) \cdot e^{-j2\pi\beta(t_1-t_2)} dt_1 dt_2$

Affine class	
$\mathbf{A}_\alpha$	$(\mathbf{A}_\alpha x)(t) = \frac{1}{\sqrt{\alpha}} x\big(\frac{t}{\alpha}\big), \quad \alpha \in (\mathbb{R}^+,\cdot)$
$\mathbf{B}_\beta$	$(\mathbf{B}_\beta x)(t) = x(t-\beta), \quad \beta \in (\mathbb{R},+)$
Composition property	$\mathbf{D}_{\alpha_2,\beta_2}\mathbf{D}_{\alpha_1,\beta_1} = \mathbf{D}_{\alpha_1\alpha_2,\beta_1\alpha_2+\beta_2}$
Covariance property	$B_{\mathbf{D}_{\alpha_1,\beta_1}x,\mathbf{D}_{\alpha_1,\beta_1}y}(\alpha,\beta) = B_{x,y}\big(\frac{\alpha}{\alpha_1},\frac{\beta-\beta_1}{\alpha_1}\big)$
Covar. bilin. θ-repres.	$B_{x,y}(\alpha,\beta) = \frac{1}{\alpha}\int_{-\infty}^{\infty}\int_{-\infty}^{\infty} x(t_1)y^*(t_2) \cdot k^*\big(\frac{t_1-\beta}{\alpha},\frac{t_2-\beta}{\alpha}\big) dt_1 dt_2$

Hyperbolic class	
$\mathbf{A}_\alpha$	$(\mathbf{A}_\alpha x)(t) = \frac{1}{\sqrt{\alpha}} x\big(\frac{t}{\alpha}\big), \quad \alpha \in (\mathbb{R}^+,\cdot)$
$\mathbf{B}_\beta$	$(\hat{\mathbf{B}}_\beta X)(f) = e^{-j2\pi\beta\ln f} X(f), \quad \beta \in (\mathbb{R},+),\ f>0$
Composition property	$\mathbf{D}_{\alpha_2,\beta_2}\mathbf{D}_{\alpha_1,\beta_1} = e^{-j2\pi\beta_1\ln\alpha_2}\mathbf{D}_{\alpha_1\alpha_2,\beta_1+\beta_2}$
Covariance property	$B_{\mathbf{D}_{\alpha_1,\beta_1}x,\mathbf{D}_{\alpha_1,\beta_1}y}(\alpha,\beta) = B_{x,y}\big(\frac{\alpha}{\alpha_1},\beta-\beta_1\big)$
Covar. bilin. θ-repres.	$B_{x,y}(\alpha,\beta) = \alpha\int_0^{\infty}\int_0^{\infty} X(f_1)Y^*(f_2)\hat{k}^*(\alpha f_1,\alpha f_2) \cdot e^{j2\pi\beta\ln(f_1/f_2)} df_1 df_2$

Power classes	
$\mathbf{A}_\alpha$	$(\mathbf{A}_\alpha x)(t) = \frac{1}{\sqrt{\alpha}} x\big(\frac{t}{\alpha}\big), \quad \alpha \in (\mathbb{R}^+,\cdot)$
$\mathbf{B}_\beta$	$(\hat{\mathbf{B}}_\beta X)(f) = e^{-j2\pi\beta\xi_\kappa(f)} X(f), \quad \beta \in (\mathbb{R},+),$ with $\xi_\kappa(f) = \mathrm{sgn}(f)\lvert f\rvert^\kappa$
Composition property	$\mathbf{D}_{\alpha_2,\beta_2}\mathbf{D}_{\alpha_1,\beta_1} = \mathbf{D}_{\alpha_1\alpha_2,\beta_1\xi_\kappa(\alpha_2)+\beta_2}$
Covariance property	$B_{\mathbf{D}_{\alpha_1,\beta_1}x,\mathbf{D}_{\alpha_1,\beta_1}y}(\alpha,\beta) = B_{x,y}\big(\frac{\alpha}{\alpha_1},\frac{\beta-\beta_1}{\xi_\kappa(\alpha_1)}\big)$
Covar. bilin. θ-rep. (extended to $X(f),Y(f)$ on $f\in\mathbb{R}$)	$B_{x,y}(\alpha,\beta) = \alpha\int_{-\infty}^{\infty}\int_{-\infty}^{\infty} X(f_1)Y^*(f_2)\hat{k}^*(\alpha f_1,\alpha f_2) \cdot e^{j2\pi\beta[\xi_\kappa(f_1)-\xi_\kappa(f_2)]} df_1 df_2$

TABLE 7.5. Some covariant bilinear/quadratic θ-representations. Note that $\theta = (\alpha,\beta)$, $\mathbf{D}_{\alpha,\beta} = \mathbf{B}_\beta\mathbf{A}_\alpha$, $X(f)$ denotes the Fourier transform of $x(t)$, and $\hat{\mathbf{B}}_\beta$ denotes the frequency-domain version of $\mathbf{B}_\beta$.

7.6 The displacement function: theory

Thus far, we have characterized all linear and bilinear θ-representations that are covariant to a given DO $\{\mathbf{D}_\theta\}_{\theta \in (\mathcal{D},\circ)}$. However, ultimately we are interested in covariant *TF representations* that are functions of (t, f) rather than of $\theta = (\alpha, \beta)$. As we shall see, covariant θ-representations can be converted into covariant TF representations via a one-to-one mapping $\theta \leftrightarrow (t, f)$ that will be termed the *displacement function* (DF) [25, 26, 28].

In this section, we shall formulate basic axioms of TF displacements and derive some remarkable consequences regarding the mathematical structure of the DF as well as the underlying group and DO. A systematic method for constructing the DF and the use of the DF for constructing covariant TF representations will be described in Sections 7.7 and 7.8, respectively.

Hereafter, similarly to our shorthand notation $\theta = (\alpha, \beta)$, we shall frequently use the shorthand notation $z \triangleq (t, f)$ for TF points. Furthermore, $\mathcal{Z} \subseteq \mathbb{R}^2$ will denote the set of TF points $z = (t, f)$ on which a covariant TF representation $L_x(z) = L_x(t, f)$ or $B_{x,y}(z) = B_{x,y}(t, f)$ is defined.

7.6.1 *Axioms of time-frequency displacement*

Let us consider the TF displacements performed by a given DO $\{\mathbf{D}_\theta\}_{\theta \in (\mathcal{D},\circ)}$. If a signal $x(t)$ is TF localized about some TF point $z_1 = (t_1, f_1) \in \mathcal{Z}$, then the transformed ("displaced") signal $(\mathbf{D}_\theta x)(t)$ will be localized about some other TF point $z_2 = (t_2, f_2) \in \mathcal{Z}$ that depends on z_1 and θ, i.e.,

$$z_2 = e(z_1, \theta).$$

The function $e(\cdot, \cdot)$ will be called the *extended DF* of the DO $\mathbf{D}_\theta$ [28] (originally termed DF in [25, 26]). Before discussing the construction of the extended DF in Section 7.7, we formulate three axioms that the extended DF is assumed to satisfy and study important consequences of these axioms.

Definition 7.16. Let $(\mathcal{D}, \circ)$ with $\mathcal{D} \subseteq \mathbb{R}^2$ be a group and let $\mathcal{Z} \subseteq \mathbb{R}^2$ be an open, simply connected set of TF points $z = (t, f)$. ("Simply connected" means that $\mathcal{Z}$ has no "holes" [53].) A mapping $e : \mathcal{Z} \times \mathcal{D} \to \mathcal{Z}$ is called an *extended DF* for the group $\mathcal{D}$ on the TF set $\mathcal{Z}$ if it satisfies the following axioms.

Axiom 1. For any fixed $\theta \in \mathcal{D}$, $z \mapsto z' = e(z, \theta)$ is an invertible, continuously differentiable, and area-preserving mapping of $\mathcal{Z}$ onto $\mathcal{Z}$.

Axiom 2. For any fixed $z \in \mathcal{Z}$, $\theta \mapsto z' = e(z, \theta)$ is an invertible, continuously differentiable mapping of $\mathcal{D}$ onto $\mathcal{Z}$ with nonzero Jacobian, i.e., a *diffeomorphism* [53]. (We note that this property presupposes that $(\mathcal{D}, \circ)$ is a *topological* group with Euclidean topology.)

Axiom 3. Composition property:

$$e\big(e(z, \theta_1), \theta_2\big) = e(z, \theta_1 \circ \theta_2) \qquad \text{for all } \theta_1, \theta_2 \in \mathcal{D}. \tag{7.6.45}$$

Furthermore, if these axioms are satisfied, the function

$$d(\theta) \triangleq e(z_0, \theta),$$

with $z_0 = (t_0, f_0) \in \mathcal{Z}$ an arbitrary but fixed reference TF point, will be called a *displacement function (DF)* associated to the extended DF e.

These axioms imply that $(\mathcal{D}, \circ)$ acts as a *transitive transformation group* [55] on $\mathcal{Z}$. According to the first two axioms, to any $\theta \in \mathcal{D}$ and $z' \in \mathcal{Z}$ there exists a unique $z \in \mathcal{Z}$ such that $z' = e(z, \theta)$; similarly, to any $z, z' \in \mathcal{Z}$ there exists a unique $\theta \in \mathcal{D}$ such that $z' = e(z, \theta)$. Axiom 3 means that displacing a signal first by θ_1 and then by θ_2 is equivalent to a single displacement by $\theta_1 \circ \theta_2$; this is the counterpart of the DO composition property (7.5.34), $\mathbf{D}_{\theta_2}\mathbf{D}_{\theta_1} = c(\theta_1, \theta_2)\mathbf{D}_{\theta_1 \circ \theta_2}$. It can also be shown that $e(z, \theta_0) = z$, i.e., the group identity element θ_0 corresponds to no displacement.

Theorem 7.17. *Consider an extended DF $e(z, \theta)$ for a group $(\mathcal{D}, \circ)$ on a TF set $\mathcal{Z}$. Any associated DF $d(\theta)$ is a diffeomorphism, i.e., an invertible, continuously differentiable mapping of $\mathcal{D}$ onto $\mathcal{Z}$ with nonzero Jacobian. Moreover, e can be written in terms of d and its inverse d^{-1} as*

$$e(z, \theta) = d\big(d^{-1}(z) \circ \theta\big). \tag{7.6.46}$$

Proof: Let $d(\theta) = e(z_0, \theta)$ with $z_0 \in \mathcal{Z}$ fixed. Axiom 2 in Definition 7.16 implies that d is a diffeomorphism from $\mathcal{D}$ onto $\mathcal{Z}$. It follows that the inverse function d^{-1} exists, so we have $z = d\big(d^{-1}(z)\big) = e\big(z_0, d^{-1}(z)\big)$, and further,

$$e(z, \theta) = e\big(e\big(z_0, d^{-1}(z)\big), \theta\big) = e\big(z_0, d^{-1}(z) \circ \theta\big) = d\big(d^{-1}(z) \circ \theta\big),$$

where (7.6.45) has been used. □

Finally, consider two different DFs $d(\theta) = e(z_0, \theta)$ and $\tilde{d}(\theta) = e(\tilde{z}_0, \theta)$ associated to the same extended DF e. From Axiom 2 in Definition 7.16, there is a unique $\tilde{\theta} \in \mathcal{D}$ such that $\tilde{z}_0 = e(z_0, \tilde{\theta})$. This yields

$$\tilde{d}(\theta) = e(\tilde{z}_0, \theta) = e\big(e(z_0, \tilde{\theta}), \theta\big) = e(z_0, \tilde{\theta} \circ \theta) = d(\tilde{\theta} \circ \theta),$$

or, equivalently formulated with the inverse DFs,

$$\tilde{d}^{-1}(z) = \tilde{\theta}^{-1} \circ d^{-1}(z). \tag{7.6.47}$$

7.6.2 Lie group structure and consequences

The existence of a DF has important consequences regarding the structure of the underlying group $(\mathcal{D}, \circ)$ and DO $\{\mathbf{D}_\theta\}_{\theta \in (\mathcal{D}, \circ)}$. In fact, as shown in the proof of the following theorem, the group must be a *simply connected Lie group*, i.e., a topological group that can be viewed as a smooth surface in two-dimensional (2D) Euclidean space. Since there exist only two simply connected 2D Lie groups (up to isomorphisms), namely, the group $(\mathbb{R}^2, +)$ and the affine group [53], we obtain the following result.

Theorem 7.18. *Let* $(\mathcal{D}, \circ)$ *be a group with* $\mathcal{D} \subseteq \mathbb{R}^2$ *and let the set* $\mathcal{Z} \subseteq \mathbb{R}^2$ *be open and simply connected. If there exists an extended DF* $e : \mathcal{Z} \times \mathcal{D} \to \mathcal{Z}$*, then either* $(\mathcal{D}, \circ)$ *is isomorphic to* $(\mathbb{R}^2, +)$*, i.e., the group* $(\mathcal{D}', \circ')$ *with* $\mathcal{D}' = \mathbb{R}^2$ *and* $(\alpha_1', \beta_1') \circ' (\alpha_2', \beta_2') = (\alpha_1' + \alpha_2', \beta_1' + \beta_2')$*, or* $(\mathcal{D}, \circ)$ *is isomorphic to the affine group, i.e., the group* $(\mathcal{D}', \circ')$ *with* $\mathcal{D}' = \mathbb{R}^+ \times \mathbb{R}$ *and* $(\alpha_1', \beta_1') \circ' (\alpha_2', \beta_2') = (\alpha_1'\alpha_2', \beta_1'\alpha_2' + \beta_2')$.

Proof: According to Axiom 2 in Definition 7.16, the existence of an extended DF presupposes that $(\mathcal{D}, \circ)$ is a topological group with Euclidean topology on $\mathcal{D} \subseteq \mathbb{R}^2$. According to [53,55], this implies that $(\mathcal{D}, \circ)$ is a Lie group. Since $\mathcal{Z}$ is simply connected and the DF $d : \mathcal{D} \to \mathcal{Z}$ is a homeomorphism (see Theorem 7.17, noting that a diffeomorphism is always a homeomorphism), then $\mathcal{D}$ is also simply connected. Hence, $(\mathcal{D}, \circ)$ is a simply connected 2D Lie group. It is well known [53] that there exist only two simply connected 2D Lie groups (up to isomorphisms), namely, the group $(\mathbb{R}^2, +)$ and the affine group. This proves the theorem. □

This result has an important consequence regarding the DO $\{\mathbf{D}_\theta\}_{\theta \in (\mathcal{D}, \circ)}$. The next two corollaries state that for the two possible cases—$(\mathcal{D}, \circ)$ isomorphic to $(\mathbb{R}^2, +)$ or the affine group—the DO has a *separable* structure. (Note the consistency of these corollaries with Theorems 7.6 and 7.9.)

Corollary 7.19. *Consider a group* $(\mathcal{D}, \circ)$ *that is isomorphic to* $(\mathbb{R}^2, +)$ *with isomorphism* $\psi : \mathcal{D} \to \mathbb{R}^2$*. If* $\{\mathbf{D}'_{\theta'}\}_{\theta' \in (\mathcal{D}, \circ)}$ *is a DO for* $(\mathcal{D}, \circ)$*, with cocycle* $c'\big((\alpha_1', \beta_1'), (\alpha_2', \beta_2')\big)$*, then* $\mathbf{D}_\theta \triangleq \mathbf{D}'_{\psi^{-1}(\theta)}$ *is a DO for* $(\mathbb{R}^2, +)$*, with cocycle* $c\big((\alpha_1, \beta_1), (\alpha_2, \beta_2)\big) = c'\big(\psi^{-1}(\alpha_1, \beta_1), \psi^{-1}(\alpha_2, \beta_2)\big)$*, and it is* separable *(up to a phase factor) according to*

$$\mathbf{D}_\theta \equiv \mathbf{D}_{\alpha,\beta} = c^*\big((\alpha, 0), (0, \beta)\big)\mathbf{B}_\beta \mathbf{A}_\alpha \qquad \text{with } \mathbf{A}_\alpha \triangleq \mathbf{D}_{\alpha,0},\ \mathbf{B}_\beta \triangleq \mathbf{D}_{0,\beta}. \tag{7.6.48}$$

Here, $\mathbf{A}_\alpha$ *and* $\mathbf{B}_\beta$ *are faithful projective representations of* $(\mathbb{R}, +)$*; they satisfy*

$$\mathbf{A}_0 = \mathbf{I}, \quad \mathbf{A}_{\alpha_2}\mathbf{A}_{\alpha_1} = c_{\mathbf{A}}(\alpha_1, \alpha_2)\mathbf{A}_{\alpha_1+\alpha_2} \qquad \text{for all } \alpha_1, \alpha_2 \in \mathbb{R}$$
$$\mathbf{B}_0 = \mathbf{I}, \quad \mathbf{B}_{\beta_2}\mathbf{B}_{\beta_1} = c_{\mathbf{B}}(\beta_1, \beta_2)\mathbf{B}_{\beta_1+\beta_2} \qquad \text{for all } \beta_1, \beta_2 \in \mathbb{R},$$

with $c_{\mathbf{A}}(\alpha_1, \alpha_2) = c\big((\alpha_1, 0), (\alpha_2, 0)\big)$ *and* $c_{\mathbf{B}}(\beta_1, \beta_2) = c\big((0, \beta_1), (0, \beta_2)\big)$*. If* $c\big((\alpha_1, \beta_1), (\alpha_2, \beta_2)\big) = e^{-j2\pi\alpha_2\beta_1}$*, then* $\mathbf{D}_{\alpha,\beta} = \mathbf{B}_\beta \mathbf{A}_\alpha$*, and* $\mathbf{A}_\alpha$ *and* $\mathbf{B}_\beta$ *are unitary representations of* $(\mathbb{R}, +)$ *with* $\mathbf{B}_\beta$ *dual to* $\mathbf{A}_\alpha$.

Proof: If $\mathbf{D}'_{\theta'}$ is a DO for the group $(\mathcal{D}, \circ)$, then $\mathbf{D}_\theta = \mathbf{D}'_{\psi^{-1}(\theta)}$ satisfies $\mathbf{D}_{0,0} = \mathbf{D}'_{\psi^{-1}(0,0)} = \mathbf{D}'_{\theta_0'} = \mathbf{I}$ and

$$\begin{aligned}
\mathbf{D}_{\alpha_2,\beta_2}\mathbf{D}_{\alpha_1,\beta_1} &= \mathbf{D}'_{\psi^{-1}(\alpha_2,\beta_2)}\mathbf{D}'_{\psi^{-1}(\alpha_1,\beta_1)} \\
&= c'\big(\psi^{-1}(\alpha_1, \beta_1), \psi^{-1}(\alpha_2, \beta_2)\big)\mathbf{D}'_{\psi^{-1}(\alpha_1,\beta_1)\circ\psi^{-1}(\alpha_2,\beta_2)} \\
&= c'\big(\psi^{-1}(\alpha_1, \beta_1), \psi^{-1}(\alpha_2, \beta_2)\big)\mathbf{D}'_{\psi^{-1}(\alpha_1+\alpha_2,\beta_1+\beta_2)} \\
&= c\big((\alpha_1, \beta_1), (\alpha_2, \beta_2)\big)\mathbf{D}_{\alpha_1+\alpha_2,\beta_1+\beta_2},
\end{aligned}$$

where $c\big((\alpha_1,\beta_1),(\alpha_2,\beta_2)\big) = c'\big(\psi^{-1}(\alpha_1,\beta_1),\psi^{-1}(\alpha_2,\beta_2)\big)$. The cocycle properties are easily shown to be satisfied. Hence, $\mathbf{D}_\theta$ is a projective representation of $(\mathbb{R}^2,+)$. It is irreducible and faithful because $\mathbf{D}'_{\psi^{-1}(\theta)}$ is; therefore, it is a DO for $(\mathbb{R}^2,+)$. The separability of $\mathbf{D}_{\alpha,\beta}$ in (7.6.48) follows with (7.5.34) and (7.2.10):

$$\begin{aligned} c^*\big((\alpha,0),(0,\beta)\big)\mathbf{B}_\beta\mathbf{A}_\alpha &= c^*\big((\alpha,0),(0,\beta)\big)\mathbf{D}_{0,\beta}\mathbf{D}_{\alpha,0} \\ &= c^*\big((\alpha,0),(0,\beta)\big)c\big((\alpha,0),(0,\beta)\big)\mathbf{D}_{\alpha,\beta} = \mathbf{D}_{\alpha,\beta}. \end{aligned}$$

We next show that $\mathbf{A}_\alpha$ is a faithful projective representation of $(\mathbb{R},+)$: We have $\mathbf{A}_0 = \mathbf{D}_{0,0} = \mathbf{I}$ and

$$\mathbf{A}_{\alpha_2}\mathbf{A}_{\alpha_1} = \mathbf{D}_{\alpha_2,0}\mathbf{D}_{\alpha_1,0} = c\big((\alpha_1,0),(\alpha_2,0)\big)\mathbf{D}_{\alpha_1+\alpha_2,0} = c_{\mathbf{A}}(\alpha_1,\alpha_2)\mathbf{A}_{\alpha_1+\alpha_2}$$

with $c_{\mathbf{A}}(\alpha_1,\alpha_2) = c\big((\alpha_1,0),(\alpha_2,0)\big)$. $\mathbf{A}_\alpha$ is faithful since $\mathbf{A}_\alpha = \mathbf{D}_{\alpha,0}$ and $\mathbf{D}_{\alpha,\beta}$ is faithful. The proof that $\mathbf{B}_\beta$ is a faithful projective representation of $(\mathbb{R},+)$ is analogous.

Finally, if $c\big((\alpha_1,\beta_1),(\alpha_2,\beta_2)\big) = e^{-j2\pi\alpha_2\beta_1}$, then $c\big((\alpha,0),(0,\beta)\big) = 1$ so that $\mathbf{D}_{\alpha,\beta} = \mathbf{B}_\beta\mathbf{A}_\alpha$. Furthermore, $c_{\mathbf{A}}(\alpha_1,\alpha_2) = c\big((\alpha_1,0),(\alpha_2,0)\big) = 1$ and similarly $c_{\mathbf{B}}(\beta_1,\beta_2) = 1$, and hence $\mathbf{A}_\alpha$ and $\mathbf{B}_\beta$ are unitary representations of $(\mathbb{R},+)$. We have

$$\mathbf{A}_\alpha\mathbf{B}_\beta = \mathbf{D}_{\alpha,0}\mathbf{D}_{0,\beta} = c\big((0,\beta),(\alpha,0)\big)\mathbf{D}_{\alpha,\beta} = e^{-j2\pi\alpha\beta}\mathbf{D}_{\alpha,\beta} = e^{-j2\pi\alpha\beta}\mathbf{B}_\beta\mathbf{A}_\alpha,$$

so that $\mathbf{B}_\beta\mathbf{A}_\alpha = e^{j2\pi\alpha\beta}\mathbf{A}_\alpha\mathbf{B}_\beta$, which shows that $\mathbf{B}_\beta$ is dual to $\mathbf{A}_\alpha$ (see Definition 7.5). □

Corollary 7.20. *Consider a group $(\mathcal{D},\circ)$ that is isomorphic to the affine group (i.e., the group $(\mathbb{R}^+\times\mathbb{R},\diamond)$ with $(\alpha_1,\beta_1)\diamond(\alpha_2,\beta_2) = (\alpha_1\alpha_2,\beta_1\alpha_2+\beta_2)$) with isomorphism $\psi:\mathcal{D}\to\mathbb{R}^+\times\mathbb{R}$. If $\{\mathbf{D}'_{\theta'}\}_{\theta'\in(\mathcal{D},\circ)}$ is a DO for $(\mathcal{D},\circ)$, with cocycle $c'\big((\alpha'_1,\beta'_1),(\alpha'_2,\beta'_2)\big)$, then $\mathbf{D}_\theta \triangleq \mathbf{D}'_{\psi^{-1}(\theta)}$ is a DO for the affine group, with cocycle $c\big((\alpha_1,\beta_1),(\alpha_2,\beta_2)\big) = c'\big(\psi^{-1}(\alpha_1,\beta_1),\psi^{-1}(\alpha_2,\beta_2)\big)$, and it is* separable *(up to a phase factor) according to*

$$\mathbf{D}_\theta \equiv \mathbf{D}_{\alpha,\beta} = c^*\big((\alpha,0),(1,\beta)\big)\mathbf{B}_\beta\mathbf{A}_\alpha \quad \text{with } \mathbf{A}_\alpha \triangleq \mathbf{D}_{\alpha,0},\ \ \mathbf{B}_\beta \triangleq \mathbf{D}_{1,\beta}.$$

Here, $\mathbf{A}_\alpha$ and $\mathbf{B}_\beta$ are faithful projective representations of $(\mathbb{R}^+,\cdot)$ and $(\mathbb{R},+)$, respectively; they satisfy

$$\begin{aligned} &\mathbf{A}_1 = \mathbf{I}, \quad \mathbf{A}_{\alpha_2}\mathbf{A}_{\alpha_1} = c_{\mathbf{A}}(\alpha_1,\alpha_2)\mathbf{A}_{\alpha_1\alpha_2} \qquad \text{for all } \alpha_1,\alpha_2\in\mathbb{R}^+ \\ &\mathbf{B}_0 = \mathbf{I}, \quad \mathbf{B}_{\beta_2}\mathbf{B}_{\beta_1} = c_{\mathbf{B}}(\beta_1,\beta_2)\mathbf{B}_{\beta_1+\beta_2} \qquad \text{for all } \beta_1,\beta_2\in\mathbb{R}, \end{aligned}$$

with $c_{\mathbf{A}}(\alpha_1,\alpha_2) = c\big((\alpha_1,0),(\alpha_2,0)\big)$ and $c_{\mathbf{B}}(\beta_1,\beta_2) = c\big((1,\beta_1),(1,\beta_2)\big)$. If $c\big((\alpha_1,\beta_1),(\alpha_2,\beta_2)\big)\equiv 1$, then $\mathbf{D}_{\alpha,\beta} = \mathbf{B}_\beta\mathbf{A}_\alpha$, and $\mathbf{A}_\alpha$ and $\mathbf{B}_\beta$ are unitary representations of $(\mathbb{R}^+,\cdot)$ and $(\mathbb{R},+)$, respectively, with $\mathbf{B}_\beta$ affine to $\mathbf{A}_\alpha$.

Proof: The proof is essentially analogous to that of Corollary 7.19. □

The operator families $\mathbf{A}_\alpha$ and $\mathbf{B}_\beta$ will be called *partial DOs* [26, 28, 30]. The last two corollaries state that a DO, after transforming the displacement parameter as $\theta' = \psi^{-1}(\theta)$, can be factored (possibly up to a phase factor) into the partial DOs which are projective or unitary representations of the 1-parameter groups $(\mathbb{R}, +)$ or $(\mathbb{R}^+, \cdot)$. (Note that $(\mathbb{R}^+, \cdot)$ is isomorphic to $(\mathbb{R}, +)$, so the respective partial DO could be reparameterized with $(\mathbb{R}, +)$.) This separable structure of DOs will be exploited in Section 7.7.

7.6.3 Unitary equivalence results

Two basic DOs are the operators $\mathbf{S}_{\tau,\nu} = \mathbf{F}_\nu \mathbf{T}_\tau$ defined on $L_2(\mathbb{R})$ and $\mathbf{R}_{\sigma,\tau} = \mathbf{T}_\tau \mathbf{C}_\sigma$ defined on the Hardy space $H^2(\mathbb{R})$. Whereas $\mathbf{S}_{\tau,\nu}$ is a projective representation of the group $(\mathbb{R}^2, +)$, $\mathbf{R}_{\sigma,\tau}$ is a unitary representation of the affine group. In Subsection 7.6.2, we showed that all groups $(\mathcal{D}, \circ)$ underlying "proper" DOs (i.e., DOs for which there exists a DF) are isomorphic to $(\mathbb{R}^2, +)$ or the affine group. This suggests that proper DOs might be related to one of the basic DOs $\mathbf{S}_{\tau,\nu}$ and $\mathbf{R}_{\sigma,\tau}$. The next two theorems consider the two cases that are possible, namely, $(\mathcal{D}, \circ)$ isomorphic to $(\mathbb{R}^2, +)$ or the affine group.

Theorem 7.21. *Let $(\mathcal{D}, \circ)$ be isomorphic to $(\mathbb{R}^2, +)$ with isomorphism ψ. Let $\{\mathbf{D}'_{\theta'}\}_{\theta' \in (\mathcal{D}, \circ)}$ be a DO for $(\mathcal{D}, \circ)$. If the cocycle of $\mathbf{D}_\theta \triangleq \mathbf{D}'_{\psi^{-1}(\theta)}$ (recall from Corollary 7.19 that $\mathbf{D}_\theta$ is a DO for $(\mathbb{R}^2, +)$) is $c\big((\alpha_1, \beta_1), (\alpha_2, \beta_2)\big) = e^{-j2\pi\alpha_2\beta_1}$, then $\mathbf{D}_\theta$ is unitarily equivalent to $\mathbf{S}_{\tau,\nu} = \mathbf{F}_\nu \mathbf{T}_\tau$, i.e.,*

$$\mathbf{D}_{\alpha,\beta} = \mathbf{U}\mathbf{S}_{\alpha,\beta}\mathbf{U}^{-1},$$

with $\mathbf{U}$ an invertible and isometric operator.

Conversely, let $\mathbf{D}'_{\theta'}$ with $\theta' \in (\mathcal{D}, \circ)$ be unitarily equivalent to $\mathbf{S}_{\tau,\nu}$ up to a phase factor and a parameter transformation by ψ, i.e.,

$$\mathbf{D}'_{\theta'} = e^{j\sigma(\psi(\theta'))}\mathbf{U}\mathbf{S}_{\psi(\theta')}\mathbf{U}^{-1},$$

with $\mathbf{U}$ an invertible and isometric operator and $\sigma(\cdot)$ continuous and real valued with $\sigma(0,0) = 0 \bmod 2\pi$. Then $\mathbf{D}'_{\theta'}$ is a DO for $(\mathcal{D}, \circ)$. Furthermore, the cocycle of $\mathbf{D}'_{\theta'}$ is given by $c'(\theta'_1, \theta'_2) = e^{j\phi(\theta'_1, \theta'_2)}$ with

$$\phi(\theta'_1, \theta'_2) = \sigma\big(\psi(\theta'_1)\big) + \sigma\big(\psi(\theta'_2)\big) - \sigma\big(\psi(\theta'_1 \circ \theta'_2)\big) - 2\pi p_2\big(\psi(\theta'_1)\big)p_1\big(\psi(\theta'_2)\big), \tag{7.6.49}$$

where the "projection" mappings $p_i : \mathbb{R}^2 \to \mathbb{R}$ $(i = 1, 2)$ are defined as $p_1(\alpha, \beta) = \alpha$ and $p_2(\alpha, \beta) = \beta$.

Proof: Assume $(\mathcal{D}, \circ)$ is isomorphic to $(\mathbb{R}^2, +)$, $\mathbf{D}'_{\theta'}$ is a DO for $(\mathcal{D}, \circ)$, and the cocycle of $\mathbf{D}_\theta = \mathbf{D}'_{\psi^{-1}(\theta)}$ is $c\big((\alpha_1, \beta_1), (\alpha_2, \beta_2)\big) = e^{-j2\pi\alpha_2\beta_1}$. According to Corollary 7.19, $\mathbf{D}_\theta$ is a DO for $(\mathbb{R}^2, +)$, and $\mathbf{D}_{\alpha,\beta} = \mathbf{B}_\beta \mathbf{A}_\alpha$ with $\mathbf{B}_\beta$ dual to $\mathbf{A}_\alpha$. According to Definition 7.11, the underlying Hilbert space is minimal invariant under $\mathbf{D}_\theta$ and hence also under both $\mathbf{A}_\alpha = \mathbf{D}_{\alpha,0}$ and $\mathbf{B}_\beta = \mathbf{D}_{0,\beta}$. According to Theorem 7.7, if $\mathbf{B}_\beta$ is dual to $\mathbf{A}_\alpha$ on a minimal invariant Hilbert space, with $\alpha, \beta \in (\mathbb{R}, +)$, then there hold unitary equivalence relations $\mathbf{A}_\alpha = \mathbf{U}\mathbf{T}_\alpha\mathbf{U}^{-1}$ and $\mathbf{B}_\beta = \mathbf{U}\mathbf{F}_\beta\mathbf{U}^{-1}$; hence,

$$\mathbf{D}_{\alpha,\beta} = \mathbf{B}_\beta \mathbf{A}_\alpha = \mathbf{U}\mathbf{F}_\beta\mathbf{U}^{-1}\mathbf{U}\mathbf{T}_\alpha\mathbf{U}^{-1} = \mathbf{U}\mathbf{S}_{\alpha,\beta}\mathbf{U}^{-1}.$$

Conversely, assume that $\mathbf{D}'_{\theta'} = e^{j\sigma(\psi(\theta'))}\mathbf{U}\mathbf{S}_{\psi(\theta')}\mathbf{U}^{-1}$ with $\sigma(0,0) = 0 \bmod 2\pi$. Then we have $\mathbf{D}'_{\theta'_0} = e^{j\sigma(0,0)}\mathbf{U}\mathbf{S}_{0,0}\mathbf{U}^{-1} = \mathbf{U}\mathbf{I}\mathbf{U}^{-1} = \mathbf{I}$ and

$$\begin{aligned}
\mathbf{D}'_{\theta'_2}\mathbf{D}'_{\theta'_1} &= e^{j\sigma(\psi(\theta'_2))}e^{j\sigma(\psi(\theta'_1))}\mathbf{U}\mathbf{S}_{\psi(\theta'_2)}\mathbf{U}^{-1}\mathbf{U}\mathbf{S}_{\psi(\theta'_1)}\mathbf{U}^{-1} \\
&= e^{j\left[\sigma(\psi(\theta'_2))+\sigma(\psi(\theta'_1))\right]}e^{-j2\pi p_2(\psi(\theta'_1))p_1(\psi(\theta'_2))}\mathbf{U}\mathbf{S}_{\psi(\theta'_1\circ\theta'_2)}\mathbf{U}^{-1} \\
&= e^{j\left[\sigma(\psi(\theta'_1))+\sigma(\psi(\theta'_2))-2\pi p_2(\psi(\theta'_1))p_1(\psi(\theta'_2))\right]}e^{-j\sigma(\psi(\theta'_1\circ\theta'_2))}\mathbf{D}'_{\theta'_1\circ\theta'_2}.
\end{aligned}$$

Thus, $\mathbf{D}'_{\theta'}$ is a projective representation of $(\mathcal{D}, \circ)$ with cocycle phase (7.6.49). It is irreducible and faithful because $\mathbf{S}_{\alpha,\beta}$ is. Therefore, it is a DO. □

Theorem 7.22. *Let $(\mathcal{D}, \circ)$ be isomorphic to the affine group with isomorphism ψ. Let $\{\mathbf{D}'_{\theta'}\}_{\theta' \in (\mathcal{D},\circ)}$ be a DO for $(\mathcal{D}, \circ)$ with cocycle $c'(\theta'_1, \theta'_2) \equiv 1$. Then $\mathbf{D}_\theta \triangleq \mathbf{D}'_{\psi^{-1}(\theta)}$ is unitarily equivalent to $\mathbf{R}_{\sigma,\tau} = \mathbf{T}_\tau\mathbf{C}_\sigma$, i.e.,*

$$\mathbf{D}_{\alpha,\beta} = \mathbf{V}\mathbf{R}_{\alpha,\pm\beta}\mathbf{V}^{-1},$$

with $\mathbf{V}$ an invertible and isometric operator.

Conversely, let $\mathbf{D}'_{\theta'}$ with $\theta' \in (\mathcal{D}, \circ)$ be unitarily equivalent to $\mathbf{R}_{\sigma,\tau}$ up to a parameter transformation, i.e.,

$$\mathbf{D}'_{\theta'} = \mathbf{V}\mathbf{R}_{\psi(\theta')}\mathbf{V}^{-1},$$

with $\mathbf{V}$ an invertible and isometric operator. Then $\mathbf{D}'_{\theta'}$ is a DO for $(\mathcal{D}, \circ)$, with cocycle $c'(\theta'_1, \theta'_2) \equiv 1$.

Proof: The proof is essentially analogous to that of Theorem 7.21. □

7.6.4 Relation between extended displacement functions

Next, we show an important relation between different extended DFs based on the same group.

Theorem 7.23. *Let $\mathcal{Z}$ and $\hat{\mathcal{Z}}$ be open, simply connected subsets of $\mathbb{R}^2$, $(\mathcal{D}, \circ)$ a simply connected 2D Lie group, and $\mathrm{e} : \mathcal{Z} \times \mathcal{D} \to \mathcal{Z}$ an extended DF for $(\mathcal{D}, \circ)$ acting on $\mathcal{Z}$. If $v : \mathcal{Z} \to \hat{\mathcal{Z}}$ is a diffeomorphism (i.e., an invertible, continuously differentiable function that maps $\mathcal{Z}$ onto $\hat{\mathcal{Z}}$ with nonzero Jacobian) with constant Jacobian, then the function*

$$\hat{\mathrm{e}} : \hat{\mathcal{Z}} \times \mathcal{D} \to \hat{\mathcal{Z}}, \qquad \hat{\mathrm{e}}(\hat{z}, \theta) = v\big(\mathrm{e}(v^{-1}(\hat{z}), \theta)\big) \tag{7.6.50}$$

is an extended DF for $(\mathcal{D}, \circ)$ acting on $\hat{\mathcal{Z}}$.

Conversely, let $\mathrm{e} : \mathcal{Z} \times \mathcal{D} \to \mathcal{Z}$ and $\hat{\mathrm{e}} : \hat{\mathcal{Z}} \times \mathcal{D} \to \hat{\mathcal{Z}}$ be extended DFs for $(\mathcal{D}, \circ)$ acting on $\mathcal{Z}$ and $\hat{\mathcal{Z}}$, respectively. Then e and $\hat{\mathrm{e}}$ are related according to (7.6.50) with $v : \mathcal{Z} \to \hat{\mathcal{Z}}$ given by

$$v(z) = \hat{d}\big(d^{-1}(z)\big), \tag{7.6.51}$$

where $d(\theta) = \mathrm{e}(z_0, \theta)$ and $\hat{d}(\theta) = \hat{\mathrm{e}}(\hat{z}_0, \theta)$ with arbitrary but fixed $z_0 \in \mathcal{Z}$ and $\hat{z}_0 \in \hat{\mathcal{Z}}$. This function v is a diffeomorphism with constant Jacobian.

Proof: We first show that $\hat{\mathrm{e}}(\hat{z}, \theta)$ in (7.6.50) satisfies the axioms of Definition 7.16. Axiom 1 is satisfied since a composition of invertible and continuously differentiable functions is invertible and continuously differentiable. Furthermore, $\hat{\mathrm{e}}$ is area-preserving since (J denotes the Jacobian; furthermore, note that $J_v(z)$ is constant by assumption and $J_{\mathrm{e}(\cdot,\theta)}(z) = \pm 1$ due to area preservation)

$$J_{\hat{\mathrm{e}}(\cdot,\theta)}(\hat{z}) = \frac{J_v\big(\mathrm{e}(v^{-1}(\hat{z}), \theta)\big) J_{\mathrm{e}(\cdot,\theta)}\big(v^{-1}(\hat{z})\big)}{J_v\big(v^{-1}(\hat{z})\big)} = J_{\mathrm{e}(\cdot,\theta)}\big(v^{-1}(\hat{z})\big) = \pm 1.$$

Axiom 2 is satisfied since a composition of diffeomorphisms is a diffeomorphism. Axiom 3, the composition property (7.6.45), can be verified in a straightforward manner. Hence, $\hat{\mathrm{e}}(\hat{z}, \theta)$ is an extended DF.

Next, we prove the converse statement. As a composition of diffeomorphisms (see Theorem 7.17), v in (7.6.51) is a diffeomorphism. Inserting $\hat{\mathrm{e}}(\hat{z}, \theta) = \hat{d}(\hat{d}^{-1}(\hat{z}) \circ \theta)$ (see (7.6.46)) into the left-hand side of (7.6.50), as well as (7.6.51) and $\mathrm{e}(z, \theta) = d\big(d^{-1}(z) \circ \theta\big)$ into the right-hand side of (7.6.50), we see that the two sides are equal and, hence, that (7.6.50) is true. Finally, setting $\hat{z} = v(z_1)$ in (7.6.50), with $z_1 \in \mathcal{Z}$ arbitrary but fixed, we obtain $\hat{\mathrm{e}}(v(z_1), \theta) = v\big(\mathrm{e}(z_1, \theta)\big)$ and hence $J_{\hat{\mathrm{e}}(\cdot,\theta)}(v(z_1)) J_v(z_1) = J_v(\mathrm{e}(z_1, \theta)) J_{\mathrm{e}(\cdot,\theta)}(z_1)$. Since $J_{\hat{\mathrm{e}}} = \pm 1$ and $J_{\mathrm{e}} = \pm 1$ due to area preservation (see Axiom 1 in Definition 7.16), this yields $J_v(z_1) = \pm J_v(\mathrm{e}(z_1, \theta))$. This equation holds for all $\theta \in \mathcal{D}$. Since to every $z \in \mathcal{Z}$ we can find a $\theta \in \mathcal{D}$ such that $\mathrm{e}(z_1, \theta) = z$ (see Axiom 2 in Definition 7.16), we get $J_v(z) = \pm J_v(z_1)$ for all $z \in \mathcal{Z}$. But since $J_v(z)$ is continuous and $\mathcal{Z}$ is connected, we have in fact $J_v(z) = J_v(z_1)$ for all $z \in \mathcal{Z}$, i.e., $J_v(z)$ is constant. □

Hence, the extended DFs of different DOs for the *same* group are related through diffeomorphisms v. (Note that the function v in Theorem 7.23 is not unique since z_0 and $\hat{z}_0$ are arbitrary.)

For $(\mathbb{R}^2,+)$ and the affine group (and all isomorphic groups), Kirillov's coadjoint orbit theory [53] allows the construction of a unique function e, called *coadjoint action* of the group, that satisfies the DF axioms in Definition 7.16. This function acts on sets called *coadjoint orbits*, some of which can be identified with open, simply connected subsets $\mathcal{Z} \subseteq \mathbb{R}^2$ [15,17,53,54,56]. Thus, the coadjoint orbit theory allows the construction of a *particular* extended DF for $(\mathbb{R}^2,+)$ and the affine group. It has been conjectured [12] that extended DFs are generally identical to the coadjoint actions. However, the coadjoint action of a given group is unique, whereas there exist different extended DFs for this group (the extended DFs of different DOs based on this group; see Theorem 7.23). In fact, an extended DF is associated to a specific DO $\mathbf{D}_\theta$, and not generically to the underlying group $(\mathcal{D},\circ)$. Thus, in general the coadjoint orbit theory does *not* allow the construction of the extended DF of a given DO.

7.7 The displacement function: construction

We shall now present a systematic method for constructing the DF of a given DO. This method avoids certain theoretical inconveniences of previous approaches [25, 28].

7.7.1 General construction

Consider a DO $\{\mathbf{D}'_{\theta'}\}_{\theta' \in (\mathcal{D},\circ)}$ whose DF is assumed to exist. According to Corollaries 7.19 and 7.20, the associated DO $\mathbf{D}_\theta \triangleq \mathbf{D}'_{\psi^{-1}(\theta)}$ is separable, i.e., $\mathbf{D}_{\alpha,\beta} = e^{j\sigma(\alpha,\beta)}\mathbf{B}_\beta \mathbf{A}_\alpha$, where the *partial DOs* $\{\mathbf{A}_\alpha\}_{\alpha \in (\mathcal{A},\bullet)}$ and $\{\mathbf{B}_\beta\}_{\beta \in (\mathcal{B},*)}$ are projective representations of groups $(\mathcal{A},\bullet)$ and $(\mathcal{B},*)$ that are $(\mathbb{R},+)$ or isomorphic to $(\mathbb{R},+)$. This separability will allow us to construct the extended DF of $\mathbf{D}_{\alpha,\beta}$ by composing the individual extended DFs of the partial DOs $\mathbf{A}_\alpha$ and $\mathbf{B}_\beta$. The extended DF of $\mathbf{D}'_{\theta'}$ can finally be obtained by a reparameterization of the extended DF of $\mathbf{D}_\theta$ using $\theta = \psi(\theta')$.

We first consider $\mathbf{A}_\alpha$. If a signal $x(t)$ is localized about some TF point $z_1 = (t_1, f_1)$, then $(\mathbf{A}_\alpha x)(t)$ will be localized about some other TF point $z_2 = e_{\mathbf{A}}(z_1, \alpha)$. We shall call the function $e_{\mathbf{A}} : \mathcal{Z} \times \mathcal{A} \to \mathcal{Z}$ the extended DF of the partial DO $\mathbf{A}_\alpha$, even though Axiom 2 in Definition 7.16 is not satisfied. To find $z_2 = (t_2, f_2)$, we use the (generalized) eigenfunctions $u_b^{\mathbf{A}}(t)$ of $\mathbf{A}_\alpha$ that are defined by the eigenvalue relation

$$\big(\mathbf{A}_\alpha u_b^{\mathbf{A}}\big)(t) = \lambda_{\alpha,b}^{\mathbf{A}} u_b^{\mathbf{A}}(t), \qquad \alpha \in (\mathcal{A},\bullet),\ b \in (\tilde{\mathcal{A}},\tilde{\bullet}). \tag{7.7.52}$$

Here, the eigenfunctions $u_b^{\mathbf{A}}(t)$ are indexed by $b \in (\tilde{\mathcal{A}},\tilde{\bullet})$, with $(\tilde{\mathcal{A}},\tilde{\bullet})$ again isomorphic to $(\mathbb{R},+)$ [36].

The TF locus of $u_b^{\mathbf{A}}(t)$ is defined by the instantaneous frequency $\nu\{u_b^{\mathbf{A}}\}(t)$ or the group delay $\tau\{u_b^{\mathbf{A}}\}(f)$, whichever exists. (The instantaneous frequency of a signal $x(t)$ is $\nu\{x\}(t) = (1/2\pi)(d/dt)\arg\{x(t)\}$; it exists if $\arg\{x(t)\}$ is differentiable and $x(t) \neq 0$ almost everywhere. The group delay of $x(t)$ is $\tau\{x\}(f) = -(1/2\pi)(d/df)\arg\{X(f)\}$ with $X(f) = \int_t x(t)e^{-j2\pi ft}dt$; it exists if $\arg\{X(f)\}$ is differentiable and $X(f) \neq 0$ almost everywhere.) Here, e.g., we shall assume existence of $\nu\{u_b^{\mathbf{A}}\}(t)$. Let us choose b_1 such that the TF curve defined by $\nu\{u_{b_1}^{\mathbf{A}}\}(t)$ passes through (t_1, f_1), i.e.,

$$\nu\{u_{b_1}^{\mathbf{A}}\}(t_1) = f_1. \tag{7.7.53}$$

This is shown in Figure 7.1. Since the instantaneous frequency is invariant to constant factors of the signal, (7.7.52) implies

$$\nu\{\mathbf{A}_\alpha u_{b_1}^{\mathbf{A}}\}(t) = \nu\{u_{b_1}^{\mathbf{A}}\}(t).$$

This shows that $\mathbf{A}_\alpha$ preserves the TF locus of $u_{b_1}^{\mathbf{A}}(t)$, in the sense that *all TF points on the curve* $\nu\{u_{b_1}^{\mathbf{A}}\}(t)$*—including* (t_1, f_1)*—are mapped again onto TF points on* $\nu\{u_{b_1}^{\mathbf{A}}\}(t)$. Hence, $(t_2, f_2) = e_{\mathbf{A}}\big((t_1, f_1), \alpha\big)$ must lie on $\nu\{u_{b_1}^{\mathbf{A}}\}(t)$ (see Figure 7.1), i.e., there must be

$$\nu\{u_{b_1}^{\mathbf{A}}\}(t_2) = f_2. \tag{7.7.54}$$

To find the exact position of (t_2, f_2) on the TF curve defined by $\nu\{u_{b_1}^{\mathbf{A}}\}(t)$, we consider the operator $\{\tilde{\mathbf{A}}_{\tilde{\alpha}}\}_{\tilde{\alpha}\in(\tilde{\mathcal{A}},\tilde{\bullet})}$ that is dual to $\{\mathbf{A}_\alpha\}_{\alpha\in(\mathcal{A},\bullet)}$ (see Definition 7.5). We note that the (generalized) eigenfunctions $\{u_{\tilde{b}}^{\tilde{\mathbf{A}}}(t)\}_{\tilde{b}\in(\mathcal{A},\bullet)}$ of $\tilde{\mathbf{A}}_{\tilde{\alpha}}$ can be derived from the eigenfunctions $\{u_b^{\mathbf{A}}(t)\}_{b\in(\tilde{\mathcal{A}},\tilde{\bullet})}$ of $\mathbf{A}_\alpha$ as [30]

$$u_{\tilde{b}}^{\tilde{\mathbf{A}}}(t) = \int_{\tilde{\mathcal{A}}} u_b^{\mathbf{A}}(t)e^{-j2\pi\psi_{\mathcal{A}}(\tilde{b})\psi_{\tilde{\mathcal{A}}}(b)}d\psi_{\tilde{\mathcal{A}}}(b).$$

(Note also that $\tilde{\mathbf{A}}_{\tilde{\alpha}}$ and $u_{\tilde{b}}^{\tilde{\mathbf{A}}}(t)$ are assumed to be indexed by the group $(\tilde{\mathcal{A}}, \tilde{\bullet})$ and $(\mathcal{A}, \bullet)$, respectively; such an indexing is always possible since all these groups are isomorphic to $(\mathbb{R}, +)$, and hence isomorphic to each other.) The TF locus of $u_{\tilde{b}}^{\tilde{\mathbf{A}}}(t)$ is defined by the instantaneous frequency $\nu\{u_{\tilde{b}}^{\tilde{\mathbf{A}}}\}(t)$ or the group delay $\tau\{u_{\tilde{b}}^{\tilde{\mathbf{A}}}\}(f)$,

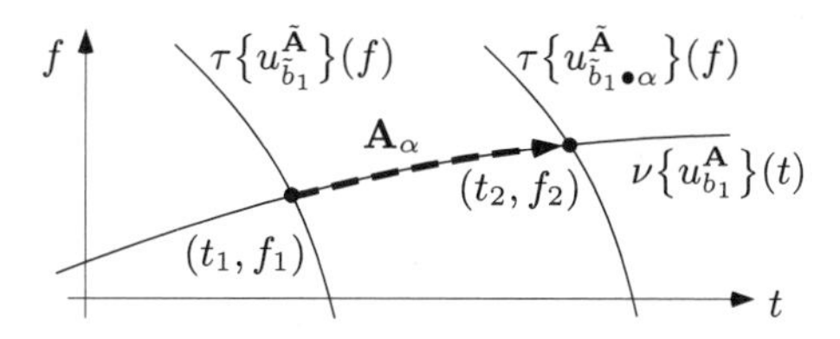

FIGURE 7.1. Construction of the extended DF of $\mathbf{A}_\alpha$.

whichever exists. Here, e.g., we assume existence of $\tau\{u_{\tilde{b}}^{\tilde{\mathbf{A}}}\}(f)$. Let us choose $\tilde{b}_1$ such that the TF curve defined by $\tau\{u_{\tilde{b}_1}^{\tilde{\mathbf{A}}}\}(f)$ passes through (t_1, f_1), i.e.,

$$\tau\{u_{\tilde{b}_1}^{\tilde{\mathbf{A}}}\}(f_1) = t_1, \tag{7.7.55}$$

as shown in Figure 7.1. According to (7.3.27), the eigenfunctions $u_{\tilde{b}}^{\tilde{\mathbf{A}}}(t)$ (suitably parameterized) satisfy

$$\big(\mathbf{A}_\alpha u_{\tilde{b}}^{\tilde{\mathbf{A}}}\big)(t) = u_{\tilde{b}\bullet\alpha}^{\tilde{\mathbf{A}}}(t), \qquad \alpha, \tilde{b} \in (\mathcal{A}, \bullet), \tag{7.7.56}$$

and hence there must be

$$\tau\{\mathbf{A}_\alpha u_{\tilde{b}_1}^{\tilde{\mathbf{A}}}\}(f) = \tau\{u_{\tilde{b}_1\bullet\alpha}^{\tilde{\mathbf{A}}}\}(f).$$

This shows that $\mathbf{A}_\alpha$ *maps all TF points on* $\tau\{u_{\tilde{b}_1}^{\tilde{\mathbf{A}}}\}(f)$*—including* (t_1, f_1)*—onto TF points on* $\tau\{u_{\tilde{b}_1\bullet\alpha}^{\tilde{\mathbf{A}}}\}(f)$. Hence, $(t_2, f_2) = e_{\mathbf{A}}\big((t_1, f_1), \alpha\big)$ must lie on $\tau\{u_{\tilde{b}_1\bullet\alpha}^{\tilde{\mathbf{A}}}\}(f)$ (see Figure 7.1), i.e., there must be

$$\tau\{u_{\tilde{b}_1\bullet\alpha}^{\tilde{\mathbf{A}}}\}(f_2) = t_2. \tag{7.7.57}$$

The relations (9.2.1) and (9.2.2) constitute a (generally nonlinear) system of equations in t_2 and f_2 whose solution for given $b_1, \tilde{b}_1$ (which follow from t_1, f_1 according to (7.7.53) and (7.7.55), respectively) and for given α defines the extended DF $e_{\mathbf{A}}$ via the identity $(t_2, f_2) \equiv e_{\mathbf{A}}\big((t_1, f_1), \alpha\big)$, provided that this solution exists and is unique. The construction of $e_{\mathbf{A}}$ can now be summarized as follows (see Figure 7.1):

1. For any given $(t_1, f_1) \in \mathcal{Z}$, we calculate associated eigenfunction parameters $b_1 \in (\tilde{\mathcal{A}}, \tilde{\bullet})$, $\tilde{b}_1 \in (\mathcal{A}, \bullet)$ as the solution to the equations (see (7.7.53), (7.7.55))

$$\nu\{u_{b_1}^{\mathbf{A}}\}(t_1) = f_1, \quad \tau\{u_{\tilde{b}_1}^{\tilde{\mathbf{A}}}\}(f_1) = t_1. \tag{7.7.58}$$

2. The extended DF $e_{\mathbf{A}}$ is defined by $(t_2, f_2) \equiv e_{\mathbf{A}}\big((t_1, f_1), \alpha\big)$, where (t_2, f_2) is obtained as the solution to the system of equations (see (9.2.1), (9.2.2))

$$\nu\{u_{b_1}^{\mathbf{A}}\}(t_2) = f_2, \quad \tau\{u_{\tilde{b}_1\bullet\alpha}^{\tilde{\mathbf{A}}}\}(f_2) = t_2. \tag{7.7.59}$$

This construction is easily modified for the case where, e.g., $\tau\{u_b^{\mathbf{A}}\}(f)$ and $\nu\{u_{\tilde{b}}^{\tilde{\mathbf{A}}}\}(t)$ exist instead of $\nu\{u_b^{\mathbf{A}}\}(t)$ and $\tau\{u_{\tilde{b}}^{\tilde{\mathbf{A}}}\}(f)$.

The extended DF $e_{\mathbf{B}}(z, \beta)$ of $\mathbf{B}_\beta$ can be constructed by an analogous procedure, using the eigenfunctions $\{u_a^{\mathbf{B}}(t)\}_{a\in(\tilde{\mathcal{B}},\tilde{*})}$ and $\{u_{\tilde{a}}^{\tilde{\mathbf{B}}}(t)\}_{\tilde{a}\in(\mathcal{B},*)}$ of $\{\mathbf{B}_\beta\}_{\beta\in(\mathcal{B},*)}$ and of the dual operator $\{\tilde{\mathbf{B}}_{\tilde{\beta}}\}_{\tilde{\beta}\in(\tilde{\mathcal{B}},\tilde{*})}$, respectively. Next, the extended DF $e(z, \theta)$ of the DO $\mathbf{D}_{\alpha,\beta} = e^{j\sigma(\alpha,\beta)}\mathbf{B}_\beta\mathbf{A}_\alpha$ follows upon composing $e_{\mathbf{A}}(z, \alpha)$ and $e_{\mathbf{B}}(z, \beta)$,

$$e\big((t,f),(\alpha,\beta)\big) = e_{\mathbf{B}}\big(e_{\mathbf{A}}\big((t,f),\alpha\big),\beta\big). \tag{7.7.60}$$

Finally, the extended DF $e'(z,\theta')$ of $\mathbf{D}'_{\theta'}$ is obtained by reparameterizing $e(z,\theta)$ according to $e'(z,\theta') \triangleq e\,(z,\psi(\theta'))$. The DF of $\mathbf{D}'_{\theta'}$ is given by $d'(z) = e'(z_0,\theta')$ with some $z_0 \in \mathcal{Z}$.

This construction of the (extended) DF assumes that the instantaneous frequencies or group delays of the operators $\mathbf{A}_\alpha$, $\tilde{\mathbf{A}}_{\tilde{\alpha}}$, $\mathbf{B}_\beta$, and $\tilde{\mathbf{B}}_{\tilde{\beta}}$ exist and that the equations (7.7.58) and (7.7.59) have unique solutions. While the construction is intuitively reasonable, there is no general proof that the resulting function $e'(z,\theta')$ satisfies the axioms of Definition 7.16. However, we will show in Subsections 7.7.3 and 7.7.4 that for the important and fairly extensive cases of dual and affine modulation and warping operators, all assumptions and axioms are satisfied and the extended DF is readily obtained in closed form.

7.7.2 Warping operator and modulation operator

By way of example and for later use, let us calculate the extended DF of a *warping operator* $\{\mathbf{W}_\alpha\}_{\alpha\in(\mathcal{A},\bullet)}$ with warping function $w_\alpha(t)$ defined on a set $\Theta \subseteq \mathbb{R}$ (see Definition 7.2). According to the second part of Theorem 7.3, under specific assumptions there exists a subset $\Omega \subseteq \Theta$ such that

$$w_\alpha(t) = n^{-1}\big(n(t) \bullet \alpha^{-1}\big), \qquad t \in \Omega, \tag{7.7.61}$$

with an invertible function $n : \Omega \to \mathcal{A}$. We recall from (7.2.21) that the eigenfunctions of $\mathbf{W}_\alpha$ are given by $u_b^{\mathbf{W}}(t) = \sqrt{|\tilde{n}'(t)|}e^{j2\pi\psi_{\tilde{\mathcal{A}}}(b)\tilde{n}(t)}$, with $b \in \tilde{\mathcal{A}}$ and $\tilde{n}(t) = \psi_{\mathcal{A}}(n(t))$, where $\tilde{\mathcal{A}}$ is some group isomorphic to $(\mathbb{R},+)$. The instantaneous frequency of $u_b^{\mathbf{W}}(t)$ is obtained as $\nu\{u_b^{\mathbf{W}}\}(t) = \psi_{\tilde{\mathcal{A}}}(b)\tilde{n}'(t)$. According to Subsection 7.3.2, the dual operator $\{\widetilde{\mathbf{W}}_{\tilde{\alpha}}\}_{\tilde{\alpha}\in\tilde{\mathcal{A}}}$ is the modulation operator with modulation function $n(t)$. We recall from (7.2.15) that the eigenfunctions of $\widetilde{\mathbf{W}}_{\tilde{\alpha}}$ are given by $u_{\tilde{b}}^{\widetilde{\mathbf{W}}}(t) = r(t)\delta\big(\tilde{n}(t) - \psi_{\mathcal{A}}(\tilde{b})\big)$, $\tilde{b} \in \mathcal{A}$ (assuming parameterization by the group $(\mathcal{A},\bullet)$). The group delay of $u_{\tilde{b}}^{\widetilde{\mathbf{W}}}(t)$ is obtained as $\tau\{u_{\tilde{b}}^{\widetilde{\mathbf{W}}}\}(f) \equiv n^{-1}(\tilde{b})$. Evaluating (7.7.58) yields $\tilde{b}_1 = n(t_1)$ and $b_1 = \psi_{\tilde{\mathcal{A}}}^{-1}\big(f_1/\tilde{n}'(t_1)\big)$. Inserting into (7.7.59) and solving for t_2 and f_2 yields $t_2 = w_{\alpha^{-1}}(t_1)$ and $f_2 = f_1/w'_{\alpha^{-1}}(t_1)$, and hence the extended DF of $\mathbf{W}_\alpha$ is obtained as

$$e_{\mathbf{W}}\big((t,f),\alpha\big) = \left(w_{\alpha^{-1}}(t), \frac{f}{w'_{\alpha^{-1}}(t)}\right), \qquad (t,f) \in \Omega\times\mathbb{R},\ \alpha \in \mathcal{A}. \tag{7.7.62}$$

Next, we calculate the extended DF of a *modulation operator* $\{\mathbf{M}_\beta\}_{\beta\in(\mathcal{B},*)}$ with modulation function $m : \Omega \to \tilde{\mathcal{B}}$ defined on a set $\Omega \subseteq \mathbb{R}$ (see Definition 7.1). The group delay of the eigenfunctions of $\mathbf{M}_\beta$ is given by (cf. the previous calculation for $\widetilde{\mathbf{W}}_{\tilde{\alpha}}$) $\tau\{u_a^{\mathbf{M}}\}(f) \equiv m^{-1}(a)$, $a \in (\tilde{\mathcal{B}},\tilde{*})$. The dual operator $\{\widetilde{\mathbf{M}}_{\tilde{\beta}}\}_{\tilde{\beta}\in\tilde{\mathcal{B}}}$ can be shown to be the warping operator with warping function $w_{\tilde{\beta}}(t) = m^{-1}\big(m(t)\tilde{*}\tilde{\beta}\big)$. Assuming parameterization of the eigenfunctions of $\widetilde{\mathbf{M}}_{\tilde{\beta}}$ according to $u_{\tilde{a}}^{\widetilde{\mathbf{M}}}(t) =$

$\sqrt{|\tilde{m}'(t)|}e^{j2\pi\psi_{\mathcal{B}}(\tilde{a})\tilde{m}(t)}$ (cf. (7.2.21)), with $\tilde{a} \in (\mathcal{B}, *)$ and $\tilde{m}(t) = \psi_{\tilde{\mathcal{B}}}(m(t))$, the instantaneous frequency of $u_{\tilde{a}}^{\widetilde{\mathbf{M}}}(t)$ is given by $\nu\{u_{\tilde{a}}^{\widetilde{\mathbf{M}}}\}(t) = \psi_{\mathcal{B}}(\tilde{a})\tilde{m}'(t)$. Evaluating (7.7.58) (with the roles of instantaneous frequency and group delay interchanged) yields $a_1 = m(t_1)$ and $\tilde{a}_1 = \psi_{\mathcal{B}}^{-1}(f_1/\tilde{m}'(t_1))$. Inserting into (7.7.59) (again with the roles of instantaneous frequency and group delay interchanged) and solving for (t_2, f_2), we obtain the extended DF of $\mathbf{M}_\beta$ as

$$e_{\mathbf{M}}\big((t,f),\beta\big) = \big(t, f + \psi_{\mathcal{B}}(\beta)\tilde{m}'(t)\big), \quad (t,f) \in \Omega \times \mathbb{R},\ \beta \in \mathcal{B}. \tag{7.7.63}$$

7.7.3 Dual modulation and warping operators

We now consider $\mathbf{D}_{\alpha,\beta} = \mathbf{M}_\beta \mathbf{W}_\alpha$, where $\{\mathbf{M}_\beta\}_{\beta\in(\mathcal{B},*)}$ is a modulation operator with modulation function $m : \Omega \to \mathcal{A}$, $m(\Omega) = \mathcal{A}$ and $\{\mathbf{W}_\alpha\}_{\alpha\in(\mathcal{A},\bullet)}$ is the warping operator with warping function $w_\alpha(t) = m^{-1}(m(t)\bullet\alpha^{-1})$. According to Subsection 7.3.2, $\mathbf{M}_\beta$ is *dual* to $\mathbf{W}_\alpha$. Hence, due to Theorem 7.6, $\mathbf{D}_{\alpha,\beta}$ is a DO for a group $(\mathcal{A}\times\mathcal{B}, \circ)$ that is isomorphic to $(\mathbb{R}^2, +)$. Composing the extended DFs $e_{\mathbf{W}}\big((t,f),\alpha\big)$ in (7.7.62) and $e_{\mathbf{M}}\big((t,f),\beta\big)$ in (7.7.63) according to (7.7.60), the extended DF of $\mathbf{D}_{\alpha,\beta}$ is obtained as

$$e\big((t,f),(\alpha,\beta)\big) = e_{\mathbf{M}}\big(e_{\mathbf{W}}\big((t,f),\alpha\big),\beta\big) = \left(w_{\alpha^{-1}}(t), \frac{f + \psi_{\mathcal{B}}(\beta)\tilde{m}'(t)}{w'_{\alpha^{-1}}(t)}\right), \tag{7.7.64}$$

with $(t,f) \in \Omega\times\mathbb{R}$, $(\alpha,\beta) \in \mathcal{A}\times\mathcal{B}$, $\tilde{m}(t) = \psi_{\mathcal{A}}\big(m(t)\big)$, and $w_\alpha(t) = m^{-1}(m(t)\bullet\alpha^{-1})$. Setting $(t,f) = (t_0, f_0)$ with $t_0 = \tilde{m}^{-1}(0)$ and $f_0 = 0$ for simplicity, the DF of $\mathbf{D}_{\alpha,\beta}$ results as

$$d(\alpha,\beta) = \big(m^{-1}(\alpha), \psi_{\mathcal{B}}(\beta)\tilde{m}'(m^{-1}(\alpha))\big). \tag{7.7.65}$$

It can be verified that the extended DF satisfies all axioms of Definition 7.16, provided that $\tilde{m}(t)$ is twice continuously differentiable and $\psi_{\mathcal{A}}(\alpha)$ and $\psi_{\mathcal{B}}(\beta)$ are continuously differentiable.

As an example, let $\mathbf{M}_\nu = \mathbf{F}_\nu$ and $\mathbf{W}_\tau = \mathbf{T}_\tau$, so that $\mathbf{D}_{\tau,\nu} = \mathbf{S}_{\tau,\nu} = \mathbf{F}_\nu\mathbf{T}_\tau$. Here, $m(t) = \tilde{m}(t) = t$, $w_\tau(t) = t - \tau$, $\Omega = \mathbb{R}$, and $(\mathcal{A},\bullet) = (\mathcal{B},*) = (\mathbb{R},+)$, so that (7.7.64) and (7.7.65) become

$$e\big((t,f),(\tau,\nu)\big) = (t+\tau, f+\nu), \quad d(\tau,\nu) = (\tau,\nu), \qquad t,f,\tau,\nu \in \mathbb{R}.$$

7.7.4 Affine modulation and warping operators

Next, let $\{\mathbf{M}_\beta\}_{\beta\in(\mathcal{B},*)}$ be a modulation operator with modulation function $m : \Omega \to \mathcal{A}$, with $\tilde{m}(\Omega) = \mathbb{R}^+$ or $\tilde{m}(\Omega) = \mathbb{R}^-$, where $\tilde{m}(t) = \psi_{\mathcal{A}}\big(m(t)\big)$. Furthermore, let $\{\mathbf{W}_\alpha\}_{\alpha\in(\mathcal{A},\bullet)}$ be the warping operator whose warping function is related to $m(t)$ according to (7.4.32), i.e., $w_\alpha(t) = \tilde{m}^{-1}\big(\tilde{m}(t)\exp\big(\psi_{\mathcal{A}}(\alpha)\big)\big)$. The construction of the extended DF of $\mathbf{W}_\alpha$ requires the dual operator of $\mathbf{W}_\alpha$, which is a modulation operator with modulation function $n(t)$ related to $w_\alpha(t)$ according to $w_\alpha(t) = n^{-1}\big(n(t) \bullet \alpha^{-1}\big)$ (see (7.7.61)). Hence, hereafter we restrict all functions and operators to a subset $\Omega' \subseteq \Omega$ on which $w_\alpha(t)$ can be represented in the form $w_\alpha(t) = n^{-1}\big(n(t) \bullet \alpha^{-1}\big)$ with an invertible function $n : \Omega' \to \mathcal{A}$.

Consider $\mathbf{D}_{\alpha,\beta} = \mathbf{M}_\beta \mathbf{W}_\alpha$. With $w_\alpha(t) = \tilde{m}^{-1}\big(\tilde{m}(t)\exp\big(\psi_\mathcal{A}(\alpha)\big)\big)$, it follows from Subsection 7.4.2 that $\mathbf{M}_\beta$ is *affine* to $\mathbf{W}_\alpha$. Hence, due to Theorem 7.9, $\mathbf{D}_{\alpha,\beta}$ is a DO for a group $(\mathcal{A} \times \mathcal{B}, \circ)$ that is isomorphic to the affine group. Using (7.7.62) and (7.7.63), the extended DF of $\mathbf{D}_{\alpha,\beta}$ is obtained as

$$\begin{aligned} e\big((t,f),(\alpha,\beta)\big) &= e_\mathbf{M}\big(e_\mathbf{W}\big((t,f),\alpha\big),\beta\big) \\ &= \left(w_{\alpha^{-1}}(t), \frac{f + \psi_\mathcal{B}(\beta)\tilde{m}'(t)\exp\big(-\psi_\mathcal{A}(\alpha)\big)}{w'_{\alpha^{-1}}(t)} \right), \end{aligned} \tag{7.7.66}$$

with $(t,f) \in \Omega' \times \mathbb{R}$, $(\alpha,\beta) \in \mathcal{A} \times \mathcal{B}$, $\tilde{m}(t) = \psi_\mathcal{A}\big(m(t)\big)$, and $w_\alpha(t) = \tilde{m}^{-1}\big(\tilde{m}(t)\exp\big(\psi_\mathcal{A}(\alpha)\big)\big)$. Setting $(t,f) = (t_0, f_0)$ with $t_0 = \tilde{m}^{-1}(\pm 1)$ (the sign depends on whether $\tilde{m}(\Omega) = \mathbb{R}^+$ or $\tilde{m}(\Omega) = \mathbb{R}^-$) and $f_0 = 0$ for simplicity, the DF of $\mathbf{D}_{\alpha,\beta}$ results as

$$d(\alpha,\beta) = \big(\xi(\alpha), \psi_\mathcal{B}(\beta)\tilde{m}'\big(\xi(\alpha)\big)\big), \qquad \text{with } \xi(\alpha) \triangleq \tilde{m}^{-1}\big(\pm\exp(-\psi_\mathcal{A}(\alpha))\big). \tag{7.7.67}$$

Again, it can be verified that the extended DF satisfies all axioms of Definition 7.16, provided that $\tilde{m}(t)$ is twice continuously differentiable and $\psi_\mathcal{A}(\alpha)$ and $\psi_\mathcal{B}(\beta)$ are continuously differentiable.

As an example, let $\mathbf{M}_\nu = \mathbf{F}_\nu$ and $\mathbf{W}_\sigma = \mathbf{C}_\sigma$ with $\sigma \in \mathbb{R}^+$, i.e., $\mathbf{D}_{\sigma,\nu} = \mathbf{F}_\nu \mathbf{C}_\sigma$. Here, $\tilde{m}(t) = t$, $w_\sigma(t) = t/\sigma$, $n(t) = t$, $\psi_\mathcal{A}(\sigma) = -\ln\sigma$, $\psi_\mathcal{B}(\nu) = \nu$, and $\Omega = \Omega' = \mathbb{R}^+$, so that (7.7.66) and (7.7.67) become

$$e\big((t,f),(\sigma,\nu)\big) = \left(\sigma t, \frac{f}{\sigma} + \nu\right), \quad d(\sigma,\nu) = (\sigma,\nu), \qquad t,\sigma \in \mathbb{R}^+,\ f,\nu \in \mathbb{R}.$$

7.8 Covariant signal representations: time-frequency domain

In Section 7.5, we derived covariant linear and bilinear signal representations that were functions of the displacement parameter $\theta = (\alpha,\beta)$. Using the DF mapping $z = d(\theta)$, $\theta = d^{-1}(z)$, it is now fairly straightforward to convert these covariant θ-representations into covariant TF representations.

7.8.1 *Covariant linear time-frequency representations*

We first consider linear TF representations (LTFRs). Any LTFR can be written in the form

$$\tilde{L}_x(z) = \langle x, h_z \rangle = \int_t x(t) h_z^*(t) dt,$$

where $h_z(t)$ is a function that depends on the TF point $z = (t, f)$ but not on the signal x.

Definition 7.24 ([25]). Let $\{\mathbf{D}_\theta\}_{\theta \in (\mathcal{D}, \circ)}$ be a DO on a Hilbert space $\mathcal{X}$, with cocycle c and extended DF e acting on a TF point set $\mathcal{Z}$. An LTFR $\tilde{L}_x(z)$ defined on $\mathcal{Z}$ is called *covariant to the DO* $\mathbf{D}_\theta$ if, for all $x \in \mathcal{X}$,

$$\tilde{L}_{\mathbf{D}_{\theta_1} x}(z) = c\big(d^{-1}(z) \circ \theta_1^{-1}, \theta_1\big) \tilde{L}_x\big(e(z, \theta_1^{-1})\big), \qquad \text{for all } z \in \mathcal{Z},\ \theta_1 \in \mathcal{D}, \tag{7.8.68}$$

where $d(\theta) = e(z_0, \theta)$ with some arbitrary but fixed $z_0 \in \mathcal{Z}$.

This covariance property states that the LTFR of the displaced signal $(\mathbf{D}_{\theta_1} x)(t)$ equals (possibly up to a phase factor) the TF displaced LTFR of the original signal $x(t)$, $\tilde{L}_x\big(e(z, \theta_1^{-1})\big)$. The covariance property depends on a reference TF point $z_0 \in \mathcal{Z}$ that can be chosen arbitrarily; different choices of z_0 result in covariance properties that differ by the phase factor $c\big(d^{-1}(z) \circ \theta_1^{-1}, \theta_1\big)$ (since $d(z)$ depends on z_0). Beyond the choice of z_0, there is some arbitrariness regarding the exact definition of the phase factor in (7.8.68). Our specific choice will result in a particularly simple expression for the covariant LTFR $\tilde{L}_x(z)$ (see (7.8.69)).

The next corollary to Theorem 7.13 characterizes all covariant LTFRs.

Corollary 7.25 ([25]). *Let $\{\mathbf{D}_\theta\}_{\theta \in (\mathcal{D}, \circ)}$ be a DO on a Hilbert space $\mathcal{X}$, with extended DF e acting on a TF point set $\mathcal{Z}$. All LTFRs covariant (with reference TF point $z_0 \in \mathcal{Z}$) to $\mathbf{D}_\theta$ are given by*

$$\tilde{L}_x(z) = \big\langle x, \mathbf{D}_{d^{-1}(z)} h \big\rangle = \int_{t'} x(t') \big(\mathbf{D}_{d^{-1}(z)} h\big)^*(t') dt', \qquad z \in \mathcal{Z}, \tag{7.8.69}$$

where h is an arbitrary function in $\mathcal{X}$ and $d(\theta) = e(z_0, \theta)$.

Proof: Setting $z = d(\theta)$ in the TF covariance property (7.8.68) and using (7.6.46) yields

$$L_{\mathbf{D}_{\theta_1} x}(\theta) = c(\theta \circ \theta_1^{-1}, \theta_1) L_x(\theta \circ \theta_1^{-1}), \qquad \text{with } L_x(\theta) \triangleq \tilde{L}_x\big(d(\theta)\big). \tag{7.8.70}$$

This is the θ-domain covariance property in (7.5.38). Since the DF mapping d is invertible, this covariance property is strictly equivalent to the original TF covariance property (7.8.68). According to Theorem 7.13, all linear θ-representations satisfying (7.8.70) are given by $L_x(\theta) = \langle x, \mathbf{D}_\theta h \rangle = \int_t x(t) (\mathbf{D}_\theta h)^*(t) dt$. Since (7.8.68) and (7.8.70) are equivalent via the DF mapping d, all LTFRs satisfying (7.8.68) are then given by

$$\tilde{L}_x(z) = L_x(\theta)\Big|_{\theta = d^{-1}(z)} = \big\langle x, \mathbf{D}_{d^{-1}(z)} h \big\rangle.$$

□

As an example, consider the operator $\mathbf{S}_{\tau,\nu} = \mathbf{F}_\nu \mathbf{T}_\tau$. This DO has cocycle $c\big((\tau_1,\nu_1),(\tau_2,\nu_2)\big) = e^{-j2\pi\nu_1\tau_2}$ and DF (with $(t_0,f_0) = (0,0)$) $d(\tau,\nu) = (\tau,\nu)$, i.e., $t=\tau$ and $f=\nu$. Thus, the covariance property (7.8.68) becomes

$$\tilde{L}_{\mathbf{S}_{\tau_1,\nu_1}x}(t,f) = e^{-j2\pi\tau_1(f-\nu_1)}\tilde{L}_x(t-\tau_1, f-\nu_1),$$

and the class of all covariant LTFRs is the short-time Fourier transform in (7.1.4), i.e.,

$$\tilde{L}_x(t,f) = \langle x, \mathbf{S}_{t,f}h\rangle = \int_{t'} x(t')h^*(t'-t)e^{-j2\pi ft'}dt'.$$

This is in fact equivalent to the θ-domain covariance property in (7.5.39) and the class of covariant linear θ-representations in (7.5.41), the reason being that the DF is the identity function. Further examples of TF covariance properties and the corresponding classes of covariant LTFRs are provided in Table 7.6—in particular, the TF version of the wavelet transform is obtained for the DO $\mathbf{R}_{\sigma,\tau} = \mathbf{T}_\tau\mathbf{C}_\sigma$—and in Subsection 7.8.3.

7.8.2 *Covariant bilinear time-frequency representations*

Next, we consider bilinear TF representations (BTFRs). Any BTFR can be written in the form

$$\tilde{B}_{x,y}(z) = \langle x, \mathbf{K}_z y\rangle = \int_{t_1}\int_{t_2} x(t_1)y^*(t_2)k_z^*(t_1,t_2)dt_1dt_2,$$

where $\mathbf{K}_z$ is a linear operator with kernel $k_z(t_1,t_2)$ that depends on $z=(t,f)$ but not on x,y.

Definition 7.26 ([25,26,28]). Let $\{\mathbf{D}_\theta\}_{\theta\in(\mathcal{D},\circ)}$ be a DO on a Hilbert space $\mathcal{X}$, with extended DF e acting on a TF point set $\mathcal{Z}$. A BTFR $\tilde{B}_{x,y}(z)$ defined on $\mathcal{Z}$ is called *covariant to the DO* $\mathbf{D}_\theta$ if, for all $x,y\in\mathcal{X}$,

$$\tilde{B}_{\mathbf{D}_{\theta_1}x,\mathbf{D}_{\theta_1}y}(z) = \tilde{B}_{x,y}\big(e(z,\theta_1^{-1})\big), \qquad \text{for all } z\in\mathcal{Z},\ \theta_1\in\mathcal{D}. \tag{7.8.71}$$

Apart from the missing cocycle c (and, hence, the independence of a reference TF point z_0), this covariance property is analogous to the corresponding property (7.8.68) in the linear case. The class of all covariant BTFRs is characterized by the following corollary to Theorem 7.15.

Corollary 7.27 ([25]). *Let* $\{\mathbf{D}_\theta\}_{\theta\in(\mathcal{D},\circ)}$ *be a DO on a Hilbert space* $\mathcal{X}$, *with extended DF* e *acting on a set* $\mathcal{Z}$. *All BTFRs covariant to* $\mathbf{D}_\theta$ *are given by*

$$\begin{aligned}\tilde{B}_{x,y}(z) &= \big\langle x, \mathbf{D}_{d^{-1}(z)}\mathbf{K}\mathbf{D}^{-1}_{d^{-1}(z)}y\big\rangle \\ &= \int_{t_1}\int_{t_2} x(t_1)y^*(t_2)\big[\mathbf{D}_{d^{-1}(z)}\mathbf{K}\mathbf{D}^{-1}_{d^{-1}(z)}\big]^*(t_1,t_2)dt_1dt_2, \quad z\in\mathcal{Z},\end{aligned} \tag{7.8.72}$$

Short-time Fourier transform	
$\mathbf{A}_\alpha$	$(\mathbf{A}_\alpha x)(t) = x(t-\alpha), \quad \alpha \in (\mathbb{R},+)$
$\mathbf{B}_\beta$	$(\mathbf{B}_\beta x)(t) = e^{j2\pi\beta t}x(t), \quad \beta \in (\mathbb{R},+)$
Composition property	$\mathbf{D}_{\alpha_2,\beta_2}\mathbf{D}_{\alpha_1,\beta_1} = e^{-j2\pi\beta_1\alpha_2}\mathbf{D}_{\alpha_1+\alpha_2,\beta_1+\beta_2}$
Displacement function	$d(\alpha,\beta) = (\alpha,\beta)$
Covariance property	$\tilde{L}_{\mathbf{D}_{\alpha,\beta}x}(t,f) = e^{-j2\pi\alpha(f-\beta)}\,\tilde{L}_x(t-\alpha,f-\beta)$
Covar. lin. TF repres.	$\tilde{L}_x(t,f) = \int_{-\infty}^{\infty} x(t')h^*(t'-t)e^{-j2\pi ft'}dt'$

Wavelet transform	
$\mathbf{A}_\alpha$	$(\mathbf{A}_\alpha x)(t) = \frac{1}{\sqrt{\alpha}}x\big(\frac{t}{\alpha}\big), \quad \alpha \in (\mathbb{R}^+,\cdot)$
$\mathbf{B}_\beta$	$(\mathbf{B}_\beta x)(t) = x(t-\beta), \quad \beta \in (\mathbb{R},+)$
Composition property	$\mathbf{D}_{\alpha_2,\beta_2}\mathbf{D}_{\alpha_1,\beta_1} = \mathbf{D}_{\alpha_1\alpha_2,\beta_1\alpha_2+\beta_2}$
Displacement function	$d(\alpha,\beta) = \big(\beta,\frac{1}{\alpha}\big)$
Covariance property	$\tilde{L}_{\mathbf{D}_{\alpha,\beta}x}(t,f) = \tilde{L}_x\big(\frac{t-\beta}{\alpha},\alpha f\big), \; f>0$
Covar. lin. TF repres.	$\tilde{L}_x(t,f) = \sqrt{f}\int_{-\infty}^{\infty} x(t')h^*\big(f(t'-t)\big)dt', \; f>0$

Hyperbolic wavelet transform	
$\mathbf{A}_\alpha$	$(\mathbf{A}_\alpha x)(t) = \frac{1}{\sqrt{\alpha}}x\big(\frac{t}{\alpha}\big), \quad \alpha \in (\mathbb{R}^+,\cdot)$
$\mathbf{B}_\beta$	$(\hat{\mathbf{B}}_\beta X)(f) = e^{-j2\pi\beta\ln f}X(f), \quad \beta \in (\mathbb{R},+), \; f>0$
Composition property	$\mathbf{D}_{\alpha_2,\beta_2}\mathbf{D}_{\alpha_1,\beta_1} = e^{-j2\pi\beta_1\ln\alpha_2}\mathbf{D}_{\alpha_1\alpha_2,\beta_1+\beta_2}$
Displacement function	$d(\alpha,\beta) = \big(\alpha\beta,\frac{1}{\alpha}\big)$
Covariance property	$\tilde{L}_{\mathbf{D}_{\alpha,\beta}x}(t,f) = e^{-j2\pi(tf-\beta)\ln\alpha}\,\tilde{L}_x\big(\frac{t-\beta/f}{\alpha},\alpha f\big), \; f>0$
Covar. lin. TF repres.	$\tilde{L}_x(t,f) = \frac{1}{\sqrt{f}}\int_0^{\infty} X(f')H^*\big(\frac{f'}{f}\big)e^{j2\pi tf\ln f'}df', \; f>0$

Power wavelet transform	
$\mathbf{A}_\alpha$	$(\mathbf{A}_\alpha x)(t) = \frac{1}{\sqrt{\alpha}}x\big(\frac{t}{\alpha}\big), \quad \alpha \in (\mathbb{R}^+,\cdot)$
$\mathbf{B}_\beta$	$(\hat{\mathbf{B}}_\beta X)(f) = e^{-j2\pi\beta\xi_\kappa(f)}X(f), \quad \beta \in (\mathbb{R},+),$ with $\xi_\kappa(f) = \mathrm{sgn}(f)\lvert f\rvert^\kappa$
Composition property	$\mathbf{D}_{\alpha_2,\beta_2}\mathbf{D}_{\alpha_1,\beta_1} = \mathbf{D}_{\alpha_1\alpha_2,\beta_1\xi_\kappa(\alpha_2)+\beta_2}$
Displacement function (extended to $f \in \mathbb{R}$)	$d(\alpha,\beta) = \big(\beta\xi'_\kappa\big(\frac{1}{\alpha}\big),\pm\frac{1}{\alpha}\big) = \big(\frac{\beta\kappa}{\alpha^{\kappa-1}},\pm\frac{1}{\alpha}\big)$ (note that $\xi'_\kappa(f) = \kappa\lvert f\rvert^{\kappa-1}$)
Covariance property	$\tilde{L}_{\mathbf{D}_{\alpha,\beta}x}(t,f) = \tilde{L}_x\big(\frac{t-\beta\xi'_\kappa(f)}{\alpha},\alpha f\big)$
Covar. lin. TF repres. (extended to $f \in \mathbb{R}$)	$\tilde{L}_x(t,f) = \frac{1}{\sqrt{\lvert f\rvert}}\int_{-\infty}^{\infty} X(f')H^*\big(\frac{f'}{f}\big)e^{j2\pi t\xi_\kappa(f')/\xi'_\kappa(f)}df'$

TABLE 7.6. Some covariant linear TF representations. Note that $\theta = (\alpha,\beta)$, $\mathbf{D}_{\alpha,\beta} = \mathbf{B}_\beta\mathbf{A}_\alpha$, $X(f)$ denotes the Fourier transform of $x(t)$, and $\hat{\mathbf{B}}_\beta$ denotes the frequency-domain version of $\mathbf{B}_\beta$.

where $\mathbf{K}$ *is an arbitrary linear operator on* $\mathcal{X}$*,* $\left[\mathbf{D}_\theta \mathbf{K} \mathbf{D}_\theta^{-1}\right](t_1, t_2)$ *denotes the kernel of* $\mathbf{D}_\theta \mathbf{K} \mathbf{D}_\theta^{-1}$*, and* $d(\theta) = e(z_0, \theta)$ *with arbitrary* $z_0 \in \mathcal{Z}$*.*

Proof: The proof is completely analogous to that of Corollary 7.25. In particular, we have

$$\tilde{B}_{x,y}(z) = B_{x,y}(\theta)\Big|_{\theta=d^{-1}(z)}, \tag{7.8.73}$$

where

$$B_{x,y}(\theta) = \langle x, \mathbf{D}_\theta \mathbf{K} \mathbf{D}_\theta^{-1} y\rangle = \int_{t_1}\int_{t_2} x(t_1) y^*(t_2) \left[\mathbf{D}_\theta \mathbf{K} \mathbf{D}_\theta^{-1}\right]^*(t_1, t_2) dt_1 dt_2$$

is the class of all covariant bilinear θ-representations according to Theorem 7.15. □

Whereas it appears that the class of all covariant BTFRs $\tilde{B}_{x,y}(z)$ in (7.8.72) depends on the reference TF point z_0 via the DF $d(\theta) = e(z_0, \theta)$, this is really not the case. Indeed, if $d(\theta) = e(z_0, \theta)$ and $\tilde{d}(\theta) = e(\tilde{z}_0, \theta)$ are two different DFs of $\mathbf{D}_\theta$ obtained with two different reference TF points z_0 and $\tilde{z}_0$, respectively, then according to (7.6.47) there exists a $\theta' \in \mathcal{D}$ such that $\tilde{d}^{-1}(z) = \theta' \circ d^{-1}(z)$. Hence, we have

$$\begin{aligned}
\mathbf{D}_{\tilde{d}^{-1}(z)} \mathbf{K} \mathbf{D}_{\tilde{d}^{-1}(z)}^{-1} &= \mathbf{D}_{\theta' \circ d^{-1}(z)} \mathbf{K} \mathbf{D}_{\theta' \circ d^{-1}(z)}^{-1} \\
&= c^*\big(\theta', d^{-1}(z)\big) \mathbf{D}_{d^{-1}(z)} \mathbf{D}_{\theta'} \mathbf{K} \left[c^*\big(\theta', d^{-1}(z)\big) \mathbf{D}_{d^{-1}(z)} \mathbf{D}_{\theta'}\right]^{-1} \\
&= \mathbf{D}_{d^{-1}(z)} \mathbf{D}_{\theta'} \mathbf{K} \mathbf{D}_{\theta'}^{-1} \mathbf{D}_{d^{-1}(z)}^{-1} = \mathbf{D}_{d^{-1}(z)} \mathbf{K}' \mathbf{D}_{d^{-1}(z)}^{-1},
\end{aligned}$$

where $\mathbf{K}' = \mathbf{D}_{\theta'} \mathbf{K} \mathbf{D}_{\theta'}^{-1}$ is again an operator on $\mathcal{X}$. This shows that (7.8.72) with DF $\tilde{d}$ yields the *same* BTFR class as with DF d, albeit with a different parameterization by $\mathbf{K}$.

Examples of covariant BTFRs along with the corresponding covariance property are provided in Table 7.7—in particular, Cohen's class and the affine class of BTFRs are obtained for the DOs $\mathbf{S}_{\tau,\nu} = \mathbf{F}_\nu \mathbf{T}_\tau$ and $\mathbf{R}_{\sigma,\tau} = \mathbf{T}_\tau \mathbf{C}_\sigma$, respectively—and in Subsection 7.8.3.

7.8.3 *Dual and affine modulation and warping operators*

We shall now specialize our results for the important case where the partial DOs are dual or affine modulation and warping operators. Specifically, the DO is given by (7.5.35), i.e.,

$$\begin{aligned}
(\mathbf{D}_{\alpha,\beta} x)(t) &= (\mathbf{M}_\beta \mathbf{W}_\alpha x)(t) \\
&= e^{j2\pi\psi_{\mathcal{B}}(\beta)\tilde{m}(t)} \sqrt{|w'_\alpha(t)|}\, x(w_\alpha(t)), \quad \alpha \in (\mathcal{A}, \bullet),\ \beta \in (\mathcal{B}, *),\ t \in \Omega,
\end{aligned}$$

with $\tilde{m}(t) = \psi_{\mathcal{A}}(m(t))$. We first consider the case where $\mathbf{M}_\beta$ is *dual* to $\mathbf{W}_\alpha$ (see Subsection 7.3.2), i.e., $m(\Omega) = \mathcal{A}$ and $w_\alpha(t) = m^{-1}\big(m(t) \bullet \alpha^{-1}\big) =$

Cohen's class	
$\mathbf{A}_\alpha$	$(\mathbf{A}_\alpha x)(t) = x(t-\alpha), \quad \alpha \in (\mathbb{R},+)$
$\mathbf{B}_\beta$	$(\mathbf{B}_\beta x)(t) = e^{j2\pi\beta t} x(t), \quad \beta \in (\mathbb{R},+)$
Composition property	$\mathbf{D}_{\alpha_2,\beta_2}\mathbf{D}_{\alpha_1,\beta_1} = e^{-j2\pi\beta_1\alpha_2}\mathbf{D}_{\alpha_1+\alpha_2,\beta_1+\beta_2}$
Displacement function	$d(\alpha,\beta) = (\alpha,\beta)$
Covariance property	$\tilde{B}_{\mathbf{D}_{\alpha,\beta}x,\mathbf{D}_{\alpha,\beta}y}(t,f) = \tilde{B}_{x,y}(t-\alpha, f-\beta)$
Covar. bilin. TF repres.	$\tilde{B}_{x,y}(t,f) = \int_{-\infty}^{\infty}\int_{-\infty}^{\infty} x(t_1)y^*(t_2)\, k^*(t_1-t, t_2-t) \cdot e^{-j2\pi f(t_1-t_2)} dt_1 dt_2$

Affine class	
$\mathbf{A}_\alpha$	$(\mathbf{A}_\alpha x)(t) = \frac{1}{\sqrt{\alpha}} x\big(\frac{t}{\alpha}\big), \quad \alpha \in (\mathbb{R}^+,\cdot)$
$\mathbf{B}_\beta$	$(\mathbf{B}_\beta x)(t) = x(t-\beta), \quad \beta \in (\mathbb{R},+)$
Composition property	$\mathbf{D}_{\alpha_2,\beta_2}\mathbf{D}_{\alpha_1,\beta_1} = \mathbf{D}_{\alpha_1\alpha_2,\beta_1\alpha_2+\beta_2}$
Displacement function	$d(\alpha,\beta) = \big(\beta, \frac{1}{\alpha}\big)$
Covariance property	$\tilde{B}_{\mathbf{D}_{\alpha,\beta}x,\mathbf{D}_{\alpha,\beta}y}(t,f) = \tilde{B}_{x,y}\big(\frac{t-\beta}{\alpha}, \alpha f\big), \ f>0$
Covar. bilin. TF repres.	$\tilde{B}_{x,y}(t,f) = f\int_{-\infty}^{\infty}\int_{-\infty}^{\infty} x(t_1)y^*(t_2) \cdot k^*\big(f(t_1-t), f(t_2-t)\big) dt_1 dt_2, \ f>0$

Hyperbolic class	
$\mathbf{A}_\alpha$	$(\mathbf{A}_\alpha x)(t) = \frac{1}{\sqrt{\alpha}} x\big(\frac{t}{\alpha}\big), \quad \alpha \in (\mathbb{R}^+,\cdot)$
$\mathbf{B}_\beta$	$(\hat{\mathbf{B}}_\beta X)(f) = e^{-j2\pi\beta \ln f} X(f), \quad \beta \in (\mathbb{R},+), \ f>0$
Composition property	$\mathbf{D}_{\alpha_2,\beta_2}\mathbf{D}_{\alpha_1,\beta_1} = e^{-j2\pi\beta_1 \ln\alpha_2}\mathbf{D}_{\alpha_1\alpha_2,\beta_1+\beta_2}$
Displacement function	$d(\alpha,\beta) = \big(\alpha\beta, \frac{1}{\alpha}\big)$
Covariance property	$\tilde{B}_{\mathbf{D}_{\alpha,\beta}x,\mathbf{D}_{\alpha,\beta}y}(t,f) = \tilde{B}_{x,y}\big(\frac{t-\beta/f}{\alpha}, \alpha f\big), \ f>0$
Covar. bilin. TF repres.	$\tilde{B}_{x,y}(t,f) = \frac{1}{f}\int_0^{\infty}\int_0^{\infty} X(f_1)Y^*(f_2)\hat{k}^*\big(\frac{f_1}{f}, \frac{f_2}{f}\big) \cdot e^{j2\pi tf \ln(f_1/f_2)} df_1 df_2, \ f>0$

Power classes	
$\mathbf{A}_\alpha$	$(\mathbf{A}_\alpha x)(t) = \frac{1}{\sqrt{\alpha}} x\big(\frac{t}{\alpha}\big), \quad \alpha \in (\mathbb{R}^+,\cdot)$
$\mathbf{B}_\beta$	$(\hat{\mathbf{B}}_\beta X)(f) = e^{-j2\pi\beta\xi_\kappa(f)} X(f), \quad \beta \in (\mathbb{R},+),$ with $\xi_\kappa(f) = \text{sgn}(f)\|f\|^\kappa$
Composition property	$\mathbf{D}_{\alpha_2,\beta_2}\mathbf{D}_{\alpha_1,\beta_1} = \mathbf{D}_{\alpha_1\alpha_2,\beta_1\xi_\kappa(\alpha_2)+\beta_2}$
Displacement function (extended to $f \in \mathbb{R}$)	$d(\alpha,\beta) = \big(\beta\xi'_\kappa(\frac{1}{\alpha}), \pm\frac{1}{\alpha}\big) = \big(\frac{\beta\kappa}{\alpha^{\kappa-1}}, \pm\frac{1}{\alpha}\big)$ (note that $\xi'_\kappa(f) = \kappa\|f\|^{\kappa-1}$)
Covariance property	$\tilde{B}_{\mathbf{D}_{\alpha,\beta}x,\mathbf{D}_{\alpha,\beta}y}(t,f) = \tilde{B}_{x,y}\big(\frac{t-\beta\xi'_\kappa(f)}{\alpha}, \alpha f\big)$
Covar. bilin. TF repres. (extended to $f \in \mathbb{R}$)	$\tilde{B}_{x,y}(t,f) = \frac{1}{\|f\|}\int_{-\infty}^{\infty}\int_{-\infty}^{\infty} X(f_1)Y^*(f_2)\hat{k}^*\big(\frac{f_1}{f}, \frac{f_2}{f}\big) \cdot e^{j2\pi t[\xi_\kappa(f_1)-\xi_\kappa(f_2)]/\xi'_\kappa(f)} df_1 df_2$

TABLE 7.7. Some covariant bilinear/quadratic TF representations. Note that $\theta = (\alpha,\beta)$, $\mathbf{D}_{\alpha,\beta} = \mathbf{B}_\beta\mathbf{A}_\alpha$, $X(f)$ denotes the Fourier transform of $x(t)$, and $\hat{\mathbf{B}}_\beta$ denotes the frequency-domain version of $\mathbf{B}_\beta$.

$\tilde{m}^{-1}\big(\tilde{m}(t) - \psi_{\mathcal{A}}(\alpha)\big)$. From (7.7.65), the inverse DF is obtained as $d^{-1}(t,f) = \big(m(t), \psi_{\mathcal{B}}^{-1}(f/\tilde{m}'(t))\big)$. Evaluating (7.8.69) and (7.8.72), we obtain the covariant LTFRs and BTFRs as

$$\tilde{L}_x(t,f) = \int_{\Omega} x(t')h^*\big(w_{m(t)}(t')\big)\sqrt{\big|w'_{m(t)}(t')\big|}\, e^{-j2\pi f \tilde{m}(t')/\tilde{m}'(t)}dt'$$

$$\tilde{B}_{x,y}(t,f) = \int_{\Omega}\int_{\Omega} x(t_1)y^*(t_2)k^*\big(w_{m(t)}(t_1), w_{m(t)}(t_2)\big) \cdot \sqrt{\big|w'_{m(t)}(t_1)w'_{m(t)}(t_2)\big|}e^{-j2\pi f[\tilde{m}(t_1)-\tilde{m}(t_2)]/\tilde{m}'(t)}dt_1dt_2,$$

for $(t,f) \in \Omega \times \mathbb{R}$. These TF representations satisfy the following covariance properties (see (7.8.68), (7.8.71), and (7.7.64)),

$$\tilde{L}_{\mathbf{D}_{\alpha,\beta}x}(t,f) = \exp\left(-j2\pi\psi_{\mathcal{A}}(\alpha)\left[\frac{f}{\tilde{m}'(t)} - \psi_{\mathcal{B}}(\beta)\right]\right) \cdot \tilde{L}_x\left(w_\alpha(t), \frac{f - \psi_{\mathcal{B}}(\beta)\tilde{m}'(t)}{w'_\alpha(t)}\right)$$

$$\tilde{B}_{\mathbf{D}_{\alpha,\beta}x,\mathbf{D}_{\alpha,\beta}y}(t,f) = \tilde{B}_{x,y}\left(w_\alpha(t), \frac{f - \psi_{\mathcal{B}}(\beta)\tilde{m}'(t)}{w'_\alpha(t)}\right).$$

Next, let $\mathbf{M}_\beta$ be *affine* to $\mathbf{W}_\alpha$ (see Subsection 7.4.2), i.e., $\tilde{m}(\Omega) = \mathbb{R}^+$ or $\tilde{m}(\Omega) = \mathbb{R}^-$ and $w_\alpha(t) = \tilde{m}^{-1}\big(\tilde{m}(t)\exp\big(\psi_{\mathcal{A}}(\alpha)\big)\big)$. From (7.7.67), the inverse DF follows as $d^{-1}(t,f) = \big(\psi_{\mathcal{A}}^{-1}\big(-\ln|\tilde{m}(t)|\big), \psi_{\mathcal{B}}^{-1}(f/\tilde{m}'(t))\big)$. Evaluating (7.8.69) and (7.8.72), we obtain the covariant LTFRs and BTFRs as

$$\tilde{L}_x(t,f) = \int_{\Omega} x(t')h^*\big(w_{\chi(t)}(t')\big)\sqrt{\big|w'_{\chi(t)}(t')\big|}\, e^{-j2\pi f \tilde{m}(t')/\tilde{m}'(t)}dt'$$

$$\tilde{B}_{x,y}(t,f) = \int_{\Omega}\int_{\Omega} x(t_1)y^*(t_2)k^*\big(w_{\chi(t)}(t_1), w_{\chi(t)}(t_2)\big) \cdot \sqrt{\big|w'_{\chi(t)}(t_1)w'_{\chi(t)}(t_2)\big|}e^{-j2\pi f[\tilde{m}(t_1)-\tilde{m}(t_2)]/\tilde{m}'(t)}dt_1dt_2,$$

where $\chi(t) \triangleq \psi_{\mathcal{A}}^{-1}\big(-\ln|\tilde{m}(t)|\big)$. These TF representations satisfy the covariance properties

$$\tilde{L}_{\mathbf{D}_{\alpha,\beta}x}(t,f) = \tilde{L}_x\left(w_\alpha(t), \frac{f - \psi_{\mathcal{B}}(\beta)\tilde{m}'(t)}{w'_\alpha(t)}\right)$$

$$\tilde{B}_{\mathbf{D}_{\alpha,\beta}x,\mathbf{D}_{\alpha,\beta}y}(t,f) = \tilde{B}_{x,y}\left(w_\alpha(t), \frac{f - \psi_{\mathcal{B}}(\beta)\tilde{m}'(t)}{w'_\alpha(t)}\right).$$

7.9 The characteristic function method

Besides covariance properties, which are the main subject of this chapter, *marginal properties* are of importance in the context of BTFRs. Marginal properties state that integration of a BTFR along certain curves in the TF plane yields corresponding energy densities of the signals, thus suggesting the BTFR's interpretation as a joint TF energy distribution. The shape of the curves and the definition of the energy densities depend on a pair of unitary operator families (unitary group representations) $\mathbf{A}_\alpha$ and $\mathbf{B}_\beta$.

The *characteristic function method* [2, 34, 57–60] aims at constructing joint "(a,b)-distributions," i.e., signal representations that are functions of real variables a,b and satisfy marginal properties in the (a,b)-domain. The nature of the variables a,b depends on the operators $\mathbf{A}_\alpha$ and $\mathbf{B}_\beta$. Using a mapping between the (a,b)-plane and the TF plane (the *localization function*, see Subsection 7.9.2), the (a,b)-distributions can then be converted into BTFRs that satisfy marginal properties in the TF domain.

If we are given two unitary operators $\mathbf{A}_\alpha$ and $\mathbf{B}_\beta$ for which $\mathbf{D}_{\alpha,\beta} = \mathbf{B}_\beta \mathbf{A}_\alpha$ is a DO, then we can apply both the covariance method and the characteristic function method to construct classes of BTFRs. In general, the classes obtained will be different. However, in Subsection 7.9.3 we will show that in the special case of dual operators, the covariance method and characteristic function method yield equivalent results.

7.9.1 (a,b)-energy distributions

Let $\{\mathbf{A}_\alpha\}_{\alpha\in(\mathcal{A},\bullet)}$ and $\{\mathbf{B}_\beta\}_{\beta\in(\mathcal{B},*)}$ be unitary representations of groups $(\mathcal{A},\bullet)$ and $(\mathcal{B},*)$ that are isomorphic to $(\mathbb{R},+)$ by $\psi_\mathcal{A}$ and $\psi_\mathcal{B}$, respectively. Let $\{u_b^{\mathbf{A}}(t)\}_{b\in(\tilde{\mathcal{A}},\tilde{\bullet})}$ and $\{u_a^{\mathbf{B}}(t)\}_{a\in(\tilde{\mathcal{B}},\tilde{*})}$ denote the (generalized) eigenfunctions of $\mathbf{A}_\alpha$ and $\mathbf{B}_\beta$, respectively, where $(\tilde{\mathcal{A}},\tilde{\bullet})$ and $(\tilde{\mathcal{B}},\tilde{*})$ are again isomorphic to $(\mathbb{R},+)$ by $\psi_{\tilde{\mathcal{A}}}$ and $\psi_{\tilde{\mathcal{B}}}$, respectively. According to the characteristic function method, (a,b)-distributions (with $a\in\tilde{\mathcal{B}}$ and $b\in\tilde{\mathcal{A}}$ the parameters of the eigenfunctions $u_a^{\mathbf{B}}(t)$ and $u_b^{\mathbf{A}}(t)$, respectively) are constructed as

$$P_{x,y}(a,b) = \int_\mathcal{A}\int_\mathcal{B} Q_{x,y}(\alpha,\beta)\, e^{-j2\pi[\psi_\mathcal{A}(\alpha)\psi_{\tilde{\mathcal{A}}}(b)-\psi_\mathcal{B}(\beta)\psi_{\tilde{\mathcal{B}}}(a)]}d\psi_\mathcal{A}(\alpha)d\psi_\mathcal{B}(\beta)$$

$$(7.9.74)$$

$$= \int_\mathbb{R}\int_\mathbb{R} Q_{x,y}\big(\psi_\mathcal{A}^{-1}(\alpha'),\psi_\mathcal{B}^{-1}(\beta')\big)e^{-j2\pi[\alpha'\psi_{\tilde{\mathcal{A}}}(b)-\beta'\psi_{\tilde{\mathcal{B}}}(a)]}d\alpha' d\beta',$$

with the "characteristic function" $Q_{x,y}(\alpha,\beta)$ defined as

$$Q_{x,y}(\alpha,\beta) \triangleq \phi(\alpha,\beta)\langle x, \mathbf{Q}_{\alpha,\beta}y\rangle = \phi(\alpha,\beta)\int_t x(t)\big(\mathbf{Q}_{\alpha,\beta}y\big)^*(t)dt. \qquad (7.9.75)$$

Here, $\phi : \mathcal{A}\times\mathcal{B}\to\mathbb{C}$ is an arbitrary "kernel" function and $\mathbf{Q}_{\alpha,\beta}$ is any operator satisfying $\mathbf{Q}_{\alpha_0,\beta} = \mathbf{B}_\beta$ and $\mathbf{Q}_{\alpha,\beta_0} = \mathbf{A}_\alpha$. Examples of such "characteristic function operators" are $\mathbf{Q}_{\alpha,\beta} = \mathbf{B}_\beta\mathbf{A}_\alpha$, $\mathbf{Q}_{\alpha,\beta} = \mathbf{A}_\alpha\mathbf{B}_\beta$, $\mathbf{Q}_{\alpha,\beta} = (1/2)\,(\mathbf{A}_\alpha\mathbf{B}_\beta + \mathbf{B}_\beta\mathbf{A}_\alpha)$, $\mathbf{Q}_{\alpha,\beta} = \mathbf{A}_{\alpha^{1/2}}\mathbf{B}_\beta\mathbf{A}_{\alpha^{1/2}}$ (where $\alpha^{1/2}$ is defined by $\psi_\mathcal{A}(\alpha^{1/2}) = \psi_\mathcal{A}(\alpha)/2$), etc. Assuming that $\phi(\alpha,\beta)$ and $\mathbf{Q}_{\alpha,\beta}$ do not depend on the signals $x(t)$ and $y(t)$, $P_{x,y}(a,b)$

is a *bilinear* signal representation. If $\phi(\alpha, \beta)$ satisfies $\phi(\alpha_0, \beta) = \phi(\alpha, \beta_0) = 1$, then it can be shown [34] that (assuming suitable parameterization of the eigenfunctions) $P_{x,x}(a, b)$ satisfies the marginal properties

$$\int_{\tilde{\mathcal{A}}} P_{x,x}(a,b) d\psi_{\tilde{\mathcal{A}}}(b) = |\langle x, u_a^{\mathbf{B}} \rangle|^2, \quad \int_{\tilde{\mathcal{B}}} P_{x,x}(a,b) d\psi_{\tilde{\mathcal{B}}}(a) = |\langle x, u_b^{\mathbf{A}} \rangle|^2. \tag{7.9.76}$$

7.9.2 Time-frequency energy distributions

Since we are primarily interested in TF distributions, we will now convert the (a, b)-distributions $P_{x,y}(a, b)$ constructed above into TF distributions. This conversion uses an invertible mapping $(a, b) \to (t, f)$ that will be termed the *localization function* (LF) [10, 26, 28, 30]. To construct the LF, we consider the eigenfunctions $\{u_b^{\mathbf{A}}(t)\}_{b \in (\tilde{\mathcal{A}}, \tilde{\bullet})}$ and $\{u_a^{\mathbf{B}}(t)\}_{a \in (\tilde{\mathcal{B}}, \tilde{*})}$ of $\mathbf{A}_\alpha$ and $\mathbf{B}_\beta$, respectively. The TF locus of $u_b^{\mathbf{A}}(t)$ is characterized by the instantaneous frequency $\nu\{u_b^{\mathbf{A}}\}(t)$ or the group delay $\tau\{u_b^{\mathbf{A}}\}(f)$, whichever exists. Here, e.g., we assume existence of the instantaneous frequency $\nu\{u_b^{\mathbf{A}}\}(t)$. Similarly, we assume existence of the group delay $\tau\{u_a^{\mathbf{B}}\}(f)$ of $u_a^{\mathbf{B}}(t)$. The TF loci of $u_b^{\mathbf{A}}(t)$ and $u_a^{\mathbf{B}}(t)$ being described by $\nu\{u_b^{\mathbf{A}}\}(t)$ and $\tau\{u_a^{\mathbf{B}}\}(f)$, respectively, the TF point (t, f) that corresponds to the eigenfunction parameter pair (a, b) is defined by the intersection of these two TF curves, i.e., (t, f) is given by the solution to the system of equations

$$\nu\{u_b^{\mathbf{A}}\}(t) = f, \quad \tau\{u_a^{\mathbf{B}}\}(f) = t. \tag{7.9.77}$$

If this system of equations has a unique solution (t, f) for each $(a, b) \in \tilde{\mathcal{B}} \times \tilde{\mathcal{A}}$, then we can write $(t, f) = l(a, b)$ with a function $l : \tilde{\mathcal{B}} \times \tilde{\mathcal{A}} \to \mathcal{Z}$ that will be termed the LF of the operators $\mathbf{A}_\alpha$ and $\mathbf{B}_\beta$. We moreover assume that, conversely, the system of equations (7.9.77) has a unique solution $(a, b) \in \tilde{\mathcal{B}} \times \tilde{\mathcal{A}}$ for each $(t, f) \in \mathcal{Z}$; the LF then is invertible and we can write $(a, b) = l^{-1}(t, f)$ for all $(t, f) \in \mathcal{Z}$.

Using the LF mapping, the (a, b)-distribution $P_{x,y}(a, b)$ in (7.9.74) can now be converted into a TF distribution $\tilde{P}_{x,y}(t, f)$ according to

$$\tilde{P}_{x,y}(t, f) \triangleq P_{x,y}(a, b)\Big|_{(a,b)=l^{-1}(t,f)}. \tag{7.9.78}$$

If the kernel $\phi(\alpha, \beta)$ in (7.9.75) satisfies $\phi(\alpha_0, \beta) = \phi(\alpha, \beta_0) = 1$, then $\tilde{P}_{x,x}(t, f)$ satisfies the following marginal properties (cf. (7.9.76)),

$$\int_{\tilde{\mathcal{A}}} \tilde{P}_{x,x}\big(l(a,b)\big) d\psi_{\tilde{\mathcal{A}}}(b) = |\langle x, u_a^{\mathbf{B}} \rangle|^2, \quad \int_{\tilde{\mathcal{B}}} \tilde{P}_{x,x}\big(l(a,b)\big) d\psi_{\tilde{\mathcal{B}}}(a) = |\langle x, u_b^{\mathbf{A}} \rangle|^2. \tag{7.9.79}$$

As an example illustrating the construction of the LF, let us consider the case where $\mathbf{A}_\alpha = \mathbf{W}_\alpha$ is a warping operator with warping function $w_\alpha(t) = n^{-1}\big(n(t) \bullet$

$\alpha^{-1})$, where $n : \Omega \to \mathcal{A}$, and $\mathbf{B}_\beta = \mathbf{M}_\beta$ is a modulation operator with modulation function $m : \Omega \to \tilde{\mathcal{B}}$. According to Subsections 7.2.3, 7.2.4, and 7.7.2, the eigenfunctions of $\mathbf{W}_\alpha$ are $u_b^{\mathbf{W}}(t) = \sqrt{|\tilde{n}'(t)|}e^{j2\pi\psi_{\tilde{\mathcal{A}}}(b)\tilde{n}(t)}$, with $b \in \tilde{\mathcal{A}}$ and $\tilde{n}(t) = \psi_{\mathcal{A}}(n(t))$, and the eigenfunctions of $\mathbf{M}_\beta$ are $u_a^{\mathbf{M}}(t) = \sqrt{|\tilde{m}'(t)|}\delta\big(\tilde{m}(t) - \psi_{\tilde{\mathcal{B}}}(a)\big)$ with $a \in \tilde{\mathcal{B}}$ and $\tilde{m}(t) = \psi_{\tilde{\mathcal{B}}}(m(t))$. We assume $m(\Omega) = \tilde{\mathcal{B}}$ and $|\tilde{n}'(t)| \neq 0$. With $\nu\{u_b^{\mathbf{W}}\}(t) = \psi_{\tilde{\mathcal{A}}}(b)\tilde{n}'(t)$ being the instantaneous frequency of $u_b^{\mathbf{W}}(t)$ and $\tau\{u_a^{\mathbf{M}}\}(f) \equiv m^{-1}(a)$ being the group delay of $u_a^{\mathbf{M}}(t)$, solution of (7.9.77) yields the LF as

$$l(a,b) = \big(m^{-1}(a), \psi_{\tilde{\mathcal{A}}}(b)\tilde{n}'\big(m^{-1}(a)\big)\big). \tag{7.9.80}$$

Inserting for $l(a,b)$, $u_a^{\mathbf{M}}(t)$, and $u_b^{\mathbf{W}}(t)$, the TF-domain marginal properties in (7.9.79) are obtained as

$$\int_{\mathbb{R}} \tilde{P}_{x,x}(t,f)df = \left|\frac{\tilde{n}'(t)}{\tilde{m}'(t)}\right| |x(t)|^2, \qquad t \in \Omega$$

$$\int_{\Omega} \tilde{P}_{x,x}\big(t, c\tilde{n}'(t)\big)|\tilde{m}'(t)|dt = \left|\int_{\Omega} x(t)e^{-j2\pi c\tilde{n}(t)}\sqrt{|\tilde{n}'(t)|}dt\right|^2, \qquad c \in \mathbb{R}.$$

In particular, if $\mathbf{M}_\beta$ is dual to $\mathbf{W}_\alpha$, then $\tilde{n}(t) = \tilde{m}(t)$ and the marginal properties simplify as

$$\int_{\mathbb{R}} \tilde{P}_{x,x}(t,f)df = |x(t)|^2, \qquad t \in \Omega$$

$$\int_{\Omega} \tilde{P}_{x,x}\big(t, c\tilde{m}'(t)\big)|\tilde{m}'(t)|dt = \left|\int_{\Omega} x(t)e^{-j2\pi c\tilde{m}(t)}\sqrt{|\tilde{m}'(t)|}dt\right|^2, \qquad c \in \mathbb{R}.$$

7.9.3 *Equivalence results for dual operators*

As before, let the groups $(\mathcal{A}, \bullet)$ and $(\mathcal{B}, *)$ be isomorphic to $(\mathbb{R}, +)$ by $\psi_{\mathcal{A}}$ and $\psi_{\mathcal{B}}$, respectively. We now consider the case where $\{\mathbf{B}_\beta\}_{\beta\in(\mathcal{B},*)}$ is *dual* to $\{\mathbf{A}_\alpha\}_{\alpha\in(\mathcal{A},\bullet)}$, i.e. (see (7.3.23)),

$$\mathbf{B}_\beta\mathbf{A}_\alpha = e^{j2\pi\psi_{\mathcal{A}}(\alpha)\psi_{\mathcal{B}}(\beta)}\mathbf{A}_\alpha\mathbf{B}_\beta. \tag{7.9.81}$$

Furthermore, according to Theorem 7.6, $\mathbf{D}_{\alpha,\beta} = \mathbf{B}_\beta\mathbf{A}_\alpha$ is a DO satisfying

$$\mathbf{D}_{\alpha_2,\beta_2}\mathbf{D}_{\alpha_1,\beta_1} = e^{-j2\pi\psi_{\mathcal{A}}(\alpha_2)\psi_{\mathcal{B}}(\beta_1)}\mathbf{D}_{\alpha_1\bullet\alpha_2,\beta_1*\beta_2}.$$

Here, the displacement parameter $\theta = (\alpha,\beta)$ belongs to the commutative group $(\mathcal{A}\times\mathcal{B}, \circ)$ with group operation $(\alpha_1,\beta_1)\circ(\alpha_2,\beta_2) = (\alpha_1\bullet\alpha_2, \beta_1 * \beta_2)$; this group is isomorphic to $(\mathbb{R}^2, +)$ by $\psi(\alpha,\beta) = \big(\psi_{\mathcal{A}}(\alpha), \psi_{\mathcal{B}}(\beta)\big)$.

Using the covariance method (see Subsection 7.5.3), we can construct the class $\{B_{x,y}(\alpha,\beta)\}$ of all bilinear (α,β)-representations $B_{x,y}(\alpha,\beta)$ that are covariant to the DO $\mathbf{D}_{\alpha,\beta} = \mathbf{B}_\beta\mathbf{A}_\alpha$; this class is parameterized by an operator $\mathbf{K}$ or, equivalently, by its kernel $k(t_1,t_2)$. On the other hand, using the characteristic function method, we can construct the class $\{P_{x,y}(a,b)\}$ of all bilinear (a,b)-distributions $P_{x,y}(a,b)$ that satisfy marginal properties related to the operators $\mathbf{A}_\alpha$ and $\mathbf{B}_\beta$; this class is parameterized by a kernel function $\phi(\alpha,\beta)$. (Here, the specific definition of the operator $\mathbf{Q}_{\alpha,\beta}$ in (7.9.75) is immaterial: due to the commutation relation (7.9.81), different operators $\mathbf{Q}_{\alpha,\beta}$ are equal up to complex factors that can be incorporated in the kernel $\phi(\alpha,\beta)$ and thus lead to the same class of distributions $\{P_{x,y}(a,b)\}$ [59].) The following theorem states that the classes $\{B_{x,y}(\alpha,\beta)\}$ and $\{P_{x,y}(a,b)\}$ are equivalent in the dual case considered.

Theorem 7.28 ([28]). *Let $\{\mathbf{B}_\beta\}_{\beta\in(\mathcal{B},*)}$ be dual to $\{\mathbf{A}_\alpha\}_{\alpha\in(\mathcal{A},\bullet)}$. Let $\{P_{x,y}(a,b)\}$ be the class of (a,b)-distributions based on $\mathbf{A}_\alpha$ and $\mathbf{B}_\beta$, as given by (7.9.74), (7.9.75) with $\mathbf{Q}_{\alpha,\beta}=\mathbf{B}_\beta\mathbf{A}_\alpha$ (in the dual case, this choice implies no loss of generality) and eigenfunction parameter groups $(\tilde{\mathcal{A}},\tilde{\bullet})=(\mathcal{B},*)$, $(\tilde{\mathcal{B}},\tilde{*})=(\mathcal{A},\bullet)$ (hence $a\in(\mathcal{A},\bullet)$, $b\in(\mathcal{B},*)$). Let $\{B_{x,y}(\alpha,\beta)\}$ be the class of covariant (α,β)-representations based on the DO $\mathbf{D}_{\alpha,\beta}=\mathbf{B}_\beta\mathbf{A}_\alpha$, as given by (7.5.44). Then these classes are identical, i.e.,*

$$\{P_{x,y}(a,b)\}=\{B_{x,y}(a,b)\}. \tag{7.9.82}$$

Moreover, specific representations of either class are identical, i.e.,

$$P_{x,y}(a,b)\equiv B_{x,y}(a,b), \tag{7.9.83}$$

if the operator $\mathbf{K}$ defining $B_{x,y}(\alpha,\beta)$ is related to the kernel $\phi(\alpha,\beta)$ defining $P_{x,y}(a,b)$ as

$$\mathbf{K}=\int_{\mathcal{A}}\int_{\mathcal{B}}\phi^*(\alpha,\beta)\mathbf{D}_{\alpha,\beta}d\psi_{\mathcal{A}}(\alpha)d\psi_{\mathcal{B}}(\beta). \tag{7.9.84}$$

Proof: With $\mathbf{Q}_{\alpha,\beta}=\mathbf{B}_\beta\mathbf{A}_\alpha=\mathbf{D}_{\alpha,\beta}$ and $(\tilde{\mathcal{A}},\tilde{\bullet})=(\mathcal{B},*)$, $(\tilde{\mathcal{B}},\tilde{*})=(\mathcal{A},\bullet)$, the (a,b)-distributions obtained by the characteristic function method (see (7.9.74), (7.9.75)) are given by

$$\begin{aligned}
&P_{x,y}(a,b)\\
&=\int_{\mathcal{A}}\int_{\mathcal{B}}\phi(\alpha,\beta)\langle x,\mathbf{D}_{\alpha,\beta}y\rangle e^{-j2\pi[\psi_{\mathcal{A}}(\alpha)\psi_{\mathcal{B}}(b)-\psi_{\mathcal{B}}(\beta)\psi_{\mathcal{A}}(a)]}d\psi_{\mathcal{A}}(\alpha)d\psi_{\mathcal{B}}(\beta)\\
&=\left\langle x,\left[\int_{\mathcal{A}}\int_{\mathcal{B}}\phi^*(\alpha,\beta)\mathbf{D}_{\alpha,\beta}e^{j2\pi[\psi_{\mathcal{A}}(\alpha)\psi_{\mathcal{B}}(b)-\psi_{\mathcal{B}}(\beta)\psi_{\mathcal{A}}(a)]}d\psi_{\mathcal{A}}(\alpha)d\psi_{\mathcal{B}}(\beta)\right]y\right\rangle\\
&=\langle x,\mathbf{K}_{a,b}y\rangle,
\end{aligned}$$

with

$$\mathbf{K}_{a,b}=\int_{\mathcal{A}}\int_{\mathcal{B}}\phi^*(\alpha,\beta)\mathbf{D}_{\alpha,\beta}e^{j2\pi[\psi_{\mathcal{A}}(\alpha)\psi_{\mathcal{B}}(b)-\psi_{\mathcal{B}}(\beta)\psi_{\mathcal{A}}(a)]}d\psi_{\mathcal{A}}(\alpha)d\psi_{\mathcal{B}}(\beta).$$

On the other hand, the (α,β)-representations covariant to $\mathbf{D}_{\alpha,\beta}=\mathbf{B}_\beta\mathbf{A}_\alpha$ are given by (see (7.5.44))

$$B_{x,y}(\alpha,\beta) = \langle x, \mathbf{D}_{\alpha,\beta}\mathbf{K}\mathbf{D}_{\alpha,\beta}^{-1} y\rangle.$$

Hence, we have $P_{x,y}(a,b) = B_{x,y}(a,b)$ if and only if $\mathbf{K}_{a,b}$ is of the form $\mathbf{D}_{a,b}\mathbf{K}\mathbf{D}_{a,b}^{-1}$, i.e., if and only if

$$\mathbf{D}_{a,b}\mathbf{K}\mathbf{D}_{a,b}^{-1} = \int_{\mathcal{A}}\int_{\mathcal{B}} \phi^*(\alpha,\beta)\mathbf{D}_{\alpha,\beta} e^{j2\pi[\psi_{\mathcal{A}}(\alpha)\psi_{\mathcal{B}}(b)-\psi_{\mathcal{B}}(\beta)\psi_{\mathcal{A}}(a)]} d\psi_{\mathcal{A}}(\alpha) d\psi_{\mathcal{B}}(\beta)$$

or, equivalently,

$$\mathbf{K} = \int_{\mathcal{A}}\int_{\mathcal{B}} \phi^*(\alpha,\beta)\mathbf{D}_{a,b}^{-1}\mathbf{D}_{\alpha,\beta}\mathbf{D}_{a,b} e^{j2\pi[\psi_{\mathcal{A}}(\alpha)\psi_{\mathcal{B}}(b)-\psi_{\mathcal{B}}(\beta)\psi_{\mathcal{A}}(a)]} d\psi_{\mathcal{A}}(\alpha) d\psi_{\mathcal{B}}(\beta). \tag{7.9.85}$$

With (7.2.13) and (7.3.24), we have (note that according to (7.3.24), the cocycle of $\mathbf{D}_{\alpha,\beta}$ is $c\big((\alpha_1,\beta_1),(\alpha_2,\beta_2)\big) = e^{-j2\pi\psi_{\mathcal{A}}(\alpha_2)\psi_{\mathcal{B}}(\beta_1)}$)

$$\begin{aligned}
&\mathbf{D}_{a,b}^{-1}\mathbf{D}_{\alpha,\beta}\mathbf{D}_{a,b}\\
&\quad= e^{j2\pi\psi_{\mathcal{A}}(a^{-1})\psi_{\mathcal{B}}(b)}\mathbf{D}_{(a,b)^{-1}}\mathbf{D}_{\alpha,\beta}\mathbf{D}_{a,b}\\
&\quad= e^{-j2\pi\psi_{\mathcal{A}}(a)\psi_{\mathcal{B}}(b)}\mathbf{D}_{a^{-1},b^{-1}} e^{-j2\pi\psi_{\mathcal{A}}(\alpha)\psi_{\mathcal{B}}(b)}\mathbf{D}_{a\bullet\alpha,b*\beta}\\
&\quad= e^{-j2\pi\psi_{\mathcal{A}}(a)\psi_{\mathcal{B}}(b)} e^{-j2\pi\psi_{\mathcal{A}}(\alpha)\psi_{\mathcal{B}}(b)} e^{-j2\pi\psi_{\mathcal{A}}(a^{-1})\psi_{\mathcal{B}}(b*\beta)}\mathbf{D}_{a\bullet\alpha\bullet a^{-1},b*\beta*b^{-1}}\\
&\quad= e^{-j2\pi\psi_{\mathcal{A}}(\alpha)\psi_{\mathcal{B}}(b)} e^{j2\pi\psi_{\mathcal{A}}(a)\psi_{\mathcal{B}}(\beta)}\mathbf{D}_{\alpha,\beta},
\end{aligned}$$

where in the last step we used the fact that $\bullet$ and $*$ are commutative. Hence, (7.9.85) becomes (7.9.84):

$$\begin{aligned}
\mathbf{K} &= \int_{\mathcal{A}}\int_{\mathcal{B}} \phi^*(\alpha,\beta) e^{-j2\pi\psi_{\mathcal{A}}(\alpha)\psi_{\mathcal{B}}(b)} e^{j2\pi\psi_{\mathcal{A}}(a)\psi_{\mathcal{B}}(\beta)}\\
&\qquad\qquad \cdot \mathbf{D}_{\alpha,\beta} e^{j2\pi[\psi_{\mathcal{A}}(\alpha)\psi_{\mathcal{B}}(b)-\psi_{\mathcal{B}}(\beta)\psi_{\mathcal{A}}(a)]} d\psi_{\mathcal{A}}(\alpha) d\psi_{\mathcal{B}}(\beta)\\
&= \int_{\mathcal{A}}\int_{\mathcal{B}} \phi^*(\alpha,\beta)\mathbf{D}_{\alpha,\beta} d\psi_{\mathcal{A}}(\alpha) d\psi_{\mathcal{B}}(\beta).
\end{aligned}$$

This shows that $P_{x,y}(a,b) = B_{x,y}(a,b)$ if and only if $\phi(\alpha,\beta)$ and $\mathbf{K}$ are related according to (7.9.84).

Finally, it is shown in [45] (see also [56]) that every $\mathbf{K}$ can be represented according to (7.9.84) with some $\phi(\alpha,\beta)$. We hence conclude that the entire classes $\{P_{x,y}(a,b)\}$ and $\{B_{x,y}(a,b)\}$ are equivalent. □

The equivalence of the characteristic function method and the covariance method in the dual case applies not only to the (a,b)- and (α,β)-representations but carries over to the BTFRs derived from them through the LF mapping and the DF mapping, respectively. This is due to the fact that *in the dual case, the LF and DF are themselves equivalent.* To show this, we recall from (7.7.60) that the extended DF of $\mathbf{D}_{\alpha,\beta} = \mathbf{B}_\beta\mathbf{A}_\alpha$ is obtained by composing the extended DF $e_{\mathbf{A}}$ of $\mathbf{A}_\alpha$ with the extended DF $e_{\mathbf{B}}$ of $\mathbf{B}_\beta$. This composition is illustrated for the dual case in Figure 7.2.

Let $\{u_b^{\mathbf{A}}(t)\}_{b\in(\mathcal{B},*)}$ and $\{u_a^{\mathbf{B}}(t)\}_{a\in(\mathcal{A},\bullet)}$ be the eigenfunctions of $\mathbf{A}_\alpha$ and $\mathbf{B}_\beta$, respectively (recall that $(\tilde{\mathcal{A}},\tilde{\bullet})=(\mathcal{B},*)$ and $(\tilde{\mathcal{B}},\tilde{*})=(\mathcal{A},\bullet)$). Furthermore let (t_0,f_0) be the intersection of the TF loci of $u_{\beta_0}^{\mathbf{A}}(t)$ and $u_{\alpha_0}^{\mathbf{B}}(t)$, with α_0 and β_0 the group identity elements of $\mathcal{A}$ and $\mathcal{B}$, respectively. Now $\mathbf{A}_\alpha$ or, equivalently, $e_{\mathbf{A}}(\cdot,\alpha)$ maps (t_0,f_0) to the intersection (t',f') of the TF loci of $u_{\beta_0}^{\mathbf{A}}(t)$ and (see (7.7.56), cf. Figure 7.1) $u_{\alpha_0\bullet\alpha}^{\tilde{\mathbf{A}}}(t)=u_{\alpha_0\bullet\alpha}^{\mathbf{B}}(t)=u_\alpha^{\mathbf{B}}(t)$, where we have used the fact that $\mathbf{B}_\beta$ is the dual operator $\tilde{\mathbf{A}}_\beta$ of $\mathbf{A}_\alpha$ and assumed appropriate parameterization of $u_\alpha^{\mathbf{B}}(t)$. Subsequently, $\mathbf{B}_\beta$ or, equivalently, $e_{\mathbf{B}}(\cdot,\beta)$ maps (t',f') to the intersection (t,f) of the TF loci of $u_\alpha^{\mathbf{B}}(t)$ and $u_{\beta_0*\beta}^{\mathbf{A}}(t)=u_\beta^{\mathbf{A}}(t)$. Thus, for given (fixed) (t_0,f_0), the composite extended DF $e\big((t_0,f_0),(\alpha,\beta)\big)=e_{\mathbf{B}}\big(e_{\mathbf{A}}((t_0,f_0),\alpha),\beta\big)=e_{\mathbf{B}}((t',f'),\beta)=(t,f)$ maps the parameters (α,β) of the eigenfunctions $u_\alpha^{\mathbf{B}}(t)$ and $u_\beta^{\mathbf{A}}(t)$ to the intersection point (t,f) of their TF loci (see Figure 7.2). But this is exactly how we defined the LF $l(\alpha,\beta)$ in Subsection 7.9.2. Hence, the DF $d(\alpha,\beta)=e\big((t_0,f_0),(\alpha,\beta)\big)$ (with the specific (t_0,f_0) defined above) and the LF $l(a,b)$ are equivalent:

$$d(\alpha,\beta)=l(\alpha,\beta). \tag{7.9.86}$$

For example, consider the case where $\mathbf{A}_\alpha$ is a warping operator and $\mathbf{B}_\beta$ is the modulation operator that is dual to $\mathbf{A}_\alpha$. In this case, the DF is

$$d(\alpha,\beta)=\big(m^{-1}(\alpha),\psi_{\mathcal{B}}(\beta)\tilde{m}'(m^{-1}(\alpha))\big)$$

(see (7.7.65)) and the LF is

$$l(a,b)=\big(m^{-1}(a),\psi_{\mathcal{B}}(b)\tilde{m}'\big(m^{-1}(a)\big)\big)$$

(see (7.9.80) with $\tilde{n}(t)=\tilde{m}(t)$ and $\tilde{\mathcal{A}}=\mathcal{B}$ due to duality). Thus, the LF and DF are recognized to be equivalent.

With this equivalence of the DF and LF in the dual case, it immediately follows from Theorem 7.28 that the TF representations corresponding to $P_{x,y}(a,b)$ and $B_{x,y}(\alpha,\beta)$ are identical. This is stated in the next corollary.

Corollary 7.29 ([26, 28, 30]). *Let $\{\mathbf{B}_\beta\}_{\beta\in(\mathcal{B},*)}$ be dual to $\{\mathbf{A}_\alpha\}_{\alpha\in(\mathcal{A},\bullet)}$. Let $\{\tilde{P}_{x,y}(t,f)\}$ denote the class of TF distributions based on $\mathbf{A}_\alpha$ and $\mathbf{B}_\beta$, as given by (7.9.78), (7.9.74) with $\mathbf{Q}_{\alpha,\beta}=\mathbf{B}_\beta\mathbf{A}_\alpha$. Let $\big\{\tilde{B}_{x,y}(t,f)\big\}$ be the class of covariant*

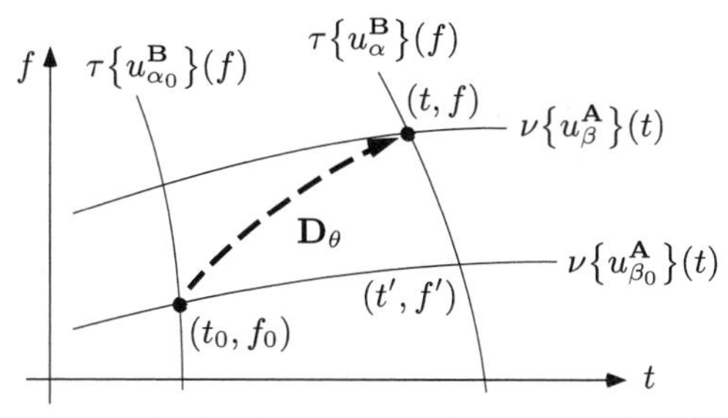

FIGURE 7.2. Equivalence of localization function and displacement function in the dual case.

BTFRs based on the DO $\mathbf{D}_{\alpha,\beta} = \mathbf{B}_\beta \mathbf{A}_\alpha$*, as given by (7.8.72). Then these classes are identical, i.e.,*

$$\{\tilde{P}_{x,y}(t,f)\} = \{\tilde{B}_{x,y}(t,f)\}.$$

Moreover, specific BTFRs of either class are identical, i.e.,

$$\tilde{P}_{x,y}(t,f) \equiv \tilde{B}_{x,y}(t,f),$$

if the operator $\mathbf{K}$ *defining* $\tilde{B}_{x,y}(t,f)$ *is related to the kernel* $\phi(\alpha,\beta)$ *defining* $\tilde{P}_{x,y}(t,f)$ *according to (7.9.84).*

Proof: Using in turn (7.9.78), (7.9.86), (7.9.83), and (7.8.73), we have

$$\begin{aligned}\tilde{P}_{x,y}(t,f) &= P_{x,y}(a,b)\big|_{(a,b)=l^{-1}(t,f)} = P_{x,y}(a,b)\big|_{(a,b)=d^{-1}(t,f)}\\ &= B_{x,y}(a,b)\big|_{(a,b)=d^{-1}(t,f)} = \tilde{B}_{x,y}(t,f),\end{aligned}$$

where due to Theorem 7.28, $\mathbf{K}$ and $\phi(\alpha,\beta)$ are related according to (7.9.84). The equivalence of the entire classes $\{\tilde{P}_{x,y}(t,f)\}$ and $\{\tilde{B}_{x,y}(t,f)\}$ then follows from (7.9.82). □

7.10 Extension to general LCA groups

From our assumption of continuous-time, nonperiodic signals and corresponding assumptions on the DF, it followed that the groups $(\mathcal{A},\bullet)$ and $(\mathcal{B},*)$ underlying the partial DOs $\mathbf{A}_\alpha$ and $\mathbf{B}_\beta$ are isomorphic to $(\mathbb{R},+)$ (see Theorem 7.18 and Corollaries 7.19 and 7.20). Such groups belong to the larger class of *locally compact abelian (LCA)* groups [36, 37]. For LCA groups one can define a *dual* group, which is unique up to isomorphisms [36, 37]. For instance, the group $(\mathbb{R},+)$ is self-dual, and the group given by the interval $[0,1)$ with addition modulo 1 is dual to the group of integers with addition. For an LCA group $(\mathcal{G},\star)$ that is isomorphic to $(\mathbb{R},+)$, the dual group $(\tilde{\mathcal{G}},\tilde{\star})$ is also isomorphic to $(\mathbb{R},+)$. Conversely, any two groups that are isomorphic to $(\mathbb{R},+)$ are dual LCA groups.

Many results of this chapter can be extended to general LCA groups. This enables the construction of, e.g., discrete-time, periodic-frequency signal representations [49, 61, 62]. In particular, the definition of modulation operators (Definition 7.1) can be extended to arbitrary LCA groups simply by replacing the complex exponential $e^{j2\pi\psi_{\mathcal{H}}(h)\tilde{m}(t)}$ in (7.2.14) with $\gamma(h, m(t))$, where $\gamma(h,g)$ is a continuous *character* [36, 37] of the LCA group $(\mathcal{H},\diamond)$. Similarly, the warping operators (Definition 7.2) can be extended to LCA groups if the normalizing factor $\sqrt{|w_g'(t)|}$ in (7.2.18) is replaced by a suitable function (since differentiation may not be defined). Theorem 7.3 and Corollary 7.4 hold—with the exception of differentiability—for general LCA groups $(\mathcal{G},\star)$ and $(\mathcal{H},\diamond)$, provided they are *dual* groups [49].

The concept of dual operators as discussed in Section 7.3 can be generalized in a very natural way to *dual* LCA groups; again, the complex exponential $e^{j2\pi\psi_{\mathcal{G}}(g)\psi_{\mathcal{H}}(h)}$ in, e.g., (7.3.23) has to be replaced with the character $\gamma(g,h)$ of the dual LCA groups $(\mathcal{G},\star)$ and $(\mathcal{H},\diamond)$. Theorem 7.6 holds, except that $(\mathcal{G},\star)\times(\mathcal{H},\diamond)$ is not always isomorphic to $(\mathbb{R}^2,+)$. Theorem 7.7 holds if the frequency-shift and time-shift operators are replaced by the group diagonal and group translation operators, respectively [29, 34, 49].

The concept of affine operators (Section 7.4) can also be generalized to arbitrary LCA groups $(\mathcal{G},\star)$, $(\mathcal{H},\diamond)$ if $\mu(g,h)$ in (7.4.29) is replaced by a more general function [49]. Theorem 7.9 essentially holds, with $\mathbf{D}_{g,h}$ being a unitary representation of a *semidirect product* [38] of $(\mathcal{G},\star)$ and $(\mathcal{H},\diamond)$. It seems that Theorem 7.10 *cannot* be generalized, since the scaling operator has no analogue in the general LCA setting.

All results of Section 7.5 can immediately be generalized to arbitrary topological groups; no further assumptions about the group are used.

In Section 7.6, we showed that under our assumptions on the TF point set $\mathcal{Z}$ and the DF, the underlying group must be isomorphic to either $(\mathbb{R}^2,+)$ or the affine group. Hence, under these assumptions, the results of Sections 7.6–7.8 cannot be extended to more general groups. However, for other sets $\mathcal{Z}$ (e.g., a grid in the TF plane), groups that are not isomorphic to $(\mathbb{R}^2,+)$ or to the affine group are appropriate [49, 62]. The extension of the DF concept to such groups is an interesting topic for future research.

The characteristic function method of Section 7.9 was generalized to LCA groups in [34]. Again, the key is to replace complex exponentials by group characters, and also Fourier transforms by *group Fourier transforms* [36, 37]. The equivalence of the covariance method and characteristic function method in the group domain (Theorem 7.28) holds also for dual operators based on dual LCA groups [49]. It is unclear whether the equivalence of these methods in the TF domain (Corollary 7.29) can be likewise extended since this would require an extension of the DF to general LCA groups.

7.11 Conclusion

The covariance theory allows one to construct the class of all linear or quadratic/bilinear time-frequency representations that are covariant to a given displacement operator, and thus it provides a theoretical basis for time-frequency analysis. Important classes of time-frequency representations—such as the short-time Fourier transform, the wavelet transform, Cohen's class, and the affine class—can all be derived from this principle.

The present development of the covariance method assumed that the groups underlying the partial displacement operators are isomorphic to $(\mathbb{R},+)$. However, many of the results can be generalized to arbitrary LCA groups [49], thus allowing for discrete-time and/or periodic signals.

Apart from the covariance method as such, we discussed several concepts that are interesting in a broader context, namely, modulation and warping operators, dual and affine pairs of operators, and a general method for constructing the displacement function that describes the action of a displacement operator in the time-frequency plane. We also showed that for dual operators the covariance method and characteristic function method—two methods based on totally different principles—are equivalent.

Acknowledgment. This work was supported by FWF grant P12228-TEC.

References

[1] P. Flandrin, *Time-Frequency/Time-Scale Analysis.* San Diego (CA): Academic Press, 1999.

[2] L. Cohen, *Time-Frequency Analysis.* Englewood Cliffs (NJ): Prentice-Hall, 1995.

[3] F. Hlawatsch and G. F. Boudreaux-Bartels, Linear and quadratic time-frequency signal representations, *IEEE Signal Processing Magazine*, vol. 9, pp. 21–67, April 1992.

[4] K. Gröchenig, *Foundations of Time-Frequency Analysis.* Boston (MA): Birkhäuser, 2001.

[5] I. Daubechies, *Ten Lectures on Wavelets.* Philadelphia (PA): SIAM, 1992.

[6] Y. Meyer, *Wavelets.* Philadelphia (PA): SIAM, 1993.

[7] O. Rioul and M. Vetterli, Wavelets and signal processing, *IEEE Signal Processing Magazine*, vol. 1, pp. 14–38, Oct. 1991.

[8] A. Papandreou, F. Hlawatsch, and G. F. Boudreaux-Bartels, The hyperbolic class of quadratic time-frequency representations—Part I: Constant-Q warping, the hyperbolic paradigm, properties, and members, *IEEE Trans. Signal Processing, Special Issue on Wavelets and Signal Processing*, vol. 41, pp. 3425–3444, Dec. 1993.

[9] F. Hlawatsch, A. Papandreou-Suppappola, and G. F. Boudreaux-Bartels, The power classes—Quadratic time-frequency representations with scale covariance and dispersive time-shift covariance, *IEEE Trans. Signal Processing*, vol. 47, pp. 3067–3083, Nov. 1999.

[10] R. G. Baraniuk and D. L. Jones, Unitary equivalence: A new twist on signal processing, *IEEE Trans. Signal Processing*, vol. 43, pp. 2269–2282, Oct. 1995.

[11] R. G. Baraniuk, Warped perspectives in time-frequency analysis, in *IEEE Int. Symp. Time-Frequency and Time-Scale Analysis*, (Philadelphia, PA), pp. 528–531, Oct. 1994.

[12] R. G. Baraniuk, Covariant time-frequency representations through unitary equivalence, *IEEE Signal Processing Letters*, vol. 3, pp. 79–81, March 1996.

[13] F. Hlawatsch, Regularity and unitarity of bilinear time-frequency signal representations, *IEEE Trans. Inf. Theory*, vol. 38, pp. 82–94, Jan. 1992.

[14] F. Hlawatsch, Duality and classification of bilinear time-frequency signal representations, *IEEE Trans. Signal Processing*, vol. 39, pp. 1564–1574, July 1991.

[15] J. Bertrand and P. Bertrand, Affine time-frequency distributions, in *Time-Frequency Signal Analysis – Methods and Applications* (B. Boashash, ed.), pp. 118–140, Melbourne: Longman Cheshire, 1992.

[16] O. Rioul and P. Flandrin, Time-scale energy distributions: A general class extending wavelet transforms, *IEEE Trans. Signal Processing*, vol. 40, pp. 1746–1757, July 1992.

[17] J. Bertrand and P. Bertrand, A class of affine Wigner functions with extended covariance properties, *J. Math. Phys.*, vol. 33, pp. 2515–2527, July 1992.

[18] P. Flandrin and P. Gonçalvès, Geometry of affine time-frequency distributions, *Applied and Computational Harmonic Analysis*, vol. 3, pp. 10–39, 1996.

[19] F. Hlawatsch, A. Papandreou, and G. F. Boudreaux-Bartels, Regularity and unitarity of affine and hyperbolic time-frequency representations, in *Proc. IEEE ICASSP-93*, (Minneapolis, MN), pp. 245–248, April 1993.

[20] F. Hlawatsch, A. Papandreou-Suppappola, and G. F. Boudreaux-Bartels, The hyperbolic class of quadratic time-frequency representations—Part II: Subclasses, intersection with the affine and power classes, regularity, and unitarity, *IEEE Trans. Signal Processing*, vol. 45, pp. 303–315, Feb. 1997.

[21] A. Papandreou-Suppappola, F. Hlawatsch, and G. F. Boudreaux-Bartels, Quadratic time-frequency representations with scale covariance and generalized time-shift covariance: A unified framework for the affine, hyperbolic, and power classes, *Digital Signal Processing—A Review Journal*, vol. 8, pp. 3–48, Jan. 1998.

[22] A. Papandreou-Suppappola, F. Hlawatsch, and G. F. Boudreaux-Bartels, Power class time-frequency representations: Interference geometry, smoothing, and implementation, in *Proc. IEEE-SP Int. Sympos. Time-Frequency Time-Scale Analysis*, (Paris, France), pp. 193–196, June 1996.

[23] A. Papandreou-Suppappola and G. F. Boudreaux-Bartels, The exponential class and generalized time-shift covariant quadratic time-frequency representations, in *Proc. IEEE-SP Int. Sympos. Time-Frequency Time-Scale Analysis*, (Paris, France), pp. 429–432, June 1996.

[24] A. Papandreou-Suppappola, R. L. Murray, B.-G. Iem, and G. F. Boudreaux-Bartels, Group delay shift covariant quadratic time-frequency representations, *IEEE Trans. Signal Processing*, vol. 49, pp. 2549–2564, Nov. 2001.

[25] F. Hlawatsch and H. Bölcskei, Unified theory of displacement-covariant time-frequency analysis, in *Proc. IEEE Int. Sympos. Time-Frequency Time-Scale Analysis*, (Philadelphia, PA), pp. 524–527, Oct. 1994.

[26] F. Hlawatsch and H. Bölcskei, Displacement-covariant time-frequency energy distributions, in *Proc. IEEE ICASSP-95*, (Detroit, MI), pp. 1025–1028, May 1995.

[27] F. Hlawatsch, Covariant time-frequency analysis: A unifying framework, in *Proc. IEEE UK Sympos. Applications of Time-Frequency and Time-Scale Methods*, (Univ. of Warwick, Coventry, UK), pp. 110–117, Aug. 1995.

[28] F. Hlawatsch and T. Twaroch, Covariant (α, β), time-frequency, and (a, b) representations, in *Proc. IEEE-SP Int. Sympos. Time-Frequency Time-Scale Analysis*, (Paris, France), pp. 437–440, June 1996.

[29] A. M. Sayeed and D. L. Jones, Integral transforms covariant to unitary operators and their implications for joint signal representations, *IEEE Trans. Signal Processing*, vol. 44, pp. 1365–1377, June 1996.

[30] F. Hlawatsch and H. Bölcskei, Covariant time-frequency distributions based on conjugate operators, *IEEE Signal Processing Letters*, vol. 3, pp. 44–46, Feb. 1996.

[31] A. M. Sayeed and D. L. Jones, A simple covariance-based characterization of joint signal representations of arbitrary variables, in *Proc. IEEE-SP Int. Sympos. Time-Frequency Time-Scale Analysis*, (Paris, France), pp. 433–436, June 1996.

[32] A. M. Sayeed and D. L. Jones, A canonical covariance-based method for generalized joint signal representations, *IEEE Signal Processing Letters*, vol. 3, pp. 121–123, April 1996.

[33] T. Twaroch and F. Hlawatsch, Modulation and warping operators in joint signal analysis, in *Proc. IEEE-SP Int. Sympos. Time-Frequency Time-Scale Analysis*, (Pittsburgh, PA), pp. 9–12, Oct. 1998.

[34] R. G. Baraniuk, Beyond time-frequency analysis: Energy densities in one and many dimensions, *IEEE Trans. Signal Processing*, vol. 46, pp. 2305–2314, Sept. 1998.

[35] R. G. Baraniuk, Marginals vs. covariance in joint distribution theory, in *Proc. IEEE ICASSP-95*, (Detroit, MI), pp. 1021–1024, May 1995.

[36] W. Rudin, *Fourier Analysis on Groups*. New York: Interscience, 1967.

[37] E. Hewitt and K. A. Ross, *Abstract Harmonic Analysis I*. Berlin, New York: Springer-Verlag, 1979.

[38] J. M. G. Fell and R. S. Doran, *Representations of *-Algebras, Locally Compact Groups, and Banach *-Algebraic Bundles*. Boston (MA): Academic Press, 1988.

[39] M. A. Naimark and A. I. Stern, *Theory of Group Representations*. New York: Springer, 1982.

[40] P. J. Higgins, *An Introduction to Topological Groups*. Cambridge (UK): Cambridge University Press, 1974.

[41] L. S. Pontryagin, *Topologische Gruppen*. Leipzig: Teubner, 1957.

[42] V. Bargmann, On unitary ray representations of continuous groups, *Annals of Mathematics*, vol. 59, pp. 1–46, Jan. 1954.

[43] O. Christensen, Atomic decomposition via projective group representations, *Rocky Mountain J. Math*, vol. 26, no. 4, pp. 1289–1312, 1996.

[44] I. M. Gelfand and G. E. Schilow, *Verallgemeinerte Funktionen (Distributionen)*. Berlin: VEB, 1960.

[45] K. Maurin, *General Eigenfunction Expansions and Unitary Representations of Topological Groups*. Warszawa: PWN–Polish Scientific Publishers, 1968.

[46] G. B. Folland, *Harmonic Analysis in Phase Space*, vol. 122 of *Annals of Mathematics Studies*. Princeton (NJ): Princeton University Press, 1989.

[47] A. Berthon, Représentations et changements d'horloge, in *Proc. Workshop Time-Frequency, Wavelets and Multiresolution: Theory, Models, and Applications*, (Lyon, France), pp. 19.1–19.4, March 1994.

[48] J. Aczel, *Vorlesungen über Funktionalgleichungen und ihre Anwendungen*. Basel (Switzerland): Birkhäuser, 1961.

[49] T. Twaroch, *Signal Representations and Group Theory*. Ph.D. thesis, Vienna University of Technology, 1999.

[50] M. H. Stone, Linear transformations in Hilbert space III, *Proc. Nat. Acad. USA*, vol. 16, pp. 172–175, 1930.

[51] J. von Neumann, Die Eindeutigkeit der Schrödingerschen Operatoren, *Math. Annalen*, vol. 104, pp. 570–578, 1931.

[52] G. W. Mackey, A theorem of Stone and von Neumann, *Duke Math. J.*, vol. 16, pp. 313–326, 1949.

[53] A. A. Kirillov, *Elements of the Theory of Representations*. Berlin, New York: Springer, 1976.

[54] R. S. Shenoy and T. W. Parks, Wide-band ambiguity functions and affine Wigner distributions, *Signal Processing*, vol. 41, no. 3, pp. 339–363, 1995.

[55] D. Montgomery and L. Zippin, *Topological Transformation Groups*. New York: Interscience, 1965.

[56] R. G. Shenoy, *Group Representations and Optimal Recovery in Signal Modeling*. Ph.D. thesis, Cornell University, 1991.

[57] L. Cohen, A general approach for obtaining joint representations in signal analysis—Part I: Characteristic function operator method, *IEEE Trans. Signal Processing*, vol. 44, pp. 1080–1090, May 1996.

[58] A. M. Sayeed and D. L. Jones, Equivalence of generalized joint signal representations of arbitrary variables, *IEEE Trans. Signal Processing*, vol. 44, pp. 2959–2970, Dec. 1996.

[59] A. M. Sayeed, On the equivalence of the operator and kernel method for joint distributions of arbitrary variables, *IEEE Trans. Signal Processing*, vol. 45, pp. 1067–1069, April 1997.

[60] F. Hlawatsch and T. Twaroch, Extending the characteristic function method for joint a-b and time-frequency analysis, in *Proc. IEEE ICASSP-97*, vol. 3, (Munich, Germany), pp. 2049–2052, April 1997.

[61] J. C. O'Neill and W. J. Williams, Shift-covariant time-frequency distributions of discrete signals, *IEEE Trans. Signal Processing*, vol. 47, pp. 133–146, Jan. 1999.

[62] M. S. Richman, T. W. Parks, and R. G. Shenoy, Discrete-time, discrete-frequency time-frequency analysis, *IEEE Trans. Signal Processing*, vol. 46, pp. 1517–1527, June 1998.

8

Time-Frequency/Time-Scale Reassignment

Eric Chassande-Mottin
Francois Auger
Patrick Flandrin

ABSTRACT This chapter reviews the reassignment principle, which aims at "sharpening" time-frequency and time-scale representations in order to improve their readability.

The basic idea, which simply consists in moving the time-frequency contributions from the point where they are computed to a more appropriate one, is presented first for the simple cases of the spectrogram and scalogram and then extended to general classes of time-frequency and time-scale energy distributions.

We further consider how the reassignment idea can be implemented efficiently and how it actually operates. Cases (with both deterministic and random signals) where closed-form expressions can be obtained offer the opportunity to better understand how reassignment works. We also give a geometrical characterization of the transform of the time-frequency plane made by the reassignment.

Finally, with two examples (signal de-noising and detection) we illustrate how the reassignment can be useful in practical signal processing applications.

8.1 Introduction

Time-frequency analysis is a topic that has been extensively studied during the past twenty years, and many tools are nowadays available [1–4]. Whereas theoretical foundations are now well established, potential users are nevertheless still faced with a number of difficulties that may limit the applicability of the techniques developed so far. First, from a qualitative point of view and for the sake of interpretation, it may seem desirable to obtain time-frequency "pictures" which would be easily *readable*, even by nonexperts. Second, from an operational perspective, getting a "picture" is generally not the very final aim of analysis, but rather a necessary step for any kind of further *processing*, expected to be more powerful in nonstationary environments, when expressed in a time-frequency setting.

These two central issues will be addressed in this paper. We will first introduce in Section 8.2 the concept of "reassignment," whose purpose is to move computed values of usual time-frequency (or time-scale) distributions so as to increase their localization properties. Section 8.3 will then detail how this theoretical idea can be turned into computationally efficient tools. Analytical examples will be derived and illustrated on both chirp-like signals and isolated singularities. A more thorough study of (spectrogram) reassignment vector fields will be conducted in Section 8.4, both statistically and geometrically. This leads us to propose in Section 8.5 two variations upon the initial method, aimed on the one hand at favoring reassignment in the vicinity of signal components while inhibiting it in "noise only" regions (*supervised reassignment*) and on the other hand at making the reassignment process continuous and governed by motion equations of time-frequency "particles" (*differential reassignment*). Finally, two signal processing applications will be presented in Section 8.6. The first one concerns the *partitioning* of the time-frequency plane, with time-varying filtering, de-noising, and signal components extraction as natural by-products. The second one is devoted to *chirp detection*, with specific emphasis on power-law chirps, in connection with the challenging problem of gravitational wave detection.

The material presented here is mostly based on [5], where detailed proofs, further extensions, algorithms, and illustrations can be found. MATLAB versions of the basic reassignment algorithms discussed in this paper are also part of the freeware [6].

8.2 The reassignment principle

The idea of reassignment was first introduced in 1976 by Kodera, Gendrin, and de Villedary [7], but the technique remained little known and rarely used until a recent past. Thanks to the many advances obtained in time-frequency analysis during the 1980s, a rebirth of reassignment has been made possible, which considerably extended its applicability, both conceptually and computationally [8]. For the sake of simplicity, we will first present basics of reassignment in the case of the spectrogram (the only case initially considered by Kodera et al.) and of the scalogram, discussing later further extensions as well as connections with other existing methods.

8.2.1 *Localization trade-offs in classical time-frequency and time-scale analysis*

The following Fourier transform (throughout this paper, we will adopt the convention of labeling time signals by lowercase symbols, and their Fourier transform, function of the angular frequency ω, by the corresponding uppercase symbols; unless otherwise specified, integrals have integration bounds running from $-\infty$ to $+\infty$):

$$X(\omega) := \int x(t)\, e^{-i\omega t}\, dt \qquad (8.1)$$

offers a natural representation for "stationary" signals, i.e., signals whose spectral properties do not vary over time. In more realistic situations, where gliding tones, transient phenomena, and abrupt changes are likely to be observed, it becomes mandatory to rather consider a mixed description in both time and frequency. The problem is therefore to properly define such a quantity, for which it is well known that no unique definition exists [1,2,4].

A first intuitive approach is to introduce some time dependence in the definition (8.1) of the Fourier transform, so as to make the computation of the transform "local" in some sense. This can be achieved by defining

$$F_x^h(t,\omega) := \int x(s)\, h^*(s-t)\, e^{-i\omega s}\, ds \times e^{it\omega/2}, \tag{8.2}$$

a quantity referred to as a *short-time Fourier transform* (STFT), because of the introduction of an arbitrary short-time window $h(t)$, aimed at limiting the evaluation of the Fourier transform to some specified neighborhood of the current date t. The purpose of introducing a pure phase factor $e^{it\omega/2}$ in the definition will be justified later by symmetry arguments (see, e.g., Eq. (8.11) and (8.12)). A closely related definition is given by

$$T_x^\psi(t,a) := \frac{1}{\sqrt{a}} \int x(s)\, \psi^*\left(\frac{s-t}{a}\right) ds, \tag{8.3}$$

referred to as a *wavelet transform* (WT) [3,9]. In (8.3), the transform is in fact a function of time t and *scale* a but, under mild conditions on the analyzing wavelet $\psi(t)$—namely that it is zero-mean, and that its spectrum is unimodal and characterized by some reference central frequency ω_0—, it can also be expressed as a function of time and (angular) frequency ω, with the identification $\omega := \omega_0/a$.

In both cases (STFT and WT), the representation is faced with a resolution trade-off between time and frequency. For the STFT, the shorter the duration of $h(t)$, the better its resolution in time, but the larger the bandwidth of its spectrum $H(\omega)$ and, henceforth, the poorer its resolution in frequency. In the WT case, a similar trade-off is observed, except that it is frequency dependent. By varying the scale a, resolutions in time and in frequency are continuously varying, but they still remain reciprocal: time resolution is increased at higher frequencies, but at the expense of a lower frequency resolution. This situation can be illustrated in the case of a "chirp" (i.e., a gliding tone), for which no perfect localization can be attained along the expected line of instantaneous frequency. This results from the fact that the choice of the window (or of the wavelet) is arbitrary and independent of the analyzed signal. An improvement can therefore be imagined by making the window (or the wavelet) signal dependent. Based on the idea of *matched filtering*, a natural choice for the STFT is to take as a window $h(t) = x_-(t) := x(-t)$, i.e., the time-reversed version of the analyzed signal. It is easy to show that such a choice leads to

$$F_x^{x_-}(t,\omega) = W_x\left(t/2, \omega/2\right)/2, \tag{8.4}$$

where

$$W_x(t,\omega) := \int x(t+s/2)\, x^*(t-s/2)\, e^{-is\omega}\, ds \tag{8.5}$$

stands for the *Wigner–Ville distribution* (WVD).

The WVD is central in time-frequency analysis, and it possesses a number of theoretical properties that make it a privileged tool [2, 4]. Among them, the WVD allows for a *perfect* localization for *linear* chirps, thus overcoming the limitations just mentioned for the STFT and the WT. However, this has to be paid at the price of interpretation difficulties: on the one hand, the WVD may take on negative values (thus forbidding any local density interpretation) and, on the other hand, the fully quadratic nature of the WVD creates spurious interference terms, characterized by oscillating contributions located in between any two interacting components [10].

Interference terms being oscillatory, they can be reduced by applying some (low-pass) smoothing to the WVD, but such a smoothing operation also has the effect of spreading out localized signal terms. This means that, whereas STFTs and WTs are faced with a trade-off between their time and frequency resolutions, smoothed WVDs are faced with a new kind of trade-off between their joint localization and the importance of their interference terms.

8.2.2 Spectrograms/scalograms revisited from a mechanical analogy

The preceding claim can be made more precise by considering *spectrograms* and *scalograms* (i.e., the energetic versions associated to STFTs and WTs) from a new perspective. Given an STFT F_x^h and a WT T_x^ψ, the corresponding spectrogram S_x^h and scalogram Σ_x^ψ are usually defined by

$$S_x^h(t,\omega) := \left|F_x^h(t,\omega)\right|^2 \tag{8.6}$$

and

$$\Sigma_x^\psi(t,a) := \left|T_x^\psi(t,a)\right|^2, \tag{8.7}$$

respectively, but they can also be expressed as [13]

$$S_x^h(t,\omega) = \iint W_x(s,\xi)\, W_h(s-t,\xi-\omega)\, \frac{ds\, d\xi}{2\pi} \tag{8.8}$$

and

$$\Sigma_x^\psi(t,a) = \iint W_x(s,\xi)\, W_\psi\left(\frac{s-t}{a}, a\xi\right) \frac{ds\, d\xi}{2\pi}, \tag{8.9}$$

thus making explicit the fact that they both result from a smoothing of the WVD.

Therefore, when we classically compute a spectrogram, the value that it takes at a given point (t,ω) of the plane cannot be considered as pointwise. In fact, because of the smoothing operation made explicit in (8.8), it rather results from the summation of all Wigner–Ville contributions within some time-frequency domain defined as the essential time-frequency support of the chosen short-time window. As a consequence, a whole *distribution* of values is summarized by a *single number*, and this number is assigned to the *geometrical center* of the domain over which the distribution is considered. Reasoning with a mechanical analogy, the situation is as if the total mass of an object were assigned to its geometrical center, an arbitrary point which—except in the very specific case of a homogeneous distribution over the domain—has no reason to suit the actual distribution. A much more meaningful choice is to assign the total mass to the *center of gravity* of the distribution within the domain, and this is precisely what reassignment performs: at each time-frequency point where a spectrogram value is computed, we also compute the local centroid $(\hat{t}_x(t,\omega), \hat{\omega}_x(t,\omega))$ of the WVD distribution W_x, as seen through the time-frequency window W_h centered in (t,ω), and the spectrogram value is *moved* from the point where it has been computed to this centroid [7, 8, 11]. This leads us to define the reassigned spectrogram as [12]

$$\check{S}_x^h(t,\omega) = \iint S_x^h(s,\xi)\, \delta\left(t - \hat{t}_x(s,\xi), \omega - \hat{\omega}_x(s,\xi)\right) \frac{ds\, d\xi}{2\pi}. \tag{8.10}$$

The centroids used in reassignment can be related to the *phase* of the STFT, a piece of information that is usually discarded when considering a spectrogram (see Eq. (8.6)). More precisely, denoting by $\varphi(t,\omega)$ the phase of the STFT (8.2), and using the simplified notation $\partial_u \varphi := \partial\varphi/\partial u$, local centroids can be shown [7, 12] to be given by

$$\hat{t}_x(t,\omega) = \frac{t}{2} - \partial_\omega \varphi(t,\omega); \tag{8.11}$$

$$\hat{\omega}_x(t,\omega) = \frac{\omega}{2} + \partial_t \varphi(t,\omega). \tag{8.12}$$

Conceptually, reassignment can be considered as a two-step process: (i) a *smoothing*, whose main purpose is to rub out oscillatory interferences, but whose drawback is to smear localized components; (ii) a *squeezing*, whose effect is to refocus the contributions that survived the smoothing.

One can therefore easily imagine that a WVD, known to perfectly localize on lines of the plane, also perfectly preserves its localization in the reassignment process, since the centroid of a line distribution necessarily belongs to its (localized) support. More generally, a somehow similar benefit is expected to be obtained as long as the analyzed signal *locally* behaves as a linear chirp, locality referring to the time-frequency support of the smoothing window. An example of this behavior is given in Figure 8.1.

8.2.3 Reassignment as a general principle

Reassignment has been considered so far for the spectrogram, and its mechanism has been explained in terms of smoothing followed by squeezing. In the specific case of the spectrogram, smoothing concerns the WVD of the analyzed signal, and it is achieved by means of a very peculiar kernel, namely the WVD of the short-time

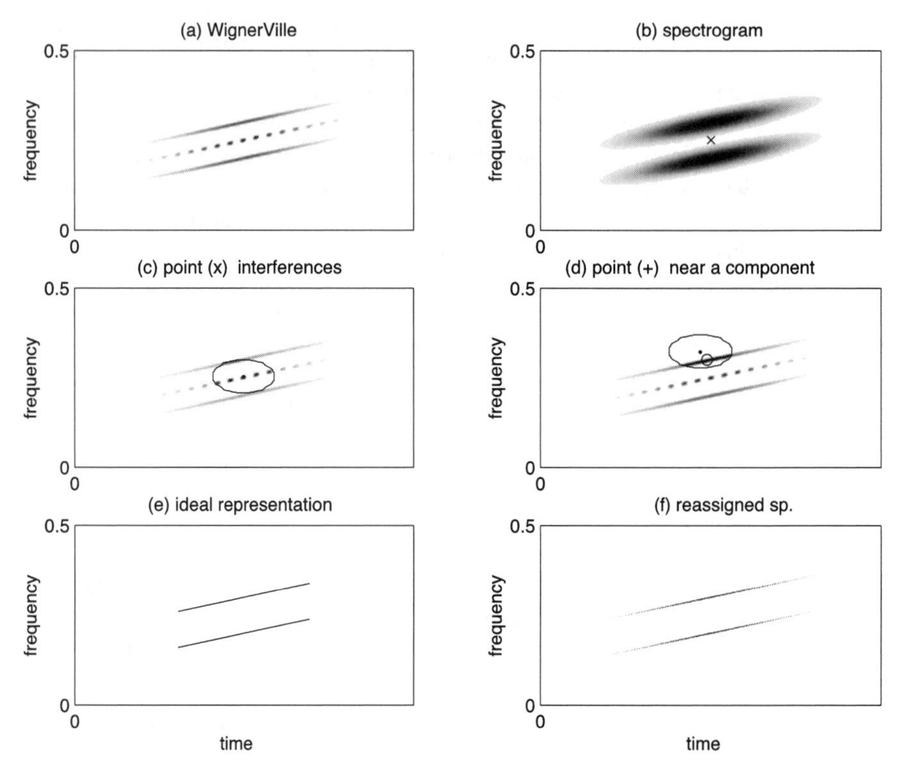

FIGURE 8.1. Principles of the reassignment method. Bilinear time-frequency distributions are faced with a trade-off between interference terms and localization. The WVD (a) is well localized on the two chirps, but it presents spurious oscillations; conversely, the spectrogram (b) is not affected by such extra terms, but it is poorly localized. As illustrated in (c), smoothing the WVD over a given time-frequency domain reduces interferences because of cancellations between positive and negative values within the domain, but this also results in a smearing of localized contributions. Whereas a spectrogram value at the time-frequency point "+" in (b) is classically obtained as the average of all Wigner–Ville values in (d), reassignment moves this value to the point "o," defined as the local centroid of all contributions in (d). This process leads to the refocused distribution (reassigned spectrogram) in (f), to be compared to the "ideal" representation (e) of the considered two-chirp model.

window. It is however straightforward to think of possible generalizations of the process, beyond the spectrogram case. The first possibility is to take for a smoothing kernel a two-dimensional (2D) (low-pass) function that does not necessarily identify to the WVD of some short-time window. Denoting such a kernel by $\Pi(t, \omega)$ and generalizing (8.8), reassignment can be applied to distributions of the form

$$C_x(t,\omega) = \iint W_x(s,\xi)\Pi(s-t,\xi-\omega)\,\frac{ds\,d\xi}{2\pi}. \tag{8.13}$$

One recognizes in (8.13) the general form of Cohen's class [2], the class of all quadratic time-frequency energy distributions that are covariant to shifts in time and frequency, and whose spectrogram is only a special case. Reassignment can therefore be equally applied to any member of Cohen's class, provided that the kernel function $\Pi(t, \omega)$ is *low-pass*, so as to guarantee that it defines an admissible smoothing operator [8].

Given a distribution $C_x(t, \omega)$ in Cohen's class, local centroids are formally given by

$$\hat{t}_x(t, \omega) = \frac{1}{C_x(t, \omega)} \iint s \, W_x(s, \xi) \, \Pi(s - t, \xi - \omega) \, \frac{ds \, d\xi}{2\pi}, \tag{8.14}$$

$$\hat{\omega}_x(t, \omega) = \frac{1}{C_x(t, \omega)} \iint \xi \, W_x(s, \xi) \, \Pi(s - t, \xi - \omega) \, \frac{ds \, d\xi}{2\pi}, \tag{8.15}$$

and the corresponding reassigned distribution is expressed as

$$\check{C}_x(t, \omega) = \iint C_x(s, \xi) \, \delta\left(t - \hat{t}_x(s, \xi), \omega - \hat{\omega}_x(s, \xi)\right) \frac{ds \, d\xi}{2\pi}. \tag{8.16}$$

Figure 8.2 illustrates the extension of reassignment to the entire Cohen's class with two examples where the smoothing kernel is a separable function.

In the spectrogram case, the expressions (8.14), (8.15) can be reduced to (8.11), (8.12). We will see in Subsection 8.3.1 that implicit ways of computing local centroids—which turn out to be much more efficient than (8.14), (8.15)—are possible for a large number of the most commonly used distributions.

A first generalization of spectrogram reassignment was to switch to more general distributions within Cohen's class. A different kind of generalization can also be obtained when considering scalograms instead of spectrograms. In fact, if we go back to the scalogram definition (8.9) and compare it to the spectrogram form (8.8), it is clear that the smoothing structure is very similar—albeit frequency dependent—paving the road for the reassignment of scalograms. More generally, scalograms are themselves only a special case of a more general class of time-scale energy distributions, referred to as the *affine class* [13]. This class, which encompasses all quadratic time-scale distributions that are covariant under shifts and dilations, parallels Cohen's class and is obtained as

$$\Omega_x(t, a) = \iint W_x(s, \xi) \, \Pi\left(\frac{s - t}{a}, a\xi\right) \frac{ds \, d\xi}{2\pi}, \tag{8.17}$$

i.e., as (8.9), with the smoothing kernel W_ψ replaced by some (band-pass) kernel function Π. Within this generalized framework, the reassignment operator in time is directly given by

$$\hat{t}_x(t, a) = \frac{1}{\Omega_x(t, a)} \iint s \, W_x(s, \xi) \, \Pi\left(\frac{s - t}{a}, a\xi\right) \frac{ds \, d\xi}{2\pi}, \tag{8.18}$$

whereas the associated operator in scale $\hat{a}_x(t, a)$ needs some intermediate step in frequency. Precisely, it is necessary to first compute the quantity

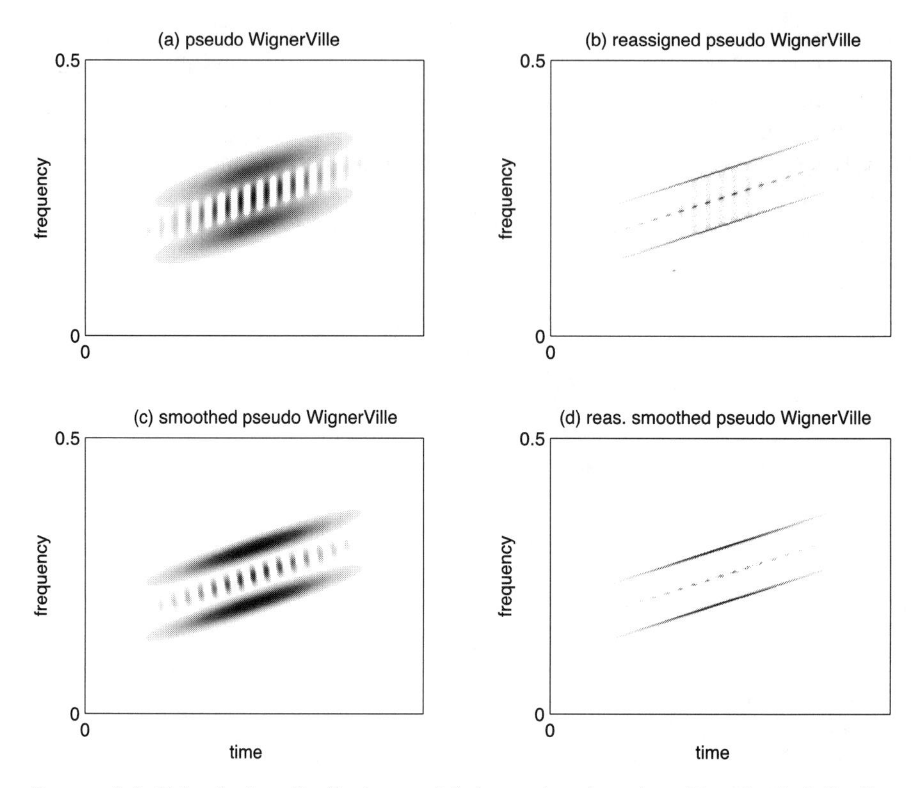

FIGURE 8.2. Cohen's class distributions and their reassigned versions. Two classical distributions within Cohen's class are represented in the left column, with their associated reassigned versions in the right column. (a), (b) Pseudo WVD: $\Pi(t,\omega) = \delta(t)H(\omega)$. (c), (d) Smoothed pseudo WVD: $\Pi(t,\omega) = g(t)H(\omega)$.

$$\hat{\omega}_x(t,a) = \frac{1}{\Omega_x(t,a)} \iint \xi \, W_x(s,\xi) \, \Pi\left(\frac{s-t}{a}, a\xi\right) \frac{ds \, d\xi}{2\pi}, \tag{8.19}$$

from which a frequency-to-scale conversion is then achieved according to

$$\hat{a}_x(t,a) = \frac{\omega_0}{\hat{\omega}_x(t,a)}, \tag{8.20}$$

where

$$\omega_0 := \iint \omega \, \Pi(t,\omega) \frac{dt \, d\omega}{2\pi} \tag{8.21}$$

is the reference frequency of the kernel $\Pi(t,\omega)$, assumed to be of unit integral over $\mathbb{R}^2$.

Reasoning along these lines, further generalizations are possible as well, beyond Cohen's class and the affine class. In fact, any distribution that results from the smoothing of some mother distribution with localization properties for specific signals can be reassigned. This is the case for, e.g., the *hyperbolic class* [14, 15], the *power class* [16], and S-distributions [17]. We will not detail further such generalizations [18, 19].

8.2.4 Connections with related approaches

Although original in many respects, the concept of reassignment is—at least in its restricted spectrogram/scalogram form—more or less closely connected with a number of other approaches that have been proposed independently. In most cases, those related approaches make use, implicitly or explicitly, of some "sinusoidal" signal model [20] of the form

$$x(t) = \sum_{n=1}^{N} A_n(t) \exp\{i\theta_n(t)\}, \tag{8.22}$$

with further assumptions of slow variations of (i) the envelopes with respect to phase evolutions, and (ii) the instantaneous frequencies. More precisely, it is generally assumed that $|\dot{A}_n(t)/A_n(t)| \ll |\dot{\theta}_n(t)|$ and $|\ddot{\theta}_n(t)/\dot{\theta}_n^2(t)| \ll 1$, where " $\dot{}$ " and " $\ddot{}$ " denote first and second time derivatives, respectively. Such models are most useful in the many situations (including, in particular, speech) where signals can be represented as the sum of quasi-monochromatic components. The underlying idea is to get a simplified representation of such signals in terms of a collection of frequency lines and amplitudes, both varying over time. Many attempts have been observed in this direction, among which it is worth mentioning the pioneering work [21]. More recently, techniques referred to as "differential spectrum analysis" [22] or "ridge and skeleton" [23–25] have surfaced, whose principle is to track fine frequency evolutions on the basis of stationary phase arguments. Such approaches have of course much to do with reassignment, in the sense that, e.g., "ridges" can be related to fixed points of reassignment operators. Even more closely related to reassignment, two other techniques, referred to as "instantaneous frequency density" [26] and "squeezing" [27] have been proposed, in which not only are lines identified but neighboring frequency components are also grouped together. A more detailed presentation of these techniques, as well as elements of comparison, can be found in [5].

8.3 Reassignment in action

After having presented the principle of reassignment, in this section, we discuss the practical implementation of the reassignment operators and the resulting computational burdens. We will also concentrate on different situations where complete analytical calculations of the reassigned spectrogram and scalogram are possible. Whereas quite general results can be obtained [8], we will focus mainly here on the spectrogram and scalogram cases.

8.3.1 Efficient algorithms

We have already seen that the reassignment operators of the spectrogram can be expressed either as mean values of weighted Wigner distributions (in Eq. (8.14) and (8.15)) or as partial derivatives of the phase of the STFT (see Eq. (8.11) and (8.12)). It has been proved in [8] that a third formulation involving ratios of STFTs still exists:

$$\hat{t}_x(t,\omega) = t + \mathrm{Re}\left\{\frac{F_x^{th}}{F_x^h}\right\}(t,\omega); \tag{8.23}$$

$$\hat{\omega}_x(t,\omega) = \omega - \mathrm{Im}\left\{\frac{F_x^{dh/dt}}{F_x^h}\right\}(t,\omega). \tag{8.24}$$

These expressions are very important for the implementation of the reassigned spectrogram. They are a convenient and efficient alternative to the direct calculation of local centroids (untractable because of their large computational cost) or to the evaluation of phase derivatives (unstable because of the difficulties encountered in the phase unwrapping).

The complete algorithm using these expressions can be summarized as follows [6]:

1. compute three STFTs based on the three windows $h(t)$, $t\,h(t)$, and dh/dt,
2. if $F_x^h(t,\omega) \neq 0$, then
   (a) combine them together through Eq. (8.23) and (8.24),
   (b) compute the spectrogram according to Eq. (8.6),
   (c) reassign the spectrogram values according to Eq. (8.16).

The final step left apart, the total computational cost is $O(NM\log M)$, when sampling the time-frequency plane with N points in time and M points in frequency. The practical use of reassigned spectrograms in effective signal processing problems is therefore conceivable. Moreover, a recursive implementation of this algorithm has been recently proposed [28], allowing the reassigned spectrogram to be part of real-time application schemes.

Similar expressions also exist for the scalogram [8]:

$$\hat{t}_x(t,a) = t + a\,\mathrm{Re}\left\{\frac{T_x^{t\psi}}{T_x^{\psi}}\right\}(t,a); \tag{8.25}$$

$$\hat{a}_x(t,a) = -\frac{a\omega_0}{\mathrm{Im}\left\{T_x^{d\psi/dt}/T_x^{\psi}\right\}(t,a)}, \tag{8.26}$$

where ω_0 is the reference central frequency of the wavelet ψ (see Eq. (8.21) with $\Pi = W_\psi$). These equations lead to an algorithm similar to the one for the reassigned spectrogram.

As will be used further in this chapter, a convenient way to handle reassignment vector fields is to map them onto the complex plane after normalization. This leads to the *normalized displacement vector*:

$$r_x^h = \frac{\hat{t}_x - t}{\Delta t_h} + i\frac{\hat{\omega}_x - \omega}{\Delta\omega_h} = \frac{1}{\Delta t_h}\mathrm{Re}\left\{\frac{F_x^{th}}{F_x^h}\right\} - \frac{i}{\Delta\omega_h}\mathrm{Im}\left\{\frac{F_x^{dh/dt}}{F_x^h}\right\}, \tag{8.27}$$

where the normalization factors are, respectively, the time duration Δt_h and the frequency bandwidth $\Delta\omega_h$ of the window (assumed to be of unit energy):

$$\Delta t_h := \left(\int t^2|h(t)|^2\,dt\right)^{1/2}; \quad \Delta\omega_h := \left(\int \omega^2|H(\omega)|^2\,\frac{d\omega}{2\pi}\right)^{1/2}.$$

In the following sections, we will use Gaussian windows,

$$h(t) = \pi^{-1/4}\lambda^{-1/2}e^{-t^2/(2\lambda^2)}, \tag{8.28}$$

since in that case only two STFTs are necessary (because the windows $th(t)$ and dh/dt are indeed proportional: $dh/dt = -(1/\lambda^2)\,t\,h(t)$), and the normalized reassignment vector field in Eq. (8.27) admits the particularly compact expression

$$r_x^h = \frac{\sqrt{2}}{\lambda}\frac{F_x^{th}}{F_x}, \tag{8.29}$$

given $\Delta t_h = \lambda/\sqrt{2}$ and $\Delta\omega_h = 1/(\sqrt{2}\lambda)$.

8.3.2 *Analysis of AM-FM signals*

This section gathers a collection of situations where it is possible to obtain a closed-form expression for the reassignment vectors. We will restrict ourselves to Gaussian analysis windows (see Eq. (8.28)).

We intend to give, for any of the following six test signals, the closed-form expression of the spectrogram $S_x^h(t,\omega)$, of the normalized reassignment vector field $r_x^h(t,\omega)$ through its image in the complex plane defined in Eq. (8.27) and of the reassigned spectrogram $\check{S}_x^h(t,\omega)$. The first results are reshaped versions of ones given in [11, 29], whereas the last calculation (two neighboring impulses) is an original contribution [5].

8.3.2.1 Impulses and complex exponentials

Complete calculations can be easily done for a Dirac impulse in t_0: $x(t) = \delta(t - t_0)$. The result is

$$\begin{aligned} S_x^h(t,\omega) &= \pi^{-1/2}\lambda^{-1}e^{-(t_0-t)^2/\lambda^2}, \\ r_x^h(t,\omega) &= \sqrt{2}(t_0 - t)/\lambda, \\ \check{S}_x^h(t,\omega) &= \delta(t - t_0). \end{aligned} \tag{8.30}$$

The dual case, a pure tone of frequency ω_0, defined by the complex exponential $x(t) = e^{i\omega_0 t}$, leads to

$$
\begin{aligned}
S_x^h(t,\omega) &= 2^{-1}\pi^{1/2}\lambda e^{-(\omega_0-\omega)^2\lambda^2}, \\
r_x^h(t,\omega) &= i\sqrt{2}\lambda(\omega_0-\omega), \\
\check{S}_x^h(t,\omega) &= \delta(\omega-\omega_0)/(2\pi).
\end{aligned} \tag{8.31}
$$

As illustrated in Figure 8.3, the reassignment vectors are all pointing toward the time location t_0 (resp. frequency ω_0) of the impulse (resp. complex exponential), yielding a perfect localization for the reassigned spectrogram.

8.3.2.2 Linear chirp with Gaussian envelope

Complete results can also be obtained if the signal is a linear *chirp* (i.e., a signal whose frequency evolves linearly with time) with a Gaussian envelope. If we let

$$
x(t) = \exp\left(-(1/T^2 - i\beta)t^2/2\right),
$$

we get

$$
\begin{aligned}
S_x^h(t,\omega) &= 2\sqrt{\pi k}\,\lambda \exp\left(-k\left((t/\sigma_t)^2 - 2\sigma_{t\omega}t\omega + (\omega/\sigma_\omega)^2\right)\right), \\
r_x^h(t,\omega) &= \sqrt{2}k\left(\left(-(\lambda/\sigma_t)^2 + i\sigma_{t\omega}\right)t/\lambda + \left(\sigma_{t\omega} - i/(\sigma_\omega\lambda)^2\right)\lambda\omega\right),
\end{aligned} \tag{8.32}
$$

with

$$
\begin{aligned}
\sigma_{t\omega} &= \lambda^2\beta, \\
\delta_{t\omega} &= \lambda^2(T^{-2}+\lambda^{-2}), \\
1/\sigma_\omega^2 &= \lambda^2\delta_{t\omega}, \\
1/\sigma_t^2 &= \sigma_{t\omega}^2/\lambda^2 + \delta_{t\omega}/T^2, \\
k^{-1} &= \sigma_{t\omega}^2 + \delta_{t\omega}^2.
\end{aligned} \tag{8.33}
$$

One can single out some interesting cases when the signal parameters take on particular values. For example, if we let $\beta = 0$ and $T = \lambda = 1$, the signal $x(t) = e^{-t^2/2}$ is a *Gaussian logon* centered at the origin $(0,0)$ of the time-frequency plane. This leads to $\sigma_{t\omega} = 0$, $\delta_{t\omega} = 2$, $\sigma_t = \sigma_\omega = 1/\sqrt{2}$, $k^{-1} = 4$, and

$$
\begin{aligned}
S_x^h(t,\omega) &= \sqrt{\pi}\exp\left(-(t^2+\omega^2)/2\right), \\
r_x^h(t,\omega) &= -\left(t+i\omega\right)/\sqrt{2}, \\
\check{S}_x^h(t,\omega) &= 4\sqrt{\pi}\exp\left(-2(t^2+\omega^2)\right).
\end{aligned} \tag{8.34}
$$

The resulting spectrogram is radially symmetric in a time-frequency plane with normalized axes. The reassignment shrinks the time-frequency plane homogeneously in every direction by a homothetic transformation centered on $(0,0)$.

When $T \to \infty$, the signal $x(t)$ tends to the *linear chirp* $x(t) = e^{i\beta t^2/2}$. Setting $\lambda = \beta = 1$ for simplicity, we get $\sigma_{t\omega} = 1$, $\delta_{t\omega} = 1$, $\sigma_t = \sigma_\omega = 1$, and $k^{-1} = 2$, whence

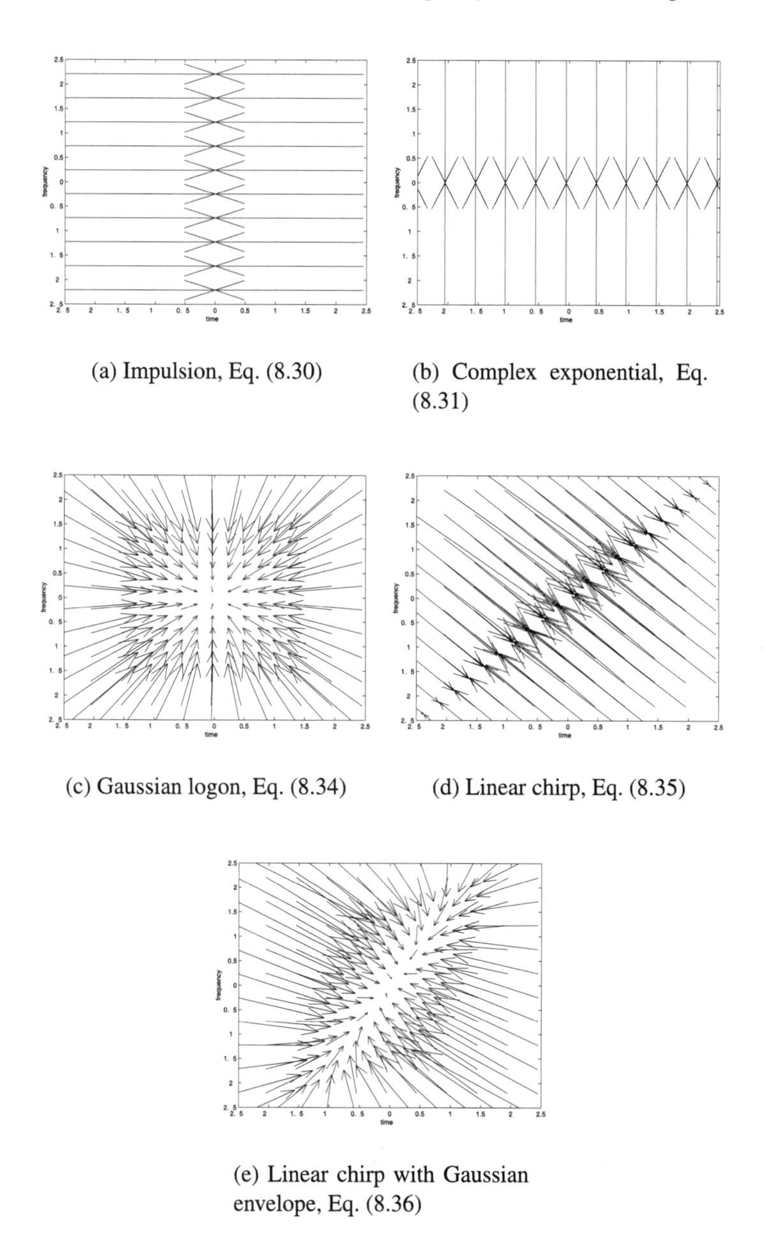

(a) Impulsion, Eq. (8.30)

(b) Complex exponential, Eq. (8.31)

(c) Gaussian logon, Eq. (8.34)

(d) Linear chirp, Eq. (8.35)

(e) Linear chirp with Gaussian envelope, Eq. (8.36)

FIGURE 8.3. Reassignment vector field for several test signals. We have gathered here the graphs of the reassignment vector fields for the set of test signals chosen in Subsection 8.3.2. Each arrow links the calculation point (t, ω) (on the tail extremity) to the reassignment point $(\hat{t}, \hat{\omega})$ (on the head extremity).

$$\begin{aligned} S_x^h(t,\omega) &= \sqrt{2\pi}\exp\left(-(\omega-t)^2/2\right), \\ r_x^h(t,\omega) &= (1-i)(\omega-t)/\sqrt{2}, \\ \check{S}_x^h(t,\omega) &= \delta(\omega-t)/(2\pi). \end{aligned} \tag{8.35}$$

Similarly to the impulse/tone case, the reassignment vectors are all pointing exactly onto the instantaneous frequency line $\omega = \beta t$ (with $\beta = 1$). This is a direct proof (in the special case of a Gaussian window) of the rationale mentioned in Subsection 8.2.2, stating that the reassigned spectrogram is perfectly localized for signals on which the Wigner distribution has itself a perfect localization.

Figure 8.3(e) illustrates the case of a linear chirp with a Gaussian envelope, where all parameters have been set to 1. Equations (8.32) simplify to

$$\begin{aligned} S_x^h(t,\omega) &= 2\sqrt{\pi/5}\exp\left(-(3t^2-2t\omega+2\omega^2)/5\right), \\ r_x^h(t,\omega) &= \sqrt{2}\left((-3+i)t+(1-2i)\omega\right)/5, \\ \check{S}_x^h(t,\omega) &= 2\sqrt{5\pi}\exp\left(-(7t^2-8t\omega+3\omega^2)\right). \end{aligned} \tag{8.36}$$

The geometrical transformation achieved by the reassignment method now combines the two transformations previously described in the logon and chirp cases.

8.3.2.3 Sum of two neighboring impulses

It is also interesting to treat the case of a signal composed of two successive Dirac impulses (and, using the duality between time and frequency, two neighboring complex exponentials). This example will give us information about the resolution (i.e., the ability of distinguishing two components) that we can expect from a reassigned spectrogram.

Let us consider a signal $x(t)$, composed of two distinct components: $x(t) = x_1(t) + x_2(t)$. We will give to each quantity a subscript corresponding to the index of the component it has been computed with. Then, from Eq. (8.29) and using the linearity of the STFT, the normalized reassignment vector field of $x(t)$ can be rewritten as the weighted sum of r_1^h and r_2^h:

$$r_x^h = \frac{F_1^h}{F_1^h + F_2^h} r_1^h + \frac{F_2^h}{F_1^h + F_2^h} r_2^h, \tag{8.37}$$

where the weights are complex valued (this corresponds to a modification in both modulus and phase) and depend on the signal. These weights sum up to 1, showing that Eq. (8.37) can be interpreted as an arithmetic mean.

Let us consider now that $x_1(t)$ and $x_2(t)$ are two Dirac impulses separated from the origin of the time axis by the same distance t_0. Combining the results given in Eq. (8.30) and (8.37) leads to the following expression for the reassignment vector field:

$$r_x^h(t,\omega) = \frac{\sqrt{2}(t_0-t)/\lambda}{1+\exp\left((-2t_0/\lambda)(t/\lambda+i\lambda\omega)\right)} + \frac{-\sqrt{2}(t_0+t)/\lambda}{1+\exp\left((2t_0/\lambda)(t/\lambda+i\lambda\omega)\right)}. \tag{8.38}$$

It is important to mention that the ratios in Eq. (8.38) are not defined for every time t and every frequency ω. In between the two impulses, namely at time $t = 0$, we have

$$r(0, \omega) = \sqrt{2}\frac{t_0}{\lambda}\tan(t_0\omega), \tag{8.39}$$

and the reassignment vector field diverges periodically when the frequency is equal to $\omega_k = k\pi/(2t_0)$ with $k \in \mathbb{Z}$ (this divergence has to be put together with the one observed when evaluating the instantaneous frequency of two beating frequencies [2]). When computing the spectrogram at these points, a direct calculation shows that its value vanishes. Therefore, the impossibility of reassigning the singular points $(0, \omega_k)_{k\in\mathbb{Z}}$ is unimportant because nothing has to be reassigned. It remains that, around all singular points, the reassignment vector attains large values, a situation that is unsatisfactory from a conceptual and algorithmic viewpoint.

Figure 8.4 presents several configurations with respect to the total distance $d = 2t_0$ between the two impulses. When the impulses are close enough (relatively to the window time duration λ, here fixed to 1), the resulting spectrogram, reassignment vector field, and reassigned spectrogram are approximately equal to the sum of the spectrograms, reassignment vector fields, and reassigned spectrograms of each isolated component. If d is of the same order as λ, the singular points mentioned above appear. The distance d has been chosen so that these points (marked with a circle) are separated by a distance of 2 using normalized units. Small spectrogram values in between the two impulses (the dynamic scale on this plot is logarithmic as opposed to the others) may be interpreted as interferences. Finally, when d is large enough so that interactions between impulses vanish, the reassignment vector field can be approximated by the sum of the vector fields obtained with each impulse taken separately.

8.3.3 Singularity characterization from reassigned scalograms

Concerning the scalogram, similar analytical results can also be obtained for the same set of test signals [29] with a Morlet wavelet. Here we are interested in another type of canonical signal.

Time-scale techniques—including scalograms—are known to provide a tool simultaneously adapted to the detection and to the characterization of singularities [3, 30]. It is therefore natural to consider reassigned scalograms of singularities. The underlying motivation is that reassignment methods may improve the contrast in the representation of singularities and therefore their detection, but a question remains: Is it still possible to characterize singularities with a reassigned scalogram?

We will limit ourselves to the family of isolated Hölder singularities and to a certain wavelet, that we will define later on. Any isolated Hölder singularity can be written as (details about the construction of this family can be found in [5])

$$X(\omega) = A_\nu |\omega|^{-\nu-1}, \tag{8.40}$$

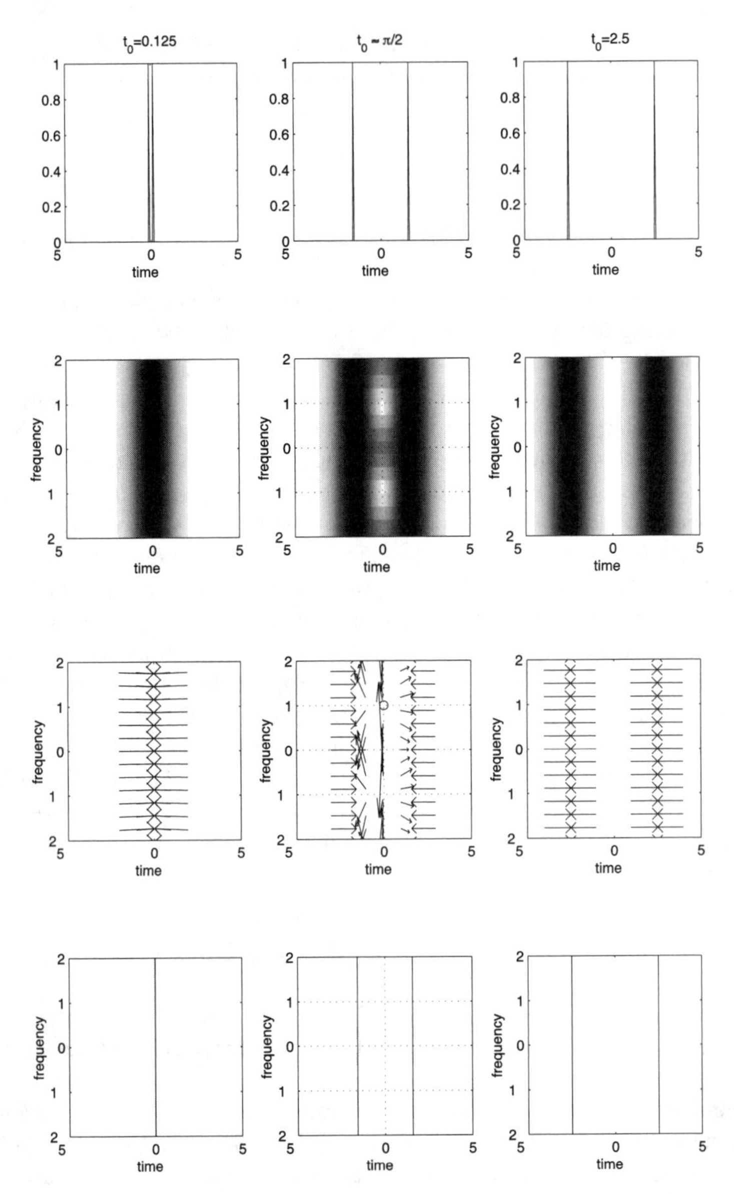

FIGURE 8.4. Spectrogram reassignment of two neighboring impulses. Waveforms, spectrograms, reassignment vector fields, and reassigned spectrograms (from top to bottom, respectively) are considered as functions of spacing (increasing from left to right). Extreme situations of closely spaced (left column) or well-separated (right column) impulses end up in representations formed of one or two lines, respectively. The intermediate situation (middle column) evidences the existence of a "beating" effect (see text).

where the amplitude function A_ν is parameterized by the Hölder exponent ν according to

$$A_\nu = \begin{cases} 2\Gamma(\nu+1)\,(-\sin(\nu\pi/2)) & \text{if } \nu \in \mathbb{R}-\mathbb{Z}, \\ 2(\nu\,!)(-1)^{(\nu+1)/2} & \text{if } \nu \in \mathbb{N}, \\ \pi(-1)^{\nu/2}/|\nu+1|\,! & \text{if } \nu \in \mathbb{Z}^*_-. \end{cases} \tag{8.41}$$

These functions are obviously a particular case of singularities. Nevertheless, they are good local approximations for a wide variety of observable singular behaviors.

Defined as is, the isolated Hölder singularities have the property [31] that their wavelet transform is equal to a rescaled version of their fractional derivative of order $\alpha = -\nu - 1$ (we define the fractional derivative of order α of g as $g^{(\alpha)}(x) = \int_0^{+\infty} (i\omega)^\alpha G(\omega) e^{i\omega x}\, d\omega/(2\pi)$). More precisely, let ψ be a wavelet and $x(t)$ an isolated Hölder singularity. Then the wavelet transform of $x(t)$ is given by

$$T_x^\psi(t,a) = A_{-\alpha-1} a^{-(\alpha+1/2)} i^\alpha \psi^{(\alpha)*}(-t/a). \tag{8.42}$$

One can see in eq. (8.42) two important characteristics of the scalogram (squared modulus of T_x^ψ) structure of a Hölder singularity. First, the energy is almost entirely concentrated in the support of $|\psi^{(\alpha)}(-t/a)|$, which defines in the time-scale plane a cone-shaped domain centered around $t = 0$, and referred to as the *influence cone*. Second, from the restriction of Eq. (8.42) at time $t = 0$

$$\log\left(|T^\psi(0,a)|^2\right) = \log\left(|A_\nu \psi^{-\nu-1}(0)|^2\right) + (2\nu+1)\log(a), \tag{8.43}$$

one can get a simple estimate of ν by measuring the scalogram slope along the scale axis in a log-log diagram [30, 32].

We have chosen the wavelet to be a *Klauder wavelet* [33], defined as

$$\kappa_{\beta,\gamma}(t) = \frac{C_{\beta,\gamma}}{(\gamma - it)^{\beta+1}}, \tag{8.44}$$

where the constant $C_{\beta,\gamma} = (2\gamma)^{\beta+1/2}\Gamma(\beta+1)/\sqrt{2\pi\Gamma(2\beta+1)}$ normalizes the wavelet to unit energy. From the Fourier transform of $\kappa_{\beta,\gamma}(t)$,

$$K_{\beta,\gamma}(\omega) = C_{\beta,\gamma}\frac{2\pi}{\Gamma(\beta+1)}\omega^\beta e^{-\gamma\omega} U(\omega), \tag{8.45}$$

with the convergence conditions $\beta > -1/2$ and $\gamma > 0$, and where $U(\cdot)$ is the Heaviside step function, one can see that the Klauder wavelet family is covariant to fractional differentiation:

$$\kappa_{\beta,\gamma}^{(\alpha)}(t) = \left(\frac{i}{2\gamma}\right)^\alpha \sqrt{\frac{\Gamma(2(\alpha+\beta)+1)}{\Gamma(2\beta+1)}}\,\kappa_{\alpha+\beta,\gamma}(t). \tag{8.46}$$

Coming back to our problem, this last equation gives us the possibility of obtaining closed-form expressions for both the wavelet transform and the scalogram of a Hölder singularity using a Klauder wavelet.

For the computation of the reassignment operators in Eq. (8.25) and (8.26), two wavelet transforms $T^{t\psi}$ and $T^{d\psi/dt}$ need to be expressed. For this, we can use the property that the Klauder wavelet is stable by multiplication by t and by differentiation (the second property resulting from Eq. (8.46) with $\alpha = 1$):

$$\begin{aligned} d\kappa_{\beta,\gamma}/dt &= i/(2\gamma)\sqrt{(2\beta+3)(2\beta+2)}\kappa_{\beta+1,\gamma}(t), \\ t\,\kappa_{\beta,\gamma}(t) &= i\gamma\sqrt{(2\beta)/(2\beta-1)}\kappa_{\beta-1,\gamma}(t) - i\gamma\kappa_{\beta,\gamma}(t). \end{aligned} \tag{8.47}$$

Combining Eq. (8.42), (8.46) and (8.47), we get the algebraic form of the three wavelet transforms involved in Eq. (8.25) and (8.26), leading finally to (see [5] for details)

$$\hat{t}(a,b) = \frac{\alpha}{\alpha+\beta}b; \tag{8.48}$$

$$\hat{a}(a,b) = \frac{\omega_0}{\alpha+\beta+1}\,\frac{(\gamma a)^2+b^2}{\gamma a}, \tag{8.49}$$

where the reference central frequency of the Klauder wavelet is equal to $\omega_0 = (\beta + 1/2)/\gamma$.

We now have all the elements to compute the reassigned scalogram, and only the effective analytical reassignment of each scalogram value still remains. This step essentially consists in inverting the reassignment operators in Eq. (8.48) and (8.49) so as to indicate which quantities are arriving in a given reassignment point $(\hat{t}, \hat{a})$. A first result is that all scalogram points are reassigned between two lines of equations $\hat{a} = \pm(2\beta+1)(\alpha+\beta)/\left(\alpha\gamma(\alpha+\beta+1)\right)\hat{t}$, defining the influence cone. Under the assumptions that the domain defined by these two lines is reasonably sampled, the reassigned scalogram is given, within its support, by

$$\check{\Sigma}_x^{\kappa}(\hat{t},\hat{a}) = \frac{\hat{C}\,\hat{a}^{-(\alpha+\beta-1)}}{(2C)^{\alpha+\beta+1}} \sum_{\epsilon=\pm 1}\left(C\hat{a} + \epsilon\sqrt{(C\hat{a})^2 - (1+\beta/\alpha)^2\hat{t}^2}\right)^{-\alpha+\beta-2} \tag{8.50}$$

where $C = (\alpha+\beta+1)/(2\omega_0)$ and $\hat{C} = A^2_{-\alpha-1}2^{2\beta+1}\gamma^{\alpha+\beta+3}\Gamma^2(\alpha+\beta+1)/\Gamma(2\beta+1)$.

At the time of occurrence of the singularity, i.e., in $\hat{t} = 0$, Eq. (8.50) simplifies to

$$\check{\Sigma}_x^{\kappa}(0,\hat{a}) = \frac{\hat{C}}{(2C)^{2\alpha-3}}\hat{a}^{-(2\alpha+1)}, \tag{8.51}$$

from which we conclude that, (i) as for the scalogram, the reassigned scalogram undergoes a power law with respect to scales and (ii) the exponent of this law is the same as in the scalogram case (see Eq. (8.43)). This means that the measurement

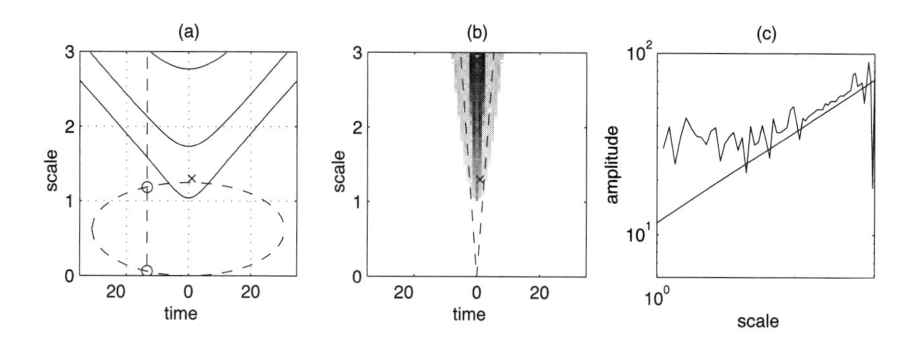

FIGURE 8.5. (a) Scalogram computed with a Klauder wavelet ($\beta = 20$, $\gamma = 50$) of an isolated Hölder singularity at time $t = 0$. (b) The support of the reassigned scalogram of a Hölder singularity with exponent $\nu = 1/3$ ($\alpha = -4/3$) is a cone-shaped domain pointing at $t = 0$ and defined by two lines (dashed lines in the diagram). (c) At the time of the singularity, we verify that the reassigned scalogram undergoes the evolution predicted in Eq. (8.50) (e.g., the reassigned scalogram is linear in a log-log plot). The deviations observed at small and large scales are due to border effects.

of the Hölder exponent ν can be possibly done with a reassigned scalogram. The evolution obtained in Eq. (8.50) is confirmed by the simulations (see Figure 8.5).

In the cases where the influence cone is not sufficiently sampled (e.g., the angle that separates the two border lines is too narrow, yielding some degeneracy in the transformation), then the reassignment has to be handled differently. We obtain in this way another form for the reassigned scalogram:

$$\check{\Sigma}_x^{\kappa}(0,\hat{a}) = (2\gamma)^{-(\alpha+\beta+1)}\hat{C}C^{-2\alpha}\hat{a}^{-2\alpha} \times \int_0^{\pi} \left((1+\sin\theta)^{-2\alpha-3} + (1-\sin\theta)^{-2\alpha-3}\right)\sin\theta\, d\theta. \qquad (8.52)$$

In this situation, the reassigned scalogram varies as $\check{\Sigma}_x^{\kappa}(0,\hat{a}) \sim \hat{a}^{-2\alpha}$.

8.4 More on spectrogram reassignment vector fields

8.4.1 Statistics of spectrogram reassignment vectors

As shown in Section 8.3, reassigned spectrograms and scalograms can be completely characterized for many deterministic signals. For noise corrupted signals, some statistical results also exist for the reassignment operators [34,35]. Although they could also be presented for the scalogram, they will be presented here for the case of the spectrogram, using the Gaussian analysis window defined in Eq. (8.28). In this situation, the normalized reassignment vector r_x^h in (8.29) essentially reduces to the ratio of F_x^{th}/F_x^h, up to a constant factor.

We will first consider the "noise-only case," where the analyzed signal $x(t)$ is a zero-mean noise $n(t)$ assumed to be white, Gaussian, and analytic (hence circular [36]), such that $\forall\,(t,s)\,\in\,\mathbb{R}^2$,

$$n(t) = n_r(t) + i\, n_i(t),$$

with

$$\begin{aligned}\mathbb{E}[n_r(t)\, n_r(s)] &= \mathbb{E}[n_i(t)\, n_i(s)] = (\sigma^2/2)\ \delta(t-s),\\ \mathbb{E}[n(t)\, n(s)] &= 0.\end{aligned}$$

In that case, the properties of both linear filtering and analytic circularity imply that F_x^h and F_x^{th} are two zero-mean circular and statistically independent Gaussian variables, with joint probability density function (pdf)

$$f\left(F_x^h, F_x^{th}\right) = \frac{1}{\pi^2\sigma_1^2\sigma_2^2}\exp\left(-\frac{|F_x^h|^2}{\sigma_1^2} - \frac{|F_x^{th}|^2}{\sigma_2^2}\right), \tag{8.53}$$

with $\sigma_1^2 := \mathbb{E}|F_x^h|^2 = 2\sigma^2$ and $\sigma_2^2 := \mathbb{E}|F_x^{th}|^2 = \sigma^2\,\lambda^2$. A change of variable, $F_x^{th} = \left(\lambda/\sqrt{2}\right)\, r_x^h\, F_x^h$, whose Jacobian is $\left(\lambda^2/2\right)\,|F_x^h|^2$, then yields (after some straightforward computations) a simple expression for the pdf of r_x^h:

$$\begin{aligned}f\left(r_x^h\right) &= \iint f\left(F_x^h, \frac{\lambda}{\sqrt{2}}\, r_x^h\, F_x^h\right)\frac{\lambda^2}{2}\,|F_x^h|^2\, d\mathrm{Re}F_x^h\, d\mathrm{Im}F_x^h\\ &= \frac{1}{\pi\left(1+|r_x^h|^2\right)^2}.\end{aligned} \tag{8.54}$$

We emphasize that this pdf depends neither depend on the noise variance σ^2, nor on the window length λ, nor on the time and frequency variables. Since this pdf has a radial symmetry, we can conclude that the real and imaginary parts of r_x^h are centered and uncorrelated, and their pdf have the same form:

$$\mathbb{E}[r_r\, r_i] = 0, \quad f(r_r) = \frac{1}{2\,(1+r_r^2)^{3/2}}, \quad f(r_i) = \frac{1}{2\,(1+r_i^2)^{3/2}}, \tag{8.55}$$

with $r_x^h = r_r + i\, r_i$. They have infinite variance, but it can be shown that $\mathbb{E}|r_r| = \mathbb{E}|r_i| = 1$. Using polar coordinates, we may also write $r_x^h = \rho\,\exp i\theta$, and the pdf expressions of ρ and θ are, respectively,

$$f(\rho) = \frac{2\rho}{(1+\rho^2)^2}, \quad f(\theta) = \frac{1}{2\pi}, \tag{8.56}$$

which means that θ is a uniform random variable (i.e., the reassignment vector does not have any preferred direction), and that for a given probability P, the maximum modulus $\rho_{\max}$ such that $\mathrm{prob}(0 \le \rho \le \rho_{\max}) = P$ is $\rho_{\max} = \sqrt{P/(1-P)}$. This value can be interpreted as the radius of the disc containing the reassignment vectors with a given probability P.

This approach can be carried over to a "signal plus noise" case, where the analyzed signal is $x(t) = s(t) + n(t)$, $s(t)$ being a deterministic component. The aim of this study is to know how much the reassignment number r_x^h differs from the noise-free value $r_s^h = (\sqrt{2}/\lambda)\,(F_s^{th}/F_s^h)$. In that case, the joint pdf of F_x^h and F_x^{th} is

$$f\left(F_x^h, F_x^{th}\right) = \frac{1}{\pi^2\sigma_1^2\sigma_2^2}\exp\left(-\frac{|F_x^h - F_s^h|^2}{\sigma_1^2} - \frac{|F_x^{th} - F_s^{th}|^2}{\sigma_2^2}\right), \tag{8.57}$$

and the same change of variable as in the noise-only case leads to a more complicated expression:

$$f\left(r_x^h\right) = \frac{1}{\pi\left(1+|r_x^h|^2\right)^2}\left(1 + \frac{S}{2\sigma^2}\,\frac{|1 + r_x^h (r_s^h)^*|^2}{1+|r_x^h|^2}\right)\exp\left(-\frac{S}{2\sigma^2}\,\frac{|r_x^h - r_s^h|^2}{1+|r_x^h|^2}\right), \tag{8.58}$$

with $S = |F_x^h|^2$. This pdf now depends on both the noiseless reassignment vector r_s^h and some kind of "local" signal-to-noise ratio $(\mathrm{SNR}) := S/(2\sigma^2)$. One may check that $\lim_{\mathrm{SNR}\to 0} f(r_x^h)$ equals the "noise-only" pdf (8.54), and $\lim_{\mathrm{SNR}\to +\infty} f(r_x^h) = \delta(r_x^h - r_s^h)$. This expression also shows that, when the local SNR is large $(S/(2\sigma^2) \gg 1)$ and the reassignment vector is small $(|r_s^h| \ll 1)$, then the pdf of r_x^h is approximately Gaussian, with variance $2\sigma^2/S = 1/\mathrm{SNR}$.

All these results should provide a statistical basis for the extraction of useful information from the reassignment vector field of noisy signals. They have been presented for the spectrogram with a Gaussian analysis window, but it has been shown in [34, 35] that the pdf of the "noise-only case" (8.54) does not depend on the choice of the window (under mild conditions that are almost always fulfilled). A similar result, obtained by a very different approach, was published in [37]. In the "signal plus noise case," several computer simulations did not evidence important differences between the pdf (8.58) and the histograms of reassignment vector fields computed with non-Gaussian analysis windows.

8.4.2 *Geometric phase and level curves*

For now, we have been characterizing the reassignment vector field within a fixed family of signals. More results can be given concerning geometrical features shared by reassignment vector fields on a general basis (i.e., no matter what signal is being picked up).

8.4.2.1 Geometric phase

When the STFT is defined as in (8.2), the quasi-symmetric form of the displacement vector field,

$$r_x^h(t,\omega) = \left(\hat{t}_x - t,\, \hat{\omega}_x - \omega\right)^t = \left(-t/2 - \partial_\omega \varphi,\, -\omega/2 - \partial_t \varphi\right)^t,$$

first suggests that this field follows the level curves of some 2D function. But unlike the phase $\varphi(t,\omega)$ of the STFT, this 2D function must be covariant to time and frequency shifts, since $\hat{t}$ and $\hat{\omega}$ both satisfy this property. This means that this function shall not depend on the origin of the time and frequency plane. For this, we introduce a new function $\Phi_{(t_0,\omega_0)}(t,\omega)$, referred to as the *geometric phase* [38]. When the origin of the time-frequency plane is chosen at the (t_0,ω_0) point, $\Phi_{(t_0,\omega_0)}(t,\omega)$ is defined as the phase of the STFT computed at the point whose coordinates are (t,ω) in the new reference frame. If we define the Weyl operator $W(t,\omega)$ as $[W(t,\omega)h](s) = h(s-t)\ \exp i(\omega\, s - \omega t/2)$, then $F_x^h = \langle x, W(t,\omega)h\rangle$, where $\langle\cdot,\cdot\rangle$ stands for the usual inner product of $L^2(\mathbb{R})$, and

$$\begin{aligned}\Phi_{(t_0,\omega_0)}(t,\omega) &= \arg\langle W(-t_0,-\omega_0)x, W(t,\omega)h\rangle \\ &= \arg\langle x, W(t_0,\omega_0)W(t,\omega)h\rangle \\ &= \varphi(t+t_0,\omega+\omega_0) + (\omega_0 t - \omega t_0)/2.\end{aligned}$$

It can be easily shown [39] that the reassignment vector at the (t_0,ω_0) point is tangent to the level curve (and perpendicular to the gradient) of $\Phi_{(t_0,\omega_0)}(t=0,\omega=0)$:

$$r_x^h(t_0,\omega_0) = \left(\partial_\omega \Phi_{(t_0,\omega_0)}(t=0,\omega=0), -\partial_t \Phi_{(t_0,\omega_0)}(t=0,\omega=0)\right)^t. \quad (8.59)$$

This first result shows that the reassignment vector r_x^h has a local geometric interpretation.

8.4.2.2 Level curves

A second result, which we now present, binds the reassignment vector to a scalar potential. As proposed by Bargmann [40], if we write the STFT as $F_x^h(t,\omega) = \mathcal{F}(z,z^*)\ \exp(-|z|^2/4)$, with $z = \omega + i\,t$, then both F_x^h and $\mathcal{F}(z,z^*)$ have the same phase. The reassignment displacements can then be expressed with the partial derivatives of $\mathcal{F}$:

$$\begin{aligned}\hat{t}_x - t &= -t/2 - \mathrm{Im}(\partial_\omega \mathcal{F}/\mathcal{F}) \\ &= -t/2 - \mathrm{Im}\left(\{\partial_z \mathcal{F} + \partial_{z^*}\mathcal{F}\}/\mathcal{F}\right); \quad (8.60)\\ \hat{\omega}_x - \omega &= -\omega/2 + \mathrm{Im}(\partial_t \mathcal{F}/\mathcal{F}) \\ &= -\omega/2 + \mathrm{Re}\left(\{\partial_z \mathcal{F} - \partial_{z^*}\mathcal{F}\}/\mathcal{F}\right). \quad (8.61)\end{aligned}$$

Differentiating $\log F_x^h$ yields another pair of equations:

$$\begin{aligned}\mathrm{Re}(\partial_t F_x^h/F_x^h) &= \partial_t |F_x^h|/|F_x^h| = -t/2 - \mathrm{Im}\left(\{\partial_z \mathcal{F} - \partial_{z^*}\mathcal{F}\}/\mathcal{F}\right),\\ \mathrm{Re}(\partial_\omega F_x^h/F_x^h) &= \partial_\omega |F_x^h|/|F_x^h| = -\omega/2 + \mathrm{Re}\left(\{\partial_z \mathcal{F} + \partial_{z^*}\mathcal{F}\}/\mathcal{F}\right),\end{aligned}$$

which, when combined with Eq. (8.60) and (8.61), leads to

$$\hat{t}_x - t = \partial_t \log|F_x^h| - 2\,\mathrm{Im}(\partial_{z^*} \log \mathcal{F}), \quad (8.62)$$

$$\hat{\omega}_x - \omega = \partial_\omega \log |F_x^h| - 2\,\mathrm{Re}(\partial_{z^*} \log \mathcal{F}). \tag{8.63}$$

The reassignment vector is therefore the sum of the gradient of a scalar potential, $\log |F_x^h|$, and of a second term that is related to the nonanalyticity of $\mathcal{F}$ [39]. When $h(t)$ is a Gaussian window with unit variance ($\lambda = 1$), then $\mathcal{F}$ is an entire function of z, and $\partial_{z^*} \log \mathcal{F} = 0$. In such a case, the reassignment vector field is the gradient of the scalar potential $\log |F_x^h|$, and the reassignment displacements are colinear to the direction of the maxima of the STFT modulus. Up to a constant, the phase and modulus of the STFT are related to each other (i.e., they bear the same information). When $h(t)$ is an arbitrary window, it has been shown in [5] that

$$\begin{cases} -2\,\mathrm{Im}(\partial_{z^*} \log \mathcal{F}) = \mathrm{Re}(F_x^{dh/dt+th}/F_x^h) \\ -2\,\mathrm{Re}(\partial_{z^*} \log \mathcal{F}) = -\mathrm{Im}(F_x^{dh/dt+th}/F_x^h) \end{cases} . \tag{8.64}$$

The importance of this correction term is related to the similarity of the time and frequency resolutions of the STFT: the more Δt_h and $\Delta\omega_h$ differ, the larger $|F_x^{dh/dt+th}|^2$, and the more the reassignment vector field deviates from a gradient field.

8.5 Two variations

Increasing the readability of a time-frequency distribution is not the only purpose of reassignment. As shown in the two examples that follow, useful information brought by reassignment vector fields can be extracted to yield a more intelligent signal characterization.

8.5.1 Supervised reassignment

As shown in [8], the reassigned spectrogram of a chirp signal does not depend on the analysis window. But this situation does not hold anymore when noise is added, and the representation becomes window dependent. Moreover, when the analyzed signal includes broadband noise, the squeezing process performed by reassignment yields peaked areas in noise-only regions, whereas a rather flat energy distribution is expected here. To overcome these drawbacks, an improved reassignment process referred to as *supervised reassignment* was proposed in [41]. This method first attempts to discriminate between "signal + noise" and "noise-only" regions in order to perform reassignment only in the signal + noise regions. Second, it reduces the dependence of the analysis on the window length by a multiwindow procedure.

This first step relies on the observation that in the neighborhood of a deterministic component, the reassignment vector obeys a specific evolution when the length of the window is changed, whereas it undergoes an erratic evolution in noise-only regions. This observation is strengthened by the results presented in Subsection 8.4.1. If the signal can be approximated in the time-frequency neighborhood of any (t, ω)

point as a linear chirp of the form $x(t) = A \exp i\,\beta t^2$, the normalized reassignment displacements are

$$\frac{\hat{t}_x - t}{\Delta t_h} = \frac{\sqrt{2}\,\beta\,(\omega - \beta t)}{\lambda(1/\lambda^4 + \beta^2)}, \quad \frac{\hat{\omega}_x - \omega}{\Delta\omega_h} = \frac{\sqrt{2}\,(\omega - \beta t)}{\lambda^3(1/\lambda^4 + \beta^2)},$$

and the reassignment direction θ satisfies

$$\tan\theta = \frac{\Delta t_h\,(\hat{\omega}_x - \omega)}{\Delta\omega_h\,(\hat{t}_x - t)} = -\frac{1}{\lambda^2\beta}.$$

Assuming that λ is a random variable uniformly distributed between $\lambda_{\min}$ and $\lambda_{\max}$, the pdf of θ is

$$f(\theta) = \frac{C}{2\sqrt{|\beta|}}\left|\frac{1+\tan^2\theta}{\tan^{3/2}\theta}\right|, \qquad \forall\theta \in [\theta_{\min}, \theta_{\max}],$$

where C is a normalization constant. This pdf can be compared to the histogram $\hat{f}(\theta)$ of the reassignment directions, obtained with a collection of N window lengths, by means of a Kullback–Leibler distance [42]:

$$d(\hat{f}, f) = \int_{\theta_{\min}}^{\theta_{\max}} \hat{f}(\theta)\,\log\frac{\hat{f}(\theta)}{f(\theta)}\,d\theta.$$

When this distance exceeds some threshold, the (t, ω) point is considered to belong to a noise-only region, and reassignment is inhibited.

This first step requires the computation of N spectrograms, using Gaussian windows with uniformly distributed window lengths and their associated N reassignment vector fields. What can be done to merge these spectrograms and these reassignment vector fields in a coherent way, so as to reinforce their common features? Answering this question requires us to find the optimal choice of a distance measure between time-frequency distributions, leading to a specific averaging rule. As this optimal choice is not known yet (see [43,44] for preliminary attempts), a pragmatic approach has been followed, and an arithmetic average of both the spectrograms and the reassignment vectors is performed. This choice corresponds to an L^2-distance between distributions.

The complete algorithm is detailed in [41], where satisfying results are also presented.

8.5.2 *Differential reassignment*

The existence of a link between reassignment vector fields and scalar potentials, shown in Subsection 8.4.2, suggests that we consider the dynamical systems governed by these potentials. The resulting dynamical processes perform what we call *differential reassignment* [38]. Whereas the original reassignment proposed by

Kodera et al. [7] moves each value by one finite *dash*, differential reassignment considers each time-frequency point as the starting point of a continuous path defined by the 2D differential equation

$$\begin{cases} t(0) = t, \\ \omega(0) = \omega, \\ dt/ds(s) = \hat{t}_x\left(t(s), \omega(s)\right) - t(s), \\ d\omega/ds(s) = \hat{\omega}_x\left(t(s), \omega(s)\right) - \omega(s), \end{cases} \tag{8.65}$$

where s is a dummy variable defining a curvilinear coordinate along the resulting path.

The reassignment vector field is then considered as a *velocity field* governing the motion of each time-frequency point, considered as an elementary particle. When the analysis window is a Gaussian function with unit variance ($\lambda = 1$), the reassignment vector field has been shown to be a gradient field (see Subsection 8.4.2) attached to the scalar potential $\log |F_x^h|$. Therefore, the dynamical equations (8.65) define a fully dissipative system so that each energy "particle" converges to a local maximum of the potential. For an arbitrary window, the differential reassignment still yields a signal description in terms of *attractors*, *basins of attraction*, and *watersheds*, that can be used for a large variety of applications. As an example, a signal partitioning (and de-noising) algorithm will be presented in Section 8.6.

8.6 Two signal processing applications

For now, we have considered reassigned distributions used for the purpose of (time-frequency) signal *analysis*. The objective of this section is to show, with two application examples, that reassignment methods can also be useful for signal *processing* tasks. In the first example, the reassignment method is used for partitioning the time-frequency plane, through the variation we have introduced in Subsection 8.5.2. The second example illustrates, with an application to the detection of gravitational waves emitted from the coalescence of astrophysical compact bodies, why reassigned distributions are good candidates for chirp detection.

8.6.1 Time-frequency partitioning and filtering

Numerous signal processing problems such as de-noising and signal classification, can be rephrased in the context of time-frequency analysis via a suitable tiling of the time-frequency plane, in which each adjacent cell indicates a signal component, thus allowing for an extraction and/or a processing of each component, independently. Drawing such a time-frequency map is essentially equivalent to decomposing a signal into amplitude and/or frequency modulated components. Strictly speaking, this problem is an ill-posed one, since no clear and universal definition can be given for the concept of a "component," therefore leading to a nonunique decomposition. Nevertheless, several methods exist, overcoming the ambiguity in the notion of components by specifying it with some additional hypotheses. These methods can be divided into two families: the first one defines a signal as a waveform that exhibits some phase coherence under a slowly varying envelope: methods based on the sinusoidal model [20] or the "ridge and skeleton" methods [23, 31] are members of this family. The second group is based on a time-frequency bulk measurement [43, 44]: it consists in counting the signal components by evaluating the time-frequency volume occupied by the observed signal and comparing it with a reference given by the signal of minimum bulk.

Our contribution [39] to this problem is different from the others in that it defines the notion of component by a certain representation of the signal, namely the reassignment vector field, instead of using physical arguments.

Given the results of Subsection 8.4.2, it becomes natural to describe and parameterize a signal in terms of attractors, basins of attraction, and watersheds. From this parameterization, one can think of a variety of solutions to characterize the signal. In accordance with intuition, a component may be described by a *time-frequency center* and a *domain*—namely, the basin of attraction—where the component essentially exists. This way, this method shares a common philosophy with the one exposed in [45].

The algorithm that gives the partition is composed of four steps:

1. *Differential reassignment: Asymptotic reassignment points*, i.e., the coordinates of the "time-frequency particle" location at the end of its trajectory, are computed using the numerical integration (with a fixed-step Runge–Kutta method of order 2) of the equations of motion (8.65). The iteration is stopped by a specific criterion which ensures that the point we are obtaining is in a square of side length d (d has to be fixed to a small value, compared to the step of the time-frequency grid in use) and centered upon the true asymptotic value. This criterion is obtained through a local approximation of the potential by a quadratic form.
2. *Preprocessing:* This consists in sorting the large number of asymptotic reassignment points that have been computed, a part of which is indicating the same time-frequency bins, and in summarizing the useful information into a smaller set of points. The principle is to gather all points converging at a distance less than $\sqrt{2}d$ (the diagonal length of the convergence square mentioned in the previous step) of each other. As a consequence, all "particles" converging to the same asymptotic value are assigned the same location, which we fix equal to their centroid. Let us remark that this operation amounts to first processing the asymptotic reassignment points belonging to an amplitude modulated signal (for which one can prove that they are all converging to a unique isolated point).
3. *Ascending hierarchical classification:* In order to keep only the points coming from a same frequency-modulated signal, we call for a more complicated algorithm: the *ascending hierarchical classification* [46] (AHC). The AHC provides a tree structure where the points are sorted according to their distance. We finally apply a threshold (which we fix to $d+D$, the largest gap between two neighbor-

ing attractors belonging to the same instantaneous frequency path, where D is the discretization step of the time-frequency grid) in the tree structure to get the sets of all points aligned on the same line.

4. *Time-frequency map:* The last step consists in choosing a label for each component detected and in assigning this label to each antecedent of the corresponding asymptotic reassignment points.

If the application requires the extraction of any of the detected components, a nonstationary filter can be set up with the method introduced in [47], according to a template given by the restriction of the time-frequency map to the domain attached to the considered component.

Comparisons have been made in [5] between this partition algorithm and other similar methods. For a canonical situation involving a signal composed of Gaussian logons only, one can analytically prove that the resulting partition is identical to the one we would obtain when applying a Voronoi tessellation [48] to the set of local maxima of the spectrogram. We have also shown that the algorithm presented here and methods based on time-frequency bulk measurements (mentioned before) essentially agree when answering the question "how many components in a signal?". (The test signal was composed of two logons separated by a varying distance. These two logons can be identified either as two components when the distance is large enough, or as one component, e.g., when the distance vanishes.)

As an example, let us consider the extraction of a linear chirp with a Gaussian envelope, embedded in white and Gaussian noise. For this, we use the time-frequency map to select the time-frequency domain that we believe belongs to the signal, and reject all the others. In other words, we design a nonstationary filter adapted to the extraction of the signal. We compare the gain in SNR between the input and the output. This is illustrated in Figure 8.6 where the input SNR varies between -15 dB and 10 dB. These simulations indicate that for an SNR of the order of -5 to 0 dB, the gain is about 5 dB.

However, the question of distinguishing a "signal + noise" region from a "noise-only" region remains open. The strategy we have adopted was to pick up the domain containing the largest energy among all the domains of the map. This simple criterion given here as an illustration could be made more sophisticated (using entropic criteria), in order to enlarge the region where the gain is positive.

8.6.2 Chirp detection

We will be interested in this section in the problem of chirp detection. Although this question can be investigated on a more general basis, we will mainly concentrate on a particular situation: the detection of gravitational wave chirps. The context of gravitational wave detection implies specific notation and therefore some changes in the definitions made in the previous sections (especially, we leave pulsation ω to move to frequency f).

Within the coming years, several gravitational wave detectors will be ready to operate. The objective of these experiments is to provide the first direct measurement of gravitational radiations whose existence has been predicted long ago by the general relativity theory. To be able to measure the tiny effects that gravitational waves induce on matter, one has to obtain a sufficiently precise measurement of changes in the relative distance between two test masses. All these projects—among them, the more important ones are the American LIGO, the French/Italian VIRGO, and the German/British GEO600 projects—have adopted laser interferometry, the best telemetric technique currently available to reach the required sensitivity. Gravitational waves interact so weakly with matter that the necessary sensitivity has been estimated to 10^{-22} for an interferometer about 4 km long (typical length currently planned), reaching an absolute length precision of about a thousandth of the diameter of an atomic nucleus. Given limitations due to noise, this should nevertheless be possible in a frequency "window" between a few tens and a few hundreds of hertz.

A large variety of gravitational wave sources have been listed, but it is almost universally accepted that the most promising one in terms of detection is the coalescence of a massive binary system composed of two neutron stars, the only source we will consider here. In a first (Newtonian) approximation, when the two bodies are far apart, the explicit expression of the waveform can be given as the real part of the complex signal [49, 50]

$$x(t; t_0, \mathcal{M}_\odot) = A\,(t_0 - t)^{-\alpha}\, e^{-i2\pi d (t_0 - t)^\beta}\, U(t_0 - t), \tag{8.66}$$

where $\alpha = 1/4$ and $\beta = 5/8$. In this expression, t_0 is the coalescing time, and d and A are constants that depend upon the physical parameters characterizing the binary system. More precisely, the constant d is best expressed using the "chirp mass" $\mathcal{M} = \mu^{3/5}\, M^{2/5}$ [49], a function of the total mass $M = m_1 + m_2$ and the reduced mass $\mu^{-1} = m_1^{-1} + m_2^{-1}$, where m_1 and m_2 are the individual masses of the two

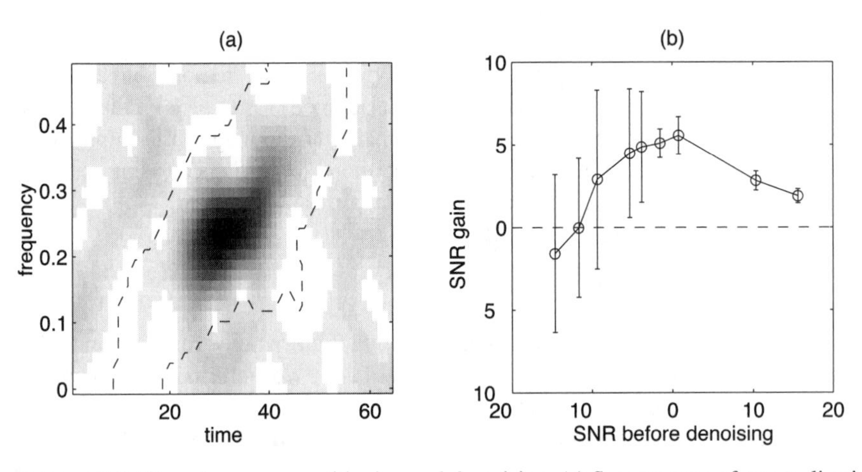

FIGURE 8.6. Time-frequency partitioning and de noising. (a) Spectrogram of one realization of a noisy linear chirp (SNR = 0 dB) and contour of the "signal region" identified by the reassignment vector field (dashed line). (b) SNR gain after de-noising (with corresponding error bars), as function of the input SNR.

bodies. We obtain [50]

$$d \approx 241\,\mathcal{M}_{\odot}^{-5/8}, \tag{8.67}$$

where the chirp mass $\mathcal{M}_{\odot} = \mathcal{M}/M_{\odot}$ is given using the solar mass $M_{\odot}$ unit. For an optimal relative orientation between the interferometer and the binary, we have [51]

$$A \approx 3.37 \times 10^{-21}\,\frac{\mathcal{M}_{\odot}^{5/4}}{r}, \tag{8.68}$$

where r is the earth-binary distance expressed in megaparsecs.

Looking to the phase in the exponential, it is clear that Eq. (8.66) defines a frequency modulated signal whose frequency evolves as a power law with respect to time. In fact, one can show that the gravitational wave chirp (8.66) belongs to the family of power law chirps defined in [52]. Any chirp within this family is characterized through their group delay (i.e., the time of arrival of a given frequency) law which must be of the following form:

$$t_X(f) = t_0 + ckf^{k-1}. \tag{8.69}$$

In the case of the chirp (8.66), the values of its power index k and modulation rate c can be obtained with a stationary phase approximation [53, 54] of its Fourier transform, leading to $k = \beta/(\beta - 1) = -5/3$ and

$$c = -\frac{\beta - 1}{\beta}\,(d\beta)^{-1/(\beta-1)} \approx 3.85 \times 10^{5}\,\mathcal{M}_{\odot}^{-5/3}. \tag{8.70}$$

Although much effort will be done on the interferometer to reduce measurement noise to a minimum, the gravitational wave data will still be corrupted by a large amount of perturbations, mainly of seismic origin at low frequencies and coming from the shot noise in the laser diodes at high frequencies. One can assume [55,56] in a first approximation that, within the measurement frequency bandwidth, these perturbations can be modeled as an additive stationary Gaussian noise of power spectral density $\Gamma_n(f) = \sigma^2\, f^{-\epsilon}$, with $\epsilon \approx 1$ and $\sigma^2 = 0.7 \times 10^{-42}/\text{Hz}$.

For a typical $1.4M_{\odot}$–$1.4M_{\odot}$ neutron star binary located at a distance of 200 Mpc from earth, the expected SNR (the signal to noise ratio is defined as the ratio between the signal and noise energy within the measurement frequency bandwidth) is about 25 ($\approx$ 14 dB). Because the signal energy is distributed into a large number of cycles, the gravitational wave signal will not be visible directly in the interferometer output data. Therefore, specific detection procedures have to designed: the matched filter technique is a natural candidate since this method is optimal under the Gaussian assumption in the sense of the "maximum likelihood ratio test" (known to be a pertinent statistical criterion).

The matched filter method relies on the correlation measurement—understood as the square modulus of a scalar product—between the observed data $r(t)$ and the signal to be detected $x(t; t_0, \mathcal{M}_{\odot})$, taken as a reference. This correlation can be expressed in frequency [57]:

$$\Lambda^w(r; t_0, \mathcal{M}_\odot) = \left| \int_0^{+\infty} \frac{R(f)\, X^*(f; t_0, \mathcal{M}_\odot)}{\Gamma_n(f)}\, df \right|^2, \tag{8.71}$$

but also equivalently in time (the Fourier transform being an isometry). This naturally suggest a third possible approach: the one according to which the scalar product in (8.71) could be written in time and frequency by means of a suitable time-frequency distribution $\rho_X(t, f)$, which we can think of as a well-adapted "signature" of non-stationary signals.

This equivalence is obviously not verified by all time-frequency distributions, but only by time-frequency distributions having the unitary property for which it is possible to write

$$\Lambda^w(r; t, \mathcal{M}_\odot) = \iint_0^{+\infty} \rho_{X_1}(\tau, f)\, \rho_{X_2}(\tau, f)\, d\tau df, \tag{8.72}$$

where $X_1(f) = R(f)/\Gamma_n(f)$ and $X_2(f) = X(f; t_0, \mathcal{M}_\odot)$.

Within the family of unitary time-frequency distributions, we choose the one that best simplifies our detector. Considering that approach, when $\rho_{X_2}(t, f)$ is perfectly localized along the group delay line $(t_X(f), f)$, the double integral in Eq. (8.72) simplifies, leading to a path integration in the time-frequency plane:

$$\Lambda^w(r; t_0, \mathcal{M}_\odot) = \int_0^{+\infty} \rho_{X_1}\left(t_X(f), f\right)\, df. \tag{8.73}$$

The optimal detector can therefore be reformulated by a *path integration* in the time-frequency plane if there exists a time-frequency distribution simultaneously *unitary* and *perfectly localized* on the signal we want to detect [54,58]. In the context of gravitational wave detection, one can prove that such a distribution exists (see [54] for details): the *Bertrand distribution* of index $k = -5/3$.

Strictly speaking, the optimal time-frequency detector (8.73) requires the computation of a filtered version of the Bertrand distribution of the data, thus implying a heavy computational burden. To obtain a more efficient solution, it is mandatory to consider instead a simpler, yet precise, time-frequency representation. Because of the values taken by the physical parameters involved, some approximations can be made in the specific case of gravitational waves. Moreover, since the crucial characteristic of the Bertrand distribution is its perfect localization on the reference signal, we propose to replace it by a reassigned spectrogram $\check{S}_x^h(t, f)$.

Finally, the amplitude of the reassigned spectrogram along the ridge has to be corrected (so as to mimic the unitary property), ending up with the final detector structure

$$\Lambda^w(r; t_0, \mathcal{M}_\odot) \propto \int \check{S}_x^h(t_X(f), f) f^{-2/3}\, df, \tag{8.74}$$

with

$$t_X(f) := t_0 + 3 \times 100^{8/3}\, \mathcal{M}_\odot^{-5/3}\, f^{-8/3}. \tag{8.75}$$

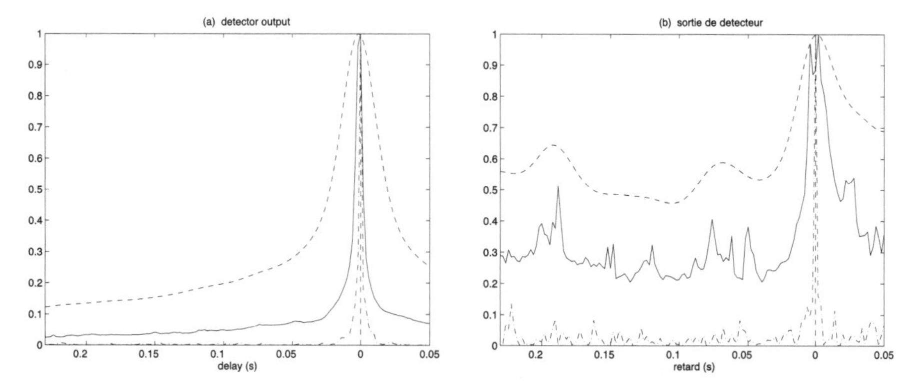

FIGURE 8.7. Gravitational wave detection. The two diagrams display the output of various detectors in the (simulated) case of a coalescing binary system composed of two objects of masses $1\mathcal{M}_{\odot}$ and $10\mathcal{M}_{\odot}$ at a distance of 200 Mpc (left) and 1 Gpc (right) from earth, with coalescence time fixed at $t = 0$. Dashed line: path integration onto a conventional spectrogram; full line: path integration onto a reassigned spectrogram; dashed-dotted line: quadrature matched filtering.

To illustrate the efficiency of the proposed approach, we present in Figure 8.7 two examples based on a typical situation discussed in [55,56]. The strategy based on the reassigned spectrogram does not reach the optimal performance of the matched filter because of the limited precision of the different approximations involved. However, this figure shows that this approach clearly allows the chirp to be detected with a performance (i.e., a contrast in the detector output) much larger than the one obtained with crude path integration onto the standard spectrogram.

8.7 Conclusion

The first natural motivation for introducing reassignment methods was time-frequency analysis. Reassignment relies on a very general principle that applies to a wide class of time-frequency distributions. Because of their limited amount of interference (as compared to the WVD) and their enhanced contrast (as compared to smoothed distributions such as spectrograms and scalograms), reassigned distributions have been shown to give an easy-to-read time-frequency description of a signal, especially in situations involving arbitrary chirp signals at favorable SNRs. Furthermore, efficient algorithms are available, which make reassigned distributions of practical use in effective signal analysis applications.

Beyond time-frequency analysis, reassignment offers other possibilities, too. On the one hand, it can be applied to the chirp detection problem (we have shown that, because of its localization properties, a reassigned spectrogram can be used as a key ingredient in the time-frequency formulation of the optimal detector). On the other hand, a time-frequency partitioning can be based on the reassignment vector field, thus allowing for an easy manipulation of multicomponent nonstationary signals, and paving the road for further processing such as signal characterization, extraction, and de-noising. However, note that, while theoretically attractive, this approach still remains practically limited because of (i) the large computational burden involved, and (ii) its sensitivity to perturbations (contaminating components and noise). It is nevertheless believed that, because of a better knowledge of the structural properties of reassignment vector fields, further developments (e.g., regularization procedures) will be possible, thus turning an appealing concept into an effective signal processing tool.

Acknowledgment. Professor Ingrid Daubechies (Princeton University) is gratefully acknowledged for many discussions and pointwise collaborations during the development of the work reported here.

References

[1] F. Hlawatsch and G. F. Boudreaux-Bartels, Linear and quadratic time-frequency signal representations. *IEEE Signal Proc. Magazine*, 21–67, 1992.

[2] L. Cohen, *Time-Frequency Analysis*. Prentice-Hall, Englewoods Cliffs (NJ), 1995.

[3] S. Mallat, *A Wavelet Tour of Signal Processing*. Academic Press, New York (NY), 1998.

[4] P. Flandrin, *Time-Frequency/Time-Scale Analysis*. Academic Press, San Diego (CA), 1999.

[5] E. Chassande-Mottin, *Méthodes de réallocation dans le plan temps-fréquence pour l'analyse et le traitement de signaux non stationnaires*. Thèse de Doctorat, Université de Cergy-Pontoise (France), 1998.

[6] F. Auger, P. Flandrin, P. Gonçalvès, and O. Lemoine, Time-Frequency Toolbox for MATLAB, Users' Guide and Reference Guide. Available at http://iut-saint-nazaire.univ-nantes.fr/~auger/tftb.html.

[7] K. Kodera, C. De Villedary, and R. Gendrin, A new method for the numerical analysis of nonstationary signals. *Phys. Earth and Plan. Int.*, 12:142–150, 1976.

[8] F. Auger and P. Flandrin, Improving the readability of time-frequency and time-scale representations by the reassignment method. *IEEE Trans. Signal Proc.*, SP-43(5):1068–1089, 1995.

[9] O. Rioul and M. Vetterli, Wavelets and signal processing. *IEEE Signal Proc. Magazine*, 14–38, 1991.

[10] F. Hlawatsch and P. Flandrin, The interference structure of Wigner distributions and related time-frequency signal representations. In *The Wigner Distribution: Theory and Applications in Signal Processing* (W. Mecklenbräuker, F. Hlawatsch, *eds.*), 59–133, Elsevier, Amsterdam, 1998.

[11] K. Kodera, R. Gendrin, and C. De Villedary, Analysis of time-varying signals with small BT values. *IEEE Trans. on Acoust., Speech, and Signal Proc.*, ASSP-26(1):64–76, 1978.

[12] K. Kodera, *Analyse numérique de signaux géophysiques non-stationnaires*. Thèse de Doctorat, Université de Paris VI (France), 1976.

[13] O. Rioul and P. Flandrin, Time-scale energy distributions: a general class extending wavelet transforms. *IEEE Trans. on Signal Proc.*, SP-40(7):1746–1757, 1992.

[14] A. Papandreou, F. Hlawatsch, and G. F. Boudreaux-Bartels, The hyperbolic class of time-frequency representations—Part I: constant Q warping, the hyperbolic paradigm, properties and members. *IEEE Trans. on Signal Proc.*, SP-41(12):3425–3444, 1993.

[15] F. Hlawatsch, A. Papandreou-Suppapola, and G. F. Boudreaux-Bartels, The hyperbolic class of quadratic time-frequency distributions—Part II: subclasses, intersection with the affine and power classes, regularity and unitarity. *IEEE Trans. on Signal Proc.*, SP-45(2):303–315, 1997.

[16] A. Papandreou, F. Hlawatsch, and G. F. Boudreaux-Bartels, Power class time-frequency representations: interference geometry, smoothing and implementation. In *Proc. of the IEEE Int. Symp. on Time-Frequency and Time-Scale Analysis*, 193–196, Paris (France), 1996.

[17] L. J. Stankovic, A method for time-frequency signal analysis. *IEEE Trans. on Signal Proc.*, SP-42(1):225–229, 1994.

[18] F. Auger and P. Flandrin, La réallocation: une méthode générale d'amélioration de la lisibilité des représentations temps-fréquence bilinéaires. In *Proc. Journées GdR TdSI, Temps-Fréquence, Ondelettes et Multirésolution*, 15.1–15.7, Lyon, 1994.

[19] I. Djurovic and L.J. Stankovic, Time-frequency representation based on the reassigned S-method. *Sig. Proc.*, 77(1):115–120, 1999.

[20] R. J. McAulay and T. F. Quatieri, Speech analysis–synthesis based on a sinusoidal representation. *IEEE Trans. on Acoust., Speech, and Signal Proc.*, ASSP-34(4):744–754, 1986.

[21] C. Berthomier, Sur une méthode d'analyse de signaux. *Ann. Geophys.*, 31(2):239–252, 1975.

[22] V. Gibiat, F. Wu, P. Perio, and S. Chantreuil, Analyse spectrale différentielle (A.S.D.). *C. R. Acad. Sc. Paris, série II*, 294:633–636, 1982.

[23] N. Delprat, B. Escudié, P. Guillemain, R. Kronland-Martinet, P. Tchamitchian, and B. Torrésani, Asymptotic wavelet and Gabor analysis: extraction of instantaneous frequencies. *IEEE Trans. on Info. Theory*, IT-38(2):644–673, 1992.

[24] N. Delprat, *Analyse temps-fréquence de sons musicaux: exploration d'une nouvelle méthode d'extraction de données pertinentes pour un modèle de synthèse.* Thèse de Doctorat, Université d'Aix-Marseille II (France), 1992.

[25] P. Guillemain and R. Kronland-Martinet, Horizontal and vertical ridges associated to continuous wavelet transforms. In *Proc. of the IEEE Int. Symp. on Time-Frequency and Time-Scale Analysis*, 63–66, Victoria (Canada), 1992.

[26] D. Friedman, Instantaneous frequency distribution vs. time: an interpretation of the phase structure of speech. In *Proc. of the IEEE Int. Conf. on Acoust., Speech, and Signal Proc.*, 1121–1124, Tampa (FL), 1985.

[27] S. Maes, The synchrosqueezed representation yields a new reading of the wavelet transform. In *Proc. SPIE 95 on OE/Aerospace Sensing and Dual Use Photonics*, 532–559, Orlando (FL), 1995.

[28] C. Richard and R. Lengellé, Joint recursive implementation of time-frequency representations and their modified version by the reassignment method. *Sig. Proc.*, 60(2):163–179, 1997.

[29] F. Auger and P. Flandrin, Improving the readability of the time-frequency and time-scale representations by the reassignment method. Technical Report LAN 93-05, Laboratoire d'Automatique de Nantes, Nantes (France), 1993.

[30] S. Mallat and W. L. Hwang, Singularity detection and processing with wavelets. *IEEE Trans. on Info. Theory*, IT-38(2):617–643, 1992.

[31] P. Guillemain and R. Kronland-Martinet, Characterization of acoustic signals through continuous linear time-frequency representations. *Proc. IEEE*, 84(4):561–587, 1996.

[32] P. Gonçalvès, *Représentations temps-fréquence et temps-échelle — Synthèse et contributions.* Thèse de Doctorat, Inst. National Polytechnique de Grenoble (France), 1993.

[33] J. R. Klauder, Path integrals for affine variables. In *Functional Integration: Theory and Applications* (J.-P. Antoine, E. Tirapegui, *eds.*), 101–119, Plenum Press, New York (NY), 1980.

[34] E. Chassande-Mottin, F. Auger, and P. Flandrin, Statistiques des vecteurs de réallocation du spectrogramme. Technical Report 96-01, Laboratoire de Physique, ENS-Lyon (URA 1325 CNRS), Lyon (France), 1996.

[35] E. Chassande-Mottin, F. Auger, and P. Flandrin, On the statistics of spectrogram reassignment. *Multidim. Syst. and Signal Proc.*, 9(4):355–362, 1998.

[36] B. Picinbono, On circularity. *IEEE Trans. on Signal Proc*, SP-42(12):3473–3482, 1994.

[37] M. Dechambre and J. Lavergnat, Statistical properties of the instantaneaous frequency for a noisy signal. *Sig. Proc.*, 2:137–150, 1980.

[38] E. Chassande-Mottin, I. Daubechies, F. Auger, and P. Flandrin, Differential reassignment. *IEEE Signal Proc. Lett.*, SPL-4(10):293–294, 1997.

[39] E. Chassande-Mottin, F. Auger, I. Daubechies, and P. Flandrin, Partition du plan temps-fréquence et réallocation. In *Proc. 16ème Colloque GRETSI*, 1447–1450, Grenoble (France), 1997.

[40] V. Bargmann, On a Hilbert space of analytic functions and an associated integral transform. *Comm. on Pure and Appl. Math.*, 14:187–214, 1961.

[41] E. Chassande-Mottin, F. Auger, and P. Flandrin, Supervised time-frequency reassignment. In *Proc. of the IEEE Int. Symp. on Time-Frequency and Time-Scale Analysis*, 517–520, Paris (France), 1996.

[42] M. Basseville, Distance measures for signal processing and pattern recognition. *Sig. Proc.* 18:349–369, 1989.

[43] R. G. Baraniuk, P. Flandrin, and O. Michel, Information and complexity on the time-frequency plane. In *Proc. 14ème Colloque GRETSI*, 359–362, Juan-Les-Pins (France), 1993.

[44] R. G. Baraniuk, P. Flandrin, A. J. E. M. Janssen, and O. Michel, Measuring time-frequency information content using the Rényi entropies. *IEEE Trans. on Info. Theory*, Vol. 47, No. 4, pp. 1391–1409.

[45] V. Pierson and N. Martin, Watershed segmentation of time-frequency images. In *Proc. IEEE Workshop on Nonlinear Signal and Image Proc.*, 1003–1006, Halkidiki (Greece), 1995.

[46] J. P. Benzecri, *L'Analyse de Données. Tome 1: La Taxinomie.* Dunod, Paris (France), 1973.

[47] W. Kozek and F. Hlawatsch, A comparative study of linear and nonlinear time-frequency filters. In *Proc. of the IEEE Int. Symp. on Time-Frequency and Time-Scale Analysis*, 163–166, Victoria (Canada), 1992.

[48] F. Preparata and M. Shamos, *Computational Geometry. An Introduction.* Springer-Verlag, New York (NY), 1985.

[49] K. S. Thorne. Gravitational radiation. In *300 Years of Gravitation* (S. W. Hawking, W. Israel, *eds.*), 330–458, Cambridge Univ. Press, Cambridge (UK), 1987.

[50] B. S. Sathyaprakash and D. V. Dhurandhar, Choice of filters for the detection of gravitational waves from coalescing binaries. *Phys. Rev. D*, 44(12):3819–3834, 1991.

[51] S. D. Mohanty and S. V. Dhurandhar, Hierarchical search strategy for the detection of gravitational waves from coalescing binaries. *Phys. Rev. D*, 54(12):7108–7128, 1996.

[52] J. Bertrand and P. Bertrand, A class of affine Wigner distributions with extended covariance properties. *J. Math. Phys.*, 33(7):2515–2527, 1992.

[53] E. Chassande-Mottin and P. Flandrin, On the stationary phase approximation of chirp spectra. In *Proc. of the IEEE Int. Symp. on Time-Frequency and Time-Scale Analysis*, 117–120, Pittsburgh (PA), 1998.

[54] E. Chassande-Mottin and P. Flandrin, On the time-frequency detection of chirps. *Appl. Comp. Harm. Anal.*, 6(9):252–281, 1999.

[55] J.-M. Innocent and B. Torrésani, A multiresolution strategy for detecting gravitational waves generated by binary coalescence. Technical Report CPT-96/P.3379, CPT-CNRS, Marseille (France), 1996.

[56] J.-M. Innocent and B. Torrésani, Wavelets and binary coalescences detection. *Appl. Comp. Harm. Anal.*, 4(2):113–116, 1997.

[57] A. D. Whalen, *Detection of Signals in Noise*. Academic Press, San Diego (CA), 1971.

[58] E. Chassande-Mottin and P. Flandrin, On the time-frequency detection of chirps and its application to gravitational waves. In *Proc. of the Second Workshop on Grav. Waves Data Analysis* (M. Davier, P. Hello, *eds.*), 47–52, Editions Frontières, Gif-sur-Yvette (France), 1997.

9

Spatial Time-Frequency Distributions: Theory and Applications

Moeness G. Amin
Yimin Zhang
Gordon J. Frazer
Alan R. Lindsey

ABSTRACT This chapter presents a comprehensive treatment of the hybrid area of time-frequency distributions (TFDs) and array signal processing. The application of quadratic TFDs to sensor signal processing has recently become of interest, and it was necessitated by the need to address important problems related to processing nonstationary signals incident on multiantenna receivers. Over the past few years, major contributions have been made to improve direction finding and blind source separation using time-frequency signatures. This improvement has cast quadratic TFDs as a key tool for source localization and signal recovery, and put bilinear transforms at equal footing with second-order and higher-order statistics as bases for effective spatial-temporal signal processing. This chapter discusses the advances made through time-frequency analysis in direction-of-arrival estimation, signal synthesis, and near-field source characterization.

9.1 Introduction

Time-frequency distributions (TFDs) are used in various applications, including speech, biomedicine, sonar, radar, global positioning systems, and geophysics. Over the past two decades, most of the work on quadratic TFDs has focused on the mono-component and multi-component temporal signal structures and the corresponding time-frequency (t-f) signatures. This work has led to major advances in nonstationary signal analysis and processing. Information on the signal instantaneous frequency and instantaneous bandwidth obtained from the t-f domain has allowed improved separation, suppression, classification, identification, and detection of signals with time-varying spectra [1–6].

Applications of the quadratic distributions to array signal processing began to flourish in the mid-1990s. The main objective was to enhance direction finding and blind source separation of nonstationary signals using their t-f signatures. Another important but different objective was to characterize near-field and far-field emitters based on their spatial signatures. In order to achieve both objectives, new definitions and generalization of quadratic distributions were in order.

The spatial time-frequency distribution (STFD) has been introduced to describe the mixing of nonstationary signals at the different sensors of the array. The relationship between the TFDs of the sensors to the TFDs of the individual source waveforms is defined by the steering, or the array, matrix, and was found to be similar to that encountered in the traditional data covariance matrix approach to array processing. This similarity has allowed a rapid progress in nonstationary array processing from the TFD perspective [7].

This chapter discusses two fundamental formulations to incorporate the spatial information into quadratic distributions. One formulation is based on STFDs and the localization of the signal arrivals in the t-f domain. The corresponding analysis, theory, and applications are covered in the first six sections of this chapter. Section 9.7 deals with another formulation, in which the quadratic distribution of the spatial signal across the array is computed. This sensor-angle distribution (SAD) localizes the source angle at each sensor and is the dual in sensor number and angle to Cohen's class of TFDs. The SAD is particularly appropriate for characterizing sources and scatter in the near field of an array. Sources arriving from the far field have the same angle at each sensor. In contrast, sources in the near field have differing angles at each sensor, and a full SAD provides a complete characterization of the near-field source and scatter environment. Knowledge of this characterization can be important when calibrating arrays of sensors placed in a nonhomogeneous environment, such as a radar or communications array deployed in a built-up environment and surrounded by other metallic structures that are not part of the array.

9.2 STFDs

9.2.1 *Signal model*

In narrowband array processing, when n signals arrive at an m-element (sensor) array, the linear data model

$$\mathbf{x}(t) = \mathbf{y}(t) + \mathbf{n}(t) = \mathbf{A}\mathbf{d}(t) + \mathbf{n}(t) \tag{9.2.1}$$

is commonly assumed, where the $m \times n$ spatial matrix $\mathbf{A} = [\mathbf{a}_1, \dots, \mathbf{a}_n]$ represents the mixing matrix or the steering matrix. In direction finding problems, we require $\mathbf{A}$ to have a known structure, and each column of $\mathbf{A}$ corresponds to a single arrival and carries a clear bearing. For blind source separation problems, $\mathbf{A}$ is a mixture of several steering vectors, due to multipaths, and its columns may assume any structure.

The mixture of the signals at each sensor renders the elements of the $m \times 1$ data vector $\mathbf{x}(t)$ to be multicomponent signals, whereas each source signal $d_i(t)$ of the $n \times 1$ signal vector $\mathbf{d}(t)$ is typically a monocomponent signal. $\mathbf{n}(t)$ is an additive noise vector whose elements are modeled as stationary, spatially and temporally white, zero-mean complex Gaussian random processes, independent of the source signals. That is,

$$E[\mathbf{n}(t+\tau)\mathbf{n}^H(t)] = \sigma\delta(\tau)\mathbf{I} \quad \text{and} \quad E[\mathbf{n}(t+\tau)\mathbf{n}^T(t)] = \mathbf{0} \qquad \text{for any } \tau \tag{9.2.2}$$

where $\delta(\tau)$ is the delta function, $\mathbf{I}$ denotes the identity matrix, σ is the noise power at each sensor, superscripts ^{H} and ^{T}, respectively, denote conjugate transpose and transpose, and $E(\cdot)$ is the statistical expectation operator.

9.2.2 STFDs

We first review the definition and basic properties of the STFDs. STFDs based on Cohen's class of TFD were introduced in [8], and their applications to direction finding and blind source separation have been discussed in [9, 10] and [8, 11], respectively.

The discrete form of TFD of a signal $x(t)$ is given by

$$D_{xx}(t,f) = \sum_{v=-\infty}^{\infty} \sum_{l=-\infty}^{\infty} \phi(v,l)x(t+v+l)x^*(t+v-l)e^{-j4\pi fl}, \tag{9.2.3}$$

where $\phi(v,l)$ is a kernel function and * denotes the complex conjugate. The STFD matrix is obtained by replacing $x(t)$ by the data snapshot vector $\mathbf{x}(t)$,

$$\mathbf{D_{xx}}(t,f) = \sum_{v=-\infty}^{\infty} \sum_{l=-\infty}^{\infty} \phi(v,l)\mathbf{x}(t+v+l)\mathbf{x}^H(t+v-l)e^{-j4\pi fl}. \tag{9.2.4}$$

Substituting (9.2.1) into (9.2.4), we obtain

$$\mathbf{D_{xx}}(t,f) = \mathbf{D_{yy}}(t,f) + \mathbf{D_{yn}}(t,f) + \mathbf{D_{ny}}(t,f) + \mathbf{D_{nn}}(t,f). \tag{9.2.5}$$

We note that $\mathbf{D_{xx}}(t,f)$, $\mathbf{D_{yy}}(t,f)$, $\mathbf{D_{yn}}(t,f)$, $\mathbf{D_{ny}}(t,f)$, and $\mathbf{D_{nn}}(t,f)$ are matrices of dimension $m \times m$. Under the uncorrelated signal and noise assumption and the zero-mean noise property, the expectation of the cross-term STFD matrices between the signal and noise vectors is zero, i.e., $E[\mathbf{D_{yn}}(t,f)] = E[\mathbf{D_{ny}}(t,f)] = \mathbf{0}$. Accordingly,

$$\begin{aligned} E[\mathbf{D_{xx}}(t,f)] &= \mathbf{D_{yy}}(t,f) + E[\mathbf{D_{nn}}(t,f)] \\ &= \mathbf{A}\mathbf{D_{dd}}(t,f)\mathbf{A}^H + E[\mathbf{D_{nn}}(t,f)], \end{aligned} \tag{9.2.6}$$

where the source TFD matrix

$$\mathbf{D_{dd}}(t,f) = \sum_{v=-\infty}^{\infty} \sum_{l=-\infty}^{\infty} \phi(v,l)\mathbf{d}(t+v+l)\mathbf{d}^H(t+v-l)e^{-j4\pi fl} \tag{9.2.7}$$

is of dimension $n \times n$. For narrowband array signal processing applications, the mixing matrix $\mathbf{A}$ holds the spatial information and maps the auto- and cross-TFDs of the source signals into auto- and cross-TFDs of the data.

Equation (9.2.6) is similar to the formula that has been commonly used in direction finding and blind source separation problems, relating the signal correlation matrix to the data spatial correlation matrix. In the preceding formulation, however, the correlation matrices are replaced by the STFD matrices. The well-established results in conventional array signal processing could, therefore, be utilized and key problems in various applications of array processing, specifically those dealing with nonstationary signal environments, can be approached using bilinear transformations.

Initially, only the t-f points in the auto-term regions of the TFD are considered for STFD matrix construction. The auto-term region refers to the t-f points along the true instantaneous frequency (IF) of each signal. The cross-terms, which can intrude on the auto-terms through the power in their mainlobes or/and sidelobes, were avoided. This intrusion depends on the signal temporal structures and the window size. Recently, the cross-terms have also been utilized and integrated into STFDs. The effect of cross-terms on direction finding and blind source separation will be discussed in Subsections 9.4.3 and 9.5.3. In the other parts of this chapter, it is assumed that the t-f points reside in an auto-term region, which has negligible cross-term effect.

9.2.3 *Joint diagonalization and time-frequency averaging*

In the rest of this chapter, we will address the application of STFDs to direction finding and blind source separation. These applications are based on the eigendecomposition of the STFD matrix. In direction finding, the source TFD matrix must be full rank (Section 9.5), whereas, to perform blind source separation, the source TFD matrix must be diagonal (Section 9.4). For either case, the STFD matrix of the data vector should be full column rank.

It is noted that the relationship (9.2.6) holds true for every (t, f) point. In order to ensure the full column rank property of the STFD matrix as well as to reduce the effect of noise, we consider multiple t-f points, instead of a single one. This allows more information of the source signal t-f signatures to be included into their respective eigenstructure formulation and, as such, enhances direction finding and source separation performance. Joint diagonalization and t-f averaging are the two main approaches that have been used for this purpose [8, 9, 12].

9.2.3.1 Joint diagonalization

Joint diagonalization (JD) can be explained by first noting that the problem of the diagonalization of a single $n \times n$ normal matrix $\mathbf{M}$ is equivalent to the minimization of the criterion [13]

$$C(\mathbf{M}, \mathbf{V}) = -\sum_i \left|\mathbf{v}_i^H \mathbf{M} \mathbf{v}_i\right|^2 \tag{9.2.8}$$

over the set of unitary matrices $\mathbf{V} = [\mathbf{v}_1, \ldots, \mathbf{v}_n]$. Hence, the JD of a set $\{\mathbf{M}_k | k = 1, \ldots, K\}$ of K arbitrary $n \times n$ matrices is defined as the minimization of the following JD criterion:

$$C(\mathbf{V}) = -\sum_k C(\mathbf{M}, \mathbf{V}) = -\sum_k \sum_i \left|\mathbf{v}_i^H \mathbf{M}_k \mathbf{v}_i\right|^2 \tag{9.2.9}$$

under the same unitary constraint. It is important to note that the preceding definition of JD does not require the matrix set under consideration to be exactly and simultaneously diagonalized by a single unitary matrix. This is so because we do not require the off-diagonal elements of all the matrices to be cancelled by a unitary transform; a joint diagonalizer is simply a minimizer of the criterion. If the matrices in $\mathbf{M}$ are not exactly joint diagonalizable, the criterion cannot be zeroed, and the matrices can only be approximately joint diagonalized. Hence, an (approximate) joint diagonalizer defines a kind of average eigenstructure.

9.2.3.2 Joint block-diagonalization

For direction finding methods such as t-f MUSIC, the source TFD matrix should not be singular but not necessarily diagonal. In this case, the joint block-diagonalization (JBD) is used to incorporate multiple t-f points rather than JD [9]. The JBD is achieved by the maximization under unitary transform of the following criterion:

$$C(\mathbf{U}) = \sum_k \sum_{i,l} \left|\mathbf{u}_i^H \mathbf{M} \mathbf{u}_l\right|^2 \tag{9.2.10}$$

over the set of unitary matrices $\mathbf{U} = [\mathbf{u}_1, \ldots, \mathbf{u}_n]$.

9.2.3.3 Time-frequency averaging

Time-frequency averaging is a linear operation that adds the STFDs over a t-f region where, typically, the desired signal is highly localized and the cross-terms are negligible. The averaged STFD is defined as

$$\mathbf{D} = \frac{1}{A} \sum_{(t,f)\in\Omega} \mathbf{D}_{\mathbf{xx}}(t, f), \tag{9.2.11}$$

where Ω is the t-f region of interest and A is a normalization constant that, for example, can be chosen as the total number of (t, f) points in the region Ω. The eigendecomposition of $\mathbf{D}$ is addressed in Subsection 9.3.3.

9.3 Properties of STFDs

To understand the properties of STFDs, we consider the case of frequency modulated (FM) signals and the simplest form of TFD, namely, the pseudo Wigner–Ville distribution (PWVD) [14]. The consideration of FM signals is motivated by the fact that these signals are uniquely characterized by their IFs, and therefore, they have clear t-f signatures that can be utilized by the STFD approach. Also, FM signals have constant amplitudes.

The FM signals can be modeled as

$$\mathbf{d}(t) = [d_1(t), \ldots, d_n(t)]^T = [D_1 e^{j\psi_1(t)}, \ldots, D_n e^{j\psi_n(t)}]^T, \tag{9.3.1}$$

where D_i and $\psi_i(t)$ are the fixed amplitude and time-varying phase of the ith source signal. For each sampling time t, $d_i(t)$ has an IF $f_i(t) = d\psi_i(t)/(2\pi dt)$.

The discrete form of the PWVD of a signal $x(t)$, using a rectangular window of odd length L, is a special case of (9.2.3) and is given by

$$D_{xx}(t, f) = \sum_{\tau=-(L-1)/2}^{(L-1)/2} x(t+\tau)x^*(t-\tau)e^{-j4\pi f\tau}. \tag{9.3.2}$$

Similarly, the spatial PWVD (SPWVD) matrix is obtained by replacing $x(t)$ by the data snapshot vector $\mathbf{x}(t)$,

$$\mathbf{D}_{\mathbf{xx}}(t, f) = \sum_{\tau=-(L-1)/2}^{(L-1)/2} \mathbf{x}(t+\tau)\mathbf{x}^H(t-\tau)e^{-j4\pi f\tau}. \tag{9.3.3}$$

9.3.1 Subspace analysis for FM signals

Analysis of the eigendecomposition of the STFD matrix is closely related to the analysis of subspace decomposition of the covariance matrix [15]. Before elaborating on this relationship, we present the case of FM signals using the conventional covariance matrix approach.

In Eq. (9.2.1), it is assumed that the number of sensors is greater than the number of sources, i.e., $m > n$. Further, matrix $\mathbf{A}$ is full column rank. We further assume that the correlation matrix

$$\mathbf{R}_{\mathbf{xx}} = E[\mathbf{x}(t)\mathbf{x}^H(t)] \tag{9.3.4}$$

is nonsingular, and the observation period consists of N snapshots with $N > m$. Under these assumptions, the correlation matrix is given by

$$\mathbf{R}_{\mathbf{xx}} = E[\mathbf{x}(t)\mathbf{x}^H(t)] = \mathbf{A}\mathbf{R}_{\mathbf{dd}}\mathbf{A}^H + \sigma\mathbf{I}, \tag{9.3.5}$$

where $\mathbf{R}_{\mathbf{dd}} = E[\mathbf{d}(t)\mathbf{d}^H(t)]$ is the source correlation matrix.

Let $\lambda_1 > \lambda_2 > \cdots > \lambda_n > \lambda_{n+1} = \lambda_{n+2} = \cdots = \lambda_m = \sigma$ denote the eigenvalues of $\mathbf{R}_{\mathbf{xx}}$. It is assumed that λ_i, $i = 1, \ldots, n$, are distinct. The unit-norm eigenvectors associated with $\lambda_1, \ldots, \lambda_n$ constitute the columns of matrix $\mathbf{S} = [\mathbf{s}_1, \ldots, \mathbf{s}_n]$, and those corresponding to $\lambda_{n+1}, \ldots, \lambda_m$ make up matrix $\mathbf{G} = [\mathbf{g}_1, \ldots, \mathbf{g}_{m-n}]$. Since the columns of $\mathbf{A}$ and $\mathbf{S}$ span the same subspace, then $\mathbf{A}^H\mathbf{G} = \mathbf{0}$.

In practice, $\mathbf{R_{xx}}$ is unknown and therefore should be estimated from the available data samples (snapshots) $\mathbf{x}(i)$, $i = 1, 2, \ldots, N$. The estimated correlation matrix is given by

$$\hat{\mathbf{R}}_{\mathbf{xx}} = \frac{1}{N} \sum_{i=1}^{N} \mathbf{x}(i)\mathbf{x}^H(i). \tag{9.3.6}$$

Let $\{\hat{\mathbf{s}}_1, \ldots, \hat{\mathbf{s}}_n, \hat{\mathbf{g}}_1, \ldots, \hat{\mathbf{g}}_{m-n}\}$ denote the unit-norm eigenvectors of $\hat{\mathbf{R}}_{\mathbf{xx}}$, arranged in the descending order of the associated eigenvalues, and let $\hat{\mathbf{S}}$ and $\hat{\mathbf{G}}$ denote the matrices defined by the set of vectors $\{\hat{\mathbf{s}}_i\}$ and $\{\hat{\mathbf{g}}_i\}$, respectively. The statistical properties of the eigenvectors of the sample covariance matrix $\hat{\mathbf{R}}_{\mathbf{xx}}$ for signals modeled as independent processes with additive white noise are given in [15].

We assume that the transmitted signals propagate in a stationary environment and are mutually uncorrelated over the observation period $1 \leq t \leq N$. Subsequently, the corresponding covariance matrices are time independent. Under these assumptions,

$$\frac{1}{N} \sum_{k=1}^{N} d_i(k) d_j^*(k) = 0 \qquad \text{for } i \neq j,\ i, j = 1, \ldots, n. \tag{9.3.7}$$

In this case, the signal correlation matrix is

$$\mathbf{R_{dd}} = \lim_{T \to \infty} \frac{1}{T} \sum_{t=1}^{T} \mathbf{d}(t)\mathbf{d}^H(t) = \text{diag}\left[D_i^2,\ i = 1, 2, \ldots, n\right], \tag{9.3.8}$$

where diag$[\cdot]$ is the diagonal matrix formed with the elements of its vector-valued arguments. From these assumptions, we have the following lemma.

Lemma 9.1 ([14]). *For uncorrelated FM signals with additive white Gaussian noise,*

(a) *The estimation errors $(\hat{\mathbf{s}}_i - \mathbf{s}_i)$ are asymptotically (for large N) jointly Gaussian distributed with zero means and covariance matrices given by*

$$E\left[(\hat{\mathbf{s}}_i - \mathbf{s}_i)(\hat{\mathbf{s}}_j - \mathbf{s}_j)^H\right] = \frac{\sigma}{N}\left[\sum_{\substack{k=1 \\ k \neq i}}^{n} \frac{\lambda_i + \lambda_k - \sigma}{(\lambda_k - \lambda_i)^2} \mathbf{s}_k \mathbf{s}_k^H + \sum_{k=1}^{m-n} \frac{\lambda_i}{(\sigma - \lambda_i)^2} \mathbf{g}_k \mathbf{g}_k^H\right] \delta_{i,j}, \tag{9.3.9}$$

$$E\left[(\hat{\mathbf{s}}_i - \mathbf{s}_i)(\hat{\mathbf{s}}_j - \mathbf{s}_j)^T\right] = -\frac{\sigma}{N} \frac{(\lambda_i + \lambda_j - \sigma)}{(\lambda_j - \lambda_i)^2} \mathbf{s}_j \mathbf{s}_i^T (1 - \delta_{i,j}), \tag{9.3.10}$$

where

$$\delta_{i,j} = \begin{cases} 1, & i = j, \\ 0, & i \neq j. \end{cases}$$

(b) *The orthogonal projections of $\{\hat{\mathbf{g}}_i\}$ onto the column space of $\mathbf{S}$ are asymptotically (for large N) jointly Gaussian distributed with zero means and covariance matrices given by*

$$E\left[\left(\mathbf{S}\mathbf{S}^H\hat{\mathbf{g}}_i\right)\left(\mathbf{S}\mathbf{S}^H\hat{\mathbf{g}}_j\right)^H\right] = \frac{\sigma}{N}\left[\sum_{k=1}^{n}\frac{\lambda_k}{(\sigma-\lambda_k)^2}\mathbf{s}_k\mathbf{s}_k^H\right]\delta_{i,j} \stackrel{\text{def}}{=} \frac{1}{N}\mathbf{U}\delta_{i,j}, \tag{9.3.11}$$

$$E\left[\left(\mathbf{S}\mathbf{S}^H\hat{\mathbf{g}}_i\right)\left(\mathbf{S}\mathbf{S}^H\hat{\mathbf{g}}_j\right)^T\right] = 0 \qquad \textit{for all } i, j. \tag{9.3.12}$$

Equations (9.3.9) and (9.3.10) hold a strong similarity to those of [15]. The only difference is that the term $(\lambda_i\lambda_k)$ in [15] is replaced by $\sigma(\lambda_i + \lambda_k - \sigma)$ in (9.3.9) and (9.3.10), due to the uncorrelation property (9.3.7). Equations (9.3.11) and (9.3.12) are identical to those derived in reference [15].

9.3.2 Signal-to-noise ratio enhancement

The TFD maps one-dimensional (1D) signals in the time domain into two-dimensional (2D) signals in the t-f domain. The TFD property of concentrating the input signal around its IF while spreading the noise over the entire t-f domain increases the effective signal-to-noise ratio (SNR) and proves valuable in the underlying problem.

The ith diagonal element of PWVD matrix $\mathbf{D}_{\mathbf{dd}}(t, f)$ is given by

$$D_{d_id_i}(t, f) = \sum_{\tau=-(L-1)/2}^{(L-1)/2} D_i^2 e^{j[\psi_i(t+\tau)-\psi_i(t-\tau)]-j4\pi f\tau}. \tag{9.3.13}$$

Assume that the third-order derivative of the phase is negligible over the window length L. Then along the true t-f points of the ith signal, $f_i(t) = d\psi_i(t)/(2\pi dt)$, and $\psi_i(t+\tau)-\psi_i(t-\tau)-4\pi f_i(t)\tau = 0$. Accordingly, for $(L-1)/2 \le t \le N-(L-1)/2$,

$$D_{d_id_i}(t, f_i(t)) = \sum_{\tau=-(L-1)/2}^{(L-1)/2} D_i^2 = LD_i^2. \tag{9.3.14}$$

Similarly, the noise SPWVD matrix $\mathbf{D}_{\mathbf{nn}}(t, f)$ is

$$\mathbf{D}_{\mathbf{nn}}(t, f) = \sum_{\tau=-(L-1)/2}^{(L-1)/2} \mathbf{n}(t+\tau)\mathbf{n}^H(t-\tau)e^{-j4\pi f\tau}. \tag{9.3.15}$$

Under the spatially and temporally white assumptions, the statistical expectation of $\mathbf{D}_{\mathbf{nn}}(t, f)$ is given by

$$E\left[\mathbf{D}_{\mathbf{nn}}(t, f)\right] = \sum_{\tau=-(L-1)/2}^{(L-1)/2} E\left[\mathbf{n}(t+\tau)\mathbf{n}^H(t-\tau)\right]e^{-j4\pi f\tau} = \sigma\mathbf{I}. \tag{9.3.16}$$

Therefore, when we select the t-f points along the t-f signature or the IF of the ith FM signal, the SNR in the model (9.2.6) becomes LD_i^2/σ, which has an improved factor L over the one associated with model (9.3.5). The IF of the FM signals can be estimated from the employed TFD, or using any appropriate IF estimator. It is noted, however, that the STFD equation (9.2.6) provides a natural platform for the direct incorporation of any *a priori* information or estimates of the IF into direction-of-arrival (DOA) estimation.

The PWVD of each FM source has a constant value over the observation period, providing that we leave out the rising and falling power distributions at both ends of the data record. For convenience of analysis, we select those $N' = N - L + 1$ t-f points of constant distribution value for each source signal. In the case where the STFD matrices are averaged over the t-f signatures of n_o sources, i.e., a total of $n_o N'$ t-f points, the result is given by

$$\hat{\mathbf{D}} = \frac{1}{n_o N'} \sum_{q=1}^{n_o} \sum_{i=1}^{N'} \mathbf{D}_{\mathbf{xx}}(t_i, f_{q,i}(t_i)), \tag{9.3.17}$$

where $f_{q,i}(t_i)$ is the IF of the qth signal at the ith time sample. $\mathbf{x}(t)$ is an instantaneous mixture of the FM signals $d_i(t), i = 1, \ldots, n$, and hence features the same IFs. The expectation of the averaged STFD matrix is

$$\begin{aligned} \mathbf{D} = E\left[\hat{\mathbf{D}}\right] &= \frac{1}{n_o N'} \sum_{q=1}^{n_o} \sum_{i=1}^{N'} E\left[\mathbf{D}_{\mathbf{xx}}(t_i, f_{q,i}(t_i))\right] \\ &= \frac{1}{n_o} \sum_{q=1}^{n_o} \left[L D_q^2 \mathbf{a}_q \mathbf{a}_q^H + \sigma \mathbf{I}\right] = \frac{L}{n_o} \mathbf{A}^o \mathbf{R}_{\mathbf{dd}}^o (\mathbf{A}^o)^H + \sigma \mathbf{I}, \end{aligned} \tag{9.3.18}$$

where $\mathbf{R}_{\mathbf{dd}}^o = \text{diag}\left[D_i^2, i = 1, 2, \ldots, n_o\right]$ and $\mathbf{A}^o = \left[\mathbf{a}_1, \mathbf{a}_2, \ldots, \mathbf{a}_{n_o}\right]$ represent the signal correlation matrix and the mixing matrix formulated by considering n_o signals out of the total number of n signal arrivals, respectively.

It is clear from (9.3.18) that the SNR improvement $G = L/n_o$ (we assume $L > n_o$) is inversely proportional to the number of sources contributing to matrix $\mathbf{D}$. Therefore, from the SNR perspective, it is best to set $n_o = 1$, i.e., to select the sets of N' t-f points that belong to individual signals one set at a time, and then separately evaluate the respective STFD matrices.

This procedure is made possible by the fact that STFD-based array processing is, in essence, a discriminatory technique in the sense that it does not require simultaneous localization and extraction of all unknown signals received by the array. With STFDs, array processing can be performed using STFDs of a subclass of the impinging signals with specific t-f signatures. In this respect, the t-f based blind source separation and direction finding techniques have implicit spatial filtering, removing the undesired signals from consideration. It is also important to note that with the ability to construct the STFD matrix from one or few signal arrivals, the well-known $m > n$ condition on source localization using arrays can be relaxed to $m > n_o$, i.e., we can perform direction finding or source separation with the number of array sensors smaller than the number of impinging signals. Further, from the angular resolution perspective, closely spaced sources with different t-f signatures can be resolved by constructing two separate STFDs, each corresponding to one source, and then proceed with subspace decomposition for each STFD matrix, followed by an appropriate source localization method (MUSIC, for example). The drawback of using different STFD matrices separately is, of course, the need for repeated computations.

9.3.3 Signal and noise subspaces using STFDs

The following lemma provides the relationship between the eigendecompositions of the STFD matrices and the data covariance matrices used in conventional array processing.

Lemma 9.2 ([14]). *Let* $\lambda_1^o > \lambda_2^o > \cdots > \lambda_{n_o}^o > \lambda_{n_o+1}^o = \lambda_{n_o+2}^o = \cdots = \lambda_m^o = \sigma$ *denote the eigenvalues of* $\mathbf{R}_{\mathbf{xx}}^o = \mathbf{A}^o \mathbf{R}_{\mathbf{dd}}^o (\mathbf{A}^o)^H + \sigma \mathbf{I}$, *which is defined from a data record of a mixture of the* n_o *selected FM signals. Denote the unit-norm eigenvectors associated with* $\lambda_1^o, \ldots, \lambda_{n_o}^o$ *by the columns of* $\mathbf{S}^o = [\mathbf{s}_1^o, \ldots, \mathbf{s}_{n_o}^o]$, *and those corresponding to* $\lambda_{n_o+1}^o, \ldots, \lambda_m^o$ *by the columns of* $\mathbf{G}^o = [\mathbf{g}_1^o, \ldots, \mathbf{g}_{m-n_o}^o]$. *We also denote* $\lambda_1^{tf} > \lambda_2^{tf} > \cdots > \lambda_{n_o}^{tf} > \lambda_{n_o+1}^{tf} = \lambda_{n_o+2}^{tf} = \cdots = \lambda_m^{tf} = \sigma^{tf}$ *as the eigenvalues of* $\mathbf{D}$ *defined in (9.3.18). The superscript* ^{tf} *denotes that the associated term is derived from the STFD matrix* $\mathbf{D}$. *The unit-norm eigenvectors associated with* $\lambda_1^{tf}, \ldots, \lambda_{n_o}^{tf}$ *are represented by the columns of* $\mathbf{S}^{tf} = [\mathbf{s}_1^{tf}, \ldots, \mathbf{s}_{n_o}^{tf}]$, *and those corresponding to* $\lambda_{n_o+1}^{tf}, \ldots, \lambda_m^{tf}$ *are represented by the columns of* $\mathbf{G}^{tf} = [\mathbf{g}_1^{tf}, \ldots, \mathbf{g}_{m-n_o}^{tf}]$. *Then,*

(a) *The signal and noise subspaces of* $\mathbf{S}^{tf}$ *and* $\mathbf{G}^{tf}$ *are the same as* $\mathbf{S}^o$ *and* $\mathbf{G}^o$, *respectively.*

(b) *The eigenvalues have the following relationship:*

$$\lambda_i^{tf} = \begin{cases} \dfrac{L}{n_o}(\lambda_i^o - \sigma) + \sigma = \dfrac{L}{n_o}\lambda_i^o + \left(1 - \dfrac{L}{n_o}\right)\sigma & i \le n_o \\ \sigma^{tf} = \sigma & n_o < i \le m. \end{cases} \tag{9.3.19}$$

An important conclusion from Lemma 9.2 is that, the largest n_o eigenvalues are amplified using STFD analysis. The amplification of the largest n_o eigenvalues improves detection of the number of the impinging signals on the array, as it widens the separation between dominant and noise-level eigenvalues. Determination of the number of signals is key to establishing the proper signal and noise subspaces and subsequently plays a fundamental role in subspace-based applications. When the input SNR is low, or the signals are closely spaced, the number of signals may often be underdetermined. When the STFD is applied, the SNR threshold level and/or angle separation necessary for the correct determination of the number of signals are greatly reduced.

Next we consider the signal and noise subspace estimates from a finite number of data samples. We form the STFD matrix based on the true (t, f) points along the IF of the n_o FM signals.

Lemma 9.3 ([10, 14]). *If the third-order derivative of the phase of the FM signals is negligible over the time period* $[t-L+1, t+L-1]$, *then*

(a) *The estimation errors in the signal vectors are asymptotically (for* $N \gg L$*) jointly Gaussian distributed with zero means and covariance matrices given by*

$$\begin{aligned}
&E\left(\hat{\mathbf{s}}_i^{tf}-\mathbf{s}_i^{tf}\right)\left(\hat{\mathbf{s}}_j^{tf}-\mathbf{s}_j^{tf}\right)^H \\
&\quad=\frac{\sigma L}{n_o N'}\left[\sum_{\substack{k=1\\k\neq i}}^{n_o}\frac{\lambda_i^{tf}+\lambda_k^{tf}-\sigma}{(\lambda_k^{tf}-\lambda_i^{tf})^2}\mathbf{s}_k^{tf}\left(\mathbf{s}_k^{tf}\right)^H+\sum_{k=1}^{m-n_o}\frac{\lambda_i^{tf}}{(\sigma-\lambda_i^{tf})^2}\mathbf{g}_k^{tf}\left(\mathbf{g}_k^{tf}\right)^H\right]\delta_{i,j} \\
&\quad=\frac{\sigma}{N'}\left[\sum_{\substack{k=1\\k\neq i}}^{n_o}\frac{(\lambda_i^o-\sigma)+(\lambda_k^o-\sigma)+\frac{n_o}{L}\sigma}{(\lambda_k^o-\lambda_i^o)^2}\mathbf{s}_k^o\left(\mathbf{s}_k^o\right)^H\right. \\
&\qquad\left.+\sum_{k=1}^{m-n_o}\frac{(\lambda_i^o-\sigma)+\frac{n_o}{L}\sigma}{(\sigma-\lambda_i^o)^2}\mathbf{g}_k^o\left(\mathbf{g}_k^o\right)^H\right]\delta_{i,j},
\end{aligned} \tag{9.3.20}$$

and

$$\begin{aligned}
&E\left(\hat{\mathbf{s}}_i^{tf}-\mathbf{s}_i^{tf}\right)\left(\hat{\mathbf{s}}_j^{tf}-\mathbf{s}_j^{tf}\right)^T \\
&\quad=-\frac{\sigma L}{n_o N'}\frac{(\lambda_i^{tf}+\lambda_j^{tf}-\sigma)}{(\lambda_j^{tf}-\lambda_i^{tf})^2}\mathbf{s}_j^{tf}\left(\mathbf{s}_i^{tf}\right)^T(1-\delta_{i,j}) \\
&\quad=-\frac{\sigma}{N'}\cdot\frac{(\lambda_k^o-\sigma)+(\lambda_i^o-\sigma)+\frac{n_o}{L}\sigma}{(\lambda_k^o-\lambda_i^o)^2}\mathbf{s}_j^o\left(\mathbf{s}_i^o\right)^T(1-\delta_{i,j}).
\end{aligned} \tag{9.3.21}$$

(b) *The orthogonal projections of* $\left\{\hat{\mathbf{g}}_i^{tf}\right\}$ *onto the column space of* $\mathbf{S}^{tf}$ *are asymptotically (for* $N \gg L$*) jointly Gaussian distributed with zero means and covariance matrices given by*

$$\begin{aligned}
&E\left(\mathbf{S}^{tf}\left(\mathbf{S}^{tf}\right)^H\hat{\mathbf{g}}_i^{tf}\right)\left(\mathbf{S}^{tf}\left(\mathbf{S}^{tf}\right)^H\hat{\mathbf{g}}_j^{tf}\right)^H \\
&\quad=\frac{\sigma L}{n_o N'}\left[\sum_{k=1}^{n_o}\frac{\lambda_k^{tf}}{(\sigma-\lambda_k^{tf})^2}\mathbf{s}_k^{tf}\left(\mathbf{s}_k^{tf}\right)^H\right]\delta_{i,j} \\
&\quad=\frac{\sigma}{N'}\left[\sum_{k=1}^{n_o}\frac{(\lambda_k^o-\sigma)+\frac{n_o}{L}\sigma}{(\sigma-\lambda_k^o)^2}\mathbf{s}_k^o\left(\mathbf{s}_k^o\right)^H\right]\delta_{i,j} \\
&\quad\overset{\text{def}}{=}\frac{1}{N'}\mathbf{U}^{tf}\delta_{i,j},
\end{aligned} \tag{9.3.22}$$

$$E\left(\mathbf{S}^{tf}\left(\mathbf{S}^{tf}\right)^H\hat{\mathbf{g}}_i^{tf}\right)\left(\mathbf{S}^{tf}\left(\mathbf{S}^{tf}\right)^H\hat{\mathbf{g}}_j^{tf}\right)^T=\mathbf{0} \quad \textit{for all } i,j. \tag{9.3.23}$$

From (9.3.20)–(9.3.23), two important observations are in order. First, if the signals are both localizable and separable in the t-f domain, then the reduction of the number of signals from n to n_o greatly reduces the estimation error, specifically when the signals are closely spaced. The second observation relates to SNR enhancements. The preceding equations show that error reductions using STFDs are more pronounced for the cases of low SNR and/or closely spaced signals. It is clear from (9.3.20)–(9.3.23) that, when $\lambda_k^o \gg \sigma$ for all $k = 1, 2, \ldots, n_o$, the results are almost independent of L (suppose $N \gg L$ so that $N' = N - L + 1 \simeq N$), and therefore there would be no obvious improvement in using the STFD over conventional array processing. On the other hand, when some of the eigenvalues are close to σ ($\lambda_k^o \simeq \sigma$, for some $k = 1, 2, \ldots, n_o$), which is the case of weak or closely spaced signals, all the results of the preceding equations are reduced by a factor of up to $G = L/n_o$, respectively. This factor represents, in essence, the gain achieved from using STFD processing.

9.4 Blind source separation

9.4.1 Source separation based on STFDs

Blind source separation based on STFDs was first considered by Belouchrani and Amin [8]. The first step of STFD-based blind source separation is the whitening of the signal part $\mathbf{y}(t)$ of the observation. This is achieved by applying a whitening matrix $\mathbf{W}$ to $\mathbf{y}(t)$, i.e., an $n \times N$ matrix satisfying

$$\lim_{T \to \infty} \frac{1}{T} \sum_{t=1}^{T} \mathbf{W}\mathbf{y}(t)\mathbf{y}^H(t)\mathbf{W}^H = \mathbf{W}\mathbf{R}_{\mathbf{yy}}\mathbf{W}^H = \mathbf{W}\mathbf{A}\mathbf{A}^H\mathbf{A}^H = \mathbf{I}. \quad (9.4.1)$$

$\mathbf{WA}$ is an $n \times n$ unitary matrix $\mathbf{U}$, and matrix $\mathbf{A}$ can be written as

$$\mathbf{A} = \mathbf{W}^{\#}\mathbf{U} \quad (9.4.2)$$

where the superscript $\#$ denotes pseudo-inverse. The whitened process $\mathbf{z}(t) = \mathbf{W}\mathbf{x}(t)$ still obeys a linear model,

$$\mathbf{z}(t) = \mathbf{W}\mathbf{x}(t) = \mathbf{W}\left[\mathbf{A}\mathbf{s}(t) + \mathbf{n}(t)\right] = \mathbf{U}\mathbf{s}(t) + \mathbf{W}\mathbf{n}(t). \quad (9.4.3)$$

By pre- and postmultiplying the STFD matrices $\mathbf{D}_{\mathbf{xx}}(t, f)$ by $\mathbf{W}$, we obtain

$$\mathbf{D}_{\mathbf{zz}}(t, f) = \mathbf{W}\mathbf{D}_{\mathbf{xx}}(t, f)\mathbf{W}^H, \quad (9.4.4)$$

which is, in essence, the STFD of the whitened data vector $\mathbf{z}(t)$. From the definitions of $\mathbf{W}$ and $\mathbf{U}$,

$$\mathbf{D}_{\mathbf{zz}}(t, f) = \mathbf{U}\mathbf{D}_{\mathbf{ss}}(t, f)\mathbf{U}^H. \quad (9.4.5)$$

Equation (9.4.5) shows that if $\mathbf{D}_{\mathbf{ss}}(t, f)$ is diagonal, which is the case for auto-term points, then any whitened data STFD matrix is diagonal in the basis of the columns of the matrix $\mathbf{U}$, and the eigenvalues of $\mathbf{D_{zz}}(t, f)$ are the diagonal entries of $\mathbf{D_{ss}}(t, f)$. An estimate $\hat{\mathbf{U}}$ of the unitary matrix $\mathbf{U}$ may be obtained as a unitary diagonalizing matrix of a whitening STFD matrix for some t-f points corresponding to the signal auto-term. The source signals can then be estimated as $\hat{\mathbf{s}}(t) = \hat{\mathbf{U}}\hat{\mathbf{W}}\mathbf{x}(t)$, and the mixing matrix $\mathbf{A}$ is estimated by $\hat{\mathbf{A}} = \hat{\mathbf{W}}^{\#}\hat{\mathbf{U}}$.

In order to reduce the noise effect as well as the possibility of having degenerate eigenvalues and subsequently nonunique solutions, the JD and t-f averaging, both discussed in Subsection 9.2.3, can be used to incorporate multiple t-f points.

The method discussed above uses STFD matrices to estimate the unitary matrix $\mathbf{U}$, but the covariance matrix is still used for whitening. Therefore, the advantages of STFD matrices are not fully utilized. Using the STFD matrix $\mathbf{D}$, instead of the covariance matrix $\mathbf{R_{xx}}$, to perform whitening is a reasonable alternative [11]. To avoid degenerate eigenvalues, the STFD matrices used for prewhitening and unitary matrix estimation should be different.

Example. Figure 9.1 shows an example of the application of STFDs to the blind source separation problem. A three-element equispaced linear array is considered where the interelement spacing is half a wavelength. Two chirp signals arrive at -10° and 10°, respectively. The number of data samples used to compute the STFD is 128. The number of t-f points employed in the JD is $p = 128$, with an equal number of points on each signature. Figure 9.1(b) shows the Choi–Williams distributions, in which an exponential kernel is applied [16] of two linear mixtures of the original chirp signals depicted in Figure 9.1(a), corresponding to the data at the first and the second sensors. Using JD of the STFDs, we are able to recover the original signals from their observed mixture, as shown in Figure 9.1(c).

9.4.2 Source separation based on spatial averaging

Source separation based on spatial averaging is proposed by Mu, Amin, and Zhang [17]. This method first performs array averaging of the TFDs of the data across the array, permitting the spatial signature of sources to play a fundamental role in improving the synthesis performance.

This method is philosophically different from the one described in the previous section, as it applies the opposite order of operations. It first synthesizes the source signals from the t-f domain, then proceeds to estimate their respective mixing matrix.

The WVD-based synthesis techniques can be found in [18, 19]. Herein, we apply the method of extended discrete-time Wigner distribution (EDTWD), introduced in [19], to the output of the array-averaged WVD. The advantage of using the EDTWD is that it does not require *a priori* knowledge of the source waveform and thereby avoids the problem of matching the two "uncoupled" vectors (even-indexed and odd-indexed vectors).

The overall synthesis procedure is summarized in the following steps.

1. Given the received data of the ith sensor $x_i(t)$, compute the EDTWD

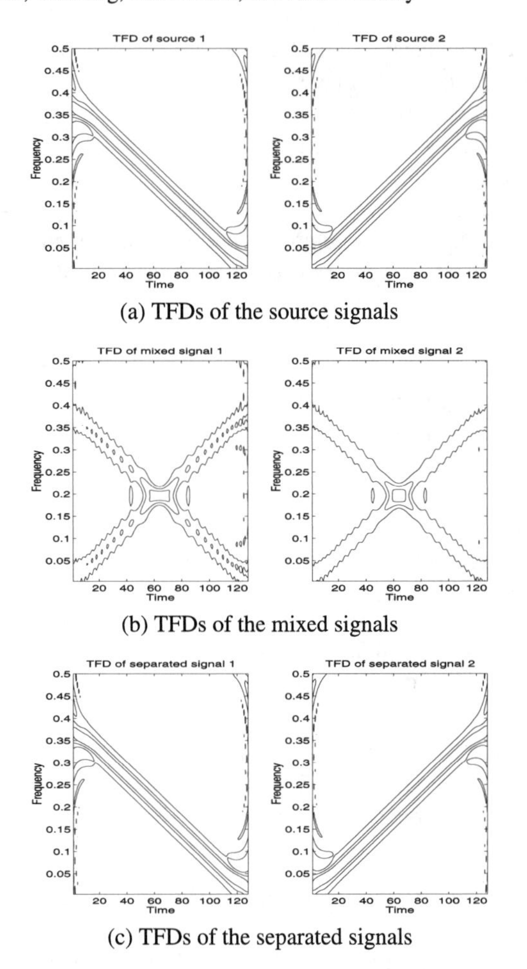

(a) TFDs of the source signals

(b) TFDs of the mixed signals

(c) TFDs of the separated signals

FIGURE 9.1. Blind source separation based on STFDs.

$$W_{x_i x_i}(t,f) = \sum_{k:(t+k/2)\in Z} x_i(t+k/2)x_i^*(t-k/2)e^{-j2\pi kf},$$

$$t = 0, \pm 0.5, \pm 1, \ldots. \quad (9.4.6)$$

2. Apply the averaging process, that is, summing the EDTWD across the array

$$\bar{W}(t,f) = \frac{1}{m}\sum_{k=1}^{m} W_{x_k x_k}(t,f). \quad (9.4.7)$$

3. Place an appropriate t-f mask on $\bar{W}(t,f)$ such that only the desired signal auto-terms are retained.
4. Take the inverse fast Fourier transform (IFFT) of the masked WVD $\bar{W}(t,f)$

$$p(t,\tau) = \sum_f \bar{W}(t,f)e^{j2\pi\tau f}. \tag{9.4.8}$$

5. Construct the matrix $\mathbf{Q} = [q_{il}]$ with

$$q_{il} = p\left(\frac{i+l}{2}, i-l\right). \tag{9.4.9}$$

6. Apply eigendecomposition to the Hermitian matrix $[\mathbf{Q} + \mathbf{Q}^H]$ and obtain the maximum eigenvalue $\lambda_{\max}$ and the associated eigenvector $\mathbf{u}$. The desired signal is given by

$$\hat{\mathbf{s}}_{opt} = e^{j\alpha}\sqrt{2\lambda_{\max}}\ \mathbf{u}, \tag{9.4.10}$$

where α is an unknown value representing the phase.
7. Repeat steps 3 through 6 until all source signals $\hat{d}_1(t), \hat{d}_2(t), \ldots, \hat{d}_L(t)$ are retrieved.

The averaging in step 2 mitigates the cross-terms and enforces the auto-terms. As such, the source t-f signatures become easier to identify, mask, and synthesize. It is noteworthy that (9.4.7) will completely suppress the cross-TFDs for sources with orthogonal spatial signatures.

Upon synthesizing all the source signals, we could utilize these signal waveforms to estimate the mixing, or array, matrix $\mathbf{A}$ through the minimization of the mean-squared error (MSE),

$$\varepsilon = \sum_{t=1}^{N} \|\mathbf{x}(t) - \mathbf{A}\hat{\mathbf{s}}(t)\|^2. \tag{9.4.11}$$

This results in

$$\hat{\mathbf{A}} = \hat{\mathbf{r}}\hat{\mathbf{R}}^{-1}, \tag{9.4.12}$$

where $\hat{\mathbf{R}} = \sum_{t=1}^{N} \hat{\mathbf{d}}(t)\hat{\mathbf{d}}^H(t)$ represents the estimated signal source covariance matrix, and $\hat{\mathbf{r}} = [\hat{\mathbf{r}}_1, \ldots, \hat{\mathbf{r}}_L]$, with $\hat{\mathbf{r}}_i = \sum_{t=1}^{N} \mathbf{x}(t)\hat{d}_i^*(t)$ being the correlation vector between the data vector received across the array and the ith source signal $\hat{d}_i(t)$.

Example. We consider three parallel chirp signals. The signals arrive with DOAs of $-20°$, $0°$, and $20°$, with the respective start and end frequencies given by $(0.9\pi, 0.5\pi)$, $(0.66\pi, 0.26\pi)$, and $(0.5\pi, 0.1\pi)$, respectively. The length of the signal sequence is set to $N = 128$. The input SNR is -5 dB. The cross-term of $d_1(t)$ and $d_3(t)$ lies closely to the t-f signature of $d_2(t)$.

We first consider the single antenna case. Figure 9.2 depicts both the WVD of the signal arrival and the WVD of the synthesized $\hat{d}_2(t)$. The signal is significantly corrupted by the cross-term of $d_1(t)$ and $d_3(t)$ as well as by the noise components.

Figure 9.3 depicts, for 16 sensor scenarios, the array-averaged WVD and the respective $\hat{d}_2(t)$. Upon averaging, both noise and cross-terms are sufficiently reduced to clearly manifest the individual source t-f signatures. The signals could, therefore, be individually recovered by placing appropriate masks in the t-f region. The significance of using array sensors is evident in Figure 9.3.

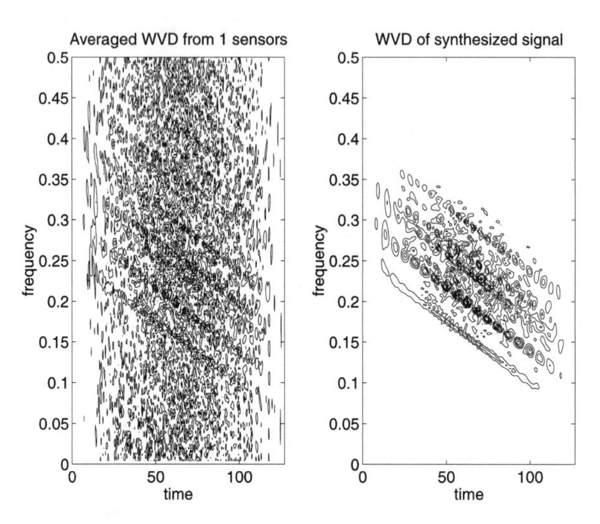

FIGURE 9.2. WVD and the synthesized signal ($M = 1$).

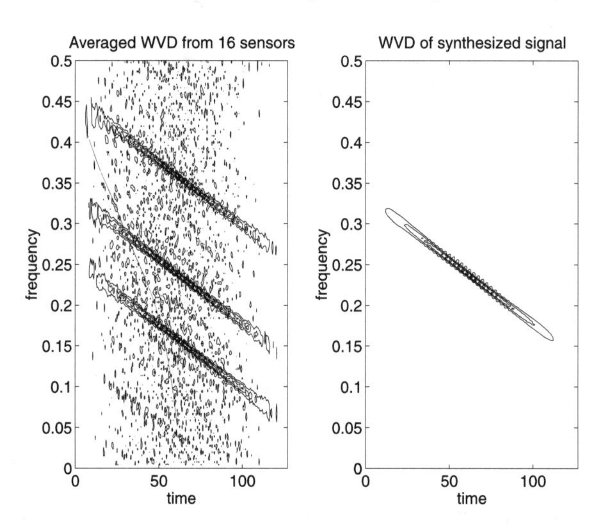

FIGURE 9.3. WVD and the synthesized signal ($M = 16$).

9.4.3 Effect of cross-terms between source signals

In this section, we examine the effect of the t-f cross-terms on source separation performance [20]. To simplify the problem, we assume that $\mathbf{R_{dd}}$ is an identity matrix. When cross-terms are present at the off-diagonal elements of the TFD matrix $\mathbf{D_{dd}}(t, f)$, then

$$\mathbf{D_{dd}}(t, f) = \mathbf{P}(t, f)\mathbf{G}(t, f)\mathbf{P}^H(t, f), \tag{9.4.13}$$

where $\mathbf{G}(t, f)$ is the diagonal matrix with the eigenvalues at the diagonal elements, and $\mathbf{P}(t, f)$ is the matrix whose columns are the corresponding eigenvectors. Note that all these matrices depend on the selected (t, f) point. From (9.4.13), the STFD matrix of the data vector under noise-free conditions becomes

$$\mathbf{D_{xx}}(t,f) = \mathbf{A}\mathbf{D_{ss}}(t,f)\mathbf{A}^H = \mathbf{A}\mathbf{P}(t,f)\mathbf{G}(t,f)\mathbf{P}^H(t,f)\mathbf{A}^H \tag{9.4.14}$$

and the STFD matrix of the whitened array signal vector is

$$\mathbf{D_{zz}}(t,f) = \mathbf{W}\mathbf{A}\mathbf{P}(t,f)\mathbf{G}(t,f)\mathbf{P}^H(t,f)\mathbf{A}^H\mathbf{W}^H. \tag{9.4.15}$$

Since $\mathbf{G}(t, f)$ is diagonal, $\mathbf{WAP}(t, f)$ is unitary. Therefore, the source separation method will assume $\mathbf{WAP}(t, f)$ as the unitary matrix and estimates the mixing matrix as

$$\hat{\mathbf{A}} = \mathbf{W}^{\#}\mathbf{W}\mathbf{A}\mathbf{P}(t,f) = \mathbf{A}\mathbf{P}(t,f), \tag{9.4.16}$$

which is dependent on the unitary matrix $\mathbf{P}(t, f)$. Furthermore,

$$\hat{\mathbf{A}}^{\#}\mathbf{A} = [\mathbf{A}\mathbf{P}(t,f)]^{\#}\,\mathbf{A} = \mathbf{P}^H(t,f). \tag{9.4.17}$$

Matrix $\hat{\mathbf{A}}$ should be close to the true one $\mathbf{A}$ so that $\hat{\mathbf{A}}^{\#}\mathbf{A}$ well approximates the identity matrix. The following variable measures, at separations, the ratio of the power of interference of the qth source signal to the power of the pth source signal [8]

$$I_{pq} = E\left|\left(\hat{\mathbf{A}}^{\#}\mathbf{A}\right)_{pq}\right|^2, \tag{9.4.18}$$

where $(\hat{\mathbf{A}}^{\#}\mathbf{A})_{pq}$ denotes the (p, q)th element of matrix $\hat{\mathbf{A}}^{\#}\mathbf{A}$. The following global rejection level is used to evaluate the overall performance of a blind source separation system

$$I_{perf} = \sum_{q \neq p} I_{pq}. \tag{9.4.19}$$

For the mixing matrix estimation given in (9.4.17), the global rejection level is approximated by the following normalized global rejection level [20]

$$I'_{perf} = \left[\text{diagonal}\left(\hat{\mathbf{A}}^{\#}\mathbf{A}\right)\right]\hat{\mathbf{A}}^{\#}\mathbf{A} = \sum_{q=1}^{n} |p_{qq}(t,f)|^{-2} - n, \tag{9.4.20}$$

where diagonal$(\mathbf{F})$ denotes the matrix formed by the diagonal elements of $\mathbf{F}$. In general, since the absolute values of $p_{qq}(t, f)$ are always equal to or smaller than 1, the global rejection level I_{perf} takes a positive value. It is clear that $I_{perf} = 0$ only when $p_{qq}(t, f) = 1$ holds true for all q. That is, $\mathbf{P}$ is an identity matrix, which implies that there are no off-diagonal nonzero elements in matrix $\mathbf{D_{dd}}(t, f)$, i.e., no cross-terms.

9.4.4 Source separation based on joint diagonalization and joint anti-diagonalization

From the previous subsection, it is clear that care must be exercised when dealing with cross-terms. The method proposed by Belouchrani, Abed-Meraim, Amin, and Zoubir [21] carefully and properly exploits both auto-terms and cross-terms of the TFDs for improved source separation. This approach is based on the simultaneous diagonalization and anti-diagonalization of a combined set of auto-term and cross-term TFD matrices, respectively.

The auto-STFD and cross-STFD are defined as

$$\mathbf{D}^a_{\mathbf{ss}}(t,f) = \mathbf{D}_{\mathbf{ss}}(t,f) \text{ for auto-term t-f points} \tag{9.4.21}$$

and

$$\mathbf{D}^c_{\mathbf{ss}}(t,f) = \mathbf{D}_{\mathbf{ss}}(t,f) \text{ for cross-term t-f points.} \tag{9.4.22}$$

Since the off-diagonal elements of $\mathbf{D}_{\mathbf{ss}}(t,f)$ are cross-terms, the auto-STFD matrix is quasi-diagonal for each t-f point that corresponds to a true power concentration, i.e., signal auto-term. Similarly, since the diagonal elements of $\mathbf{D}_{\mathbf{ss}}(t,f)$ are auto-terms, the cross-STFD matrix is quasi-anti-diagonal (i.e., its diagonal entries are close to zero) for each t-f point that corresponds to a cross-term.

As discussed earlier, JD can be used to incorporate multiple auto-term t-f points. Similarly, the joint anti-diagonalization (JAD) is appropriate to incorporate multiple cross-term t-f points. By selecting cross-term t-f points, the data cross-STFD will have the following structure:

$$\mathbf{D}^c_{\mathbf{xx}}(t,f) = \mathbf{U}\mathbf{D}^c_{\mathbf{ss}}(t,f)\mathbf{U}^H, \tag{9.4.23}$$

where $\mathbf{D}^c_{\mathbf{ss}}(t,f)$ is anti-diagonal. The JAD searches for the unitary matrix that anti-diagonalizes a combined set $\{\mathbf{D}^c_{\mathbf{xx}}(t_i,f_i)|i=1,\dots,q\}$ of q STFD matrices. The procedure for anti-diagonalization of a single $m \times m$ matrix $\mathbf{N}$ is explained in [21] and is equivalent to the maximization of the criterion

$$C(\mathbf{N},\mathbf{V}) \stackrel{\text{def}}{=} -\sum_{i=1}^{m} \left|\mathbf{v}_i^H \mathbf{N} \mathbf{v}_i\right|^2 \tag{9.4.24}$$

over the set of unitary matrices $\mathbf{V} = [\mathbf{v}_1, \dots, \mathbf{v}_m]$.

The combined JD and JAD of two sets $\{\mathbf{M}_k|k=1..p\}$ and $\{\mathbf{N}_k|k=1..q\}$ of $m \times m$ matrices is defined as the maximization of the JD/JAD criterion:

$$C(\mathbf{V}) \stackrel{\text{def}}{=} \sum_{i=1}^{m}\left(\sum_{k=1}^{p}\left|\mathbf{v}_i^H \mathbf{M}_k \mathbf{v}_i\right|^2 - \sum_{k=1}^{q}\left|\mathbf{v}_i^H \mathbf{N}_k \mathbf{v}_i\right|^2\right) \tag{9.4.25}$$

over the set of unitary matrices $\mathbf{V} = [\mathbf{v}_1, \dots, \mathbf{v}_m]$.

9.4.4.1 Selection of auto-term and cross-term points

The success of the JD or JAD of STFD matrices in determining the unitary matrix $\mathbf{U}$ depends strongly on the correct selection of the auto-term and cross-term points. Therefore, it is crucial to have a selection procedure that is able to distinguish between auto-term and cross-term points based only on the STFD matrices of the observation. A selection approach was proposed in [21] to exploit the anti-diagonal structure of the cross-term STFD matrices. More precisely,

$$\text{trace}(\mathbf{D}^c_{\mathbf{xx}}(t,f)) = \text{trace}(\mathbf{U}\mathbf{D}^c_{\mathbf{ss}}(t,f)\mathbf{U}^H) = \text{trace}(\mathbf{D}^c_{\mathbf{ss}}(t,f)) \approx 0.$$

Based on this observation, the following testing procedure can be defined:

$$\begin{aligned}
&\text{if } \frac{\text{trace}(\mathbf{D}_{\mathbf{xx}}(t,f))}{\text{norm}(\mathbf{D}_{\mathbf{xx}}(t,f))} < \epsilon \longrightarrow \text{ decide that } (t,f) \text{ is a cross-term} \\
&\text{if } \frac{\text{trace}(\mathbf{D}_{\mathbf{xx}}(t,f))}{\text{norm}(\mathbf{D}_{\mathbf{xx}}(t,f))} > \epsilon \longrightarrow \text{ decide that } (t,f) \text{ is an auto-term}
\end{aligned}$$

where ϵ is a “small” positive real scalar.

Example. We consider a uniform linear array of $m = 3$ sensors having half-wavelength spacing and receiving signals from $n = 2$ sources in the presence of white Gaussian noise. The sources arrive from different directions $\theta_1 = 10$ and $\theta_2 = 20$ degrees. The emitted signals are two chirps. The WVD is computed over 1024 samples, and eight STFD matrices are considered.

We compare in Figure 9.4 the performance of the JD-based algorithm, introduced in Subsection 9.4.1, and the JD/JAD algorithm, for SNRs in the range [5 – 20 dB]. The mean rejection levels are evaluated over 100 Monte Carlo runs. In this case, the new algorithm performs slightly better than the JD-based algorithm.

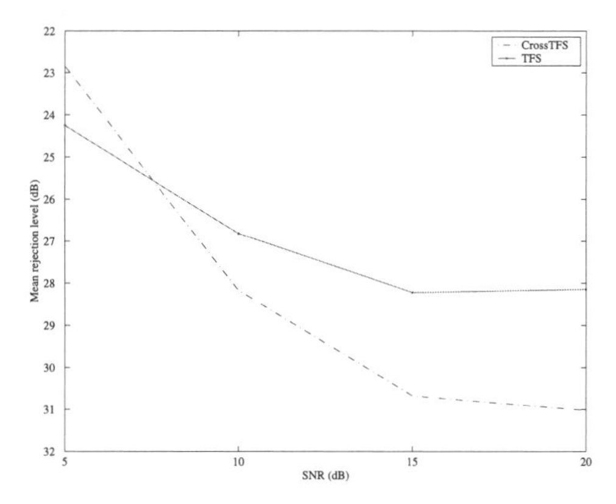

FIGURE 9.4. Mean rejection level versus input SNR for JD- and JD/JAD-based source separation methods.

9.5 Direction finding

9.5.1 Time-frequency MUSIC

The t-f MUSIC algorithm was proposed by Belouchrani and Amin [9], and its performance is analyzed by Zhang, Mu, and Amin [14]. Without loss of generality, we consider 1-D direction finding where the DOAs are described by θ. First, recall that the DOAs are estimated in the MUSIC technique by determining the n values of θ for which the following spatial spectrum is maximized [22]:

$$f_{MU}(\theta) = \left[\mathbf{a}^H(\theta)\hat{\mathbf{G}}\hat{\mathbf{G}}^H\mathbf{a}(\theta)\right]^{-1} = \left[\mathbf{a}^H(\theta)\left(\mathbf{I} - \hat{\mathbf{S}}\hat{\mathbf{S}}^H\right)\mathbf{a}(\theta)\right]^{-1}, \qquad (9.5.1)$$

where $\mathbf{a}(\theta)$ is the steering vector corresponding to θ. The variance of those estimates in the MUSIC technique, assuming white noise processes, is given by [15]

$$E\left(\hat{\omega}_i - \omega_i\right)^2 = \frac{1}{2N}\frac{\mathbf{a}^H(\theta_i)\mathbf{U}\mathbf{a}(\theta_i)}{h(\theta_i)}, \qquad (9.5.2)$$

where $\omega_i = (2\pi d/\lambda)\sin\theta_i$ is the spatial frequency associated with DOA θ_i, and $\hat{\omega}_i$ is its estimate obtained from MUSIC. Moreover, $\mathbf{U}$ is defined in (9.3.11), and

$$h(\theta_i) = \mathbf{d}^H(\theta_i)\mathbf{G}\mathbf{G}^H\mathbf{d}(\theta_i), \quad \text{with} \quad \mathbf{d}(\theta_i) = d\mathbf{a}(\theta_i)/d\omega. \qquad (9.5.3)$$

Similarly, for t-f MUSIC with n_o signals selected, the DOAs are determined by locating the n_o peaks of the spatial spectrum defined from the n_o signals' t-f regions,

$$f_{MU}^{tf}(\theta) = \left[\mathbf{a}^H(\theta)\hat{\mathbf{G}}^{tf}\left(\hat{\mathbf{G}}^{tf}\right)^H\mathbf{a}(\theta)\right]^{-1} = \left[\mathbf{a}^H(\theta)\left(\mathbf{I} - \hat{\mathbf{S}}^{tf}\left(\hat{\mathbf{S}}^{tf}\right)^H\right)\mathbf{a}(\theta)\right]^{-1}. \qquad (9.5.4)$$

$\hat{\mathbf{G}}^{tf}$ and $\hat{\mathbf{S}}^{tf}$ can be obtained by using either JBD or t-f averaging (Subsection 9.2.3). When t-f averaging is used, using the results of Lemmas 9.2 and 9.3, the variance of the DOA estimates based on t-f MUSIC is obtained as [14]

$$E\left(\hat{\omega}_i^{tf} - \omega_i\right)^2 = \frac{1}{2N'}\frac{\mathbf{a}^H(\theta_i)\mathbf{U}^{tf}\mathbf{a}(\theta_i)}{h^{tf}(\theta_i)}, \qquad (9.5.5)$$

where $\hat{\omega}_i^{tf}$ is the estimate of ω_i, $\mathbf{U}^{tf}$ is defined in (9.3.22), and

$$h^{tf}(\theta_i) = \mathbf{d}^H(\theta_i)\mathbf{G}^{tf}\left(\mathbf{G}^{tf}\right)^H\mathbf{d}(\theta_i). \qquad (9.5.6)$$

Note that $h^{tf}(\theta) = h(\theta_i)$ if $n_o = n$.

Examples. Consider a uniform linear array of eight sensors spaced by half a wavelength, and an observation period of 1024 samples. Two chirp signals are emitted from two sources positioned at angles θ_1 and θ_2. The start and end frequencies of

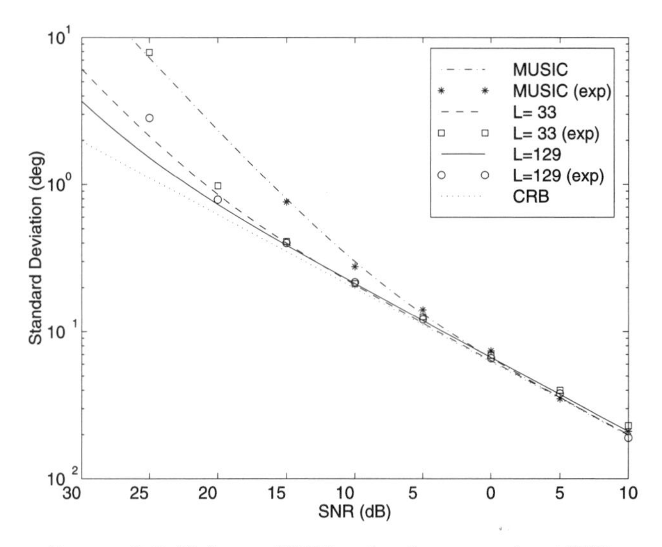

FIGURE 9.5. Variance of DOA estimation versus input SNR.

the signal source at θ_1 are $\omega_{s1} = 0$ and $\omega_{e1} = \pi$, while the corresponding two frequencies for the other source at θ_2 are $\omega_{s2} = \pi$ and $\omega_{e2} = 0$, respectively.

Figure 9.5 displays the variance of the estimated DOA $\hat{\theta}_1$ versus SNR for the case $(\theta_1, \theta_2) = (-10^\circ, 10^\circ)$. The curves in this figure show the theoretical and experimental results of the conventional MUSIC and t-f MUSIC techniques (for $L = 33$ and 129). The Cramer–Rao bound (CRB) is also shown in Figure 9.5 for comparison.

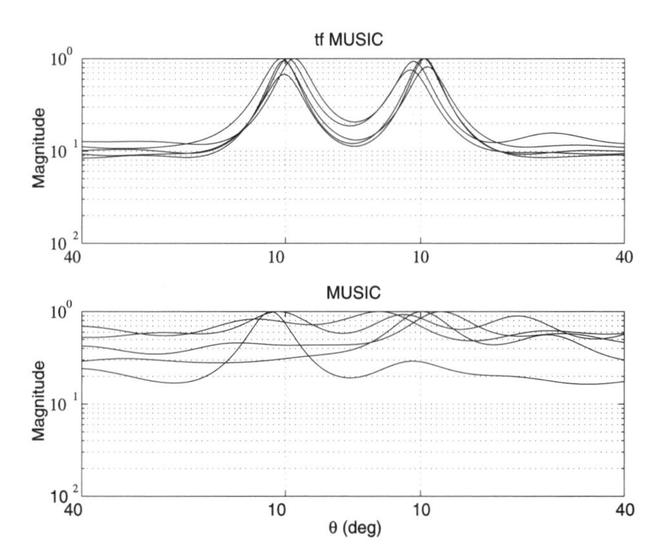

FIGURE 9.6. Estimated spatial spectra of MUSIC and t-f MUSIC.

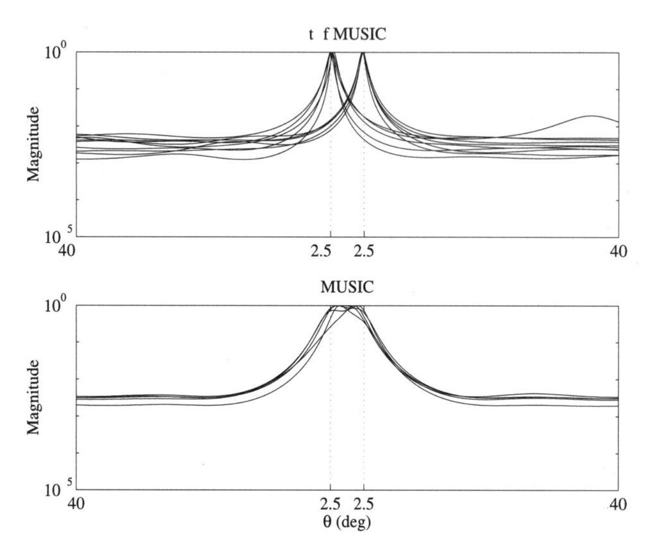

FIGURE 9.7. Estimated spatial spectra of MUSIC and t-f MUSIC for closely spaced signals.

Both signals were selected when performing t-f MUSIC ($n_o = n = 2$). Simulation results were averaged over 100 independent Monte Carlo runs. The advantages of t-f MUSIC in low SNR cases are evident from this figure. The experiment results deviate from the theoretical results for low SNR, since we only considered the lowest order of the coefficients of the perturbation expansion in deriving the theoretical results [14]. Figure 9.6 shows estimated spatial spectra at SNR $= -20$ dB based on t-f MUSIC ($L = 129$) and conventional MUSIC. The t-f MUSIC spectral peaks are clearly resolved.

Figure 9.7 shows examples of the estimated spatial spectrum based on t-f MUSIC ($L = 129$) and conventional MUSIC where the angle separation is small ($\theta_1 = -2.5^\circ$, $\theta_2 = 2.5^\circ$). The input SNR is -5 dB. Two t-f MUSIC algorithms are performed using two sets of t-f points, and each set belongs to the t-f signature of one source ($n_o = 1$). It is evident that the two signals cannot be resolved when the conventional MUSIC technique is applied, whereas by utilizing the signals' distinct t-f signatures and applying t-f MUSIC separately for each signal, the two signals become clearly separated and reasonable DOA estimation is achieved. It is noted that there is a small bias in the estimates of t-f MUSIC due to the imperfect separation of the two signals in the t-f domain.

9.5.2 *Time-frequency maximum likelihood method*

In this section, we introduce the time-frequency maximum likelihood (t-f ML) methods. This method was proposed by Zhang, Mu, and Amin [10] to deal with coherent nonstationary sources. For conventional ML methods, the joint density function of the sampled data vectors $\mathbf{x}(1), \mathbf{x}(2), \ldots, \mathbf{x}(N)$, is given by [23]

$$f(\mathbf{x}(1),\ldots,\mathbf{x}(N))$$
$$= \prod_{i=1}^{N} \frac{1}{\pi^m \det[\sigma \mathbf{I}]} \exp\left(-\frac{1}{\sigma}\left[\mathbf{x}(i) - \mathbf{Ad}(i)\right]^H \left[\mathbf{x}(i) - \mathbf{Ad}(i)\right]\right), \qquad (9.5.7)$$

where $\det[\cdot]$ denotes the matrix determinant. It follows from (9.5.7) that the log-likelihood function of the observations $\mathbf{x}(1), \mathbf{x}(2), \ldots, \mathbf{x}(N)$, is given by

$$L = -mN\ln\sigma - \frac{1}{\sigma}\sum_{i=1}^{N}\left[\mathbf{x}(i) - \mathbf{Ad}(i)\right]^H \left[\mathbf{x}(i) - \mathbf{Ad}(i)\right]. \qquad (9.5.8)$$

To carry out this minimization, we fix $\mathbf{A}$ and minimize (9.5.8) with respect to $\mathbf{d}$. This yields the well-known solution

$$\hat{\mathbf{d}}(i) = \left[\mathbf{A}^H\mathbf{A}\right]^{-1}\mathbf{A}^H\mathbf{x}(i). \qquad (9.5.9)$$

We can obtain the concentrated likelihood function as [23]

$$F_{ML}(\Theta) = \text{tr}\left\{\left[\mathbf{I} - \hat{\mathbf{A}}(\hat{\mathbf{A}}^H\hat{\mathbf{A}})^{-1}\hat{\mathbf{A}}^H\right]\hat{\mathbf{R}}_{\mathbf{xx}}\right\}, \qquad (9.5.10)$$

where $\text{tr}(\mathbf{A})$ denotes the trace of $\mathbf{A}$. The ML estimate of Θ is obtained as the minimizer of (9.5.10). Let ω_i and $\hat{\omega}_i$, respectively, denote the spatial frequency and its ML estimate associated with θ_i. Then the estimation errors $(\hat{\omega}_i - \omega_i)$ are asymptotically (for large N) jointly Gaussian distributed with zero means and the covariance matrix [24]

$$E\left[(\hat{\omega}_i - \omega_i)^2\right] = \frac{1}{2N}\left[\text{Re}(\mathbf{H}\odot\mathbf{R}_{\mathbf{dd}}^T)\right]^{-1}$$
$$\times \text{Re}\left[\mathbf{H}\odot(\mathbf{R}_{\mathbf{dd}}\mathbf{A}^H\mathbf{UAR}_{\mathbf{dd}})^T\right]\left[\text{Re}(\mathbf{H}\odot\mathbf{R}_{\mathbf{dd}}^T)\right]^{-1}, \qquad (9.5.11)$$

where $\odot$ denotes the Hadamard product, and $\mathbf{U}$ is defined in (9.3.11). Moreover,

$$\mathbf{H} = \mathbf{C}^H\left[\mathbf{I} - \mathbf{A}(\mathbf{A}^H\mathbf{A})^{-1}\mathbf{A}^H\right]\mathbf{C}, \qquad \text{with } \mathbf{C} = d\mathbf{A}/d\omega. \qquad (9.5.12)$$

Next we consider the t-f ML method. As we discussed in the previous section, we select $n_o \leq n$ signals in the t-f domain. The concentrated likelihood function defined from the STFD matrix is similar to (9.5.10) and is obtained by replacing $\hat{\mathbf{R}}_{\mathbf{xx}}$ by $\hat{\mathbf{D}}$,

$$F_{ML}^{tf}(\Theta) = \text{tr}\left[\mathbf{I} - \hat{\mathbf{A}}^o\left((\hat{\mathbf{A}}^o)^H\hat{\mathbf{A}}^o\right)^{-1}(\hat{\mathbf{A}}^o)^H\right]\hat{\mathbf{D}}. \qquad (9.5.13)$$

Therefore, the estimation errors $(\hat{\omega}_i^{tf} - \omega_i)$ associated with the t-f ML method are asymptotically (for $N \gg L$) jointly Gaussian distributed with zero means and the covariance matrix [10]

$$
\begin{aligned}
E\left[\left(\hat{\omega}_i^{tf} - \omega_i\right)^2\right] &= \frac{\sigma}{2N'} \left[\mathrm{Re}(\mathbf{H}^o \odot \mathbf{D}_{\mathbf{dd}}^T)\right]^{-1} \\
&\times \mathrm{Re}\left[\mathbf{H}^o \odot \left(\mathbf{D}_{\mathbf{dd}}(\mathbf{A}^o)^H \mathbf{U}^{tf} \mathbf{A}^o \mathbf{D}_{\mathbf{dd}}\right)^T\right] \\
&\times \left[\mathrm{Re}(\mathbf{H}^o \odot \mathbf{D}_{\mathbf{dd}}^T)\right]^{-1} \\
&= \frac{\sigma}{2N'} \left[\mathrm{Re}\left(\mathbf{H}^o \odot (\mathbf{R}_{\mathbf{dd}}^o)^T\right)\right]^{-1} \\
&\times \mathrm{Re}\left[\mathbf{H}^o \odot \left(\mathbf{R}_{\mathbf{dd}}^o(\mathbf{A}^o)^H \mathbf{U}^{tf} \mathbf{A}^o \mathbf{R}_{\mathbf{dd}}^o\right)^T\right] \\
&\times \left[\mathrm{Re}\left((\mathbf{H}^o \odot \mathbf{R}_{\mathbf{dd}}^o)^T\right)\right]^{-1}, \qquad (9.5.14)
\end{aligned}
$$

where $\mathbf{U}^{tf}$ is defined in (9.3.22), and

$$
\mathbf{H}^o = (\mathbf{C}^o)^H \left[\mathbf{I} - \mathbf{A}^o \left((\mathbf{A}^o)^H \mathbf{A}^o\right)^{-1} (\mathbf{A}^o)^H\right] \mathbf{C}^o, \qquad \text{with } \mathbf{C}^o = d\mathbf{A}^o/d\omega. \qquad (9.5.15)
$$

In the case of $n_o = n$, then $\mathbf{H}^o = \mathbf{H}$, and $\mathbf{C}^o = \mathbf{C}$.

The signal localization in the t-f domain enables us to select fewer signal arrivals. This fact is not only important in improving the estimation performance, particularly when the signals are closely spaced, but also reduces the dimension of the optimization problem solved by the ML algorithm, and subsequently reduces the computational requirement.

Examples. To demonstrate the advantages of t-f ML over both conventional ML and the t-f MUSIC, consider a uniform linear array of eight sensors separated by half a wavelength. Two FM signals arrive from $(\theta_1, \theta_2) = (-10°, 10°)$ with the IFs $f_1(t) = 0.2 + 0.1t/N + 0.2\sin(2\pi t/N)$ and $f_2(t) = 0.2 + 0.1t/N + 0.2\sin(2\pi t/N + \pi/2), t = 1, \ldots, N$. The SNR of both signals is -20 dB, and the number of snapshots used in the simulation is $N = 1024$. We used $L = 129$ for t-f ML. Figure 9.8 shows (θ_1, θ_2) that yield the minimum values of the likelihood function of the t-f ML and the ML methods for 20 independent trials. It is evident that t-f ML provides much improved DOA estimation over conventional ML.

In the next example, the t-f ML and the t-f MUSIC methods are compared for coherent sources. The two coherent FM signals have common IFs $f_{1,2}(t) = 0.2 + 0.1t/N + 0.2\sin(2\pi t/N), t = 1, \ldots, N$, with $\pi/2$ phase difference. The signals arrive at $(\theta_1, \theta_2) = (-2°, 2°)$. The SNR of both signals is 5 dB, and the number of snapshots is 1024. Figure 9.9 shows the contour plots of the likelihood function of t-f ML and the estimated spectra of t-f MUSIC for three independent trials. It is clear that t-f ML can separate the two signals, whereas t-f MUSIC cannot.

9.5.3 Effect of cross-terms

Identifying the sources' t-f signatures often requires searching the t-f domain for peak values. In some cases, these values correspond to cross-terms. Building the

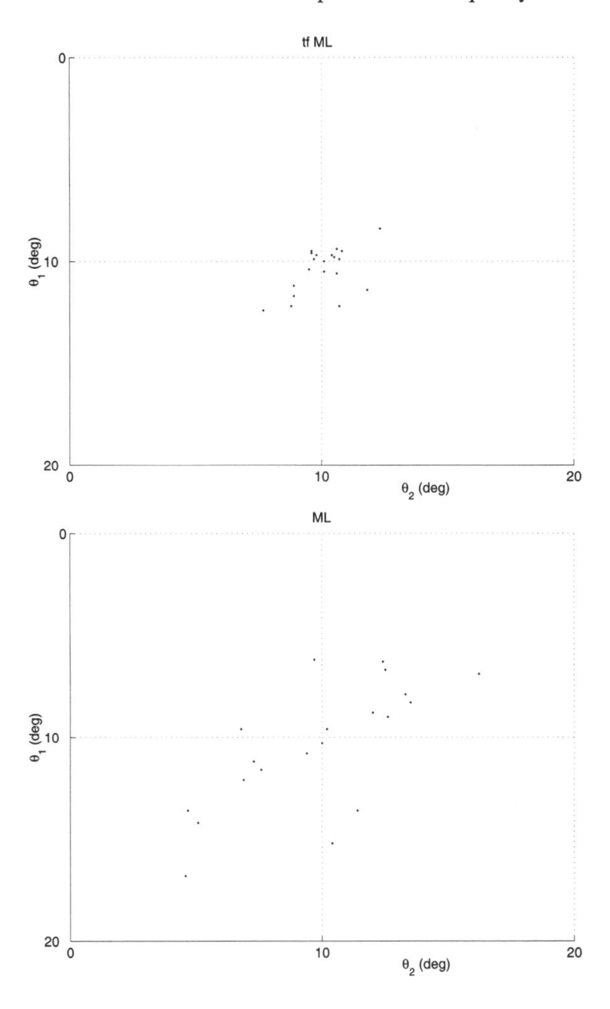

FIGURE 9.8. (θ_1, θ_2) which minimize the t-f ML (upper) and ML (lower) functions.

STFDs around only cross-terms or a mixture of auto-terms and cross-terms and its effect on the t-f MUSIC performance is considered by Amin and Zhang [25]. To understand the role of cross-terms in direction finding, it is important to compare the cross-terms to the cross-correlation between signals in conventional array processing, whose properties are familiar. The source TFD matrix takes the following general form:

$$\mathbf{D_{dd}}(t,f) = \begin{bmatrix} D_{d_1d_1}(t,f) & D_{d_1d_2}(t,f) & \cdots & D_{d_1d_n}(t,f) \\ D_{d_2d_1}(t,f) & D_{d_2d_2}(t,f) & \cdots & D_{d_2d_n}(t,f) \\ \vdots & \vdots & \ddots & \vdots \\ D_{d_nd_1}(t,f) & D_{d_nd_2}(t,f) & \cdots & D_{d_nd_n}(t,f) \end{bmatrix}. \tag{9.5.16}$$

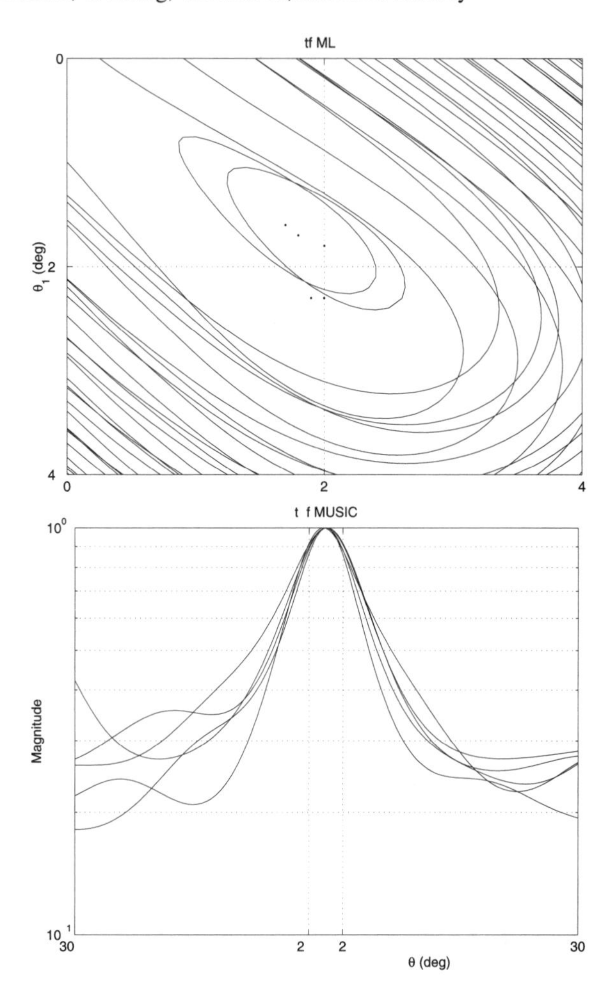

FIGURE 9.9. Contour plots of t-f ML function (upper) and spatial spectra of t-f MUSIC (lower).

On the other hand, the covariance matrix of correlated source signals is given at the form

$$\mathbf{R_{dd}} = \begin{bmatrix} R_{d_1 d_1} & R_{d_1 d_2} & \cdots & R_{d_1 d_n} \\ R_{d_2 d_1} & R_{d_2 d_2} & \cdots & R_{d_2 d_n} \\ \vdots & \vdots & \ddots & \vdots \\ R_{d_n d_1} & R_{d_n d_2} & \cdots & R_{d_n d_n} \end{bmatrix}, \tag{9.5.17}$$

where the off-diagonal element $R_{d_i d_j} = E[d_i(t) d_j^*(t)]$ represents the correlation between source signals d_i and d_j. Direction finding problems can usually be solved when the signals are partially correlated; however, full rank property of the source covariance matrix $\mathbf{R_{dd}}$ is a necessary condition.

Comparing Eq. (9.5.16) and (9.5.17), it is clear that the cross-correlation terms and the cross-terms have analogous forms. However, the correlation matrix in (9.5.17) is defined for stationary signal environments, whereas the source TFD matrix in (9.5.16) is defined at a (t, f) point and its value usually varies with respect to t and f. Detailed observations are made through the following example.

Example. Consider a six-element linear array with half-wavelength interelement spacing, and arrival of two chirp signal. The start and end frequencies of the first signal $d_1(t)$ are $f_{1s} = 0.1$ and $f_{1e} = 0.5$, and those for the second signal $d_2(t)$ are $f_{2s} = 0$ and $f_{2e} = 0.4$, respectively. The SNR is 10 dB for each signal, and the DOAs of the two signals are $\theta_1 = -5°$ and $\theta_2 = 5°$, respectively. The number of samples is 256. The PWVD is used and the window length is $N = 129$.

We consider the auto-terms and the cross-terms over the following two regions: (i) auto-term regions (t, f_1) with $f_1(t) = 0.1 + 0.4t/N$ and (t, f_2) with $f_2(t) = 0.4t/N$, where the auto-terms are dominant; and (ii) cross-term region (t, f_c) with $f_c(t) = [f_1(t) + f_2(t)]/2 = 0.05 + 0.4t/N$, where the cross-terms are dominant. Both the auto-term and cross-term regions have large peak values and are most likely to be selected.

(i) *Auto-term regions.* In the auto-term region of $d_1(t)$, (t, f_1), the auto-term of $d_1(t)$ is constant. The auto-term of $d_2(t)$ and the cross-term between $d_1(t)$ and $d_2(t)$ are relatively small. Since the STFD matrix in the auto-term region has dominant diagonal elements with constant values, incorporating only auto-term points, either by JBD or by t-f averaging, usually provides good direction finding performance.

(ii) *Cross-term regions.* In this region the cross-terms $D_{d_1 d_2}(t, f) = D^*_{d_2 d_1}(t, f)$ are dominant. Therefore, the source TFD matrix on the cross-term region is nearly anti-diagonal. Note that this source TFD matrix is still full rank (although not necessarily positive definite). Accordingly, the noise subspace can be properly estimated, even when only the cross-term points are selected. However, since the cross-terms change with time t, taking both positive and negative values, summing them at different (t, f) points yields small smoothed values. Therefore, the t-f averaging is expected to yield degraded performance in some cases. Performing JBD instead of t-f averaging avoids such risk.

Table 9.1 shows the DOA variance of signal $d_1(t)$ obtained from 100 independent Monte Carlo runs of t-f MUSIC. Both JBD and t-f averaging are considered for four cases, namely, (a) auto-term regions $f(t) = f_1(t)$ and $f(t) = f_2(t)$, (b) cross-term region $f(t) = [f_1(t) + f_2(t)]/2$, (c) auto-term and cross-term regions $f(t) = f_1(t)$, $f(t) = f_2(t)$, and $f(t) = [f_1(t) + f_2(t)]/2$, and (d) auto-term region of the first signal, $f(t) = f_1(t)$. Although both the JBD and t-f averaging resolve the signals in all the four cases, it is evident that the JBD outperforms the t-f averaging, particularly when the cross-term region is involved. Case (d), in which only one of the two signals is selected, has the best performance for both methods of JBD and t-f averaging. An interesting observation is that, in case (b), where only the cross-term region is used,

	Case (a)	*Case (b)*	*Case (c)*	*Case (d)*
JBD	0.156°	0.154°	0.180°	0.121°
T-f averaging	0.179°	0.339°	0.199°	0.161°

TABLE 9.1. Variances of DOA estimates.

the JBD yields second best performance, whereas the t-f averaging shows its worst performance.

9.6 Spatial ambiguity functions

The spatial ambiguity function is proposed by Amin, Belouchrani, and Zhang [26]. The discrete form SAF matrix of a signal vector $\mathbf{x}(t)$ is defined as

$$\mathbf{D_{xx}}(\theta, \tau) = \sum_{u=-\infty}^{\infty} \mathbf{x}(u + \tau/2)\mathbf{x}^H(u - \tau/2)e^{j\theta u}, \tag{9.6.1}$$

where θ and τ are the frequency lag and the time lag, respectively. In a noise-free environment, $\mathbf{x}(t) = \mathbf{Ad}(t)$. In this case,

$$\mathbf{D_{xx}}(\theta, \tau) = \mathbf{AD_{dd}}(\theta, \tau)\mathbf{A}^H. \tag{9.6.2}$$

Equation (9.6.2) is similar to the formula that has been commonly used in blind source separation and DOA estimation problems, relating the data correlation matrix to the signal correlation matrix. Here, these matrices are replaced by the data SAF and signal ambiguity function matrices, respectively. The two subspaces spanned by the principle eigenvectors of $\mathbf{D_{xx}}(\theta, \tau)$ and the columns of $\mathbf{A}$ are identical. This implies that array signal processing problems can be approached and solved based on the SAF.

By replacing the STFD matrix $\mathbf{D_{xx}}(t, f)$ by the SAF matrix $\mathbf{D_{xx}}(\theta, \tau)$, we can easily derive the ambiguity-domain source separation methods and the ambiguity-domain MUSIC (AD MUSIC) [26], following the same procedures described in Sections 9.4 and 9.5.

The SAFs have the following two important offerings that distinguish them from other array spatial functions.

(1) The cross-terms in between source signals reside on the off-diagonal entries of matrix $\mathbf{D_{dd}}(\theta, \tau)$, violating its diagonal structure, which is necessary to perform blind source separation. In the ambiguity domain, the signal auto-terms are positioned near and at the origin, making it easier to leave out cross-terms from matrix construction.

(2) The auto-terms of all narrowband signals, regardless of their frequencies and phases, fall on the time-lag axis ($\theta = 0$), while those of the wideband signals fall on a different (θ, τ) region or spread over the entire ambiguity domain. Therefore, the SAF is a natural choice for recovering and spatially localizing narrowband sources in broadband signal platforms.

9.7 Sensor-angle distributions

In this section, we use quadratic distributions to address the problem of characterizing the power attributed to near-field scattering local to an array of sensors. The proposed method is based on the quadratic sensor-angle distribution (SAD), previously called the spatial Wigner distribution [27]. This distribution is a characterization of the power at every angle for each sensor in the array. It is altogether different than the STFDs discussed in the previous sections. These two types of distributions have different structures and objectives.

In the SAD, near-field sources have different angles for the various array sensors. The SAD is a joint-variable distribution and a dual in sensor number and angle to Cohen's class of TFDs [1]. We use a known test source to illuminate the local scatterer distribution we wish to characterize. Orthogonal subspace projection techniques are then applied to the array data to suppress the direct propagation path from the test source so as to reveal the less powerful local scatter. An example from the area of high-frequency surface wave radar is provided for illustration.

A typical surface-wave radar receiving array may consist of between 8 and 64 sensors and can be hundreds of meters or indeed more than 1 km in total length. It is typically sited on a coastal beach which may or may not provide a uniform transition from land to sea. The coast may in fact be a bay, in which case the land sea boundaries beyond either end of the array may cause near-field scattering and distort the wave front arriving at the array. There may be other locally sited structures, such as buildings and fences, which can be the source of local scatter (consider that the wavelength of the radar signal is between 30–100 m). This makes achieving very low sidelobe spatial beams with a classical beamformer a difficult problem and can render the receiver system vulnerable to interference through beam sidelobes (possibly via skywave propagation).

The near-field scatter produced by these mechanisms is correlated with the desired direct far-field radar return from targets (and clutter). This scatter is typically approximately 20–40 dB weaker than the direct signal. Without compensating for the effects of local scatter, it is possible to achieve classical beam sidelobes of 30–35 dB. However, in general, the remaining components of the receiving system can sustain substantially higher performance [28].

The effect of local scatter on beamforming must be mitigated in order for the radar system to realize the inherent sidelobe capability as set by the radar equipment (as distinct from the sensors) [28]. A first step to achieving this is to characterize the local scatter distribution. A means of performing this characterization using techniques derived from time-frequency analysis is the focus of the remainder of this chapter.

A generalization of the spatial Wigner distribution introduced in [27] is provided and combined with orthogonal projection techniques for detection, classification, and characterization of near-field and far-field sources lying in the field of view of the multiantenna receiver.

9.7.1 Signal model

9.7.1.1 Geometry

Consider a linear, equispaced array of M sensors placed on a flat plane in a 2D surface. Assume that sensor position errors are negligible and the gain and phase of all sensors and corresponding data acquisition equipment are accurately matched. It is also assumed that the array is narrowband, i.e., the reciprocal of the bandwidth of any signals received is large compared with the propagation delay across the array. The wavelength of all sources received is λ. Let the origin of a coordinate system be at the midpoint of the array, with the sensors individually spaced by d regularly along the x-axis and indexed $i = 1, \dots, M$ from left to right. We assume that $d < \lambda/2$. Boresight is along the y-axis.

A source is placed in the near field (i.e., a circular wavefront impinges on the array) at location r_s meters from the origin and θ_s degrees from boresight. For convenience (although somewhat unconventionally), we have defined that angles are to be measured clockwise from array boresight (the y-axis). For M odd there is a sensor at the origin, whereas for M even the origin is midpoint between two sensors. The array geometry and the notation are shown in Figure 9.10 for the case of M even.

The distance from the ith sensor to the source is given by

$$r_{s,i}^2 = r_s^2 + \left[(i-1)\cdot d - \frac{M-1}{2}\cdot d\right]^2 - 2r_s \cdot d \cdot \left[(i-1) - \frac{M-1}{2}\right] \cdot \sin(\theta_s) \tag{9.7.1}$$

and the corresponding complex response at the ith sensor is

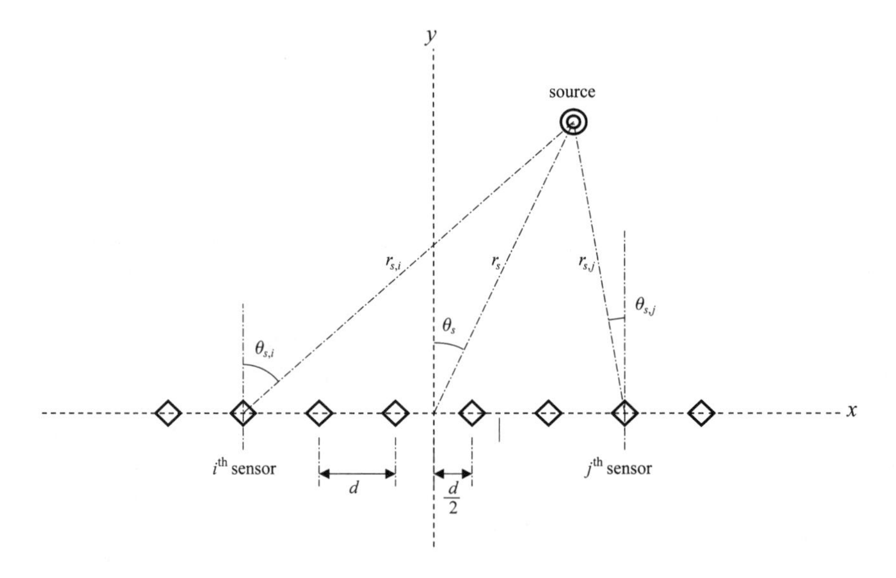

FIGURE 9.10. Sensor array geometry and notation for linear array and a near-field source.

$$a_i(r_s, \theta_s) = \frac{1}{r_{s,i}} \cdot \exp\left(-j\frac{2\pi}{\lambda} \cdot r_{s,i}\right) \tag{9.7.2}$$

assuming a normalized and equal gain for each sensor. The vector $\mathbf{a}(r_s, \theta_s) = [a_1, \ldots, a_M]^T$ is the response of the complete array to the source at (r_s, θ_s).

Likewise, the angle from the ith sensor to the source is

$$\theta_{s,i} = \cos^{-1}\left[\frac{[(i-1) - \frac{M-1}{2}]^2 d^2 + r_{s,i}^2 - r_s^2}{2d[(i-1) - \frac{M-1}{2}]r_{s,i}}\right] - \frac{\pi}{2}. \tag{9.7.3}$$

It is a characteristic of near-field sources that they are viewed at different angles by the different sensors in the array.

Given sensor-angle measurements from any two or more sensors, it is possible to determine the range and bearing (r_s, θ_s) (with respect to the origin) of a source in the near field. This is, however, subject to identifiability requirements that each sensor has a different angle to the source. Given $\theta_{s,i}$ and $\theta_{s,j}$ with $i \neq j$, one determines sensor to source ranges $r_{s,i}$ and $r_{s,j}$, respectively, using

$$r_{s,i} = [|i-j|] \cdot d \cdot \frac{\cos[\theta_{s,j}]}{\sin[\theta_{s,i} - \theta_{s,j}]} \tag{9.7.4}$$

and

$$r_{s,j} = [|i-j|] \cdot d \cdot \frac{\cos[\theta_{s,i}]}{\sin[\theta_{s,i} - \theta_{s,j}]}. \tag{9.7.5}$$

This requires that $\theta_{s,i} - \theta_{s,j} \neq 0$. The range and bearing with respect to the origin can be determined relative to any of the individual sensors using the individual sensor range and bearing. For example, for the jth sensor, we use $r_{s,j}$ and $\theta_{s,j}$ according to

$$r_s^2 = r_{s,j}^2 + \left[[\frac{M-1}{2} - [j-1]] \cdot d\right]^2 - 2 \cdot r_{s,j} \cdot \left[[\frac{M-1}{2} - [j-1]] \cdot d\right] \sin[\theta_{s,j}] \tag{9.7.6}$$

and

$$\theta_s = \frac{r_{s,j}}{r_s} \cos^{-1}[\theta_{s,j}]. \tag{9.7.7}$$

9.7.1.2 Model

Our proposed source characterization technique requires one cooperative source with complex envelope s_k^{S} in the far field of the array at known angle θ^{S}. Steering vectors for the far field $\mathbf{a}(\theta)$ and near field $\mathbf{a}(\theta, r)$ take on the standard form where θ is the angle, and r denotes range [29].

Assume that the conditions on the test source and sensor array are such that the following signal model is appropriate

$$\mathbf{z}_k = \mathbf{A}\mathbf{s}_k + \mathbf{q}_k + \mathbf{n}_k. \tag{9.7.8}$$

In this model, $\mathbf{z}_k$ is the kth snapshot of sensor data outputs (dimension M). The $\mathbf{q}_k$ represents additive spatial and temporal colored noise produced in the environment, and $\mathbf{n}_k$ represents additive white noise modeling the internal noise of the array of sensors receiving system.

The matrix $\mathbf{A}$ can take on one of two forms, depending on whether the local scatterer is best modeled as a collection of P discrete scatterers, or as a single spatially distributed scatterer. For the case of P discrete scatterers,

$$\mathbf{A} = \left[\mathbf{a}(\theta^{\mathrm{S}}), \mathbf{a}(\theta_1, r_1), \dots, \mathbf{a}(\theta_P, r_P)\right]. \tag{9.7.9}$$

In this equation, the near-field scatterer $i = 1, \dots, P$ is characterized by the angle θ_i and range r_i. For a distributed scatterer with scatter amplitude $f[\mathbf{a}(\theta, r)]$ contained in the near-field azimuth and range set Ω,

$$\mathbf{A} = \left[\mathbf{a}(\theta^{\mathrm{S}}), \int_{\theta, r \in \Omega} f[\mathbf{a}(\theta, r)] \mathrm{d}\theta \mathrm{d}r\right]. \tag{9.7.10}$$

Likewise, the signal vector $\mathbf{s}_k$ can be constructed in two ways, depending on whether the near-field scatterers are best modeled as discrete or continuous. For the discrete case,

$$\mathbf{s}_k = \left[s_k^{\mathrm{S}}, s_k^1, \dots, s_k^P\right]^T, \tag{9.7.11}$$

where the test source complex amplitude is given by s_k^{S} and the complex amplitude of the ith of P discrete scatterers is denoted as s_k^i. In the continuous scatterer case,

$$\mathbf{s}_k = \left[s_k^{\mathrm{S}}, s_k\right]^T. \tag{9.7.12}$$

The s_k^{S} and s_k^i and s_k may be uncorrelated, correlated, or coherent for each case.

The spatial covariance matrix $\mathbf{E}[\mathbf{z}\mathbf{z}^H]$ is

$$\mathbf{R} = \mathbf{A}\mathbf{S}\mathbf{A}^H + \mathbf{Q} + \sigma^2\mathbf{I}, \tag{9.7.13}$$

where $\mathbf{S}$ is the source covariance matrix. Let the elements of $\mathbf{S}$ be ρ_{ij}. We ensure that the cooperative test source has sufficient SNR (generally greater than 50 dB) to perform our measurement by requiring that

$$\rho_{\mathrm{snr}} = \frac{\rho_{11}}{(\sigma^2 + \mathrm{tr}[\mathbf{Q}])} \gg 1. \tag{9.7.14}$$

It is also expected that the direct far-field source power will be substantially greater than the total near-field power (by 20–40 dB):

$$\rho_{\mathrm{snf}} = \frac{\rho_{11}}{\sum_{k=2}^{P} \rho_{kk}} \gg 1. \tag{9.7.15}$$

9.7.1.3 Background

Breed and Posch [27] introduced the spatial Wigner distribution as a tool for determining the range and angle of a near-field source. They exploited the property that the phase front of a wave emanating from a source in the near field of an array has an approximately quadratic phase law or, equivalently, an approximately linear spatial frequency law. They then determined source location by determining the parameters governing the linear frequency law as represented by the Wigner distribution applied to the spatial signal. The true propagating wave phase front is in fact spherical and is only approximately quadratic for near-field sources some distance from the array. The method proposed in [27] breaks down for sources close to the array. Swindlehurst and Kailath [30] examined the applicability of the quadratic-spherical approximation and applied a parametric high resolution technique to determine the linear frequency law parameters (and hence the near-field source position). However, it can be seen from Eq. (9.7.4)–(9.7.7) that it is possible to determine the source position without invoking the quadratic phase approximation to the spherical phase front.

There is a substantial body of literature concerned with processing spatial signals received by an array of sensors from sources in the near field of the array. It is mostly concerned with techniques for estimating the angle and range of the source. For example, both subspace and ML algorithms are derived in [31]. Subsequently, we will present an example showing near-field characterization using both the SAD (discussed next) and the near-field MUSIC, as developed in [31].

Several authors have proposed methods for determining the angle of distributed sources located in the far field of an array [32]. These techniques address the effect of scatter local to a transmitter in the far field, not the effect of scatter that is sufficiently local to the receiving system to be in the near field of the array.

9.7.2 SADs

Our method extends the spatial Wigner distribution introduced by Breed and Posch. To avoid confusion, it has been necessary to change the name to reflect the generalization to all members of Cohen's class of quadratic distributions [1]. While the title "spatial Wigner distribution" is informative, retaining the name "spatial time-frequency distribution" for the remaining members of Cohen's class applied to spatial signals does not correctly describe the distribution we are interested in, and will be confused with STFD, as discussed in earlier sections of this chapter. Therefore, in this work we have renamed the class of quadratic distributions applied to spatial signals to be *sensor-angle distributions* (SADs). The corresponding spectra are called sensor-angle spectra (SAS).

The Cohen's class SAD for the kth snapshot is a distribution of the angle of sources impinging on the array *at each sensor*,

$$\mathbf{T}_k(i,\theta;\mathbf{z}_k) = \sum_{v=-\infty}^{\infty}\sum_{l=-\infty}^{\infty} \phi(v,l)\mathbf{z}_k(i+v+l)\mathbf{z}_k^*(i+v-l)\mathrm{e}^{-j4\pi\theta l}, \tag{9.7.16}$$

where i and θ are the sensor index and angle respectively. The kernel $\phi(v, l)$ characterizes the distribution and is a function of sensor position and sensor lag. All the standard kernel designs applied in the time-frequency literature may be used with the SAD.

The sensor-angle spectrum (SAS) is the *power* (not energy or energy density) distribution of the sources impinging on the array. The SAS is given by

$$\mathbf{T}^S(i, \theta; \mathbf{z}) = \mathbf{E}[\mathbf{T}_k(i, \theta; \mathbf{z}_k)], \tag{9.7.17}$$

where an estimate for temporally stationary sources is given by

$$\hat{\mathbf{T}}^S(i, \theta; \mathbf{z}) = \frac{1}{N} \sum_{k=0}^{N-1} \mathbf{T}_k(i, \theta; \mathbf{z}_k) \tag{9.7.18}$$

for N snapshots.

9.7.3 *Characterizing local scatter*

The objective is to use data received by the array from a test source in the far field that illuminates the local near-field scatterer distribution and to visualize and characterize this scatterer distribution using the SAS. We expect the test signal to be substantially more powerful than the local scatter that we wish to characterize (see (9.7.15)). Subspace projection is applied to the array snapshots to remove the dominant far-field component and allow a clear depiction of the near-field source in the sensor-angle (s-a) domain.

In (9.7.16) and (9.7.17), the data snaphot $\mathbf{z}_k$ is replaced by $\mathbf{P}^{\theta^S}\mathbf{z}_k$, where $\mathbf{P}^{\theta^S}$ is the orthogonal projection operator formed from the far-field test source steering vector $\mathbf{a}(\theta^S)$ as

$$\mathbf{P}^{\theta^S} = \mathbf{I} - \mathbf{a}(\theta^S)[\mathbf{a}^H(\theta^S)\mathbf{a}(\theta^S)]^{-1}\mathbf{a}^H(\theta^S). \tag{9.7.19}$$

Therefore, we compute the modified SAS

$$\hat{\mathbf{T}}^S(i, \theta; \mathbf{P}^{\theta^S}, \mathbf{z}). \tag{9.7.20}$$

In some applications, a single test angle will provide sufficient characterization using (9.7.20). In other applications, two or several test angles will be required, in which case θ^S is scanned over the required domain of angles for the test source.

9.7.4 *Simulations and examples*

The following example is used to demonstrate the proposed approach for near scattering characterizations. Consider a 32 sensor, linear, equispaced array operating at a carrier frequency of 6.41 MHz and with 15 m sensor spacing. The local scatterer distribution comprises a point scatterer in the near field at a range of 400 m and

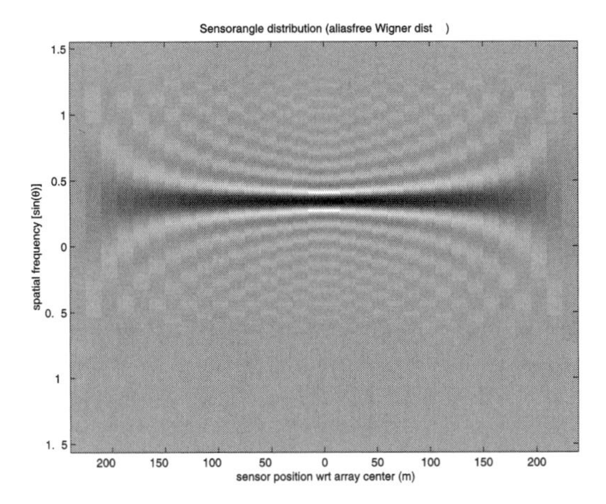

FIGURE 9.11. SAD for the received data $\mathbf{z}_k$. The far-field test source dominates the SAD characterization.

bearing of 30 degrees in front of the array (the array has a total length of 465 m). Assume that the test source is temporally stationary and located at 20 degrees angle with respect to boresight. The test source is coherent with and 20 dB stronger than the scattered source. In this example we have used the alias-free Wigner distribution [33]. Of course, other members of Cohen's class may also be used.

Figure 9.11 shows the SAD for the received data. The SAD is dominated by the substantially more powerful far-field test source, and there is no clear indication of any additional scattering. The far-field source has the same angle for every sensor, and, therefore, depicts a horizontal signature in the s-a domain. In Figure 9.12 we

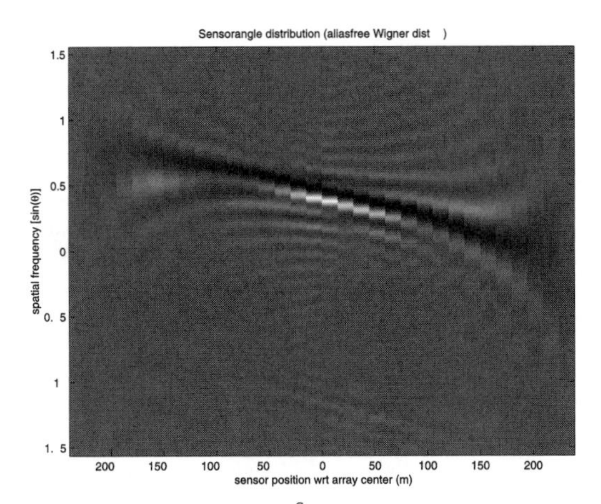

FIGURE 9.12. SAD for the received data $\mathbf{P}^{\theta^S}\mathbf{z}_k$. With the direct propagation path from the far-field test signal removed by the orthogonal projection operator, the local scatterer spectrum is now revealed.

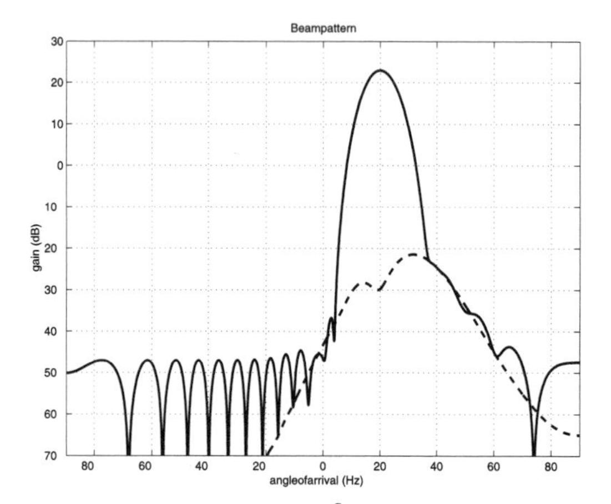

FIGURE 9.13. Beampatterns for $\mathbf{z}_k$ (—) and $\mathbf{P}^{\theta^S}\mathbf{z}_k$ (- -). Without the sensor-angle characterization it is not possible to identify perturbations from the ideal test source beampattern as being due to near-field scatter.

have applied the orthogonal projection operator and computed the SAD for $\mathbf{P}^{\theta^S}\mathbf{z}_k$. The SAD now clearly shows the presence of near-field local scatter. The location of the near-field source may be determined using equations (9.7.4)–(9.7.7).

The beampatterns for the cases of $\mathbf{z}_k$ and $\mathbf{P}^{\theta^S}\mathbf{z}_k$ are shown in Figure 9.13. The presence of near-field scatterers cannot be confirmed as compared with alternative explanations for the distorted beampatterns, such as poor array calibration.

The projection approach has also been applied to real data collected from a 16 sensor high frequency (HF) receiving array. An array calibration source was transmitted from the far-field of the array at boresight. The SAD is shown in Figure 9.14

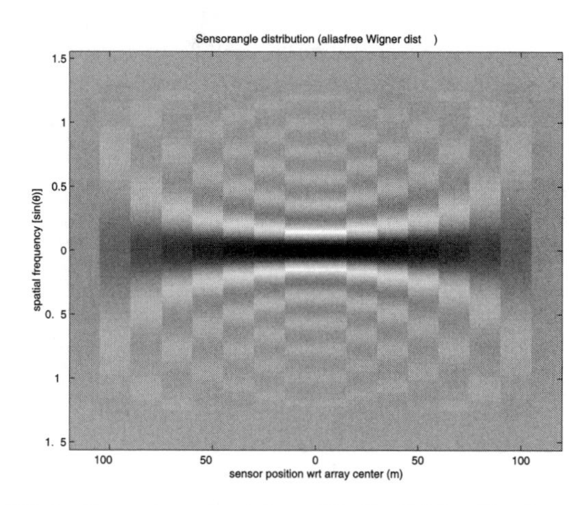

FIGURE 9.14. SAD for the real received data. The far-field calibration source at boresight dominates the SAD characterization.

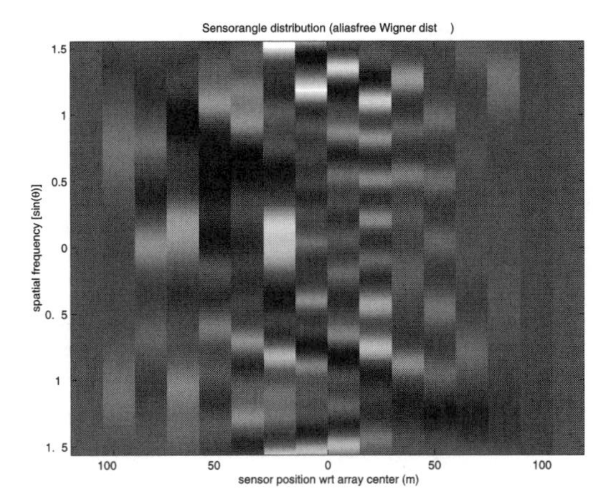

FIGURE 9.15. SAD for the real received data with the direct propagation path from the far-field calibration source removed by orthogonal projection operator. The local scatterer SAD is revealed.

and is dominated by the calibration source. Following calibration of the array using the calibration source, the received and calibrated boresight source is removed using orthogonal projection. The SAD of the residual is shown in Figure 9.15. No discrete near-field scatterers are apparent; however, there is a concentration in the SAD in the upper left region of the distribution. This indicates that there is some asymmetric local scattering near the array.

A second example is used to contrast the SAD with existing techniques for near-field sources characterization. In this case, we have chosen to compare with an implementation of near-field MUSIC as described by [31]. We consider an ideal computer

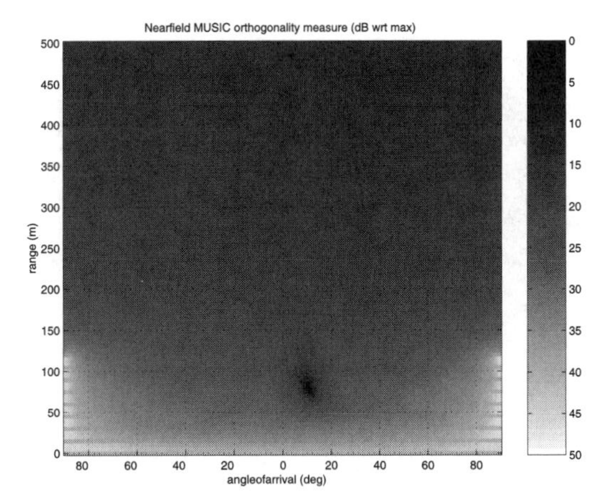

FIGURE 9.16. Near-field MUSIC diagram for an ideal source at range 80 m and bearing 10.5 degrees.

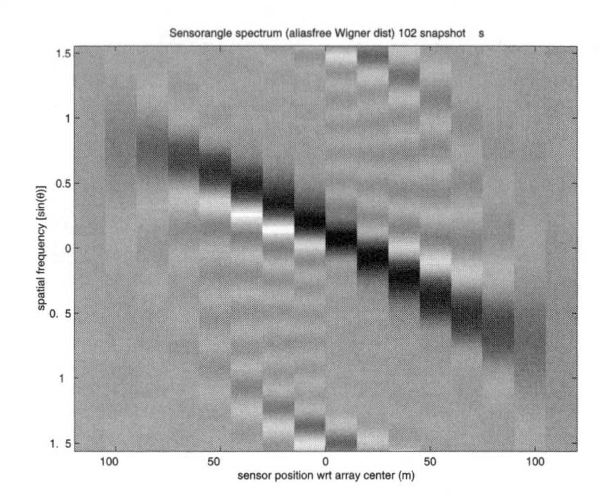

FIGURE 9.17. SAS for an ideal source at range 80 m and bearing 10.5 degrees.

generated case and an equivalent case where the data has been collected using a real HF radar array.

Consider an ideal case with a single source placed at range 80 m and bearing 10.5 degrees. Assume a 16 sensor array, that the source signal-to-noise power ratio is high, and 102 array data snapshots are available (generated using computer). Figures 9.16 and 9.17, respectively, show the near-field MUSIC diagram and the SAS, both computed using all 102 data snapshots. The source is well localized using MUSIC, and there is a characteristic structure in the SAS showing the sensor-angle for each sensor in the array.

We have repeated the analysis, but this time used 102 data snapshots collected from the HF radar receiving array. The near-field source was approximately 80 m

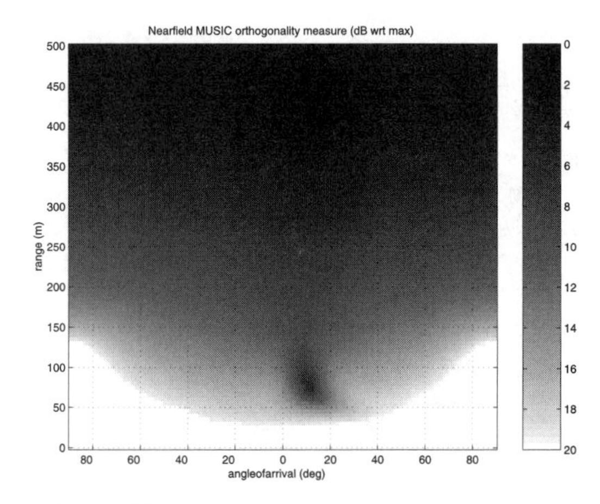

FIGURE 9.18. Near-field MUSIC diagram for a real source placed at approximate range 80 m and bearing 10.5 degrees.

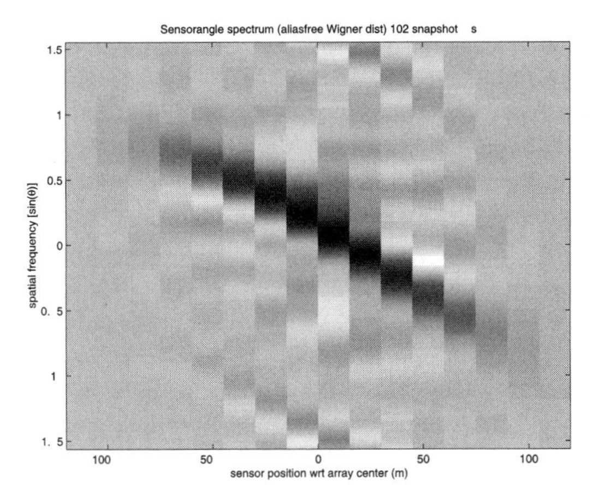

FIGURE 9.19. SAS for a real source placed at approximate range 80 m and bearing 10.5 degrees.

from the array midpoint at an angle of approximately 10 degrees. In this case, the source was behind the array. The exact location is not known precisely. Figures 9.18 and 9.19 show the near-field MUSIC diagram and the SAS, respectively. Imprecise array calibration has smeared the localization in the MUSIC diagram, while the structure in the SAS is preserved.

9.8 Conclusion

We have presented two different new perspectives of TFDs. One perspective is driven by direction finding and blind source separation problems, whereas the other stems from the need to characterize near-field sources or reflectors. The fundamental offering of quadratic distributions in both cases is the ability to discriminate between the sources based on the joint-variable signatures of their respective waveforms. This allows the enhancement of SNR as well as the consideration of only the sources of interest, and subsequently improves the estimation of the source positions and waveforms.

The first six sections presented the general framework of STFDs. The advantages of an STFD matrix over the covariance matrix-based approach to array processing are the SNR enhancement and the robustness of the eigenstructure to noise. A variety of methods have been introduced for both blind source separation and direction finding applications using STFDs. The first class of source separation methods is based on prewhitening and the recovery of a unitary matrix. Unlike similar methods based on second-order statistics, which cannot separate signals with the same spectra, the STFD-based method can separate nonstationary signals with identical spectra when they have different time-frequency signatures. The SNR enhancement and signal localization properties in the time-frequency domain can substantially improve the source separation performance. The second class of source separation methods is based on array averaging of the TFD, signal synthesis, and waveform recovery using the minimum MSE criterion. For direction finding, both the MUSIC and the ML methods have been extended and modified to incorporate the STFDs. The performance improvement, evident by both the analytical and simulation results, is most significant when the input SNR of source arrivals is low, and/or when the sources are closely spaced.

In the second part of the chapter, we have used the TFD of the spatial signal received by an array to characterize sources based on their angle at each sensor. This SAD is a tool for characterizing near-field scatter local to the receiving array. The method uses a test source in the far field to illuminate the local scatterer distribution. An orthogonal projection operator derived from the steering vector for the far-field test source is used to exclude the direct propagation path from the test source in the characterization. As part of the characterization, we exploit the spatial Wigner distribution, although we have renamed it the SAD to avoid confusion with a similarly named but differently defined STFD discussed in the first part of the chapter. We have shown the application of the method using simulation and for real data collected using an HF radar receiving array. Additional simulations and real data results contrast the SAD characterization with that of a conventional near-field localization technique (in this case near-field MUSIC).

References

[1] L. Cohen, *Time-Frequency Analysis*, Englewood Cliffs, NJ: Prentice-Hall, 1995.

[2] S. Qian and D. Chen, *Joint Time-Frequency Analysis*, Englewood Cliffs, NJ: Prentice-Hall, 1996.

[3] F. Hlawatsch and G. Boudreau-Bartles, Linear and quadratic time-frequency signal representations, *IEEE Signal Processing Mag.*, vol. 9, no. 2, pp. 21–68, April 1992.

[4] M. G. Amin, Time-frequency distributions in statistical signal and array processing, section in T. Chen (ed.), Highlights of statistical signal and array processing, *IEEE Signal Processing Mag.*, vol. 15, no. 5, Sept. 1998.

[5] M. G. Amin and A. N. Akansu, Time-frequency for interference excision in spread spectrum communications, section in G. Giannakis (ed.), Highlights of signal processing for communications, *IEEE Signal Processing Mag.*, vol. 16, no. 2, March 1999 .

[6] M. G. Amin and A. R. Lindsey, Time-frequency interference mitigation in spread spectrum communication systems, in B. Boashash (ed.), *Time-Frequency Signal Analysis and Processing*, Englewood Cliffs, NJ: Prentice-Hall, 2002.

[7] M. G. Amin and Y. Zhang, Spatial time-frequency distributions and their applications, in B. Boashash (ed.), *Time-Frequency Signal Analysis and Processing*, Englewood Cliffs, NJ: Prentice-Hall, 2002.

[8] A. Belouchrani and M. G. Amin, Blind source separation based on time-frequency signal representation, *IEEE Trans. Signal Processing*, vol. 46, no. 11, pp. 2888–2898, Nov. 1998.

[9] A. Belouchrani and M. Amin, Time-frequency MUSIC, *IEEE Signal Processing Letters*, vol. 6, no. 5, pp. 109–110, May 1999.

[10] Y. Zhang, W. Mu, and M. G. Amin, Time-frequency maximum likelihood methods for direction finding, *J. Franklin Inst.*, vol. 337, no. 4, pp. 483–497, July 2000.

[11] Y. Zhang and M. G. Amin, Blind separation of sources based on their time-frequency signatures, in *Proc. IEEE Int. Conf. Acoust., Speech, Signal Process.*, Istanbul, Turkey, pp. 3132–3135, June 2000.

[12] K. Sekihara, S. Nagarajan, D. Poeppel, and Y. Miyashita, Time-frequency MEG-MUSIC algorithm, *IEEE Trans. Medical Imaging*, vol. 18, no. 1, pp. 92–97, Jan. 1999.

[13] G. H. Golub and C. F. Van Loan, *Matrix Computations*, 3rd edition. Baltimore, MD: Johns Hopkins University Press, 1996.

[14] Y. Zhang, W. Mu, and M. G. Amin, Subspace analysis of spatial time-frequency distribution matrices, *IEEE Trans. Signal Processing*, vol. 49, no. 4, pp. 747–759, Apr. 2000.

[15] P. Stoica and A. Nehorai, MUSIC, maximum likelihood, and Cramér–Rao bound, *IEEE Trans. Acoust. Speech, Signal Process.*, vol. 37, no. 5, pp. 720–741, May 1989.

[16] H. I. Choi and W. J. Williams, Improved time-frequency representation of multicomponent signals using exponential kernels, *IEEE Trans. Acoust., Speech, Signal Process.*, vol. 37, pp. 862–871, June 1989.

[17] W. Mu, M. G. Amin, and Y. Zhang, Bilinear signal synthesis in array processing, *IEEE Trans. Signal Processing*, vol. 51, no. 1, pp. 90–100, Jan. 2003.

[18] G. Boudreaux–Bartels and T. Parks, Time–varying filtering and signal estimation using Wigner distribution synthesis techniques, *IEEE Trans. Acoust., Speech, Signal Process.*, vol. ASSP–34, no. 3, pp. 442–451, June 1986.

[19] J. Jeong and W. Williams, Time-varying filtering and signal synthesis, in B. Boashash, ed., *Time-Frequency Signal Analysis — Methods and Applications*, Melbourne, Australia: Longman Cheshire, 1995.

[20] Y. Zhang and M. G. Amin, Spatial averaging of time-frequency distributions for signal recovery in uniform linear arrays, *IEEE Trans. Signal Processing*, vol. 48, no. 10, pp. 2892–2902, Oct. 2000.

[21] A. Belouchrani, K. Abed-Meraim, M. G. Amin, and A. M. Zoubir, Joint antidiagonalization for blind source separation, in *Proc. Int. Conf. Acoust., Speech, Signal Process.*, Salt Lake City, UT, May 2001.

[22] R. O. Schmidt, Multiple emitter location and signal parameter estimation, *IEEE Trans. Antennas Propagat.*, vol. 34, no. 3, pp. 276–280, March 1986.

[23] I. Ziskind and M. Wax, Maximum likelihood localization of multiple sources by alternating projection, *IEEE Trans. Acoust. Speech Signal Process.*, vol. ASSP-36, no. 10, pp. 1553–1560, 1988.

[24] P. Stoica and A. Nehorai, MUSIC, maximum likelihood, and Cramér–Rao bound: further results and comparisons, *IEEE Trans. Acoust. Speech, Signal Process.*, vol. 38, no. 12, pp. 2140–2150, 1990.

[25] M. G. Amin and Y. Zhang, Direction finding based on spatial time-frequency distribution matrices, *Digital Signal Processing*, vol. 10, no. 4, pp. 325–339, Oct. 2000.

[26] M. G. Amin, A. Belouchrani, and Y. Zhang, The spatial ambiguity function and its applications, *IEEE Signal Processing Letters*, vol. 7, no. 6, pp. 138–140, June 2000.

[27] B. R. Breed and T. E. Posch, A range and azimuth estimator based on forming the spatial Wigner distribution, in *Proc. Int. Conf. Acoust., Speech, Signal Process.*, pp. 41B.9.1–41B.9.2, 1984.

[28] G. J. Frazer and Y. I. Abramovich, Quantifying multi-channel receiver calibration, Tech. Rep. DSTO-TR-1152, Defence Science and Technology Organisation, Edinburgh, SA, Australia, June 2001.

[29] D. H. Johnson and D. E. Dudgeon, *Array Signal Processing: Concepts and Techniques*, Englewood Cliffs, NJ: Prentice-Hall Signal Processing Series. Prentice-Hall, 1993.

[30] A. L. Swindlehurst and T. Kailath, Near-field source parameter estimation using a spatial Wigner distribution approach, in *Advanced Algorithms and Architectures for Signal Processing III*. SPIE, vol. 975, pp. 86–92, 1988.
[31] Y. D. Huang and M. Barkat, Near-field multiple source localization by passive sensor array, *IEEE Trans. Antennas Propagat.*, vol. 39, no. 7, pp. 968–975, July 1991.
[32] S. Valaee, B. Champagne, and P. Kabal, Parametric localization of distributed sources, *IEEE Trans. Signal Processing*, vol. 43, no. 9, pp. 2144–2153, Sept. 1995.
[33] J. Jeong and W. J. Williams, Alias-free generalized discrete-time time-frequency distributions, *IEEE Trans. Signal Processing*, vol. 40, no. 11, pp. 2757–2765, Nov. 1992.

10

Time-Frequency Processing of Time-Varying Signals with Nonlinear Group Delay

Antonia Papandreou-Suppappola

ABSTRACT Quadratic time-frequency representations (QTFRs) can be successful processing tools for different applications depending on the signal changes they can preserve. This chapter provides a tutorial on classes of QTFRs which are covariant to time shifts that match changes in the group delay function of the analysis signal. These changes may be constant or depend linearly or nonlinearly on frequency, and they may be the result of the signal propagating through systems with dispersive time-frequency characteristics. The unitary warping relationships of these group delay shift covariant QTFR classes to the constant time-shift covariant ones are established. Specific QTFR members are also presented together with the signal properties they satisfy. Various simulation examples are provided to demonstrate the importance of matching the time-frequency characteristics of the signal with the group delay shift of the QTFR.

10.1 Introduction

Many existing, traditional signal analysis tools such as the Fourier transform (FT) possess intrinsic limitations when the signals under consideration are nonstationary. Nonstationary signals, such as speech and biological sounds, are signals whose frequency content changes with time due to their time-varying spectral properties. The FT is of limited use for the analysis of nonstationary signals since it does not provide easily accessible information about the time localization of a given frequency component in a signal. A suitable analysis tool for nonstationary signals is provided by quadratic time-frequency representations (QTFRs), which are functions of both time and frequency and thus provide temporal localization of the signal's spectral components [1–13]. As a result, QTFRs have been successfully used in many nonstationary signal processing application areas, including speech, radar and sonar processing, image processing, and biological and biomedical signal processing.

Although many QTFRs have been proposed in the literature, no one QTFR exists that can be used effectively in all possible applications. This is so because different

QTFRs are best suited for analyzing signals with specific types of properties and time-frequency (TF) structures. For example, the well-known Wigner distribution (WD) [3, 8, 14–17] has been used to analyze speech since it preserves both constant time shifts and constant frequency shifts on a signal [3]. Preserving TF shifts is an important property since a suitable signal transformation for this application should be able to detect a change of pitch, or formant frequencies, as time changes. In order to assist the user in selecting an appropriate analysis tool, QTFRs are often classified based on the various properties they satisfy. A QTFR is then chosen that satisfies desirable properties for a given application.

There are many desirable properties for QTFRs to satisfy [1–3]. Some important QTFR properties include the covariance properties, examples of which are listed in Table 10.1. A QTFR is said to satisfy a covariance property if the QTFR preserves, or is covariant to, certain TF changes on a signal. For example, a QTFR (denoted as T) satisfies the constant frequency-shift covariance property (ii) in Table 10.1 if that QTFR preserves the constant frequency shift ν on a signal $x(t)$ with FT $X(f) = \int x(t)\, e^{-j2\pi tf}\, dt$. That is, if the signal is transformed as $Y(f) = (\mathcal{M}_\nu X)(f) = X(f-\nu)$, then the frequency-shift covariant QTFR is correspondingly transformed as $T_Y(t,f) = T_X(t, f-\nu)$. Here, the subscript in $T_X(t,f)$ indicates the signal that is being transformed, and t and f denote time and frequency, respectively. A useful QTFR classification is based on the grouping of various covariance properties that the QTFRs satisfy. For example, Cohen's class [1, 8] QTFRs (with signal-independent kernels) such as the WD and the spectrogram [18, 19] satisfy the constant time-shift and frequency-shift covariance properties. The affine class QTFRs [2, 20–23] such as the scalogram [21, 24] preserve constant time shifts and scale changes (dilations)

Covariance property	*Signal transformation*	*QTFR covariance*
(i) Scale covariance	$(\mathcal{C}_a X)(f) = \frac{1}{\sqrt{\lvert a \rvert}} X(\frac{f}{a})$	$T_{\mathcal{C}_a X}(t,f) = T_X(at, \frac{f}{a})$
(ii) Frequency-shift covariance	$(\mathcal{M}_\nu X)(f) = X(f-\nu)$	$T_{\mathcal{M}_\nu X}(t,f) = T_X(t, f-\nu)$
(iii) Time-shift covariance	$(\mathcal{S}_\tau X)(f) = e^{-j2\pi\tau f}\, X(f)$	$T_{\mathcal{S}_\tau X}(t,f) = T_X(t-\tau, f)$
(iv) Hyperbolic time-shift covariance	$(\mathcal{H}_c X)(f) = e^{-j2\pi c \ln \frac{f}{f_r}} X(f)$	$T_{\mathcal{H}_c X}(t,f) = T_X\left(t - \frac{c}{f}, f\right)$
(v) Power time-shift covariance	$(\mathcal{D}_c^{(\xi_\kappa)} X)(f) = e^{-j2\pi c\, \xi_\kappa\left(\frac{f}{f_r}\right)}\, X(f)$ where $\xi_\kappa(b) = \mathrm{sgn}(b)\, \lvert b \rvert^\kappa$	$T_{\mathcal{D}_c^{(\xi_\kappa)} X}(t,f) = T_X\big(t - c\,\tau_\kappa(f), f\big)$ where $\tau_\kappa(f) = \frac{d}{df}\, \xi_\kappa(\frac{f}{f_r})$
(vi) Exponential time-shift covariance	$(\mathcal{E}_c X)(f) = e^{-j2\pi c\, e^{f/f_r}} X(f)$	$T_{\mathcal{E}_c X}(t,f) = T_X\left(t - \frac{c}{f_r}\, e^{f/f_r}, f\right)$
(vii) Power exponential time-shift covariance	$(\mathcal{E}_c^{(\xi_\kappa)} X)(f) = e^{-j2\pi c\, e^{\kappa f/f_r}} X(f)$	$T_{\mathcal{E}_c^{(\xi_\kappa)} X}(t,f) = T_X\left(t - \kappa \frac{c}{f_r}\, e^{\kappa f/f_r}, f\right)$
(viii) Group delay shift covariance	$(\mathcal{D}_c^{(\xi)} X)(f) = e^{-j2\pi c\, \xi(\frac{f}{f_r})}\, X(f)$	$T_{\mathcal{D}_c^{(\xi)} X}(t,f) = T_X\big(t - c\,\tau(f), f\big)$ where $\tau(f) = \frac{d}{df}\, \xi(\frac{f}{f_r})$

TABLE 10.1. Covariance properties of a QTFR $T_X(t,f)$ due to various TF transformations on $X(f)$. Here, $f_r > 0$ is a reference frequency, $c, \kappa \in \Re$, $\mathrm{sgn}(b)$ yields the sign (± 1) of b, and $\xi(b)$ is a differentiable one-to-one function.

on the signal. The hyperbolic class [12, 25–28] consists of QTFRs such as the Altes Q-distribution (QD) [29, 30] that are covariant to hyperbolic time shifts and scale changes on the signal. Other covariant QTFR classes include the power classes [27, 28, 31–34], the exponential class [27, 35–37], the power exponential class [27, 38], many unitarily equivalent QTFR classes [39–42], and the displacement covariant classes [43–46]. These classes satisfy different covariance properties, and, as a result, are used for different types of applications.

In this chapter, we first review the traditional Cohen's class and affine class QTFRs, which always preserve constant time shifts on the analysis signal. Next, we study QTFR classes that always preserve nonlinear, frequency-dependent time shifts on the analysis signal that may be the result of a change in the signal's group delay when propagating through systems with dispersive TF characteristics. We show that these group delay shift covariant QTFR classes are related to the constant time-shift covariant QTFR classes through a unitary warping transformation that is fixed by the dispersive system characteristics. The warping is specified by a function $\xi(b)$ that identifies: (i) the group delay change or time shift $\tau(f) = (d/df)\xi(f/f_r)$, and (ii) the corresponding QTFR class that preserves that time shift. We provide various simulation examples to demonstrate the importance of matching the TF characteristics of the analysis signal with the group delay shift of the QTFR used in the analysis.

10.2 Constant time-shift covariant QTFRs

An important property for QTFRs to satisfy is the constant time-shift covariance property. Consider a signal $x(t)$ that is shifted in time by a constant amount τ, forming $y(t) = x(t - \tau)$. This signal transformation causes a linear change in the phase of the FT of the signal, $Y(f) = (\mathcal{S}_\tau X)(f) = e^{-j2\pi\tau f}\, X(f)$, where $\mathcal{S}_\tau$ is the constant time-shift operator defined to operate on the frequency-domain signal. A list of operators used in this chapter is provided in Table 10.2. If one wants to maintain the shift information in a particular application, then it is important for the analysis QTFR $T_X(t, f)$ to preserve that time shift. Specifically,

$$(\mathcal{S}_\tau X)(f) = e^{-j2\pi\tau f}\, X(f) \quad \Rightarrow \quad T_{\mathcal{S}_\tau X}(t, f) = T_X(t - \tau,\, f)\,. \tag{10.1}$$

For example, in shallow water sonar signal processing applications, where boundary interactions are prevalent, one may receive a bottom bounce path several milliseconds after the direct path. The bottom bounce arrival can simplistically be modeled as a delayed version of the direct path. Thus, a QTFR analyzing the received signal must preserve the delay associated with the difference in path lengths.

Next, we consider two classes of QTFRs that always satisfy the constant time-shift covariance property in (10.1) together with an additional important covariance property.

Operator name	*Operator symbol*	*Effect of operator on* $X(f)$
Identity	$\mathcal{I}$	$(\mathcal{I}X)(f) = X(f)$
Group delay shift	$\mathcal{D}_c^{(\xi)}$	$(\mathcal{D}_c^{(\xi)}X)(f) = e^{-j2\pi c\,\xi(f/f_r)}X(f)$
Constant time-shift	$\mathcal{S}_{c/f_r} = \mathcal{D}_c^{(\xi_{\mathrm{linear}})}$	$(\mathcal{S}_{c/f_r}X)(f) = e^{-j2\pi c(f/f_r)}X(f)$
Hyperbolic time-shift	$\mathcal{H}_c = \mathcal{D}_c^{(\xi_{\ln})}$	$(\mathcal{H}_cX)(f) = e^{-j2\pi c\ln(f/f_r)}X(f),\ \ f>0$
κth Power time-shift	$\mathcal{D}_c^{(\xi_\kappa)}$	$(\mathcal{D}_c^{(\xi_\kappa)}X)(f) = e^{-j2\pi c\,\xi_\kappa(f/f_r)}X(f)$
Exponential time-shift	$\mathcal{E}_c = \mathcal{D}_c^{(\xi_{\exp})}$	$(\mathcal{E}_cX)(f) = e^{-j2\pi\,c\,e^{f/f_r}}X(f)$
κth Power exponential time-shift	$\mathcal{E}_c^{(\xi_\kappa)} = \mathcal{D}_c^{(\xi_\kappa\,\exp)}$	$(\mathcal{E}_c^{(\xi_\kappa)}X)(f) = e^{-j2\pi c\,e^{\kappa f/f_r}}X(f)$
Generalized exponential time-shift	$\mathcal{E}_c^{(\xi)} = \mathcal{D}_c^{(e^{\xi})}$	$(\mathcal{E}_c^{(\xi)}X)(f) = e^{-j2\pi c\,e^{\xi(f/f_r)}}X(f)$
Generalized hyperbolic time-shift	$\mathcal{H}_c^{(\xi)} = \mathcal{D}_c^{(\ln\xi)}$	$(\mathcal{H}_c^{(\xi)}X)(f) = e^{-j2\pi c\ln\xi(f/f_r)}X(f)$
Frequency-shift	$\mathcal{M}_\nu$	$(\mathcal{M}_\nu X)(f) = X(f-\nu)$
Scale (dilation)	$\mathcal{C}_a$	$(\mathcal{C}_aX)(f) = \frac{1}{\sqrt{\lvert a\rvert}}X(\frac{f}{a})$
Dispersive warping	$\mathcal{W}_\xi$	$(\mathcal{W}_\xi X)(f) = \frac{1}{\sqrt{\lvert\xi'(\xi^{-1}(\frac{f}{f_r}))\rvert}}\,X\left(f_r\xi^{-1}(\frac{f}{f_r})\right)$
Hyperbolic warping	$\mathcal{W}_{\xi_{\ln}}$	$(\mathcal{W}_{\xi_{\ln}}X)(f) = \sqrt{e^{f/f_r}}\,X\left(f_r\,e^{f/f_r}\right)$
Power warping	$\mathcal{W}_{\xi_\kappa}$	$(\mathcal{W}_{\xi_\kappa}X)(f) = \frac{1}{\sqrt{\lvert\xi_\kappa'(\xi_\kappa^{-1}(\frac{f}{f_r}))\rvert}}X\left(f_r\xi_\kappa^{-1}(\frac{f}{f_r})\right)$
Exponential warping	$\mathcal{W}_{\xi_{\exp}}$	$(\mathcal{W}_{\xi_{\exp}}X)(f) = \sqrt{\frac{f_r}{f}}\,X\left(f_r\ln\frac{f}{f_r}\right),\ f>0$
Power exponential warping	$\mathcal{W}_{\xi_\kappa\,\exp}$	$(\mathcal{W}_{\xi_\kappa\,\exp}X)(f) = \sqrt{\frac{f_r}{\lvert\kappa\rvert f}}\,X\left(\frac{f_r}{\kappa}\ln\frac{f}{f_r}\right),\ f>0$
Dispersively warped frequency-shift	$\mathcal{Y}_\nu^{(\xi)}$	$(\mathcal{Y}_\nu^{(\xi)}X)(f) = \sqrt{\frac{\lvert\xi'(\frac{f}{f_r})\rvert}{\lvert\xi'(y(f,\nu))\rvert}}\,X(f_ry(f,\nu))$ where $y(f,\nu) = \xi^{-1}(\xi(\frac{f}{f_r}) - \frac{\nu}{f_r})$
Dispersively warped scale	$\mathcal{L}_a^{(\xi)}$	$(\mathcal{L}_a^{(\xi)}X)(f) = \sqrt{\frac{\lvert\xi'(\frac{f}{f_r})\rvert}{\lvert a\xi'(l(f,a))\rvert}}\,X(f_rl(f,a))$ where $l(f,a) = \xi^{-1}(\frac{1}{a}\xi(\frac{f}{f_r}))$

TABLE 10.2. A list of operators used in the chapter together with their effect on $X(f)$. Here, $\xi^{-1}(\xi(b)) = b$, $\xi'(b) = \frac{d}{db}\xi(b)$, $\xi_{\mathrm{linear}}(b) = b$, $\xi_{\ln}(b) = \ln b$, $\xi_\kappa(b) = \mathrm{sgn}(b)\lvert b\rvert^\kappa$, $\xi_{\exp}(b) = e^{\kappa b}$, and $\xi_{\kappa\,\exp}(b) = e^{\kappa b}$.

10.2.1 Cohen's class

Cohen's class (C) [1, 8] with signal-independent kernels (or Cohen's class for simplicity) consists of all QTFRs, $T_X^{(C)}(t,f)$, that satisfy the time-shift covariance property in (10.1) and the frequency-shift covariance property

$$(\mathcal{M}_\nu X)(f) = X(f-\nu) \quad\Rightarrow\quad T_{\mathcal{M}_\nu X}^{(C)}(t,f) = T_X^{(C)}(t,f-\nu)\,. \tag{10.2}$$

Both the time- and frequency-shift covariance properties are important in applications where the signal needs to be analyzed at various TF points using the same time resolution and the same frequency resolution. In particular, Cohen's class QTFRs feature a TF analysis resolution that is independent of the analysis time and frequency (i.e., constant-bandwidth analysis), and they maintain the signal's time and frequency shifts. This is important in applications such as speech analysis and sonar and radar signal processing.

Any QTFR of Cohen's class can be written as

$$T_X^{(C)}(t,f) = \iint \Phi_T^{(C)}(f-\hat{f},\nu) U_X(\hat{f},\nu) e^{j2\pi t\nu}\, d\hat{f}\, d\nu \tag{10.3}$$

$$= \iint \psi_T^{(C)}(t-\hat{t}, f-\hat{f}) W_X(\hat{t},\hat{f})\, d\hat{t}\, d\hat{f}, \tag{10.4}$$

where $U_X(f,\nu) = X(f+\nu/2)X^*(f-\nu/2)$, and $W_X(t,f)$ is the WD of $X(f)$,

$$W_X(t,f) = \int X\left(f+\frac{\nu}{2}\right) X^*\left(f-\frac{\nu}{2}\right) e^{j2\pi t\nu}\, d\nu. \tag{10.5}$$

Throughout this chapter, the range of integration is from $-\infty$ to ∞ unless otherwise stated. From (10.4), it can be seen that Cohen's class QTFRs can be written in terms of the WD using a two-dimensional (2D) convolution. The 2D functions $\Phi_T^{(C)}(f,\nu)$ and $\psi_T^{(C)}(t,f)$ are *kernels* that uniquely characterize any Cohen's class QTFR $T^{(C)}$. Note that the kernel functions are FT pairs $\psi_T^{(C)}(t,f) = \int \Phi_T^{(C)}(f,\nu) e^{j2\pi t\nu}\, d\nu$ [47], and they do not depend on the signal $X(f)$.

The Cohen's class formulations in (10.3) and (10.4) can be obtained using Cohen's characteristic method [1,8,48,49]. The same formulations can also be obtained by constraining the general form of any QTFR [20,47],

$$T_X(t,f) = \iint K_T(t,f;f_1,f_2) X(f_1) X^*(f_2)\, df_1 df_2, \tag{10.6}$$

to satisfy both the time-shift covariance in (10.1) and the frequency-shift covariance in (10.2). The resulting constraint simplifies the signal-independent, four-dimensional kernel $K_T(t,f;f_1,f_2)$ in (10.6) to any one of the 2D kernels in (10.3) and (10.4). In addition to the TF-shift covariance properties, Cohen's class QTFRs satisfy other desirable properties, provided that their kernels follow some simple constraints [3, 27]. The WD is an important member of Cohen's class, as it satisfies many desirable properties including preservation of signal energy, marginal distributions, and various moments such as the instantaneous frequency or group delay of a signal. Some other members of Cohen's class include the spectrogram [18, 19], the Choi–Williams exponential distribution [1, 50], and the generalized exponential distribution [51]. A list of numerous QTFR members of Cohen's class that have been developed in the literature can be found, for example, in [1–3].

10.2.2 Affine class

The affine class (A) [2, 20–23] consists of all QTFRs, $T_X^{(A)}(t,f)$, that satisfy the time-shift covariance in (10.1) and the scale covariance property

$$(\mathcal{C}_a X)(f) = \frac{1}{\sqrt{|a|}} X\left(\frac{f}{a}\right) \quad \Rightarrow \quad T_{\mathcal{C}_a X}^{(A)}(t,f) = T_X^{(A)}\left(at, \frac{f}{a}\right). \tag{10.7}$$

The scale covariance property is important for multiscale or multiresolution analysis [3], for self-similar signals [52], scale covariant systems [53], and Doppler invariant signals [54]. For example, it is desirable to maintain scale changes on the signal in applications such as image enlargement or compression and wavelet analysis. Many affine class QTFRs, like the scalogram (squared magnitude of the wavelet transform), are used for constant-Q TF analysis where the analysis bandwidth is proportional to the analysis frequency. This offers an alternative from the constant-bandwidth analysis achieved by Cohen's class QTFRs.

By constraining any QTFR in (10.6) to satisfy the time-shift covariance property in (10.1) and the scale covariance property in (10.7), any affine class QTFR can be written as [2, 21–23]

$$T_X^{(A)}(t,f) = \frac{1}{|f|} \int\int \Phi_T^{(A)}\left(-\frac{\hat{f}}{f}, \frac{\nu}{f}\right) U_X(\hat{f},\nu) e^{j2\pi t\nu} d\hat{f}\, d\nu \tag{10.8}$$

$$= \int\int \psi_T^{(A)}\left(f(t-\hat{t}), -\frac{\hat{f}}{f}\right) W_X(\hat{t},\hat{f})\, d\hat{t}\, d\hat{f}. \tag{10.9}$$

The 2D kernel functions $\Phi_T^{(A)}(b,\beta)$ and $\psi_T^{(A)}(c,b)$ uniquely characterize any affine class QTFR $T_X^{(A)}(t,f)$, and they are FT pairs, $\psi_T^{(A)}(c,b) = \int \Phi_T^{(A)}(b,\beta)\, e^{j2\pi c\beta}\, d\beta$ [2]. A table of some desirable QTFR properties and corresponding kernel constraints of the affine class can be found in [2]. The WD is a member of both the affine class, with affine class kernel $\Phi_W^{(A)}(b,\beta) = \delta(b+1)$, and Cohen's class, with Cohen's class kernel $\Phi_W^{(C)}(f,\nu) = \delta(f)$. It is a member of the intersection between the two classes [55], and, as such, it is time-shift covariant, frequency-shift covariant, and scale covariant. Other members of the affine class include the scalogram, the Bertrand P_κ-distributions, the Flandrin D-distribution, and the passive and active versions of the Unterberger distribution [2].

10.3 Group delay shift covariance property

10.3.1 Definition and special cases

The group delay shift covariance property [12, 27, 28, 33, 35, 36, 38] is an important property for analyzing signals propagating through systems with dispersive TF characteristics or, equivalently, with nonlinear group delay. The group delay $\mu(f)$ of a signal, with FT $X(f) = r(f)\, e^{-j2\pi\, \lambda(f/f_r)}$, is defined as the derivative of the phase response $\lambda(f/f_r)$ of the signal, i.e., $\mu(f) = (d/df)\lambda(f/f_r)$. Here, $r(f) \geq 0$ is the amplitude response of the signal, and $f_r > 0$ is a reference frequency. A dispersive system is one that delays in time different frequencies by different amounts. If the signal $X(f)$ is passed through an all-pass, dispersive system or operator with output $Y(f) = e^{-j2\pi\xi(f/f_r)}X(f)$, then the change in group delay, $\tau(f) = d/df\,\xi(f/f_r)$, is proportional to the derivative of the one-to-one phase function $\xi(f/f_r)$. Since group delay is a measure of the time delay introduced in each sinusoidal component of the signal at frequency f, the ideal QTFR $T_X(t, f)$ should preserve this change in group delay or frequency-dependent time shift $\tau(f)$. If the group delay shift operator of a signal $X(f)$ is given as

$$(\mathcal{D}_c^{(\xi)} X)(f) = e^{-j2\pi c\,\xi(f/f_r)} X(f), \qquad f \in \wp, \tag{10.10}$$

where $\wp$ represents the values in the domain of $\xi(\cdot)$, then a QTFR T is group delay shift covariant if the QTFR of the output, $T_{\mathcal{D}_c^{(\xi)}X}$, corresponds to the QTFR of the input, T_X, shifted in time by an amount equal to the change in group delay $c\tau(f) = c(d/df)\xi(f/f_r)$ that is introduced in (10.10), i.e.,

$$T_{\mathcal{D}_c^{(\xi)}X}(t, f) = T_X\big(t - c\tau(f), f\big), \qquad f \in \wp. \tag{10.11}$$

The QTFR transformation in (10.11), due to the signal transformation in (10.10) resulting in a nonlinear change in group delay, is demonstrated in Figure 10.1. In particular, if the function $\xi(b)$ is nonlinear, then the frequency-dependent time shift $\tau(f) = (d/df)\xi(f/f_r)$ is dispersive in the sense that high frequency components are delayed by a different amount in time from low frequency components. The real-valued parameter c is a measure of the amount of dispersion (see Figure 10.2). For example, when an object immersed in water, a dispersive medium, is illuminated by a plane wave [3, 56–58], $\tau(f)$ depends on the properties of the medium, whereas c depends on various factors such as the object's circumference, its distance from the

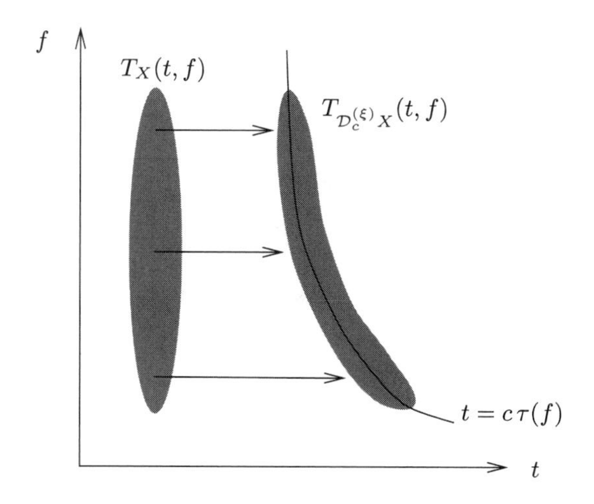

FIGURE 10.1. The QTFR preserves a dispersive change $c\tau(f)$ in a signal's group delay as a shift along the time axis of the QTFR.

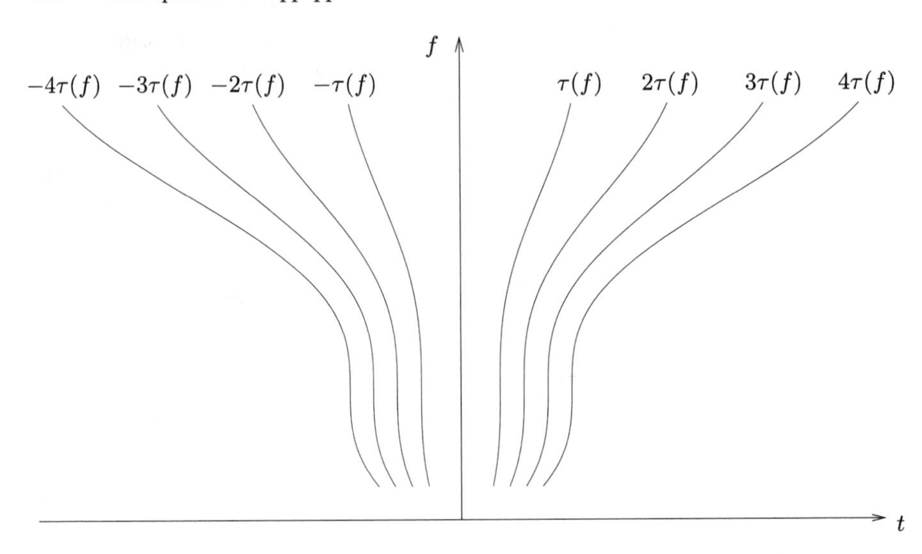

FIGURE 10.2. The group delay shift operator $\mathcal{D}_c^{(\xi)}$ in (10.10) corresponds to a group delay curve $t = c\tau(f)$ in the TF plane.

transmitter or receiver, and the number of times the wave oscillates around the object before it is reflected back to the transmitter or receiver [32].

The group delay shift operator $\mathcal{D}_c^{(\xi)}$ in (10.10) is unitarily equivalent to the constant time-shift operator $\mathcal{S}_\tau$ in (10.1) [12,27,41]. Specifically, the following relationship holds between the two operators:

$$\mathcal{D}_c^{(\xi)} = \mathcal{W}_\xi^{-1} \mathcal{S}_{c/f_r} \mathcal{W}_\xi, \tag{10.12}$$

where the warping operator is given by

$$(\mathcal{W}_\xi X)(f) = \left|\xi'\left(\xi^{-1}(f/f_r)\right)\right|^{-1/2} X\left(f_r \xi^{-1}(f/f_r)\right), \qquad f \in \aleph. \tag{10.13}$$

Here, $\xi^{-1}(b)$ denotes the inverse function, $\xi'(b) = (d/db)\xi(b)$, $\aleph$ is the range of $\xi(\cdot)$, and $\mathcal{W}_\xi^{-1}$ is the inverse of the warping operator. The warping operator in (10.13) (which we call the "dispersive warping operator" due to the possible dispersive nature of $\xi(b)$), is unitary [59] since it preserves inner products, i.e., $\int_\aleph (\mathcal{W}_\xi X)(f)\,(\mathcal{W}_\xi X)^*(f)\,df = \int_\wp X(f)\,X^*(f)\,df$.

The group delay shift covariance property in (10.10) and (10.11) simplifies to a particular covariance property (satisfied by a different class of QTFRs) when the one-to-one function $\xi(b)$ and the time shift $\tau(f) = (d/df)\xi(f/f_r)$ are fixed. For example,

- when $\xi(b) = \xi_{\text{linear}}(b) = b$ and $\tau(f) = 1/f_r$, the group delay shift covariance in (10.11) simplifies to the constant time-shift covariance in (10.1) of Cohen's class or the affine class [3];

- when $\xi(b) = \xi_{\ln}(b) = \ln b$ and $\tau(f) = 1/f$, $f > 0$, the covariance in (10.11) is the (dispersive) hyperbolic time-shift covariance (iv) in Table 10.1 of the hyperbolic class [25,26,60];
- when $\xi(b) = \xi_{\kappa}(b) = \text{sgn}(b)\,|b|^{\kappa}$ and $\tau_{\kappa}(f) = (d/df)\xi_{\kappa}(f/f_r) = (\kappa/f_r) \times |f/f_r|^{\kappa-1}$, the covariance in (10.11) is the (dispersive) κth power time-shift covariance (v) in Table 10.1 of the κth power class [27,31–34];
- when $\xi(b) = \xi_{\exp}(b) = e^{b}$ and $\tau(f) = (1/f_r)e^{f/f_r}$, the covariance in (10.11) is the (dispersive) exponential time-shift covariance (vi) in Table 10.1 of the exponential class [27,35–37];
- and when $\xi(b) = \xi_{\kappa\ \exp}(b) = e^{\kappa b}$ and $\tau(f) = (\kappa/f_r)e^{\kappa f/f_r}$, the covariance in (10.11) is the (dispersive) κth power exponential time-shift covariance (vii) in Table 10.1 of the κth power exponential class [27,38].

10.3.2 QTFR time shift and signal group delay

For successful TF analysis, it is advantageous to match the group delay shift of a QTFR in (10.11) with (changes in) the group delay of a signal. Thus, a hyperbolic QTFR that always satisfies the hyperbolic time-shift covariance would be chosen to analyze a signal with hyperbolic TF characteristics. For example, linear phase (constant group delay) systems are matched to the constant time-shift covariance of Cohen's class. Cohen's class QTFRs such as the WD are well matched to complex, frequency-domain sinusoids (i.e., time-domain Dirac impulses) $X(f) = (1/\sqrt{f_r})e^{-j2\pi(c/f_r)f}$ with constant group delay $c\mu(f) = c/f_r$, as shown in Figure 10.3(a). This is because Cohen's class QTFRs satisfy the properties (10.10) and (10.11) with $\xi(b) = b$ and constant time shift $c\tau(f) = c/f_r$. On the other hand, logarithmic phase ($1/f$ group delay) systems are matched to the hyperbolic time-shift covariance property of the hyperbolic class. Thus, hyperbolic class QTFRs such as the QD [29, 30] are well-matched to hyperbolic impulses $X(f) = (1/\sqrt{f})e^{-j2\pi c \ln(f/f_r)}$, $f > 0$, with group delay $c\mu(f) = c/f$, as shown in Figure 10.3(d). This follows since hyperbolic class QTFRs satisfy the property of (10.10) and (10.11) with $\xi(b) = \ln b$ and hyperbolic time shift $c\tau(f) = c/f$. Note, however, that when the signal group delay $\mu(f)$ does *not* match the group delay shift $\tau(f)$ of the QTFR, then significant distortion may occur that could impede analysis [61,62]. This is demonstrated by analyzing the complex sinusoid using the hyperbolic class QD in Figure 10.3(c) and the hyperbolic impulse using Cohen's class WD in Figure 10.3(b). The mismatched QTFRs suffer from inner interference terms [17].

10.3.3 Generalized impulse and signal expansion

As demonstrated in the previous section, QTFRs covariant to group delay time shifts in (10.11) are ideally suited to analyze signals with group delay equal to that time shift. That is, QTFRs that preserve time shifts $\tau(f)$ are matched analysis tools for signals with group delay $\tau(f)$. Thus, the TF geometry underlying the group delay shift covariance property is related to the frequency-domain *generalized impulse* defined as

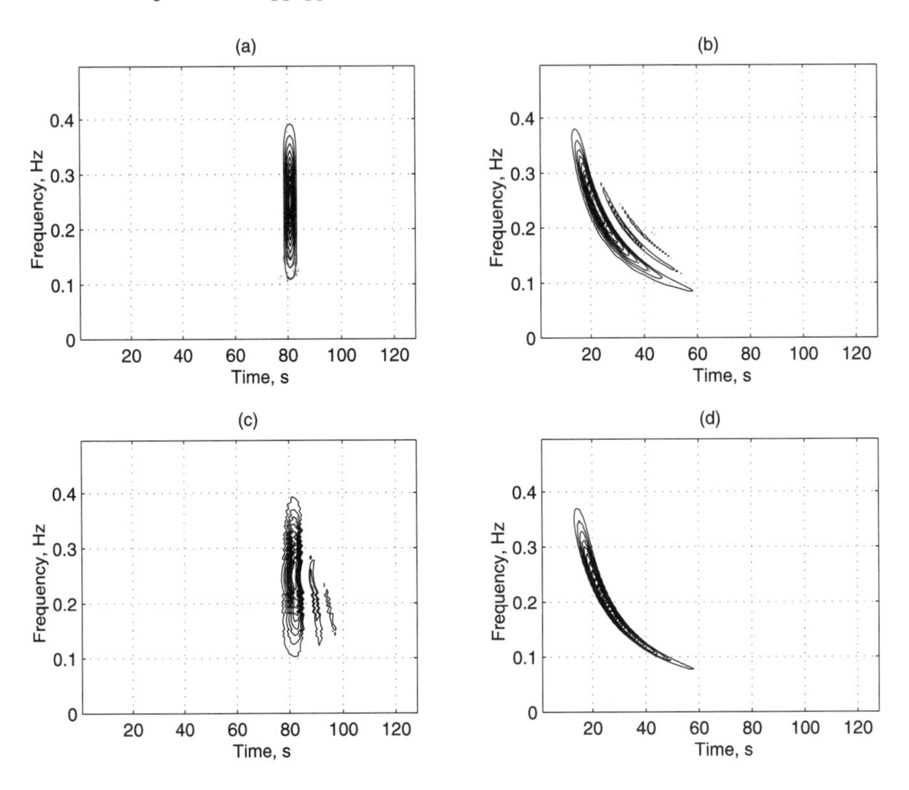

FIGURE 10.3. TF analysis of a (windowed) impulse using (a) the WD and (c) the QD, and TF analysis of a windowed hyperbolic impulse using (b) the WD and (d) the QD. In (a), Cohen's class WD is well matched to constant group delay, while in (d), the hyperbolic class QD is well matched to hyperbolic group delay.

$$I_c^{(\lambda)}(f) \triangleq \sqrt{|\mu(f)|}e^{-j2\pi c\lambda(f/f_r)}, \qquad f \in \wp, \tag{10.14}$$

where $\wp$ is the domain of the phase response $\lambda(b)$. In particular, the group delay $\mu(f) = (d/df)\lambda(f/f_r)$ reflects the dispersion characteristics of the QTFR class which is covariant to group delay shifts $\tau(f)$ in (10.11) only when $\mu(f) = \tau(f)$ (or, equivalently, when $\lambda(b) = \xi(b)$ in (10.10) and (10.11)).

The generalized impulse in (10.14) simplifies to known signals when $\lambda(b)$ and $\mu(f)$ are fixed, as summarized in Table 10.3. An important property of the generalized impulse follows when the signal is transformed using the group delay shift operator in (10.10). Specifically, group delay shifting the impulse in (10.14) simply shifts its (subscript) parameter value, $(\mathcal{D}_{c_0}^{(\xi)} I_c^{(\lambda)})(f) = I_{c+c_0}^{(\lambda)}(f)$, provided $\xi(b) = \lambda(b)$ in (10.10).

The generalized impulse plays an important role in a generalized signal expansion for any finite-energy signal $X(f)$, since the parameter c in (10.14) can vary from $-\infty$ to ∞, and thus the family of generalized impulses covers the entire TF plane.

Phase response $\lambda(b)$	*Group delay function* $\mu(f) = \frac{d}{df}\lambda\left(\frac{f}{f_r}\right)$	*Impulse* $I_c^{(\lambda)}(f)$ in (10.14)	*Coefficient function* $\rho_X^{(\lambda)}(c)$ in (10.16)
b	$1/f_r$	$\frac{1}{\sqrt{f_r}}\, e^{-j2\pi c \frac{f}{f_r}}$	$\frac{1}{\sqrt{f_r}} \int X(f)\, e^{j2\pi c \frac{f}{f_r}}\, df$ (Inverse FT)
$\ln b$	$1/f$	$\frac{1}{\sqrt{f}}\, e^{-j2\pi c \ln \frac{f}{f_r}}, \quad f > 0$	$\int_0^\infty X(f) e^{j2\pi c \ln \frac{f}{f_r}} \frac{df}{\sqrt{f}}$ (Mellin transform)
$\mathrm{sgn}(b)\, \lvert b\rvert^\kappa$	$\frac{\kappa}{f_r}\left\lvert\frac{f}{f_r}\right\rvert^{\kappa-1}$	$\sqrt{\frac{\lvert\kappa\rvert}{f_r}}\, \lvert\frac{f}{f_r}\rvert^{\frac{\kappa-1}{2}}\, e^{-j2\pi c\, \mathrm{sgn}(f)\, \lvert\frac{f}{f_r}\rvert^\kappa}$	$\int X(f) \sqrt{\frac{\lvert\kappa\rvert}{f_r}}\, \lvert\frac{f}{f_r}\rvert^{\frac{\kappa-1}{2}}\, e^{j2\pi c\, \mathrm{sgn}(f)\, \lvert\frac{f}{f_r}\rvert^\kappa}\, df$
e^b	$\frac{1}{f_r}\, e^{f/f_r}$	$\frac{1}{\sqrt{f_r}} e^{\frac{f}{2f_r}} e^{-j2\pi c\, e^{f/f_r}}$	$\int X(f) \frac{1}{\sqrt{f_r}} e^{\frac{f}{2f_r}} e^{j2\pi c\, e^{f/f_r}}\, df$
$e^{\kappa b}$	$\frac{\kappa}{f_r}\, e^{\kappa f/f_r}$	$\frac{1}{\sqrt{\lvert\kappa\rvert f_r}} e^{\frac{\kappa f}{2f_r}} e^{-j2\pi c\, e^{\kappa f/f_r}}$	$\int X(f) \frac{1}{\sqrt{\lvert\kappa\rvert f_r}} e^{\frac{\kappa f}{2f_r}} e^{j2\pi c\, e^{\kappa f/f_r}}\, df$

TABLE 10.3. The generalized impulse in (10.14) and coefficient function in (10.16) for various choices of the phase response $\lambda(b)$ of the impulse.

If the phase response $\lambda(b)$ is one-to-one with range $\Re$, then any finite-energy signal $X(f)$ can be expanded as

$$X(f) = \int \rho_X^{(\lambda)}(c)\, I_c^{(\lambda)}(f)\, dc = \sqrt{|\mu(f)|} \int \rho_X^{(\lambda)}(c) e^{-j2\pi c\, \lambda(f/f_r)}\, dc, \qquad (10.15)$$

where the generalized coefficient function, $\rho_X^{(\lambda)}(c)$, is the inner product of $X(f)$ with $I_c^{(\lambda)}(f)$ [32,33,35],

$$\rho_X^{(\lambda)}(c) = \int_{\wp} X(f) I_c^{(\lambda)*}(f)\, df = \int_{\wp} X(f) \sqrt{|\mu(f)|} e^{j2\pi c\, \lambda(f/f_r)}\, df. \qquad (10.16)$$

The validity of the expansion in (10.15) and (10.16) follows from the completeness property $\int I_c^{(\lambda)}(f_1)\, I_c^{(\lambda)\,*}(f_2)\, dc = \delta(f_1 - f_2)$ [25]. The generalized impulse also satisfies the orthogonality property [25] if $\lambda(b)$ is one-to-one with range $\Re$, that is, $\int_{\wp} I_{c_1}^{(\lambda)}(f)\, I_{c_2}^{(\lambda)\,*}(f)\, df = \delta(c_1 - c_2)$. The expansion in (10.15) is a unitary linear transform $X(f) \leftrightarrow \rho_X^{(\lambda)}(c)$ that generalizes both the FT and the Mellin transform (cf. Table 10.3). In particular, as shown in the last column of Table 10.3, the generalized coefficient function corresponds to the inverse FT when $\lambda(b) = b$, and to the Mellin transform when $\lambda(b) = \ln b$. A group delay shift of the signal, $Y(f) = (\mathcal{D}_{c_0}^{(\xi)} X)(f)$ in (10.10), simply translates the argument of the generalized coefficient function by c_0, i.e., $\rho_Y^{(\lambda)}(c) = \rho_X^{(\lambda)}(c - c_0)$ when $\lambda(b) = \xi(b)$ in (10.10). Note also that $\rho_X^{(\lambda)}(c)$ is unitary since it preserves inner products, i.e., $\int_{\wp} X(f) X^*(f) df = \int \rho_X^{(\lambda)}(c) \rho_X^{(\lambda)\,*}(c)\, dc$.

Next we consider various group delay shift covariant QTFR classes that we summarize in Figure 10.4.

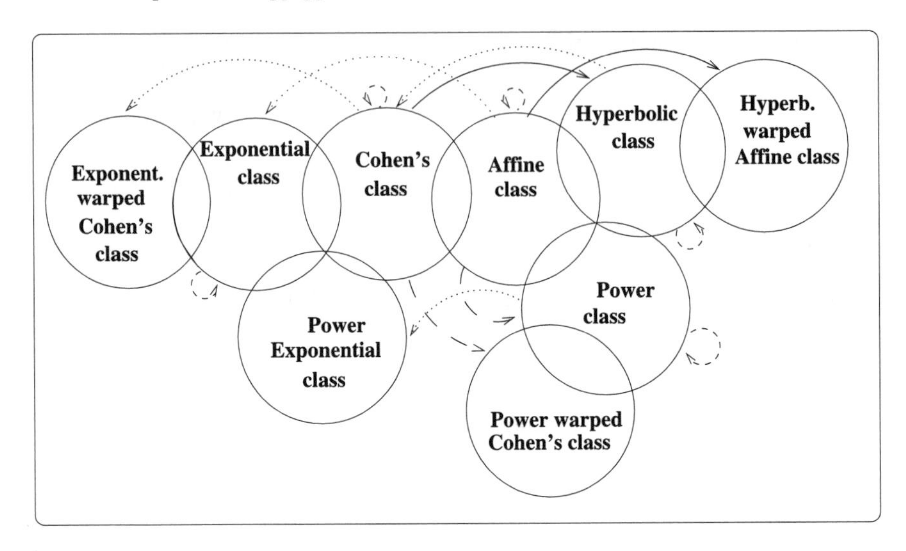

FIGURE 10.4. Some group delay shift covariant QTFR classes. The arrows point from a class being mapped to a new class formed using hyperbolic warping (——), power warping (— — —), or exponential warping (. . .). The warping (- - -) maps a class back to itself ("self-mapping").

10.4 Group delay shift covariant QTFRs

Group delay shift covariant QTFRs preserve frequency-dependent time shifts $\tau(f)$ in (10.11) that correspond to changes in the signal's group delay. These QTFRs are important for analyzing signals passing through systems with dispersive TF characteristics. They are ideally suited for signals whose group delay $\mu(f)$ is exactly equal to the group delay shift $\tau(f)$, as in the case of the generalized impulses defined in (10.14). We can obtain group delay shift covariant QTFR classes by warping Cohen's class QTFRs or affine class QTFRs using [12,27,28,35,36]

$$T_X^{(D\,class)}(t,f) = T_{\mathcal{W}_\xi X}^{(class)}\left(\frac{t}{f_r\tau(f)}, f_r\xi\left(\frac{f}{f_r}\right)\right), \qquad f \in \wp, \tag{10.17}$$

where $\wp$ is the domain of $\xi(\cdot)$, and the dispersive warping operator $\mathcal{W}_\xi$ is defined in (10.13). The superscript $(class)$ indicates which QTFR class undergoes the warping in (10.17). For example, when $(class) = (C)$ corresponding to Cohen's class, we obtain the dispersively warped Cohen's class QTFRs, $T_X^{(DC)}(t,f)$. When $(class) = (A)$ corresponding to the affine class, we obtain the dispersively warped affine class QTFRs, $T_X^{(DA)}(t,f)$. A dispersively warped affine class QTFR (e.g. the dispersively warped version of the WD) is obtained in three steps using (10.17) with a specified warping function $\xi(b)$. First, the analysis signal is transformed as $Y(f) = (\mathcal{W}_\xi X)(f)$. Second, the corresponding affine class QTFR (e.g., the WD) of the warped signal $Y(f)$ is computed. Last, the TF axes are transformed for correct TF localization using $t \to t/(f_r \tau(f))$ and $f \to f_r \xi(f/f_r)$ [35,36]. Note that arbitrary unitary warpings from Cohen's class or the affine class were considered in [40,41]. Here, we consider warpings that lead to the group delay shift covariance QTFR property in (10.11).

The warping function $\xi(b)$ is very important, as it specifies (a) the new QTFR class, $T^{(D\,class)}$, in (10.17), and (b) the group delay $\tau(f) = (d/df)\xi(f/f_r)$ shift covariance in (10.11) of the new class. Thus, fixing the warping function $\xi(b)$ (which then fixes the time shift $\tau(f)$ and the domain $\wp$ of $\xi(b)$), results in a new class that satisfies a new group delay shift covariance. In the following sections, we will provide various class examples obtained by fixing $\xi(b)$. Note that when $\xi(b) = \xi_{\text{linear}}(b) = b$, the warping operator in (10.13) simplifies to the identity operator $(\mathcal{W}_{\xi_{\text{linear}}} X)(f) = (\mathcal{I}X)(f) = X(f)$, and (10.17) simply maps a QTFR class back to itself ("self-mapping"). The group delay shift covariant QTFRs always satisfy (10.11) for a given one-to-one function $\xi(b)$. As we will demonstrate next, these QTFRs also satisfy an additional covariance property that depends on the function $\xi(b)$.

10.4.1 Dispersively warped Cohen's class

10.4.1.1 Class covariance properties and formulation

The dispersively warped Cohen's class (DC) QTFRs, $T_X^{(DC)}(t,f)$, are obtained by warping Cohen's QTFRs, $T_X^{(C)}(t,f)$, using (10.17) with $(class) = (C)$ and the function $\xi(b)$ chosen to fix a desired group delay shift in (10.10) and (10.11). Due to the warping $(\mathcal{W}_\xi X)(f)$ in (10.13), the time- and frequency-shift covariance properties in (10.1) and (10.2) in Cohen's class are transformed into two new properties in the dispersively warped Cohen's class. The time-shift operator $\mathcal{S}_\tau = \mathcal{S}_{c/f_r}$ in (10.1) maps to the group delay shift operator $\mathcal{D}_c^{(\xi)} = \mathcal{W}_\xi^{-1} \mathcal{S}_{c/f_r} \mathcal{W}_\xi$ in (10.10), and any dispersively warped Cohen's class QTFR preserves group delay shifts in (10.11). The frequency-shift operator $\mathcal{M}_\nu$ in (10.2) in Cohen's class maps to the dispersively warped frequency-shift operator $\mathcal{Y}_\nu^{(\xi)} = \mathcal{W}_\xi^{-1} \mathcal{M}_\nu \mathcal{W}_\xi$, which transforms the signal as

$$Y(f) = (\mathcal{Y}_\nu^{(\xi)} X)(f) = \frac{\left|\xi'\left(\frac{f}{f_r}\right)\right|^{1/2}}{|\xi'(y(f,\nu))|^{1/2}} X(f_r\, y(f,\nu)), \tag{10.18}$$

where $y(f,\nu) = \xi^{-1}(\xi(f/f_r) - \nu/f_r)$. The resulting covariance property is

$$T_Y^{(DC)}(t,f) = T_X^{(DC)}\left(t\,\frac{\tau(f_r y(f,\nu))}{\tau(f)}, f_r y(f,\nu)\right). \tag{10.19}$$

The dispersively warped frequency-shift covariance in (10.18) and (10.19) may or may not be useful in a particular application, depending on the warping function $\xi(b)$. If $\xi(b) = \xi_{\ln}(b) = \ln b$ and $\nu = f_r \ln a$, then (10.18) simplifies to the scaled

Characteristic class functions	*Examples of dispersively warped Cohen's classes*			
	Cohen's class	*Hyperbolic class*	*κth power warped Cohen's class*	*Exponentially warped Cohen's class*
Warping function $\xi(b)$	b	$\ln b$	$\xi_\kappa(b)$	e^b
Domain $\wp$ of $\xi(\cdot)$	$\Re$	$\Re_+$	$\Re$	$\Re$
Range $\aleph$ of $\xi(\cdot)$	$\Re$	$\Re$	$\Re$	$\Re_+$
Group delay shift $\tau(f)$	$1/f_r$	$1/f$	$\tau_\kappa(f)$	$\frac{1}{f_r} e^{f/f_r}$
Covariance operator $\mathcal{D}_c^{(\xi)}$ in (10.10)	$\mathcal{S}_{c/f_r}$	$\mathcal{H}_c = \mathcal{D}_c^{(\xi_{\ln})}$	$\mathcal{D}_c^{(\xi_\kappa)}$	$\mathcal{E}_c = \mathcal{D}_c^{(\xi_{\exp})}$
Covariance property in (10.11)	constant time-shift	hyperbolic time-shift	power time-shift	exponential time-shift
Covariance operator $\mathcal{Y}_\nu^{(\xi)}$ in (10.18)	$\mathcal{M}_\nu$	$\mathcal{C}_{e^\nu/f_r}$	$\mathcal{Y}_\nu^{(\xi_\kappa)}$	$\mathcal{Y}_\nu^{(\xi_{\exp})}$
Covariance property in (10.19)	frequency-shift	scale	power warped frequency-shift	exponentially warped frequency-shift

TABLE 10.4. Examples of classes from the dispersively warped Cohen's class in (10.20) with associated warping function $\xi(b)$, group delay shift $\tau(f)$, and covariance operators and properties (10.10), (10.11) and (10.18), (10.19), respectively. Here, $\xi_{\ln}(b) = \ln b$, $\xi_\kappa(b) = \operatorname{sgn}(b)|b|^\kappa$, $\tau_\kappa(f) = \frac{\kappa}{f_r}|\frac{f}{f_r}|^{\kappa-1}$, and $\xi_{\exp}(b) = e^b$. The effect of the various operators on $X(f)$ is described in Table 10.2.

signal $(\mathcal{C}_a X)(f) = X(f/a)/\sqrt{|a|} = (\mathcal{Y}_{f_r \ln a}^{(\xi_{\ln})} X)(f)$, and (10.19) simplifies to the QTFR scale covariance, $T_{\mathcal{C}_a X}^{(DC)}(t,f) = T_X^{(DC)}(at, f/a)$, which is important for multiresolution analysis. The corresponding dispersively warped Cohen's class is the hyperbolic class [25,26] (see Table 10.4).

Based on warping the Cohen's class formulations in (10.3) and (10.4) using (10.17), any dispersively warped Cohen's class QTFR, $T_X^{(DC)}(t,f)$, $f \in \wp$, can be expressed as

$$T_X^{(DC)}(t,f)$$
$$= \iint_{\mathcal{B}} \Phi_T^{(DC)}\left(\xi\left(\frac{f}{f_r}\right) - b, \beta\right) V_X^{(\xi)}(b,\beta) e^{j2\pi \frac{t}{\tau(f)}\beta}\, db\, d\beta \tag{10.20}$$
$$= \iint_{\wp} \psi_T^{(DC)}\left(\frac{t}{\tau(f)} - \frac{\hat{t}}{\tau(\hat{f})}, \xi\left(\frac{f}{f_r}\right) - \xi\left(\frac{\hat{f}}{f_r}\right)\right) W_X^{(\xi)}(\hat{t},\hat{f})\, d\hat{t}\, d\hat{f}, \tag{10.21}$$

where $\wp$ is the domain of $\xi(\cdot)$. The integration range $\mathcal{B}$ in (10.20) is over values of b and β such that $(b \pm \beta/2)$ is in $\aleph$, which is the range of $\xi(\cdot)$. The warped spectrum product in (10.20) is the spectrum product U in (10.3) of the warped signal $(\mathcal{W}_\xi X)(f)$ in (10.13), i.e.,

$$V_X^{(\xi)}(b,\beta) = f_r U_{\mathcal{W}_\xi X}(f_r b, f_r \beta), \qquad (b \pm \beta/2) \in \aleph. \tag{10.22}$$

The 2D functions $\Phi_T^{(DC)}(b,\beta)$ and $\psi_T^{(DC)}(c,b)$ are kernel functions that uniquely characterize any dispersively warped Cohen's class QTFR $T^{(DC)}$. The dispersively warped Cohen's class formulations in (10.20) and (10.21) generalize Cohen's class, since when $\xi(b) = b$, they simplify to the corresponding Cohen's class formulations in (10.3) and (10.4). The kernels of the dispersively warped Cohen's class and the kernels of Cohen's class are the same up to a constant scaling, i.e., $\Phi_T^{(DC)}(b, \beta) = f_r \Phi_T^{(C)}(f_r b, f_r \beta)$ and $\psi_T^{(DC)}(c, b) = \psi_T^{(C)}(c/f_r, f_r b)$. Thus, the dispersively warped Cohen's class kernels are also related via the FT. When a different one-to-one function $\xi(b)$ is chosen, the dispersively warped Cohen's class simplifies to other classes of QTFRs, as summarized in Table 10.4 and Subsection 10.4.1.3.

An important member of the dispersively warped Cohen's class is the dispersively warped version of the WD in (10.5). The dispersively warped WD [12,27,28, 32,33,35] in (10.21) is given by

$$W_X^{(\xi)}(t,f) = W_{\mathcal{W}_\xi X}\left(\frac{t}{f_r \tau(f)}, f_r \xi\left(\frac{f}{f_r}\right)\right), \qquad f \in \wp$$
$$= \int V_X^{(\xi)}\left(\xi\left(\frac{f}{f_r}\right), \beta\right) e^{j2\pi[t/\tau(f)]\beta} d\beta, \tag{10.23}$$

where the integration over β is such that $(\xi(f/f_r) \pm \beta/2) \in \aleph$.

10.4.1.2 QTFR properties and kernel constraints

All dispersively warped Cohen's class QTFRs always satisfy the covariance properties in (10.10), (10.11) and (10.18), (10.19). They will also satisfy other desirable properties, provided their kernels, $\Phi_T^{(DC)}(b, \beta)$ in (10.20), satisfy some constraints. A number of desirable QTFR properties and their corresponding kernel constraints are listed in Table 10.5. Except for [P-DC1] and [P-DC2], the kernel constraints are valid only when the range of $\xi(b)$ is $\aleph = \Re$. For example, the generalized marginal property [P-DC6] in Table 10.5 states that if a dispersively warped Cohen's class QTFR is integrated over all group delay curves $t = c\tau(f)$, the result should be equal to the squared magnitude of the generalized coefficient function in (10.16). The only QTFRs that satisfy [P-DC6] are those whose kernels $\Phi_T^{(DC)}(b, \beta)$, when integrated over all values of b, equal unity for all values of β. The dispersively warped WD $W_X^{(\xi)}(t, f)$ in (10.23) with $\aleph = \Re$ satisfies [P-DC6] since its kernel is $\Phi_{W^{(\xi)}}^{(DC)}(b, \beta) = \delta(b)$.

As the choice of $\xi(b)$ specifies a group delay shift covariant class, the properties and kernel constraints in Table 10.5 simplify for that specific class. For example, any member of Cohen's class satisfies property [P-DC8] with $\xi(b) = \xi_{\text{linear}}(b) = b$ and $\tau(f) = 1/f_r$, i.e.,

$$X(f) = I_c^{(\xi_{\text{linear}})}(f) = \frac{1}{\sqrt{f_r}} e^{-j2\pi c(f/f_r)} \quad \Rightarrow \quad T_X^{(C)}(t,f) = \frac{1}{f_r}\delta\left(t - \frac{c}{f_r}\right)$$

if the QTFR's kernel in (10.3) is constrained as $\int \Phi_T^{(C)}(f, \nu) df = 1, \forall \nu$. On the other hand, any member of the hyperbolic class (dispersively warped Cohen's class with $\xi(b) = \xi_{\ln}(b) = \ln b$) also satisfies property [P-DC8] with $\xi(b) = \xi_{\ln}(b) = \ln b$ and $\tau(f) = 1/f$, i.e.,

Dispersively warped Cohen's class property	*Kernel constraint*
[P-DC1] Group delay shift covariance Eq. (10.10), (10.11)	$\Phi_T^{(DC)}(b,\beta)$ arbitrary
[P-DC2] Dispersively warped frequency-shift covariance Eq. (10.18), (10.19)	$\Phi_T^{(DC)}(b,\beta)$ arbitrary
[P-DC3] Real-valuedness $T_X^{(DC)}(t,f) = T_X^{(DC)*}(t,f)$	$\Phi_T^{(DC)}(b,\beta) = \Phi_T^{(DC)*}(b,-\beta)$
[P-DC4] Energy distribution $\int\int_\wp T_X^{(DC)}(t,f)\,dtdf = \int_\wp \lvert X(f)\rvert^2 df$	$\int \Phi_T^{(DC)}(b,0)db = 1$
[P-DC5] Frequency marginal $\int T_X^{(DC)}(t,f)\,dt = \lvert X(f)\rvert^2$	$\Phi_T^{(DC)}(b,0) = \delta(b)$
[P-DC6] Generalized marginal $\int_\wp T_X^{(DC)}(c\tau(f),f)\,\lvert\tau(f)\rvert\,df = \lvert\rho_X^{(\xi)}(c)\rvert^2$ in (10.16)	$\int \Phi_T^{(DC)}(b,\beta)db = 1, \forall\beta$
[P-DC7] Frequency localization $X(f) = \delta(f-\hat{f}) \Rightarrow T_X^{(DC)}(t,f) = \delta(f-\hat{f})$	$\Phi_T^{(DC)}(b,0) = \delta(b)$
[P-DC8] Generalized localization $X(f) = I_c^{(\xi)}(f)$ in (10.14) $\Rightarrow T_X^{(DC)}(t,f) = \lvert\tau(f)\rvert\delta(t-c\tau(f))$	$\int \Phi_T^{(DC)}(b,\beta)db = 1, \forall\beta$
[P-DC9] Moyal's formula $\int\int_\wp T_{X_1}^{(DC)}(t,f)\,T_{X_2}^{(DC)*}(t,f)\,dt\,df =$ $\lvert\int_\wp X_1(f)X_2^*(f)\,df\rvert^2$	$\lvert\int \Phi_T^{(DC)}(b,\beta)\,e^{j2\pi b\zeta}\,db\rvert = 1$, $\forall\beta$
[P-DC10] Generalized exponential time-shift covariance $(\mathcal{E}_c^{(\xi)}X)(f) = (\mathcal{D}_c^{(e^\xi)}X)(f) = e^{-j2\pi c e^{\xi(f/f_r)}}X(f)$ $\Rightarrow T_{\mathcal{E}_c^{(\xi)}X}^{(DC)}(t,f) = T_X^{(DC)}(t-c\tau(f)e^{\xi(\frac{f}{f_r})},f)$	$\Phi_T^{(DC)}(b,\beta) = G_T^{(DC)}(\beta)\delta(b-F(\beta))$ with $F(\beta) = \ln\frac{\sinh\beta/2}{\beta/2}$ and $G_T^{(DC)}(\beta)$ arbitrary

TABLE 10.5. Desirable QTFR properties and corresponding kernel constraints of the dispersively warped Cohen's class.

$$X(f) = I_c^{(\xi_{\ln})}(f) = \frac{1}{\sqrt{f}}e^{-j2\pi c\ln(f/f_r)}$$
$$\Rightarrow \quad T_X^{(DC)}(t,f) = \frac{1}{f}\delta\left(t-\frac{c}{f}\right), \qquad f>0,$$

provided the QTFR's kernel is constrained as $\int \Phi_T^{(DC)}(b,\beta)db = 1, \forall\beta$. Note that property [P-DC8] holds only if the warping function $\xi(b)$ and the QTFR's group delay shift $\tau(f)$ equal, respectively, to the phase response $\lambda(b)$ and the group delay $\mu(f)$ of the generalized impulse in (10.14).

If a given Cohen's class QTFR, $T^{(C)}$, satisfies a particular property in Table 10.5 with $\xi(b) = b$, its dispersively warped versions, $T^{(DC)}$, must also satisfy the dispersively warped version of this same property with a given $\xi(b)$ with range $\aleph = \Re$. This is because (a) the kernel constraints in [P-DC1]–[P-DC10] are independent of the choice of the warping function $\xi(b)$, and (b) the kernel of a given Cohen's class QTFR is identical in form to the kernel of its dispersively warped Cohen's class QTFRs, up to a scaling factor f_r. It is important to note that, although the kernel constraints are essentially identical for a given Cohen's class QTFR and its generalizations, the actual form of the property does depend on $\xi(b)$. For example, the Cohen's class kernel of the WD, $\Phi_W^{(C)}(f,\nu) = \delta(f)$, is identical in form to the QD kernel, $\Phi_Q^{(DC)}(b,\beta) = \delta(b)$. Since the WD kernel, $\Phi_W^{(C)}$, satisfies the constraint $\int \Phi_W^{(C)}(f,\nu)\,df = 1$, $\forall\nu$, in the generalized marginal property [P-DC6], then the QD kernel, $\Phi_Q^{(DC)}$, must also satisfy the same constraint. Thus, if [P-DC6] with $\xi(b) = \xi_{\text{linear}}(b) = b$ holds for the WD, it must also hold for the QD with $\xi(b) = \xi_{\ln}(b) = \ln b$. However, the actual form of property [P-DC6] does differ for the two QTFRs. For the WD, [P-DC6] states that the line integral of the WD is equal to the squared magnitude of the time-domain signal,

$$\int W_X\left(\frac{c}{f_r}, f\right) df = |\rho_X^{(\xi_{\text{linear}})}(c)|^2 = \left|x\left(\frac{c}{f_r}\right)\right|^2.$$

For the QD, [P-DC6] states that the integral of the QD along hyperbolic curves is equal to the squared magnitude of the Mellin transform of the signal's frequency response,

$$\int_0^\infty Q_X\left(\frac{c}{f}, f\right)\frac{df}{f} = |\rho_X^{(\xi_{\ln})}(c)|^2 = \left|\int_0^\infty X(f)e^{j2\pi c\ln(f/f_r)}\frac{df}{\sqrt{f}}\right|^2.$$

10.4.1.3 Dispersively warped Cohen's class members

Some dispersively warped Cohen's class QTFRs with kernels $\Phi_T^{(DC)}(b,\beta)$ in (10.20) and some of the properties they satisfy from Table 10.5 are listed below. Note that these QTFRs are for any one-to-one function $\xi(b)$, and they always satisfy [P-DC1] and [P-DC2]. They will also satisfy other properties if their kernels satisfy the corresponding constraints.

- The *dispersively warped WD* is obtained by warping the WD using (10.17), and it is given in (10.23). Its kernel is $\Phi_{W^{(\xi)}}^{(DC)}(b,\beta) = \delta(b)$, and it satisfies properties [P-DC1]–[P-DC9].
- The *dispersively warped α-form WD*, defined as

$$W_T^{(\xi)(\alpha)}(t,f) = \int V_X^{(\xi)}\left(\xi\left(\frac{f}{f_r}\right) - \alpha\beta, \beta\right) e^{j2\pi(t/\tau(f))\beta} d\beta, \tag{10.24}$$

  is the warped version of the α-form WD in [3]. Here, the range of integration depends on $\aleph$, and $V^{(\xi)}$ is the warped spectrum in (10.22). The QTFR kernel is $\Phi_{W^{(\xi)(\alpha)}}^{(DC)}(b,\beta) = \delta(b-\alpha\beta)$, and the QTFR satisfies properties [P-DC1], [P-DC2], [P-DC4]–[P-DC9]. When $\alpha = 0$, (10.24) is the dispersively warped WD in (10.23).
- The *dispersively warped Cohen–Bertrand P_0-distribution* is the warped version of the Cohen–Bertrand P_0-distribution [27]. It is defined as

$$CP_{0X}^{(\xi)}(t,f) = \int V_X^{(\xi)}\left(\xi\left(\frac{f}{f_r}\right) - \ln\frac{\sinh\beta/2}{\beta/2}, \beta\right) e^{j2\pi(t/\tau(f))\beta} d\beta, \quad (10.25)$$

where the integration range depends on $\aleph$. The QTFR kernel is

$$\Phi_{CP_0^{(\xi)}}^{(DC)}(b,\beta) = \delta\left(b - \ln\frac{\sinh\beta/2}{\beta/2}\right),$$

and it satisfies [P-DC1]–[P-DC10].

- The *dispersively warped spectrogram* is defined in (10.20) with kernel $\Phi_{SPEC^{(\xi)}}^{(DC)}(b,\beta) = V_\Theta^{(\xi)}(-b,-\beta)$ where $\Theta(f)$ is an analysis window. The dispersively warped spectrogram satisfies properties [P-DC1]–[P-DC3], and [P-DC4] if $\int |\Theta(f)|^2 df = 1$.
- The *dispersively warped pseudo WD* has a kernel that depends on a window $\Theta(f)$, $\Phi_{PWD^{(\xi)}}^{(DC)}(b,\beta) = f_r W_\Theta^{(\xi)}(0, f_r\xi^{-1}(b))$, where $W^{(\xi)}$ is the dispersively warped WD in (10.23). This QTFR satisfies properties [P-DC1]–[P-DC3], and [P-DC4], [P-DC6], [P-DC8] if $f_r|\rho_\Theta^{(\xi)}(0)|^2 = 1$.

Other dispersively warped Cohen's class QTFRs include the dispersively warped versions of the smoothed pseudo WD [3], and the Choi–Williams exponential distribution [1,50].

10.4.1.4 Dispersively warped Cohen's class examples

Different dispersively warped Cohen's classes can be obtained by choosing a one-to-one function $\xi(b)$ to match the TF characteristics of the analysis signal. Once $\xi(b)$ is fixed, it can be replaced in: Eq. (10.20), (10.21) to obtain the corresponding formulations of the specified class; Eq. (10.10), (10.11) and (10.18), (10.19) to obtain the two covariance properties defining the class; Table 10.5 to obtain some properties and corresponding kernel constraints of the class; and Subsection 10.4.1.3 to obtain QTFR members of the class. Some class examples obtained by choosing $\xi(b)$ are summarized in Table 10.4 and are also listed in more detail below.

10.4.1.4.1 Cohen's class

When $\xi(b) = \xi_{\text{linear}}(b) = b$ and $\tau(f) = 1/f_r$, the warping in (10.17) maps Cohen's class (Section 10.2.1) back to itself since $\mathcal{W}_{\xi_{\text{linear}}} = \mathcal{I}$ in (10.13) becomes the identity operator.

10.4.1.4.2 Hyperbolic class

When $\xi(b) = \xi_{\ln}(b) = \ln b$ and $\tau(f) = 1/f$, $f > 0$, the dispersively warped Cohen's class in (10.20) is the hyperbolic class (H) [12, 25–28, 60, 63]. Any hyperbolic class QTFR, $T_X^{(H)}(t,f)$, $f > 0$, can be obtained by warping a corresponding Cohen's class QTFR, $T_X^{(C)}(t,f)$, using the transformation

$$T_X^{(H)}(t,f) = T_{\mathcal{W}_{\xi_{\ln}}X}^{(C)}\left(\frac{tf}{f_r},\, f_r \ln\frac{f}{f_r}\right),$$

where the warping operator $\mathcal{W}_{\xi_{\ln}}$ is defined in (10.13) with $\xi(b) = \xi_{\ln}(b) = \ln b$. Hyperbolic class QTFRs always satisfy the hyperbolic time-shift covariance and the scale covariance properties. This follows since the group delay shift covariance property in (10.10), (10.11) simplifies to the hyperbolic time-shift covariance property defined as

$$(\mathcal{H}_c X)(f) = e^{-j2\pi c \ln(f/f_r)} X(f) \quad \Rightarrow \quad T_{\mathcal{H}_c X}^{(H)}(t,f) = T_X^{(H)}\left(t - \frac{c}{f},\, f\right) \tag{10.26}$$

and $\mathcal{W}_{\xi_{\ln}}^{-1} \mathcal{S}_{c/f_r} \mathcal{W}_{\xi_{\ln}} = \mathcal{H}_c = \mathcal{D}_c^{(\xi_{\ln})}$ in (10.10) and (10.12). The dispersively warped frequency-shift covariance in (10.18), (10.19) simplifies to the scale covariance property in (10.7) since $\mathcal{W}_{\xi_{\ln}}^{-1} \mathcal{M}_\nu \mathcal{W}_{\xi_{\ln}} = \mathcal{Y}_\nu^{(\xi_{\ln})} = \mathcal{C}_{e^\nu/f_r}$ is the scale operator. The two covariance properties (10.26) and (10.7) defining the hyperbolic class are important for the analysis of Doppler invariant signals similar to the signals used by bats for echolocation [29, 64], and for the analysis of self-similar random processes [52]. Members of the hyperbolic class include the QD, the unitary Bertrand P_0-distribution, and the hyperbologram [25]. The QD [25, 29, 30] is the hyperbolically warped version of the WD in (10.5). For $f > 0$ and $Y(f) = (\mathcal{W}_{\xi_{\ln}} X)(f)$, it is given by

$$Q_X(t,f) = W_Y\left(\frac{tf}{f_r},\, f_r \ln\frac{f}{f_r}\right) = f \int X\left(f e^{\beta/2}\right) X^*\left(f e^{-\beta/2}\right) e^{j2\pi t f \beta}\, d\beta. \tag{10.27}$$

10.4.1.4.3 *Power warped Cohen's class*

When $\xi(b) = \xi_\kappa(b) = \operatorname{sgn}(b)|b|^\kappa$, and $\tau(f) = \tau_\kappa(f) = (\kappa/f_r)|f/f_r|^{\kappa-1}$, $\kappa \neq 0$, the dispersively warped Cohen's class QTFR in (10.20), (10.21) simplifies to the κth power warped Cohen's class [27, 35, 36]. The group delay shift covariance property in (10.11) simplifies to the κth power time-shift covariance, $T_{\mathcal{D}_c^{(\xi_\kappa)}X}^{(DC)}(t,f) = T_X^{(DC)}(t - c\tau_\kappa(f), f)$, since $\mathcal{W}_{\xi_\kappa}^{-1} \mathcal{S}_{c/f_r} \mathcal{W}_{\xi_\kappa} = \mathcal{D}_c^{(\xi_\kappa)}$ in (10.12); however, the dispersively warped frequency-shift covariance property in (10.19) does not reduce to any simple covariance. Members of this class include the κth power WD and the power spectrogram [35, 36]. The κth power WD is obtained by κth power warping the WD in (10.5),

$$\begin{aligned} W_X^{(\xi_\kappa)}(t,f) &= W_{\mathcal{W}_{\xi_\kappa}X}\left(\frac{t}{f_r \tau_\kappa(f)},\, f_r \xi_\kappa\left(\frac{f}{f_r}\right)\right) \\ &= \left|\frac{f}{\kappa}\right| \int \left|1 - \frac{\beta^2}{4}\right|^{(1-\kappa)/(2\kappa)} X\left(f \xi_\kappa^{-1}\left(1 + \frac{\beta}{2}\right)\right) \\ &\quad \times X^*\left(f \xi_\kappa^{-1}\left(1 - \frac{\beta}{2}\right)\right) e^{j2\pi (tf/\kappa)\beta}\, d\beta, \end{aligned} \tag{10.28}$$

where $\mathcal{W}_{\xi_\kappa}$ is the power warping operator in (10.13) with $\xi(b) = \xi_\kappa(b)$.

10.4.1.4.4 Exponentially warped Cohen's class

When $\xi(b) = \xi_{\exp}(b) = e^b$ and $\tau(f) = (1/f_r)e^{f/f_r}$, the dispersively warped Cohen's class QTFR in (10.20) is the exponentially warped Cohen's class [27, 28, 35, 36]. Since the range $\aleph$ of $\xi(b) = e^b$ is $\Re_+$, not all the property kernel constraints in Table 10.5 are valid for this class. The covariance property in (10.11) simplifies to the exponential time-shift covariance, $T^{(DC)}_{\mathcal{E}_c X}(t, f) = T^{(DC)}_X(t - (c/f_r)e^{f/f_r}, f)$, since $\mathcal{W}^{-1}_{\xi_{\exp}} \mathcal{S}_{c/f_r} \mathcal{W}_{\xi_{\exp}} = \mathcal{E}_c = \mathcal{D}^{(\xi_{\exp})}_c$ in (10.12), but the covariance property in (10.19) does not simplify to any known covariance. Here, $\mathcal{W}_{\xi_{\exp}}$ is the exponential warping operator in (10.13) with $\xi(b) = \xi_{\exp}(b) = e^b$. Members of this QTFR class include the exponential WD [37], which is the exponentially warped version of the WD in (10.5),

$$
\begin{aligned}
E_X(t, f) &= W_{\mathcal{W}_{\xi_{\exp}} X}(t e^{-f/f_r}, f_r e^{f/f_r}) \\
&= \int \rho^{(\xi_{\exp})}_X\left(\frac{t f_r}{e^{f/f_r}} + \frac{\zeta}{2}\right) \rho^{(\xi_{\exp})*}_X\left(\frac{t f_r}{e^{f/f_r}} - \frac{\zeta}{2}\right) e^{-j2\pi\zeta e^{f/f_r}} d\zeta.
\end{aligned}
\tag{10.29}
$$

Here, $\rho^{(\xi_{\exp})}_X(c) = \int X(f)\sqrt{(1/f_r)e^{f/f_r}} e^{j2\pi c e^{f/f_r}} df$ is the exponential coefficient in (10.16) when $\xi(b) = \xi_{\exp}(b) = e^b$.

10.4.2 Dispersively warped affine class

10.4.2.1 Class covariance properties and formulation

The dispersively warped affine class (DA) consists of QTFRs, $T^{(DA)}_X(t, f)$, that are group delay shift covariant. They are obtained by warping the affine class QTFRs, $T^{(A)}_X(t, f)$, using (10.17) with $(class) = (A)$. The warping transforms the two covariance properties defining the affine class, the constant time-shift covariance and the scale covariance, into two new covariance properties defining the dispersively warped affine class, the group delay shift covariance and the dispersively warped scale covariance. Specifically, constant time shifts in (10.1) are mapped to group delay shifts in (10.11) since the time-shift operator, $\mathcal{S}_{c/f_r}$ in (10.1), maps to the group delay shift operator $\mathcal{D}^{(\xi)}_c = \mathcal{W}^{-1}_\xi \mathcal{S}_{c/f_r} \mathcal{W}_\xi$ in (10.12) where $\mathcal{W}_\xi$ is defined in (10.13). The scale covariance in (10.7) is transformed to the dispersively warped scale covariance. In particular, the scale operator, $\mathcal{C}_a$, in the affine class maps to $\mathcal{W}^{-1}_\xi \mathcal{C}_a \mathcal{W}_\xi = \mathcal{L}^{(\xi)}_a$ in the dispersively warped affine class, which transforms the signal as

$$
(\mathcal{L}^{(\xi)}_a X)(f) = \frac{|\xi'(f/f_r)|^{1/2}}{|a\xi'(l(f, a))|^{1/2}} X(f_r l(f, a)),
\tag{10.30}
$$

where $l(f, a) = \xi^{-1}((1/a)\xi(f/f_r))$. The QTFR of the transformed signal is

$$T^{(DA)}_{\mathcal{L}_a^{(\xi)}X}(t,f) = T_X^{(DA)}\left(at\frac{\tau(f_r l(f,a))}{\tau(f)}, f_r\, l(f,a)\right), \qquad (10.31)$$

which defines the dispersively warped scale covariance property.

The covariance property in (10.30) and (10.31) appears complicated, but simplifies to known covariance properties for the following warping functions. When $\xi(b) = \xi_\kappa(b) = \text{sgn}(b)|b|^\kappa$, $\mathcal{L}_a^{(\xi)}$ in (10.30) simplifies to the scale operator $\mathcal{W}_{\xi_\kappa}^{-1}\mathcal{C}_a\mathcal{W}_{\xi_\kappa} = \mathcal{C}_{\xi_{1/\kappa}(a)}$ in (10.7). Consequently, (10.31) results in the scale covariance property in (10.7). The corresponding dispersively warped affine class is the κth power class [31–34]. When $\xi(b) = \xi_{\ln}(b) = \ln b$, $\mathcal{L}_a^{(\xi)}$ in (10.30) simplifies to the power warping operator $\mathcal{W}_{\xi_{\ln}}^{-1}\mathcal{C}_a\mathcal{W}_{\xi_{\ln}} = \mathcal{W}_{\xi_a}$ (see (10.13) with $\xi(b) = \xi_a(b) = \text{sgn}(b)|b|^a$). Furthermore, when $\xi(b) = \xi_{\exp}(b) = e^b$, (10.31) results in the frequency-shift covariance in (10.2) since $\mathcal{W}_{\xi_{\exp}}^{-1}\ \mathcal{C}_a\ \mathcal{W}_{\xi_{\exp}} = \mathcal{L}_a^{(\xi_{\exp})} = \mathcal{M}_{f_r \ln a}$ is the frequency-shift operator, and the corresponding class is the exponential class [27, 35–37]. These results are summarized in Table 10.6.

By warping the affine class QTFR formulations in (10.8) and (10.9) using (10.17), any dispersively warped affine class QTFR, $T_X^{(DA)}(t,f)$, $f \in \wp$, can be written as

$$T_X^{(DA)}(t,f) = \iint_{\mathcal{B}} \Phi_T^{(DA)}\left(\frac{-b}{\xi(f/f_r)}, \frac{\beta}{\xi(f/f_r)}\right) V_X^{(\xi)}(b,\beta)e^{j2\pi[t/\tau(f)]\beta}\frac{db\,d\beta}{|\xi(f/f_r)|} \qquad (10.32)$$

Characteristic class functions	*Examples of dispersively warped affine classes*				
	Affine class	*Hyperbolically warped affine class*	*κth power class*	*Exponential class*	*κth power exponential class*
Warping function $\xi(b)$	b	$\ln b$	$\xi_\kappa(b)$	e^b	$e^{\kappa b}$
Domain $\wp$ of $\xi(\cdot)$	$\Re$	$\Re_+$	$\Re$	$\Re$	$\Re$
Range $\aleph$ of $\xi(\cdot)$	$\Re$	$\Re$	$\Re$	$\Re_+$	$\Re_+$
Group delay shift $\tau(f)$	$1/f_r$	$1/f$	$\tau_\kappa(f)$	$\frac{1}{f_r}e^{f/f_r}$	$\frac{\kappa}{f_r}e^{\kappa f/f_r}$
Covariance operator $\mathcal{D}_c^{(\xi)}$ in (10.10)	$\mathcal{S}_{c/f_r}$	$\mathcal{H}_c = \mathcal{D}_c^{(\xi_{\ln})}$	$\mathcal{D}_c^{(\xi_\kappa)}$	$\mathcal{E}_c = \mathcal{D}_c^{(\xi_{\exp})}$	$\mathcal{E}_c^{(\kappa)} = \mathcal{D}_c^{(\xi_\kappa \exp)}$
Covariance property in (10.11)	constant time-shift	hyperbolic time-shift	power time-shift	exponential time-shift	power exponential time-shift
Covariance operator $\mathcal{L}_a^{(\xi)}$ in (10.30)	$\mathcal{C}_a$	$\mathcal{W}_{\xi_a}$	$\mathcal{C}_{\xi_{1/\kappa}(a)}$	$\mathcal{M}_{f_r \ln a}$	$\mathcal{M}_{\frac{f_r}{\kappa}\ln a}$
Covariance property in (10.31)	scale	power warping	scale	frequency-shift	frequency-shift

TABLE 10.6. Examples of QTFR classes from the dispersively warped affine class defined in (10.32) with associated warping function $\xi(b)$, group delay shift $\tau(f)$, and covariance operators and properties corresponding to (10.10), (10.11) and (10.30), (10.31), respectively. Here, $\xi_{\ln}(b) = \ln b$, $\xi_\kappa(b) = \text{sgn}(b)|b|^\kappa$, $\tau_\kappa(f) = (\kappa/f_r)|f/f_r|^{\kappa-1}$, $\xi_{\exp}(b) = e^b$, and $\xi_{\kappa \exp}(b) = e^{\kappa b}$.

$$= \int\int_{\wp} \psi_T^{(DA)}\left(\xi(f/f_r)\left(\frac{t}{\tau(f)} - \frac{\hat{t}}{\tau(\hat{f})}\right), -\frac{\xi(\hat{f}/f_r)}{\xi(f/f_r)}\right) W_X^{(\xi)}(\hat{t},\hat{f})\, d\hat{t}\, d\hat{f}, \tag{10.33}$$

where $\mathcal{B}$ is defined in (10.20), $V_X^{(\xi)}(b,\beta)$ in (10.22), and $W_X^{(\xi)}(t,f)$ in (10.23). The kernels $\Phi_T^{(DA)}(b,\beta)$ and $\psi_T^{(DA)}(c,b)$ uniquely characterize any dispersively warped affine class QTFR $T^{(DA)}$. These kernel functions are identical in form to the affine class kernels in (10.8) and (10.9), that is $\Phi_T^{(DA)}(b,\beta) = \Phi_T^{(A)}(b,\beta)$ and $\psi_T^{(DA)}(c,b) = \psi_T^{(A)}(c,b)$, and they are also FT related [27].

The dispersively warped affine class in (10.32) provides a generalization of the affine class since it simplifies to the affine class in (10.8) when $\xi(b) = b$. Other classes of QTFRs are obtained by choosing different warping functions $\xi(b)$ (see Subsection 10.4.2.4 and Table 10.6). Note that the dispersively warped WD, $W_X^{(\xi)}(t,f)$, in (10.33) is a member of both the dispersively warped affine class and the dispersively warped Cohen's class.

10.4.2.2 QTFR properties and kernel constraints

In addition to the two covariance properties in (10.11) and (10.31), a dispersively warped affine class QTFR satisfies other desirable properties, provided its kernel follows some constraint. Table 10.7 provides a list of desirable properties and corresponding kernel constraints. These constraints are valid only when $\aleph = \Re$, except in properties [P-DA1] and [P-DA2].

The kernel of a given affine class QTFR is identical in form to the corresponding dispersively warped affine class QTFR kernel. For example, the affine WD kernel in (10.8) is $\Phi_W^{(A)}(b,\beta) = \delta(b+1)$. The dispersively warped WD, $W_X^{(\xi)}(t,f)$, has the identical kernel $\Phi_{W^{(\xi)}}^{(DA)}(b,\beta) = \delta(b+1)$ in (10.32). Note that none of the constraints in [P-DA1]–[P-DA10] in Table 10.7 depends on $\xi(b)$, but the form of the properties changes with $\xi(b)$. As a result, if an affine class QTFR satisfies a property in Table 10.7 with $\xi(b) = b$, its dispersively warped version satisfies the dispersively warped version of the same property with the given $\xi(b)$. For example, the WD satisfies the localization property [P-DA8], $W_{I_c^{(\xi_{\text{linear}})}}(t,f) = (1/f_r)\delta(t - c/f_r)$, where $I_c^{(\xi_{\text{linear}})}(f) = (1/\sqrt{f_r})e^{-j2\pi c(f/f_r)}$, since $\int \Phi_W^{(A)}(b,\beta)db = 1, \forall\beta$. Similarly, when $\aleph = \Re$, the dispersively warped WD satisfies the generalized localization property [P-DA8] since $\int \Phi_{W^{(\xi)}}^{(DA)}(b,\beta)db = 1, \forall\beta$. In particular, when $\xi(b) = \xi_\kappa(b) = \text{sgn}(b)|b|^\kappa$ and $\tau(f) = \tau_\kappa(f) = (\kappa/f_r)|f/f_r|^{\kappa-1}$, the dispersively warped WD is the κth power WD, $W_X^{(\xi_\kappa)}(t,f)$ in (10.28). Thus, the κth power WD automatically satisfies the power localization property given by [P-DA8], $W_{I_c^{(\xi_\kappa)}}^{(\xi_\kappa)}(t,f) = |\tau_\kappa(f)|\delta(t - c\tau_\kappa(f))$, where $I_c^{(\xi_\kappa)}(f) = \sqrt{|\tau_\kappa(f)|}e^{-j2\pi c\,\xi_\kappa(f/f_r)}$ is the generalized impulse in (10.14) with phase response $\lambda(b) = \xi(b) = \xi_\kappa(b)$.

Dispersively warped affine class property	*Kernel constraint*
[P-DA1] Group delay shift covariance Eq. (10.10), (10.11)	$\Phi_T^{(DA)}(b,\beta)$ arbitrary
[P-DA2] Dispersively warped scale covariance Eq. (10.30), (10.31)	$\Phi_T^{(DA)}(b,\beta)$ arbitrary
[P-DA3] Real-valuedness $T_X^{(DA)}(t,f) = T_X^{(DA)*}(t,f)$	$\Phi_T^{(DA)}(b,\beta) = \Phi_T^{(DA)*}(b,-\beta)$
[P-DA4] Energy distribution $\int\int_\wp T_X^{(DA)}(t,f)\,dtdf = \int_\wp \lvert X(f)\rvert^2 df$	$\int \Phi_T^{(DA)}(b,0)\frac{db}{\lvert b\rvert} = 1$
[P-DA5] Frequency marginal $\int T_X^{(DA)}(t,f)\,dt = \lvert X(f)\rvert^2$	$\Phi_T^{(DA)}(b,0) = \delta(b+1)$
[P-DA6] Generalized marginal $\int_\wp T_X^{(DA)}(c\tau(f),f)\,\lvert\tau(f)\rvert\,df = \lvert\rho_X^{(\xi)}(c)\rvert^2$ in (10.16)	$\int \Phi_T^{(DA)}(b,\gamma b)\frac{db}{\lvert b\rvert} = 1,\ \forall\gamma$
[P-DA7] Frequency localization $X(f) = \delta(f-\hat{f}) \Rightarrow T_X^{(DA)}(t,f) = \delta(f-\hat{f})$	$\Phi_T^{(DA)}(b,0) = \delta(b+1)$
[P-DA8] Generalized localization $X(f) = I_c^{(\xi)}(f)$ in (10.14) $\Rightarrow T_X^{(DA)}(t,f) = \lvert\tau(f)\rvert\delta(t-c\tau(f))$	$\int\Phi_T^{(DA)}(b,\beta)db = 1, \forall\beta$
[P-DA9] Moyal's formula $\int\int_\wp T_{X_1}^{(DA)}(t,f)\,T_{X_2}^{(DA)*}(t,f)\,dt\,df =$ $\lvert\int_\wp X_1(f)X_2^*(f)\,df\rvert^2$	$\int \Phi_T^{(DA)}(\beta,\alpha\beta)$ $\cdot\Phi_T^{(DA)*}(b\beta,\alpha\beta)\,d\beta$ $=\delta(b-1),\ \forall\alpha$
[P-DA10] Generalized hyperbolic time-shift covariance $(\mathcal{H}_c^{(\xi)}X)(f) = (\mathcal{D}_c^{(\ln\xi)}X)(f) = e^{-j2\pi c\ln\xi(\frac{f}{f_r})}X(f)$ $\Rightarrow T_{\mathcal{H}_c^{(\xi)}X}^{(DA)}(t,f) = T_X^{(DA)}(t-c\tau(f)/\xi(\frac{f}{f_r}),f)$	$\Phi_T^{(DA)}(b,\beta) = G_T^{(DA)}(\beta)\delta(b+K(\beta))$ with $K(\beta) = \frac{\beta}{2}\coth\frac{\beta}{2}$ and $G_T^{(DA)}(\beta)$ arbitrary

TABLE 10.7. Desirable QTFR properties and corresponding kernel constraints of the dispersively warped affine class.

10.4.2.3 Dispersively warped affine class members

Some dispersively warped affine class QTFRs together with their kernels in (10.32) are given below. Also included are the properties from Table 10.7 that these members may satisfy.

- The *dispersively warped WD* in (10.23) satisfies properties [P-DA1]–[P-DA9] in Table 10.7 with kernel $\Phi_{W^{(\xi)}}^{(DA)}(b,\beta) = \delta(b+1)$ in (10.32). It is a member of both the dispersively warped Cohen's class (Section 10.4.1.3) and the dispersively warped affine class QTFRs.
- The *dispersively warped α-form WD* in (10.24) is also a member of both dispersively warped classes. Its kernel is $\Phi_{W^{(\xi)(\alpha)}}^{(DA)}(b,\beta) = \delta(b+1-\alpha\beta)$, and it satisfies [P-DA1], [P-DA2], [P-DA4]–[P-DA9].
- The *dispersively warped Bertrand P_0-distribution*, $P_{0_X}^{(\xi)}(t,f)$, is the dispersively warped version of the unitary Bertrand P_0-distribution. It is given in (10.32) with kernel $\Phi_{P_0^{(\xi)}}^{(DA)}(b,\beta) = [(\beta/2)/\sinh(\beta/2)]\delta(b+(\beta/2)\coth(\beta/2))$,

$$P_{0X}^{(\xi)}(t,f) = \left|\xi\left(\frac{f}{f_r}\right)\right| \int V_X^{(\xi)}\left(\xi\left(\frac{f}{f_r}\right)\frac{\beta}{2}\coth\frac{\beta}{2}, \xi\left(\frac{f}{f_r}\right)\beta\right)\frac{\beta/2}{\sinh\beta/2} \cdot e^{j2\pi[t/\tau(f)]\xi(f/f_r)\beta}\, d\beta, \quad (10.34)$$

where the range of integration depends on the range $\aleph$ of $\xi(b)$, and $V_X^{(\xi)}$ is defined in (10.22). This QTFR satisfies properties [P-DA1]–[P-DA5], [P-DA7], [P-DA9], and [P-DA10].

- The *dispersively warped scalogram* is the dispersively warped version of the scalogram (squared magnitude of the wavelet transform) with kernel $\Phi_{SCAL^{(\xi)}}^{(DA)}(b,\beta) = V_\Theta^{(\xi)}(-b,-\beta)$, where $\Theta(f)$ is the analysis wavelet. It satisfies properties [P-DA1]–[P-DA3], and property [P-DA4], provided $\int_\wp [1/|\xi(f/f_r)|] \times |\Theta(f)|^2 df = 1$.

Other dispersively warped affine class QTFRs include the dispersively warped versions of the Choi–Williams distribution, the Flandrin D-distribution, and the passive and active Unterberger distributions [2].

10.4.2.4 Dispersively warped affine class examples

As demonstrated in Table 10.6, dispersively warped affine class examples can be obtained by choosing a one-to-one function $\xi(b)$ in (10.32). Also listed in Table 10.6 are the simplified covariance properties obtained when $\xi(b)$ and $\tau(f)$ are fixed in the covariance property formulations (10.10), (10.11) and (10.30), (10.31). By fixing $\xi(b)$, members for each specified class can be obtained in Subsection 10.4.2.3., and property kernel constraints can be found in Table 10.7. The dispersively warped affine class examples are discussed in more detail below.

10.4.2.4.1 Affine class

The affine class is an example of a "self-mapping" since when $\xi(b) = \xi_{\text{linear}}(b) = b$ and $\tau(f) = 1/f_r$, the warping in (10.17) maps the affine class back to itself.

10.4.2.4.2 Hyperbolically warped affine class

When $\xi(b) = \xi_{\ln}(b) = \ln b$ and $\tau(f) = 1/f$, $f > 0$, the corresponding dispersively warped affine class in (10.32) is the hyperbolically warped affine class [27,28,35,36]. The group delay shift covariance property in (10.11) simplifies to the hyperbolic time-shift covariance property, $T_{\mathcal{H}_c X}^{(DA)}(t,f) = T_X^{(DA)}(t - c/f, f)$ in (10.26), since $\mathcal{W}_{\xi_{\ln}}^{-1} \mathcal{S}_{c/f_r} \mathcal{W}_{\xi_{\ln}} = \mathcal{H}_c = \mathcal{D}_c^{(\xi_{\ln})}$ in (10.12). The warped scale covariance results in the power warping covariance property. The QD in (10.27) is an important member of this class as well as the hyperbolic class [27,35].

10.4.2.4.3 κth power class

When $\xi(b) = \xi_\kappa(b) = \mathrm{sgn}(b)|b|^\kappa$ and $\tau(f) = \tau_\kappa(f) = (\kappa/f_r)|f/f_r|^{\kappa-1}$, the corresponding dispersively warped affine class in (10.32) is the κth power class (κP) [31–34]. A different power class is obtained by choosing a different value of $\kappa \neq 0$. Thus, any κth power class QTFR can be obtained by warping a corresponding member of the affine class (see (10.17)) using

$$T_X^{(\kappa P)}(t,f) = T_{\mathcal{W}_{\xi_\kappa} X}^{(A)}\left(\frac{t}{f_r \tau_\kappa(f)}, \, f_r \xi_\kappa\left(\frac{f}{f_r}\right)\right), \tag{10.35}$$

where $\mathcal{W}_{\xi_\kappa}$ is defined in (10.13) with $\xi(b)=\xi_\kappa(b)$ and $\tau(f)=\tau_\kappa(f)$. The group delay shift covariance in (10.10), (10.11) simplifies to the κth power time-shift covariance property. Thus, for $(\mathcal{D}_c^{(\xi_\kappa)} X)(f) = e^{-j2\pi c\,\xi_\kappa(f/f_r)} X(f)$,

$$T_{\mathcal{D}_c^{(\xi_\kappa)} X}^{(\kappa P)}(t,f) = T_X^{(\kappa P)}(t - c\tau_\kappa(f), f) \tag{10.36}$$

since $\mathcal{W}_{\xi_\kappa}^{-1} \mathcal{S}_{c/f_r} \mathcal{W}_{\xi_\kappa} = \mathcal{D}_c^{(\xi_\kappa)}$ in (10.12). The warped scale covariance in (10.31) simplifies to the scale covariance in (10.7) since $\mathcal{W}_{\xi_\kappa}^{-1} \mathcal{C}_a \mathcal{W}_{\xi_\kappa} = \mathcal{L}_a^{(\xi_\kappa)} = \mathcal{C}_{\xi_{1/\kappa}(a)}$. The two covariance properties (10.36) and (10.7) defining the κth power class are important for the analysis of systems with power dispersive TF characteristics. Power dispersive group delays occur in various application areas such as the dispersive propagation of a shock wave in a steel beam ($\kappa = 1/2$) [65–67], trans-ionospheric chirps measured by satellites ($\kappa = -1$) [68], acoustical waves reflected from a spherical shell immersed in water [56], and some cetacean mammal whistles [69]. Note that when $\kappa = 1$, the power class simplifies to the affine class. Members of the κth power class include the power WD in (10.28), the power Bertrand P_0-distribution, and the powergram [31].

10.4.2.4.4 Exponential class

When $\xi(b) = \xi_{\exp}(b) = e^b$ and $\tau(f) = (1/f_r)e^{f/f_r}$, the corresponding dispersively warped affine class in (10.32) is the exponential class (E) [27, 28, 35–37]. The group delay shift covariance in (10.11) simplifies to the exponential time-shift covariance,

$$(\mathcal{E}_c X)(f) = e^{-j2\pi c\, e^{f/f_r}} X(f) \quad \Rightarrow \quad T_{\mathcal{E}_c X}^{(E)}(t,f) = T_X^{(E)}\left(t - \frac{c}{f_r} e^{f/f_r}, f\right)$$

since (10.12) yields $\mathcal{W}_{\xi_{\exp}}^{-1} \mathcal{S}_{c/f_r} \mathcal{W}_{\xi_{\exp}} = \mathcal{E}_c = \mathcal{D}_c^{(\xi_{\exp})}$. Moreover, the dispersively warped scale covariance simplifies to the frequency-shift covariance in (10.2) since $\mathcal{W}_{\xi_{\exp}}^{-1} \mathcal{C}_a \mathcal{W}_{\xi_{\exp}} = \mathcal{L}_a^{(\xi_{\exp})} = \mathcal{M}_{f_r \ln a}$. Some important members of the exponential class include the exponential WD in (10.29), and the Cohen–Bertrand P_0-distribution [37].

10.4.2.4.5 κth power exponential class

The κth power exponential class (κPE) is obtained when $\xi(b) = \xi_{\kappa\ \exp}(b) = e^{\kappa b}$ and $\tau(f) = (\kappa/f_r)e^{\kappa f/f_r}$, $\kappa \neq 0$, in the dispersively warped affine class formulation (10.32) [27, 28, 38]. Since the range $\aleph$ of $\xi_{\kappa\ \exp}(b) = e^{\kappa b}$ is $\Re_+$, some of the property constraints in Table 10.7 do not hold for the κth power exponential class or the exponential class ($\kappa = 1$). A table of property constraints for the exponential class can be found in [37]. Any power exponential QTFR can be written as a warped affine class QTFR,

$$T_X^{(\kappa PE)}(t,f) = T^{(A)}_{\mathcal{W}_{\xi_{\kappa\ \exp}}X}\left(\frac{t}{\kappa e^{\kappa f/f_r}},\, f_r\, e^{\kappa f/f_r}\right),$$

where $\mathcal{W}_{\xi_{\kappa\ \exp}}$ is given in (10.13) with $\xi(b) = \xi_{\kappa\ \exp}(b) = e^{\kappa b}$. Note that when $\kappa = 1$, the power exponential class is the exponential class discussed in Subsection 10.4.2.4.4. The power exponential QTFRs satisfy the power exponential time-shift covariance property. That is, for $(\mathcal{E}_c^{(\xi_\kappa)}X)(f) = e^{-j2\pi c\, e^{\kappa f/f_r}}X(f)$,

$$T^{(\kappa PE)}_{\mathcal{E}_c^{(\xi_\kappa)}X}(t,f) = T_X^{(\kappa PE)}\left(t - c\frac{\kappa}{f_r}e^{\kappa f/f_r}, f\right), \tag{10.37}$$

which is the group delay shift covariance in (10.11) with $\xi(b) = \xi_{\kappa\ \exp}(b) = e^{\kappa b}$ and $\tau(f) = (\kappa/f_r)e^{\kappa(f/f_r)}$. The power exponential time-shift operator is given by $\mathcal{W}^{-1}_{\xi_\kappa\ \exp}\mathcal{S}_{c/f_r}\mathcal{W}_{\xi_\kappa\ \exp} = \mathcal{E}_c^{(\xi_\kappa)} = \mathcal{D}_c^{(\xi_{\kappa\ \exp})}$ in (10.12). The power exponential time-shift covariance in (10.37) is an important property for analyzing signals passing through exponentially dispersive systems. The power exponential QTFRs also satisfy the frequency-shift covariance property in (10.2), which is the dispersively warped scale covariance in (10.31) when $\xi(b) = e^{\kappa b}$. This follows since $\mathcal{L}_a^{(\xi)}$ in (10.30) simplifies to $\mathcal{W}^{-1}_{\xi_\kappa\ \exp}\mathcal{C}_a\mathcal{W}_{\xi_\kappa\ \exp} = \mathcal{M}_{(f_r/\kappa)\ln a}$ in (10.2). Some important members of this class include the power exponential WD in (10.23) with $\xi(b) = e^{\kappa b}$, and the power Cohen–Bertrand P_0-distribution in (10.34) with $\xi(b) = e^{\kappa b}$ [27].

10.4.3 Other dispersively warped classes

Using the warping in (10.17), it is possible to warp other QTFR classes such as the hyperbolic class [25–27], the power classes [31–34], or the exponential class [27, 35–37]. The effect of this warping is to generalize the hyperbolic, the power, and the exponential time-shift covariance properties. Note that, as we have already shown, these classes can themselves be obtained by warping either Cohen's class or the affine class. Assume (i) that we warp a Class A to obtain a Class B using (10.17) with $\xi(b)$ and $\tau(f)$, and (ii) that we also warp Class B to obtain a Class C using (10.17) with $\xi(b)$ and $\tau(f)$ replaced by $\tilde{\xi}(b)$ and $\tilde{\tau}(f) = (d/df)\tilde{\xi}(f/f_r)$, respectively. Equivalently, we can warp Class A to obtain Class C using (10.17) with $\xi(b)$ and $\tau(f)$ replaced by $\xi(\tilde{\xi}(b))$ and $\tilde{\tau}(f)\,\xi'(\tilde{\xi}(f/f_r))$, respectively. This is demonstrated graphically in Figure 10.5. This "composition" of warpings results in

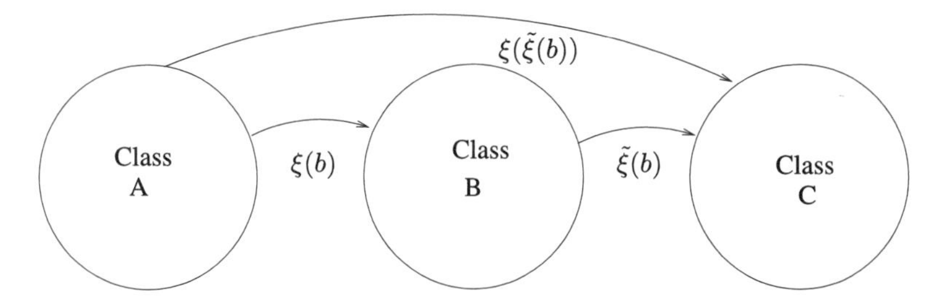

FIGURE 10.5. If Class B is obtained by warping Class A using $\xi(b)$, and Class C is obtained by warping Class B using $\tilde{\xi}(b)$, then Class C can also be obtained by warping Class A using $\xi(\tilde{\xi}(b))$.

new QTFR classes. For example, the κth power exponential QTFRs $T_X^{(\kappa\, PE)}(t, f)$ in Subsection 10.4.2.4.5 can be obtained by warping the κth power QTFRs $T_X^{(\kappa\, P)}(t, f)$ in Subsection 10.4.2.4.3 using the exponential mapping $\tilde{\xi}(b) = \tilde{\xi}_{\exp}(b) = e^b$. However, we have already shown that the κth power QTFRs are obtained by warping the affine QTFRs $T_X^{(A)}(t, f)$ using the κth power mapping $\xi(b) = \xi_\kappa(b) = \operatorname{sgn}(b)|b|^\kappa$ in (10.35). Thus, the κth power exponential class can also be obtained by warping the affine class using the κth power exponential mapping $\xi(\tilde{\xi}(b)) = \xi_\kappa(\tilde{\xi}_{\exp}(b)) = \xi_{\kappa\, \exp}(b) = e^{\kappa b}$. This mapping is the composition $\xi(\tilde{\xi}(b))$ of the κth power warping $\xi(b) = \xi_\kappa(b)$ and the exponential warping $\tilde{\xi}(b) = e^b$. Thus, two possible warping methods to obtain the κth power exponential class are

$$T_X^{(\kappa\, PE)}(t, f) = T_{\mathcal{W}_{\tilde{\xi}_{\exp}} X}^{(\kappa\, P)}\left(\frac{t}{e^{f/f_r}}, f_r e^{f/f_r}\right) \tag{10.38}$$

$$= T_{\mathcal{W}_{\xi_{\kappa\, \exp}} X}^{(A)}\left(\frac{t}{\kappa e^{\kappa f/f_r}}, f_r e^{\kappa f/f_r}\right) \tag{10.39}$$

since $T_Y^{(\kappa\, P)}(t, f) = T_{\mathcal{W}_{\xi_\kappa} Y}^{(A)}(t/(f_r \tau_\kappa(f)), f_r \xi_\kappa(f/f_r))$.

One advantage of the composition of warpings is that the implementation of the new QTFR classes could be simplified by combining existing implementation techniques. For example, the κth power exponential class can be implemented by warping the affine class as in (10.39) [12]. On the other hand, the κth power exponential class can also be implemented using the already existing implementation techniques for exponential [27, 37] and κth power mappings [34]. One only needs to apply the κth power warping to an affine QTFR in order to obtain a κth power QTFR, and then apply the exponential warping to the κth power QTFR as in (10.38) to obtain the corresponding κth power exponential QTFR. Thus, the various compositions of warpings facilitate the implementation of various QTFRs, because often no new algorithms need to be devised.

10.4.3.1 Dispersively warped hyperbolic class

The dispersively warped hyperbolic class (DH) QTFRs, $T_X^{(DH)}(t,f)$, are obtained by warping the hyperbolic class [25,26] QTFRs, $T_X^{(H)}(t,f)$, using (10.17) with $\xi(b)$ and $\tau(f)$ replaced by $\tilde{\xi}(b)$ and $\tilde{\tau}(f)$, respectively, and $(class) = (H)$, i.e.,

$$T_X^{(DH)}(t,f) = T_{\mathcal{W}_{\tilde{\xi}}X}^{(H)}\left(\frac{t}{f_r\tilde{\tau}(f)}, f_r\tilde{\xi}\left(\frac{f}{f_r}\right)\right), \tag{10.40}$$

where $(\mathcal{W}_{\tilde{\xi}}X)(f)$ is defined in (10.17), $\tilde{\xi}(b)$ is a one-to-one function with range $\Re_+$, and $\tilde{\tau}(f) = (d/df)\tilde{\xi}(f/f_r)$. The dispersively warped hyperbolic class can also be obtained by warping Cohen's class QTFRs, $T_X^{(C)}(t,f)$, using (10.17) provided $\xi(b)$ is replaced by $\ln(\tilde{\xi}(b))$ and $\tau(f)$ is replaced by $(\tilde{\tau}(f)/\tilde{\xi}(f/f_r))$ in (10.17) with $(class) = (C)$, i.e.,

$$T_X^{(DH)}(t,f) = T_{\mathcal{W}_{\ln\tilde{\xi}}X}^{(C)}\left(\frac{t\tilde{\xi}\left(\frac{f}{f_r}\right)}{f_r\tilde{\tau}(f)}, f_r\ln\left(\tilde{\xi}\left(\frac{f}{f_r}\right)\right)\right). \tag{10.41}$$

This follows since the hyperbolic class is the warped Cohen's class with $\xi(b) = \ln b$ and $(class) = (C)$ in (10.17). For example, if $\tilde{\xi}(b) = b$, $b \in \Re_+$ in (10.40) or (10.41), the dispersively warped hyperbolic class yields the hyperbolic class. However, the hyperbolic class can also be obtained by letting $\xi(b) = \ln b$ and $(class) = (C)$ for Cohen's class in (10.17). Similarly, if $\tilde{\xi}(b) = e^b$ in (10.40) or (10.41), then the dispersively warped hyperbolic class is Cohen's class; Cohen's class can also be obtained by letting $\xi(b) = \ln e^b = b$ and $(class) = (C)$ in (10.17).

The hyperbolic class QTFRs satisfy the scale covariance property in (10.7), and the hyperbolic time-shift covariance property in (10.26). Thus, the dispersively warped hyperbolic class QTFRs satisfy the dispersively warped scale covariance given in (10.31) with $\xi(b)$ replaced by $\tilde{\xi}(b)$. They also satisfy the generalized hyperbolic time-shift covariance ([P-DA10] in Table 10.7). Thus, for $(\mathcal{D}_c^{(\ln\tilde{\xi})}X)(f) = e^{-j2\pi c\ln(\tilde{\xi}(f/f_r))}X(f)$,

$$T_{\mathcal{D}_c^{(\ln\tilde{\xi})}X}^{(DH)}(t,f) = T_X^{(DH)}\left(t - c\frac{\tilde{\tau}(f)}{\tilde{\xi}(f/f_r)}, f\right). \tag{10.42}$$

Note that (10.42) is in the form of (10.11) with $\xi(b)$ in (10.11) replaced by $\ln(\tilde{\xi}(b))$. Some dispersively warped hyperbolic class QTFRs include the dispersively warped versions of the QD and the unitary Bertrand P_0-distribution [36].

10.4.3.2 Dispersively warped power classes

The dispersively warped κth power class QTFRs, $T_X^{(\kappa DP)}(t,f)$, are obtained by warping the κth power class QTFRs, $T_X^{(\kappa P)}(t,f)$, using

$$T_X^{(\kappa\, DP)}(t,f) = T_{\mathcal{W}_{\tilde{\xi}}X}^{(\kappa\, P)}\left(\frac{t}{f_r \tilde{\tau}(f)}, f_r \tilde{\xi}\left(\frac{f}{f_r}\right)\right) \tag{10.43}$$

$$= T_{\mathcal{W}_{\ln \tilde{\xi}}X}^{(A)}\left(\frac{t}{f_r \tilde{\tau}(f)\, \xi_\kappa'\left(\tilde{\xi}\left(\frac{f}{f_r}\right)\right)}, f_r \xi_\kappa\left(\tilde{\xi}\left(\frac{f}{f_r}\right)\right)\right),$$

where $\tilde{\xi}(b)$ is a one-to-one function and $\tilde{\tau}(f) = (d/df)\tilde{\xi}(f/f_r)$. As shown above, dispersively warped power class QTFRs can also be obtained by warping the affine class as in (10.17) with $\xi(b)$ replaced by $\xi_\kappa(\tilde{\xi}(b))$ and $\tau(f)$ replaced by $\tilde{\tau}(f) \times \xi_\kappa'(\tilde{\xi}(f/f_r))$. The QTFRs in (10.43) satisfy the dispersively warped scale covariance in (10.31) with $\xi(b)$ replaced by $\tilde{\xi}(b)$. They also satisfy the generalized power time-shift covariance property that is given by (10.11) with $\xi(b)$ replaced by $\xi_\kappa(\tilde{\xi}(b))$.

10.4.3.3 Dispersively warped exponential class

Exponential class QTFRs, $T_X^{(E)}(t,f)$, can also be dispersively warped using (10.17) with $(class) = (E)$ and with $\xi(b)$ and $\tau(f)$ replaced by $\tilde{\xi}(b)$ and $\tilde{\tau}(f)$, respectively, to form the dispersively warped exponential class QTFRs, $T_X^{(DE)}(t,f)$. These QTFRs can also be obtained by directly warping affine class QTFRs using (10.17) with $(class) = (A)$ and with $\xi(b)$ and $\tau(f)$ replaced by $e^{\tilde{\xi}(b)}$ and $\tilde{\tau}(f)\, e^{\tilde{\xi}(b)}$, respectively. These QTFRs satisfy the generalized exponential time-shift covariance property given by (10.11) with $\xi(b)$ replaced by $e^{\tilde{\xi}(b)}$ ([P-DC10] in Table 10.5), and the dispersively warped frequency-shift covariance property in (10.19) with $\xi(b)$ replaced by $\tilde{\xi}(b)$. Some members include the dispersively warped versions of the exponential WD and the exponogram [35–37].

10.4.4 *Group delay shift covariant intersections*

Some of the group delay shift covariant classes intersect as they share some common QTFRs as well as a common covariance property. For example, the dispersively warped WD is a member of both the dispersively warped Cohen's and affine classes, and both classes satisfy the group delay shift covariance in (10.11) for a given $\xi(b)$. Next, we discuss some specific intersection QTFR subclasses shown in Figure 10.6, and we show that their 2D kernels simplify to one-dimensional (1D) kernels.

10.4.4.1 Dispersively warped Cohen's and affine class intersection

The dispersively warped Cohen-affine intersection consists of all QTFRs satisfying three covariance properties: the group delay shift covariance in (10.11), the dispersively warped frequency-shift covariance in (10.19), and the dispersively warped scale covariance in (10.31). Thus, it contains QTFRs that are members of both the dispersively warped Cohen's class (Subsection 10.4.1) and the dispersively warped affine class (Subsection 10.4.2).

In order to find all the QTFR members in the intersection, we impose the dispersively warped scale covariance property in (10.31) on the dispersively warped Cohen's class in (10.20) [27,35,36]. The resulting condition on the QTFR kernel in (10.20) is derived as (cf. [55])

$$\Phi_T^{(DC)}(b,\beta) = s_T(b/\beta)/\beta, \qquad \beta \neq 0. \tag{10.44}$$

The 2D kernel $\Phi_T^{(DC)}(b,\beta)$ reduces to the 1D kernel $s_T(\eta)$ that uniquely characterizes any QTFR in the dispersively warped Cohen-affine intersection. As a result, the formulation of the dispersively warped Cohen-affine intersection is now simplified, and can be obtained by inserting (10.44) into (10.20). Also, the kernel constraints for desirable QTFR properties in Table 10.5 for any dispersively warped Cohen-affine QTFR are also simplified, as they can be expressed in terms of $s_T(\eta)$.

The dispersively warped version of the WD in (10.23) is an important member of the dispersively warped Cohen-affine intersection with kernel $s_{W^{(\xi)}}(\eta) = \delta(\eta)$ in (10.44). Another important member is the dispersively warped α-form WD, $W_T^{(\xi)(\alpha)}(t,f)$ in (10.24), with kernel $s_{W^{(\xi)(\alpha)}}(\eta) = \delta(\eta-\alpha)$ in (10.44). It can be shown that any dispersively warped Cohen-affine intersection QTFR, $T_X^{(DA\cap DC)}(t,f)$, can be written in terms of the dispersively warped α-form WD and the 1D kernel $s_T(\eta)$ as

$$T_X^{(DA\cap DC)}(t,f) = \int s_T(\alpha) W_X^{(\xi)(\alpha)}(t,f)\, d\alpha. \tag{10.45}$$

Note that the dispersively warped Cohen-affine intersection simplifies to known intersection subclasses when $\xi(b)$ in (10.45) is fixed [27, 35]. For example, when $\xi(b) = b$, the intersection simplifies to the one between Cohen's class and the affine class [55]. When $\xi(b) = \ln b$, the intersection simplifies to the one between the hyperbolic class and the hyperbolically warped affine class in Figure 10.4, and when

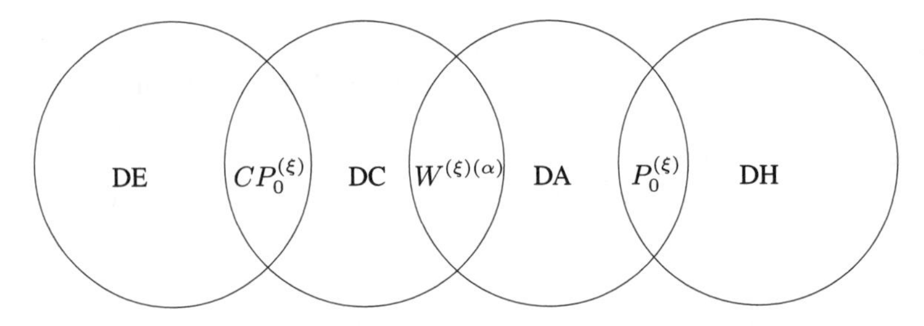

FIGURE 10.6. The intersection between the dispersively warped Cohen's class (DC) and the dispersively warped affine class (DA) includes the dispersively warped α-form WD $W_T^{(\xi)(\alpha)}(t,f)$ in (10.24). The intersection between the DA and the dispersively warped hyperbolic class (DH) includes the dispersively warped Bertrand P_0-distribution $P_{0_X}^{(\xi)}(t,f)$ in (10.34). The intersection between the DC and the dispersively warped exponential class (DE) includes the dispersively warped Cohen–Bertrand P_0-distribution $CP_{0_X}^{(\xi)}(t,f)$ in (10.25).

$\xi(b) = \xi_\kappa(b)$, it corresponds to the intersection between the κth power class and the κth power warped Cohen's class in Figure 10.4. Also, when $\xi(b) = e^b$, the intersection simplifies to the one between the exponential class and the exponentially warped Cohen's class in Figure 10.4.

10.4.4.2 Dispersively warped affine and hyperbolic class intersection

The dispersively warped affine class in Subsection 10.4.2 also intersects with the dispersively warped hyperbolic class (DH) in (10.40) (with $\xi(b)$ equal to $\tilde{\xi}(b)$) in Subsection 10.4.3.1 since both classes satisfy the dispersively warped scale covariance in (10.31). Thus, any member of the dispersively warped affine-hyperbolic intersection satisfies three covariance properties: the group delay shift covariance property in (10.11), the generalized hyperbolic time-shift covariance in (10.42), and the dispersively warped scale covariance in (10.31). If the dispersively warped affine class is constrained to satisfy the generalized hyperbolic time-shift covariance in (10.42) (with $\xi(b)$ equal to $\tilde{\xi}(b)$), then the corresponding constraint on the dispersively warped affine class kernel is given in [P-DA10] in Table 10.7 as

$$\Phi_T^{(DA)}(b,\beta) = G_T^{(DA)}(\beta)\delta\left(b + (\beta/2)\coth(\beta/2)\right). \tag{10.46}$$

Thus, the 2D kernel, $\Phi_T^{(DA)}(b,\beta)$, simplifies to a 1D kernel, $G_T^{(DA)}(\beta)$, that uniquely characterizes any intersection QTFR.

An important member of this intersection is the dispersively warped Bertrand P_0-distribution in (10.34) with kernel $G_{P_0^{(\xi)}}^{(DA)}(\beta) = (\beta/2)\sinh(\beta/2)$ in (10.46). Any member of the intersection, denoted as $T_X^{(DA\cap DH)}(t,f)$, can be written in terms of the dispersively warped Bertrand P_0-distribution as

$$T_X^{(DA\cap DH)}(t,f) = \frac{\xi(f/f_r)}{\tau(f)} \int h_T\left(\frac{\xi(f/f_r)}{\tau(f)}(t-\hat{t})\right) P_{0X}^{(\xi)}(\hat{t},f)\,d\hat{t}, \tag{10.47}$$

where $h_T(c)$ is a 1D kernel with FT $H_T(\beta) = G_T^{(DA)}(\beta)/G_{P_0^{(\xi)}}^{(DA)}(\beta)$.

The choice of $\xi(b)$ in (10.47) determines the intersection between various known classes. For example, when $\xi(b) = b$, the intersection yields the one between the affine class and the hyperbolic class [26], and when $\xi(b) = e^b$, it yields the one between Cohen's class and the exponential class [27, 35–37], as shown in Figure 10.4.

10.4.4.3 Other intersection subclasses

Some other possible intersections are the intersections between the dispersively warped power classes and the dispersively warped affine class, between the dispersively warped exponential class and the dispersively warped Cohen's class, and between the dispersively warped power classes and the dispersively warped hyperbolic class. For example, the intersection between the dispersively warped Cohen's class and the dispersively warped exponential class consists of all dispersively warped Cohen's class QTFRs that satisfy the generalized exponential time-shift covariance defined in [P-DC10] in Table 10.5. According to [P-DC10], the dispersively warped Cohen's class kernel constraint is $\Phi_T^{(DC)}(b,\beta) = G_T^{(DC)}(\beta)\delta(b - \ln(\sinh(\beta/2)/(\beta/2)))$. An important member of this dispersively warped intersection is the dispersively warped Cohen-Bertrand P_0-distribution in Subsection 10.4.1.

10.5 Simulation examples

When analyzing nonstationary signals using QTFRs, some preprocessing or *a priori* knowledge is necessary in order to identify the TF characteristics of the signal so as to match them to the group delay shift covariance of the QTFR. For example, signals with linear TF characteristics, such as impulses (the signal in (10.14) with linear phase response $\lambda(b) = b$ and constant group delay $\mu(f) = 1/f_r$) or dolphin and whale click sounds [69], are best analyzed using Cohen's class QTFRs (see Subsection 10.2.1) or affine class QTFRs (see Subsection 10.2.2) as demonstrated in Figure 10.3. These QTFRs are more suitable because they preserve constant time shifts $\tau(f) = 1/f_r$ that equal the group delay of the signal $\mu(f) = 1/f_r$. However, signals with hyperbolic TF characteristics such as hyperbolic impulses (the signal in (10.14) with $\lambda(b) = \ln b$ and $\mu(f) = 1/f$), or Doppler invariant signals, such as the signals used by bats for echolocation, are best analyzed using hyperbolic class QTFRs (see Subsection 10.4.1.4), as also demonstrated in Figure 10.3. This follows since hyperbolic class QTFRs preserve hyperbolic time shifts $\tau(f) = 1/f$. Signals with power TF characteristics, such as power impulses (the signal in (10.14) with $\lambda(b) = \text{sgn}(b)|b|^\kappa$ and $\mu(f) = (\kappa/f_r)|f/f_r|^{\kappa-1}$) or power dispersive waves, are best analyzed using power class QTFRs that preserve power time shifts $\tau(f) = (\kappa/f_r)|f/f_r|^{\kappa-1}$ (see Subsection 10.4.2.4). Furthermore, signals with power exponential TF characteristics, such as power exponential impulses (the signal in (10.14) with $\lambda(b) = e^{\kappa b}$ and $\mu(f) = (\kappa/f_r)e^{\kappa f/f_r}$), are best analyzed using power exponential QTFRs that preserve power exponential time shifts $\tau(f) = (\kappa/f_r)e^{\kappa f/f_r}$ (see Subsection 10.4.2.4). In each case, a matched TF analysis is achieved when the signal group delay $\mu(f)$ is the same or approximately the same as the QTFR time shift $\tau(f)$.

Next, we provide some simulation examples by analyzing signals using group delay shift covariant QTFRs whose time shift is matched to the signal's group delay. Furthermore, we discuss and demonstrate the effect of signal-QTFR mismatch [61, 62]. Such a mismatch occurs when the signal's group delay does not equal the group delay shift covariance of a QTFR, causing significant distortion that may impede analysis. Smoothed dispersively warped WDs are important in multicomponent analysis applications, as they reduce cross-terms. However, they reduce cross-terms with some loss of TF resolution only when (a) the QTFR's group delay shift covariance matches the signal's group delay, and (b) the QTFR's smoothing characteristics match the signal's TF structure [61, 62]. Note that distortion in the TF plane could still result if (a) holds but (b) does not hold. Thus, we also use examples to demonstrate the distortion that occurs when the signal's group delay does not match the smoothing operation of the QTFR that was used to analyze it. In addition, we provide a real-data analysis example using group delay shift covariant QTFRs. A discussion on the discrete implementation of hyperbolic class QTFRs can be found in [70, 71], of power class QTFRs in [34], of exponential QTFRs in [37], and of power exponential class QTFRs in [27]. For computational purposes, whenever we analyze a generalized impulse in (10.14), we window it in the frequency domain. Unless otherwise specified, the sampling frequency used is 1 Hz.

10.5.1 Impulse example

An impulse is defined in the frequency domain as $X(f) = (1/\sqrt{f_r})e^{-j2\pi c(f/f_r)}$. It is the generalized impulse in (10.14) with $\lambda(b) = b$ and constant group delay $\mu(f) = 1/f_r$ (see Table 10.3). Cohen's class QTFRs are well matched to analyzing this constant group delay signal since these QTFRs are constant time-shift covariant ($\tau(f) = 1/f_r$ in (10.11)). In particular, the WD in (10.5) of an impulse is a Dirac delta function perfectly concentrated along the constant group delay c/f_r of the signal

$$X(f) = \frac{1}{\sqrt{f_r}} e^{-j2\pi c(f/f_r)} \quad \Rightarrow \quad W_X(t, f) = \frac{1}{f_r}\delta(t - c/f_r).$$

Figure 10.7 analyzes the sum of two windowed impulses that, before windowing, is given by $Y(f) = (e^{-j2\pi 8(f/f_r)} + e^{-j2\pi 18(f/f_r)})/\sqrt{f_r}$. As expected, the WD of this signal in Figure 10.7(a) is well concentrated along the constant group delay of each signal term, but the WD has cross-terms between the two signal components [17]. The pseudo WD in Figure 10.7(b) removes the cross-terms with a small loss of TF concentration. This occurs because both of these QTFRs are Cohen's class QTFRs (dispersively warped Cohen's class QTFRs with $\xi(b) = b$) that are well suited for analyzing impulses as they preserve constant group delay changes. The QD in Figure 10.7(c) and the κth power WD with $\kappa = 3$ in Figure 10.7(d) both suffer from oscillatory cross-terms just like the WD in Figure 10.7(a), but they are not matched to the signal's TF structure, resulting in a loss of TF resolution. The smoothed pseudo QD in Figure 10.7(d) is not successful at removing oscillatory cross-terms, and it forces a hyperbolic structure to the second impulse. The κth power smoothed pseudo WD with $\kappa = 3$ in Figure 10.7(f) largely distorts the signal's TF structure, and does not appear to remove any cross-terms. Furthermore, the smoothing imposes a power TF structure to the resulting QTFR. Thus, hyperbolic and power QTFRs are not well suited to analyze signals with constant group delay characteristics. As we will show next, hyperbolic and power QTFRs are better matched to signals with dispersive group delay characteristics.

10.5.2 Hyperbolic impulse example

A hyperbolic impulse is the generalized impulse in (10.14) with logarithmic phase response $\lambda(b) = \ln b$ and hyperbolic group delay $\mu(f) = 1/f$, i.e., $X(f) =$

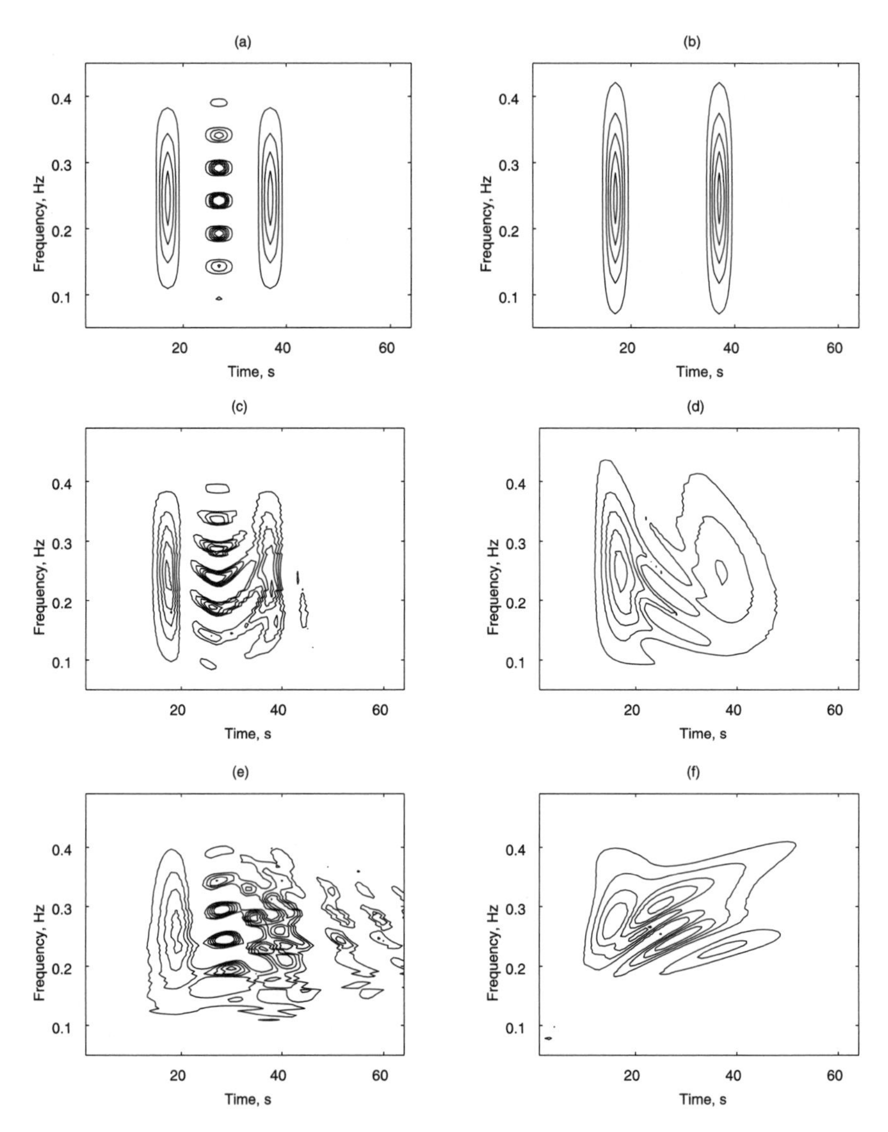

FIGURE 10.7. TF analysis of the sum of two windowed impulses. The first row shows QTFRs from Cohen's class, the second from the hyperbolic class, and the third from the power class with $\kappa = 3$. (a) WD, (b) pseudo WD, (c) QD, (d) smoothed pseudo QD, (e) κth power WD with $\kappa = 3$, (f) κth power smoothed pseudo WD with $\kappa = 3$.

$\frac{1}{\sqrt{f}} e^{-j2\pi c \ln(f/f_r)}$, $f > 0$. Since this signal has hyperbolic TF characteristics, it is well matched to QTFRs that preserve hyperbolic time shifts $\tau(f) = 1/f$, such as the QD in (10.27). Specifically, the QD of the hyperbolic impulse is perfectly concentrated along a hyperbola in the TF plane (property [P-DC8] in Table 10.5 with

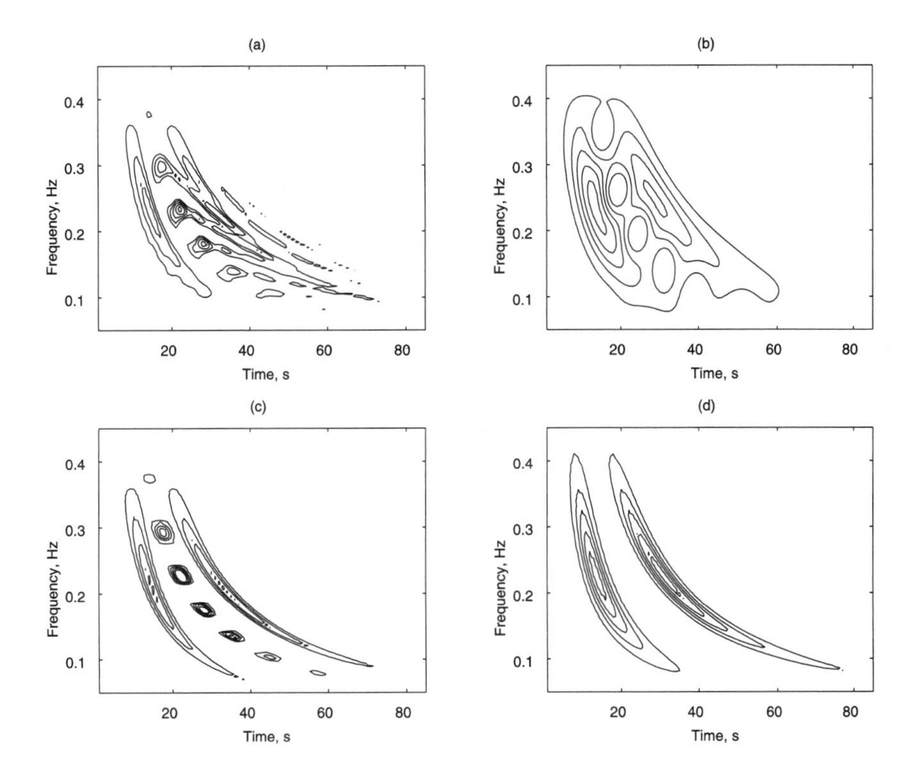

FIGURE 10.8. TF analysis of the sum of two windowed hyperbolic impulses. The first row shows QTFRs from Cohen's class, and the second from the hyperbolic class. (a) WD, (b) smoothed pseudo WD, (c) QD, (d) smoothed pseudo QD.

$\xi(b) = \ln b$), i.e.,

$$X(f) = \frac{1}{\sqrt{f}} e^{-j2\pi c \ln(f/f_r)} \quad \Rightarrow \quad Q_X(t,f) = \frac{1}{f}\delta\left(t - \frac{c}{f}\right), \qquad f > 0.$$

Figure 10.8 analyzes the sum of two windowed hyperbolic impulses given (before windowing) as $Y(f) = (e^{-j2\pi 3 \ln(f/f_r)} + e^{-j2\pi 7 \ln(f/f_r)})/\sqrt{f}$, $f > 0$. As expected, the QD in Figure 10.8(c) has good TF concentration along the two hyperbolas $t = 3/f$ and $t = 7/f$. However, it also results in cross-terms along the mean hyperbola $t = 5/f$ [25]. The smoothed pseudo QD (the dispersively warped smoothed pseudo WD with $\xi(b) = \ln b$ in Subsection 10.4.1.3) in Figure 10.8(d) removes the cross-terms with some loss of TF resolution. Cohen's class QTFRs, such as the WD in Figure 10.8(a) and the smoothed pseudo WD in Figure 10.8(b), are not well matched to hyperbolic impulses. The WD results in complicated cross-terms between the two signal components as well as inner interference terms [17]. In comparison to the smoothed pseudo QD in Figure 10.8(d), the smoothed pseudo WD in Figure 10.8(b) has a larger loss of TF resolution and it is not as successful at removing all the cross-terms.

10.5.3 Power impulse example

The generalized impulse in (10.14) simplifies to the κth power impulse, $X(f) = \sqrt{|\mu_\kappa(f)|}\, e^{-j2\pi c \lambda_\kappa(f/f_r)}$ with phase response $\lambda_\kappa(b) = \mathrm{sgn}(b)\,|b|^\kappa$, and group delay function $\mu_\kappa(f) = (\kappa/f_r)|f/f_r|^{\kappa-1}$. The κth power WD in (10.28) is ideally matched to this signal. Specifically, since the signal group delay $\mu_\kappa(f)$ equals the power time shift $\tau_\kappa(f)$ of the power WD, the power WD of the power impulse is a delta function perfectly concentrated along the group delay of the signal (property [P-DA8] in Table 10.7 with $\xi(b) = \xi_\kappa(b)$)

$$X(f) = \sqrt{|\tau_\kappa(f)|} e^{-j2\pi c \xi_\kappa(f/f_r)} \quad \Rightarrow \quad W_X^{(\xi_\kappa)}(t,f) = \tau_\kappa(f)\delta(t - c\tau_\kappa(f)). \tag{10.48}$$

The signal we analyze next is the sum of two windowed power impulses with signal power parameter $\kappa_{\mathrm{sig}} = 3$. In order to avoid any confusion between the various power parameters, we use κ_{sig} to denote the power parameter of a κth power impulse in (10.48), and κ_{distr} to denote the power parameter of a κth power distribution in (10.35). Before windowing, the signal is given by $Y(f) = \sqrt{|\tau_3(f)|}\,(e^{-j2\pi 8\, \xi_3(f/f_r)} + e^{-j2\pi 23 \xi_3(f/f_r)})$, where $\xi_3(b) = \mathrm{sgn}(b)|b|^3$ and $\tau_3(f) = (3/f_r)|f/f_r|^2$. For this signal's analysis, we use power class QTFRs with various power parameters κ_{distr}. Figures 10.9(a) and 10.9(b) show the results obtained with the power WD and the power smoothed pseudo WD [32]. These QTFRs are both members of the power class with power parameter $\kappa_{\mathrm{distr}} = 3$. Note that the power parameter κ_{distr} of the two power QTFRs is matched to the power parameter κ_{sig} of the power impulses. The power WD in Figure 10.9(a) has very good TF concentration but large cross-terms [34]. Note that the cross-term geometry of the power WD of multicomponent signals is discussed in [34]. These cross-terms are effectively suppressed in the power smoothed pseudo WD in Figure 10.9(b) with hardly any loss of TF concentration. Figures 10.9(c) and 10.9(d) show the WD and an affine smoothed pseudo WD, both members of the affine class (i.e., both have power parameter $\kappa_{\mathrm{distr}} = 1$). The WD in Figure 10.9(c) is not matched to the power impulses and, as a result, it suffers from complicated cross-terms. The affine smoothed pseudo WD in Figure 10.9(d) does not remove all the cross-terms and has a larger loss of TF concentration than does the power smoothed pseudo WD in Figure 10.9(b). Even though all QTFRs in Figure 10.9 are scale covariant, the results of the two power class QTFRs with $\kappa_{\mathrm{distr}} = 3$ in Figures 10.9(a) and 10.9(b) are better than those of the corresponding two affine class QTFRs (power class QTFRs with $\kappa_{\mathrm{distr}} = 1$) in Figures 10.9(c) and 10.9(d) because the former two are optimally matched to the $\kappa_{\mathrm{sig}} = 3$ power-law group delays of the power impulse signal components.

In order to further demonstrate the effect of mismatch in the signal and distribution power parameters κ_{sig} and κ_{distr}, Figures 10.9(e) and 10.9(f) show the results obtained when analyzing the preceding signal using the power WD and the power

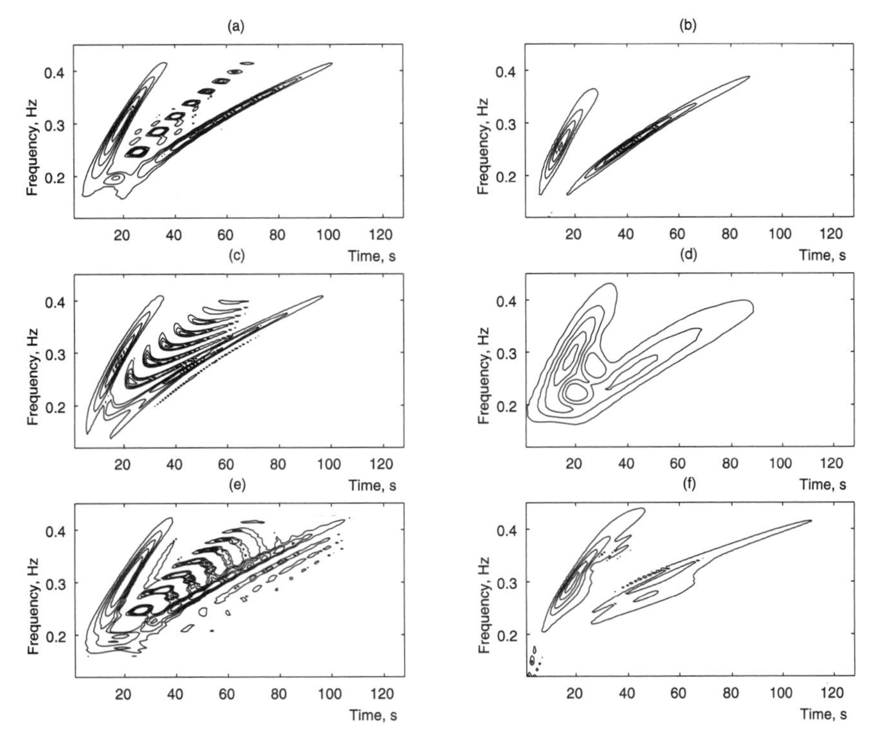

FIGURE 10.9. TF analysis of the sum of two windowed power impulses with $\kappa_{\rm sig} = 3$. The first row shows QTFRs from the power class with $\kappa_{\rm distr} = 3$, the second from the affine class ($\kappa_{\rm distr} = 1$), and the third from the power class with $\kappa_{\rm distr} = 4$. (a) Power WD with $\kappa_{\rm distr} = 3$, (b) power smoothed pseudo WD with $\kappa_{\rm distr} = 3$, (c) WD ($\kappa_{\rm distr} = 1$), (d) affine smoothed pseudo WD ($\kappa_{\rm distr} = 1$), (e) power WD with $\kappa_{\rm distr} = 4$, and (f) power smoothed pseudo WD with $\kappa_{\rm distr} = 4$.

smoothed pseudo WD with power parameter $\kappa_{\rm distr} = 4$. Note that in Figures 10.9(e) and 10.9(f) the power parameter of the power QTFRs, $\kappa_{\rm distr} = 4$, is different from that of the signal, $\kappa_{\rm sig} = 3$. The power smoothed pseudo WD in Figure 10.9(f) has better cross-term removal and better TF concentration along the true group delay than the affine smoothed pseudo WD in Figure 10.9(d) since the power parameter mismatch in Figure 10.9(f) is smaller than in Figure 10.9(d) [34, 61].

10.5.4 *Power exponential impulse example*

The κth power exponential WD is the dispersively warped WD in (10.23) when $\xi(b) = e^{\kappa b}$. Since it preserves power exponential time shifts $\tau(f) = (\kappa/f_r)e^{\kappa f/f_r}$, it is well matched to signals with power exponential TF characteristics, such as power exponential impulses. The power exponential impulse is given in (10.14) with exponential phase response $\lambda(b) = e^{\kappa b}$ and group delay $\mu(f) = (\kappa/f_r)e^{\kappa f/f_r}$, i.e.,

$X(f) = \sqrt{(|\kappa|/f_r)e^{\kappa f/f_r}}\, e^{-j2\pi c\, e^{\kappa f/f_r}}$. The power exponential WD is well matched to a power exponential impulse, and provides good TF concentration along the power exponential group delay curve. However, since the range of $\xi(b) = e^{\kappa b}$ is $\Re_+$, the power exponential WD of a power exponential impulse does not result in a perfectly concentrated delta function along the power exponential curve. For example, when $\kappa = 1$, the exponential WD, $E_X(t, f)$ in (10.29), of an exponential impulse is given by

$$X(f) = \sqrt{(1/f_r)e^{f/f_r}}\, e^{-j2\pi c\, e^{f/f_r}}$$
$$\Rightarrow \quad E_X(t, f) = \frac{1}{f_r} e^{f/f_r}\, \mathrm{sc}\left(t - \frac{c}{f_r} e^{f/f_r}; f\right),$$

where $\mathrm{sc}(t; f) = \sin\left(4\pi f_r (1 - e^{-f/f_r})\, t\right)/(\pi t)$ is a frequency-dependent sinc function. As frequency increases, the exponential WD becomes very narrow since the width of the sinc function is inversely proportional to $(1 - e^{-f/f_r})$. Thus, in the limit, the exponential WD becomes concentrated along the signal's exponential group delay and is indeed well matched to the signal.

In the following example, we demonstrate the advantage of using κth power exponential QTFRs to analyze signals with κth power exponential group delay. The test signal consists of the sum of three windowed power exponential impulses with $\kappa = \kappa_{\mathrm{sig}} = 2.3$. Specifically, before windowing, the signal is given by $Y(f) = \sqrt{(\kappa/f_r)e^{\kappa f/f_r}}\left(e^{-j2\pi(0.3)\, e^{\kappa f/f_r}} + e^{-j2\pi(1.8)\, e^{\kappa f/f_r}} + e^{-j2\pi(3.8)\, e^{\kappa f/f_r}}\right)$ with $\kappa = 2.3$. Figure 10.10 compares six different QTFRs of this signal: (i) two affine class QTFRs (dispersively warped affine class with $\xi(b) = b$), the WD and the affine smoothed pseudo WD, (ii) two power exponential class QTFRs with $\kappa_{\mathrm{distr}} = 2.3$ (dispersively warped affine class with $\xi(b) = e^{\kappa b}$), the power exponential WD and the power exponential smoothed pseudo WD, and (iii) two exponential class QTFRs (dispersively warped affine class with $\xi(b) = e^{b}$), the exponential WD and the exponential smoothed pseudo WD. Note that κth power exponential QTFRs are ideal for analyzing κth power exponential impulses because the κth power exponential time shift, $\tau(f) = (\kappa/f_r)e^{\kappa f/f_r}$, of the QTFRs is matched to the κth power exponential group delay, $\mu(f) = (\kappa/f_r)e^{\kappa f/f_r}$, of each signal component, provided that the power κ is the same for both the QTFRs and the signal. This is demonstrated in Figure 10.10(c), where the power exponential WD with $\kappa_{\mathrm{distr}} = 2.3$, chosen to match the power parameter of the input signal, shows well-concentrated cross-terms along the power exponential group delay curve. On the other hand, the WD and the exponential WD have more complicated cross-term structures, as shown in Figures 10.10(a) and Figure 10.10(e), respectively. Using a smoothing component along the power exponential group delay curve, the power exponential smoothed pseudo WD with $\kappa_{\mathrm{distr}} = 2.3$ reduces the cross-terms with only moderate loss of TF resolution, as shown in Figure 10.10(d). However, as the affine smoothed pseudo WD smoothing is not matched to the signal's power exponential group delay curve, it does not succeed in removing all of the cross-terms, as shown in Figure 10.10(b). Since the exponential smoothed pseudo WD is actually equal to the power exponential smoothed pseudo WD when $\kappa_{\mathrm{distr}} = 1$, the exponential smoothed pseudo WD in

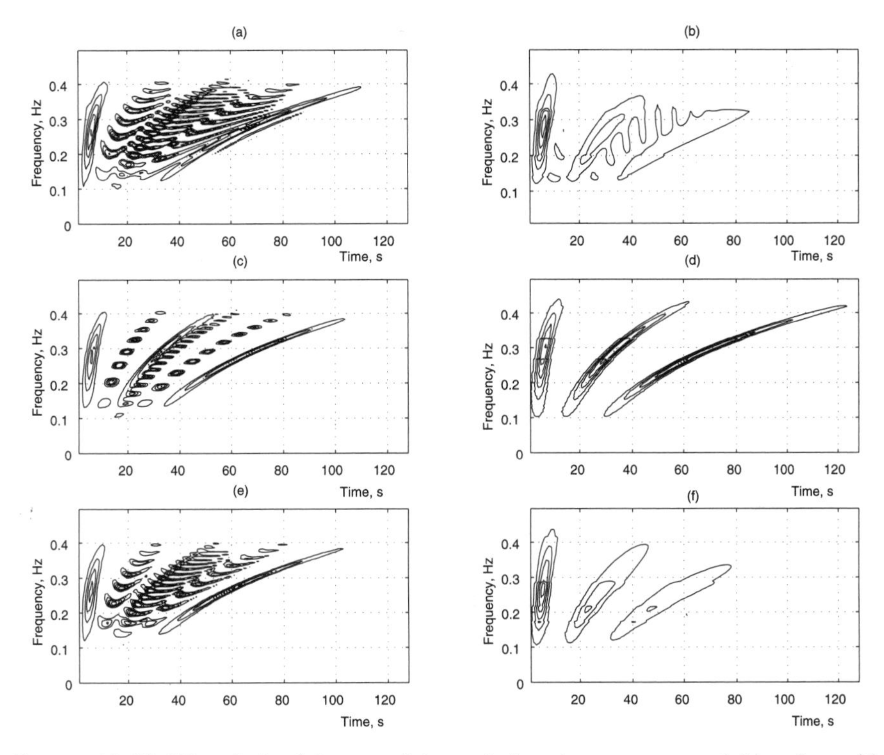

FIGURE 10.10. TF analysis of the sum of three windowed power exponential impulses with $\kappa_{\text{sig}} = 2.3$. The first row shows QTFRs from the affine class, the second from the power exponential class with $\kappa_{\text{distr}} = 2.3$, and the third from the exponential class ($\kappa_{\text{distr}} = 1$). (a) WD, (b) affine smoothed pseudo WD, (c) power exponential WD with $\kappa_{\text{distr}} = 2.3$, (d) power exponential smoothed pseudo WD with $\kappa_{\text{distr}} = 2.3$, (e) exponential WD, and (f) exponential smoothed pseudo WD.

Figure 10.10(f) removes more cross-terms [37] than the affine smoothed pseudo WD in Figure 10.10(b). On the other hand, the auto-terms of the exponential smoothed pseudo WD are distorted. This is because the power exponential time shift of the power exponential smoothed pseudo WD with $\kappa_{\text{distr}} = 1$ is not matched to the $\kappa_{\text{sig}} = 2.3$ power exponential group delay parameter of the signal, and a mismatch occurs. Note, however, that choosing the wrong power κ_{distr} of the power exponential smoothed pseudo WD in Figure 10.10(f) yields better results than choosing an affine QTFR that is covariant to constant time shifts, as in Figure 10.10(b). In this example, a mismatch occurs in Figure 10.10(a) and 10.10(b) between the group delay of the power exponential impulses and the constant time shift of the WDs, and in Figure 10.10(e) and 10.10(f) between the group delay of the power exponential impulses and the exponential time shift of the exponential WDs.

10.5.5 Hyperbolic impulse and Gaussian example

In this example, we consider dispersively warped WDs and their smoothed versions to demonstrate some cases of signal-QTFR mismatch. A mismatch will occur whenever the group delay shift in (10.11) does not match the signal's group delay [61,62]. The windowed signal consists of three hyperbolic impulses, $(1/\sqrt{f})(e^{-j2\pi 3 \ln(f/f_r)} + e^{-j2\pi 7 \ln(f/f_r)} + e^{-j2\pi 11 \ln(f/f_r)})$, $f > 0$ and two TF-shifted Gaussian signals. Only the hyperbolic class QTFRs (covariant to hyperbolic time shifts $\tau(f) = 1/f$) like the QD and its smoothed versions match the hyperbolic impulse components of this signal, whereas only Cohen's class QTFRs (covariant to constant time shifts $\tau(f) = 1/f_r$) like the WD and its smoothed versions match the Gaussian signal components (with constant support in the TF plane). Thus, the hyperbolic impulses are better matched to hyperbolic class QDs, whereas the linear, Gaussian components are better matched to Cohen's class WDs. Figure 10.11 shows the "ideal" TF representation of this signal for comparison. The ideal representation consists of the sum of the group delay functions of each of the components. Figure 10.12 shows the TF analysis. Both the WD in Figure 10.12(a) and the QD in Figure 10.12(b) suffer from oscillatory cross-terms between pairs of the five signal components. Figure 10.12(c) shows that the smoothed pseudo WD is not as successful at removing oscillatory cross-terms. The smoothed pseudo QD in Figure 10.12(d) is well matched to the three hyperbolic impulses, as the smoothing window is matched to this signal's TF structure, and thus it is better at reducing cross-terms. Due to local smoothing with negative cross-terms, the amplitude of the hyperbolic impulses in the smoothed pseudo WD and that of the Gaussian signals in the smoothed pseudo QD are significantly reduced. Also, there is distortion in the TF support of the smoothed QTFRs: the hyperbolic impulses in the smoothed pseudo WD now appear linear, and

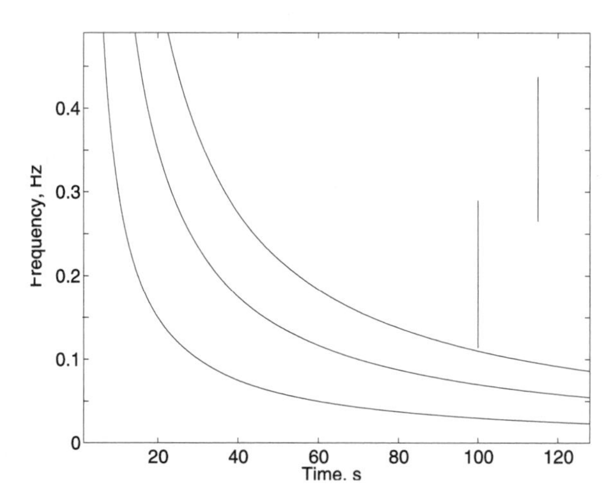

FIGURE 10.11. The ideal TF representation of the sum of three hyperbolic impulses and two TF shifted Gaussian components.

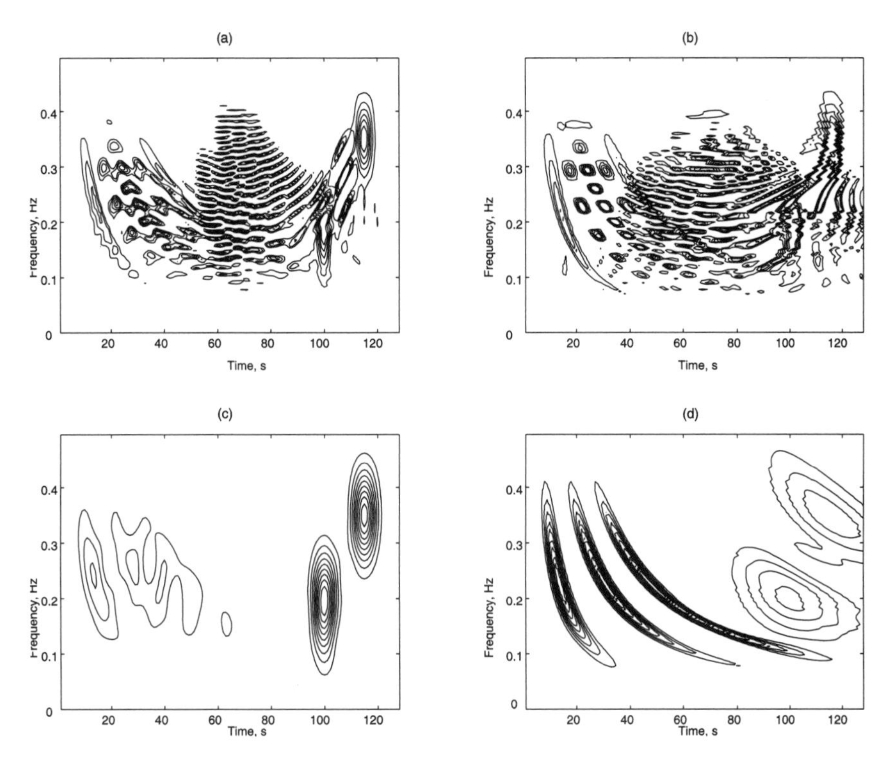

FIGURE 10.12. TF analysis of the sum of three windowed hyperbolic impulses and two TF-shifted Gaussian components. Cohen's class QTFRs are shown on the left-hand side, and hyperbolic class QTFRs are shown on the right-hand side. (a) WD, (b) QD, (c) smoothed pseudo WD, and (d) smoothed pseudo QD.

the Gaussian components in the smoothed pseudo QD are no longer vertical. Note that a mismatch occurs between the group delay of each hyperbolic impulse and the constant time shift of the WDs in Figure 10.12(a) and in Figure 10.12(c). Also, a mismatch occurs between the constant group delay of the Gaussian components and the hyperbolic time shift of the QDs in Figure 10.12(b) and in Figure 10.12(d). One should use the smoothed pseudo QD if it is more important to analyze the hyperbolic impulses in a particular application.

10.5.6 Impulse response of a steel beam example

The next example analyzes a real-data signal using group delay shift covariant QTFRs. The signal consists of the measured impulse response of a steel beam with rectangular cross section [65–67]. The impulse response was obtained by lightly tapping one end of the steel beam in the direction orthogonal to the flat side of the beam. Bending waves travel along the beam until they are reflected at the free end. They

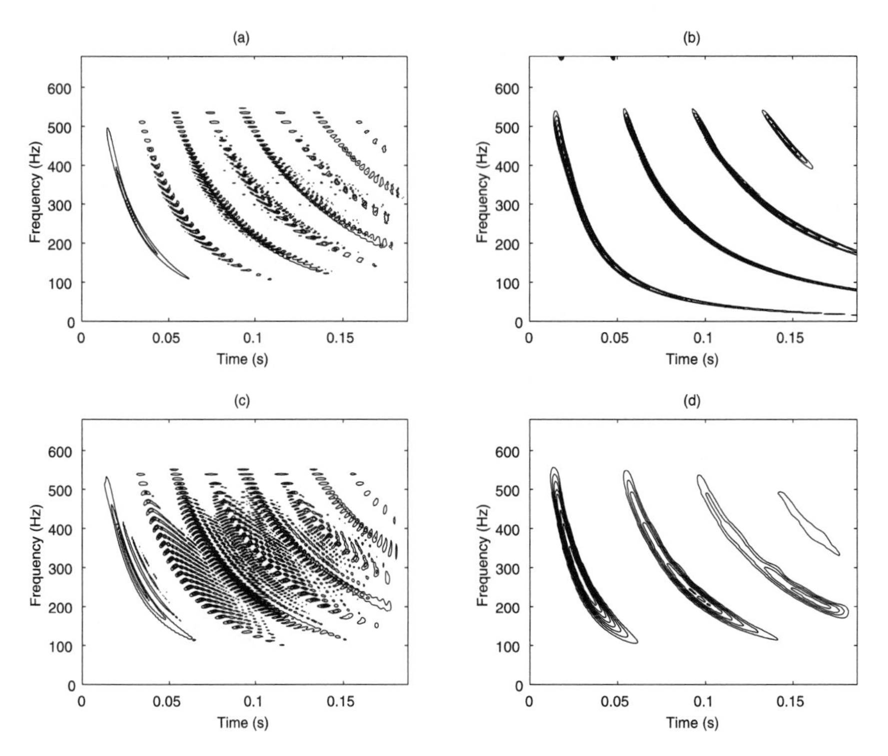

FIGURE 10.13. TF analysis of a bandpass-filtered segment of the measured impulse response of a steel beam. The first row shows QTFRs in the power class with $\kappa = 0.35$, and the second row shows QTFRs in the affine class. (a) Power WD with $\kappa = 0.35$, (b) power smoothed pseudo WD with $\kappa = 0.35$, (c) WD ($\kappa = 1$), and (d) affine smoothed pseudo WD ($\kappa = 1$). The sampling frequency of the data is 4096 Hz.

continue to return to the point of impact and are reflected again, thus producing a series of echoes with increasing dispersion. The QTFRs in Figure 10.13 display a bandpass-filtered segment of the impulse response. Specifically, Figure 10.13 shows two power class QTFRs with $\kappa = 0.35$, the power WD in (10.28) and the power smoothed pseudo WD [32], and two affine class QTFRs (that correspond to power class QTFRs with $\kappa = 1$), the WD and the affine smoothed pseudo WD of the impulse response. Both the WD in Figure 10.13(c) and the power WD with $\kappa = 0.35$ in Figure 10.13(a) suffer from oscillatory cross-terms. These cross-terms are suppressed by both the affine smoothed pseudo WD in Figure 10.13(d) and the power smoothed pseudo WD with $\kappa = 0.35$ in Figure 10.13(b). However, only the matched QTFR in Figure 10.13(b) shows better TF resolution than the other three QTFRs. Here, the specific value of $\kappa = 0.35$ was chosen empirically to match the TF curvature of the primary reflection. Short-time power QTFRs were also found to be useful in analyzing cetacean mammal whistles [69].

10.6 Conclusion

QTFRs are important tools for analyzing nonstationary signals. Many different QTFRs have been proposed in the literature, and various tutorials on QTFRs can be found in [1–6, 10–13]. Different QTFRs satisfy different sets of properties, and, as a result, they can be used more successfully in certain applications. In this chapter, we provided a tutorial on classes of QTFRs that are covariant to group delay shifts $\tau(f)$ corresponding to the group delay or changes in the group delay of the analysis signal. Depending on $\tau(f)$, these group delay shifts may be constant or dispersive (i.e., nonlinear and frequency-dependent such that high frequency components are delayed by amounts different from low frequency components). Thus, the group delay shift covariance property is important when analyzing signals propagating through systems with specific dispersive characteristics. For example, the nondispersive time-shift covariance of Cohen's class and the affine class is matched to constant group delays and is useful in many applications, such as speech analysis. The hyperbolic time-shift covariance of the hyperbolic class is matched to hyperbolic group delays and is important when analyzing Doppler-invariant signals, such as the signals used by bats for echolocation. The power time-shift covariance of the power classes is matched to power-law group delays and is useful for analyzing signals with power TF characteristics, such as the acoustical waves reflected from spherical shells immersed in water. On the other hand, the exponential time-shift covariance of the exponential and power exponential classes is matched to exponential group delays and is useful for analyzing signals passing through systems with exponentially dispersive TF characteristics.

The various group delay shift covariant QTFR classes that were discussed in this chapter are summarized in Tables 10.4 and 10.6, and are shown in Figure 10.4. We have demonstrated that, since the group delay shift operator in (10.10) is unitarily equivalent to the constant time-shift operator in (10.1), we can obtain the group delay shift covariant classes through unitary warpings of known constant time-shift covariant QTFR classes such as Cohen's class and the affine class. The warping involves a frequency-dependent axis transformation of the analysis signal combined with a transformation of the TF axes, both of which depend on a differentiable one-to-one warping function $\xi(b)$. The group delay (possibly dispersive) time shift corresponding to each class is matched to the warping function as it is chosen to be $\tau(f) = (d/df)\xi(f/f_r)$. For example, we obtain the hyperbolic class of scale covariant and hyperbolic time-shift ($\tau(f) = 1/f$) covariant QTFRs by warping Cohen's class QTFRs using $\xi(b) = \ln b$; we obtain the κth power class of scale covariant and κth power time-shift ($\tau(f) = (\kappa/f_r)|f/f_r|^{\kappa-1}$) covariant QTFRs by warping the affine class using $\xi(b) = \text{sgn}(b)|b|^{\kappa}$; and we obtain the power exponential class of frequency-shift covariant and power exponential time-shift ($\tau(f) = (\kappa/f_r)e^{\kappa f/f_r}$) covariant QTFRs by warping the affine class using $\xi(b) = e^{\kappa b}$.

For a particular application, it is important to identify the TF characteristics of the signal under analysis so as to design a QTFR to ideally analyze the signal. Thus, by matching the signal's group delay with the time-shift covariance property of a given class of QTFRs, one may avoid causing significant distortion in the TF plane, which may impede analysis. We have shown, for example, that the dispersively warped WD in (10.23) is highly concentrated for generalized impulses in (10.14) that have a phase response $\xi(b)$ with range $\Re$ (that is, it satisfies property [P-DC8] in Table 10.5). However, the high concentration is obtained only when the group delay shift $\tau(f)$ used to obtain the dispersively warped WD in (10.23) is equal to the group delay function $\mu(f)$ of the generalized impulse in (10.14). Note that, in practical applications, dispersively warped smoothed QTFRs (such as the dispersively warped pseudo WD in Subsection 10.4.1.3) are important for multicomponent signal analysis, as they reduce cross-terms. Moreover, not only does the time shift need to match the signal's group delay in order to avoid TF distortion, but also the smoothing characteristics of the QTFR need to match the signal's TF structure, as was demonstrated with the simulation examples. When a signal consists of components with different group delay or TF structures, it is difficult to find an ideal QTFR to match the signal for successful TF analysis. Depending on the application at hand, it might be necessary to decide which signal components need to be best matched, and then to design a QTFR to match closely those components with only minimum loss of TF resolution for the other components.

The QTFRs presented in this chapter are important analysis tools for analyzing nonstationary signals with dispersive TF characteristics. These dispersive TF tools have been extended to higher order TF representations in [72, 73]. Also, TF symbols have been developed in [74–77] as quadratic TF analysis tools for analyzing nonstationary random processes or linear time-varying systems, such as wireless communication, sonar, or radar channels.

Acknowledgments. The author is with the Department of Electrical Engineering, Arizona State University, P. O. Box 877206, Tempe, AZ 85287-7206 (e-mail: papandreou@asu.edu). This work was supported in part by the Office of Naval Research grant N00014-96-1-0350, by the Naval Undersea Warfare Center contract N66604-96-C-A336, and by the National Science Foundation grant NSF-EIA0074663. The author thanks G. F. Boudreaux-Bartels and F. Hlawatsch for their extensive collaboration on generalized time-shift covariant QTFRs, D. Newland and J. Woodhouse for providing the steel beam impulse response data, P. Gonçalvès for providing a part of the software used for computing the affine and power class QTFRs, and B. Iem for providing the power exponential class example. Also, special thanks to S. B. Suppappola for his insightful comments in reviewing this manuscript.

References

[1] L. Cohen, *Time-Frequency Analysis*. Englewood Cliffs, New Jersey: Prentice-Hall, 1995.

[2] P. Flandrin, *Time-Frequency/Time-Scale Analysis*. San Diego, California: Academic Press, 1999. (Translated from French, *Temps-fréquence*. Paris: Hermès, 1993).

[3] F. Hlawatsch and G. F. Boudreaux-Bartels, Linear and quadratic time-frequency signal representations, *IEEE Signal Processing Magazine*, vol. 9, pp. 21–67, April 1992.

[4] B. Boashash, ed., *Time-Frequency Signal Analysis—Methods and Applications*. Melbourne, Australia: Longman-Cheshire, 1992.

[5] S. Qian and D. Chen, *Joint Time-Frequency Analysis: Methods and Applications*. Englewood Cliffs, New Jersey: Prentice-Hall, 1996.

[6] A. Papandreou-Suppappola, ed., *Applications in Time-Frequency Signal Processing*. Boca Raton, Florida: CRC Press, 2002.

[7] W. F. G. Mecklenbräuker and F. Hlawatsch, eds., *The Wigner Distribution—Theory and Applications in Signal Processing*. Amsterdam, The Netherlands: Elsevier, 1997.

[8] L. Cohen, Generalized phase-space distribution functions, *Journal of Mathematics and Physics*, vol. 7, pp. 781–786, 1966.

[9] T. A. C. M. Claasen and W. F. G. Mecklenbräuker, The Wigner distribution—A tool for time-frequency signal analysis, Part III: Relations with other time-frequency signal transformations, *Philips Journal of Research*, vol. 35, pp. 372–389, 1980.

[10] B. Boashash, Time-frequency signal analysis, in *Advances in Spectrum Estimation* (S. Haykin, ed.), Englewood Cliffs, New Jersey: Prentice Hall, 1990.

[11] G. F. Boudreaux-Bartels, Mixed time-frequency signal transformations, in *The Transforms and Applications Handbook* (A. Poularikas, ed.), Boca Raton, Florida: CRC Press, 1996.

[12] A. Papandreou-Suppappola, F. Hlawatsch, and G. F. Boudreaux-Bartels, Quadratic time-frequency representations with scale covariance and generalized time-shift covariance: A unified framework for the affine, hyperbolic, and power classes, *Digital Signal Processing: A Review Journal*, vol. 8, pp. 3–48, January 1998.

[13] A. Papandreou-Suppappola, Time-varying processing: tutorial on principles and practice, in *Applications in Time-Frequency Signal Processing* (A. Papandreou-Suppappola, ed.), Boca Raton, Florida: CRC Press, 2002.

[14] E. P. Wigner, On the quantum correction for thermo-dynamic equilibrium, *Physics Review*, vol. 40, pp. 749–759, 1932.

[15] T. A. C. M. Claasen and W. F. G. Mecklenbräuker, The Wigner distribution—A tool for time-frequency signal analysis, Part I: Continuous-time signals, *Philips Journal of Research*, vol. 35, pp. 217–250, 1980.

[16] T. A. C. M. Claasen and W. F. G. Mecklenbräuker, The Wigner distribution—A tool for time-frequency signal analysis, Part II: Discrete-time signals, *Philips Journal of Research*, vol. 35, pp. 276–300, 1980.

[17] F. Hlawatsch and P. Flandrin, The interference structure of the Wigner distribution and related time-frequency signal representations, *The Wigner Distribution—Theory and Applications in Signal Processing* (W. Mecklenbräuker and F. Hlawatsch), Amsterdam: North Holland Elsevier Science Publishers, 1997.

[18] L. R. Rabiner and R. W. Schafer, *Digital Processing of Speech Signals*. Englewood Cliffs, New Jersey: Prentice-Hall, 1978.

[19] R. A. Altes, Detection, estimation, and classification with spectrograms, *Journal of the Acoustical Society of America*, vol. 67, pp. 1232–1246, April 1980.

[20] J. Bertrand and P. Bertrand, A class of affine Wigner functions with extended covariance properties, *Journal of Mathematics and Physics*, vol. 33, pp. 2515–2527, 1992.

[21] O. Rioul and P. Flandrin, Time-scale energy distributions: A general class extending wavelet transforms, *IEEE Transactions on Signal Processing*, vol. 40, pp. 1746–1757, July 1992.

[22] P. Flandrin and P. Gonçalvès, From wavelets to time-scale energy distributions, in *Recent Advances in Wavelet Analysis* (L. L. Schumaker and G. Webb, eds.), pp. 309–334, New York: Academic Press, 1994.

[23] P. Flandrin and P. Gonçalvès, Geometry of affine time-frequency distributions, *Applied and Computational Harmonic Analysis*, vol. 3, pp. 10–39, January 1996.
[24] O. Rioul and M. Vetterli, Wavelets and signal processing, *IEEE Signal Processing Magazine*, vol. 8, pp. 14–38, October 1991.
[25] A. Papandreou, F. Hlawatsch, and G. F. Boudreaux-Bartels, The hyperbolic class of quadratic time-frequency representations, Part I: Constant-Q warping, the hyperbolic paradigm, properties, and members, *IEEE Transactions on Signal Processing*, vol. 41, pp. 3425–3444, December 1993.
[26] F. Hlawatsch, A. Papandreou-Suppappola, and G. F. Boudreaux-Bartels, The hyperbolic class of quadratic time-frequency representations, Part II: Subclasses, intersection with the affine and power classes, regularity, and unitarity, *IEEE Transactions on Signal Processing*, vol. 45, pp. 303–315, February 1997.
[27] A. Papandreou-Suppappola, R. L. Murray, B. G. Iem, and G. F. Boudreaux-Bartels, Group delay shift covariant quadratic time-frequency representations, *IEEE Transactions on Signal Processing*, vol. 49, pp. 2549–2564, November 2001.
[28] A. Papandreou-Suppappola, Time-frequency representations covariant to group delay shifts, in *Time-Frequency Signal Analysis and Processing* (B. Boashash, ed.), Englewood Cliffs, New Jersey: Prentice-Hall, 2003.
[29] R. A. Altes, Wide-band, proportional-bandwidth Wigner–Ville analysis, *IEEE Transactions on Acoustics, Speech, and Signal Processing*, vol. 38, pp. 1005–1012, June 1990.
[30] N. M. Marinovich, *The Wigner Distribution and the Ambiguity Function: Generalizations, Enhancement, Compression and Some Applications*. Ph.D. thesis, The City University of New York, 1986.
[31] F. Hlawatsch, A. Papandreou, and G. F. Boudreaux-Bartels, The power classes of quadratic time-frequency representations: A generalization of the affine and hyperbolic classes, in *Proceedings Twenty-Seventh Asilomar Conference on Signals, Systems and Computers*, (Pacific Grove, California), pp. 1265–1270, November 1993.
[32] F. Hlawatsch, A. Papandreou-Suppappola, and G. F. Boudreaux-Bartels, The power classes—quadratic time-frequency representations with scale covariance and dispersive time-shift covariance, *IEEE Transactions on Signal Processing*, vol. 47, pp. 3067–3083, November 1999.
[33] A. Papandreou, F. Hlawatsch, and G. F. Boudreaux-Bartels, A unified framework for the scale covariant affine, hyperbolic, and power class time-frequency representations using generalized time-shifts, in *Proceedings 1995 International Conference on Acoustics, Speech and Signal Processing*, (Detroit, Michigan), May 1995.
[34] A. Papandreou-Suppappola, F. Hlawatsch, and G. F. Boudreaux-Bartels, Power class time-frequency representations: Interference geometry, smoothing, and implementation, in *Proceedings IEEE International Symposium on Time-Frequency/Time-Scale Analysis*, (Paris, France), pp. 193–196, June 1996.
[35] A. Papandreou-Suppappola, Generalized time-shift covariant quadratic time-frequency representations with arbitrary group delays, in *Proceedings Twenty-Ninth Asilomar Conference on Signals, Systems and Computers*, (Pacific Grove, California), pp. 553–557, October/November 1995.
[36] A. Papandreou-Suppappola and G. F. Boudreaux-Bartels, The exponential class and generalized time-shift covariant quadratic time-frequency representations, in *Proceedings IEEE International Symposium on Time-Frequency/Time-Scale Analysis*, (Paris, France), pp. 429–432, June 1996.
[37] A. Papandreou-Suppappola, B. G. Iem, R. L. Murray, and G. F. Boudreaux-Bartels, Properties and implementation of the exponential class of quadratic time-frequency represen-

tations, in *Proceedings Thirtieth Asilomar Conference on Signals, Systems and Computers*, (Pacific Grove, California), pp. 237–241, November 1996.

[38] A. Papandreou-Suppappola, R. L. Murray, and G. F. Boudreaux-Bartels, Localized subclasses of quadratic time-frequency representations, in *Proceedings IEEE International Conference on Acoustics, Speech, and Signal Processing*, (Munich, Germany), pp. 2041–2044, April 1997.

[39] R. G. Baraniuk and D. L. Jones, Warped wavelet bases: Unitary equivalence and signal processing, in *Proceedings IEEE International Conference on Acoustics, Speech, and Signal Processing*, vol. 3, (Minneapolis, Minnesota), pp. 320–323, April 1993.

[40] R. G. Baraniuk, Warped perspectives in time-frequency analysis, in *Proceedings IEEE-SP International Symposium on Time-Frequency and Time-Scale Analysis*, (Philadelphia, Pennsylvania), pp. 528–531, October 1994.

[41] R. G. Baraniuk and D. L. Jones, Unitary equivalence: A new twist on signal processing, *IEEE Transactions on Signal Processing*, vol. 43, pp. 2269–2282, October 1995.

[42] R. G. Baraniuk, Covariant time-frequency representations through unitary equivalence, *IEEE Signal Processing Letters*, vol. 3, pp. 79–81, March 1996.

[43] F. Hlawatsch and H. Bölcskei, Displacement-covariant time-frequency distributions based on conjugate operators, *IEEE Signal Processing Letters*, vol. 3, pp. 44–46, February 1996.

[44] F. Hlawatsch and T. Twaroch, Covariant (α, β), time-frequency, and (a, b) representations, in *Proceedings IEEE International Symposium on Time-Frequency/Time-Scale Analysis*, (Paris, France), pp. 437–440, June 1996.

[45] A. M. Sayeed and D. L. Jones, A canonical covariance-based method for generalized joint signal representations, *IEEE Signal Processing Letters*, vol. 3, pp. 121–123, April 1996.

[46] A. M. Sayeed and D. L. Jones, A simple covariance-based characterization of joint signal representations of arbitrary variables, in *Proceedings IEEE International Symposium on Time-Frequency/Time-Scale Analysis*, (Paris, France), pp. 433–436, June 1996.

[47] F. Hlawatsch, Duality and classification of bilinear time-frequency signal representations, *IEEE Transactions on Signal Processing*, vol. 39, pp. 1564–1574, July 1991.

[48] L. Cohen, Time-Frequency Distribution—A review, *Proceedings IEEE*, vol. 77, pp. 941–981, July 1989.

[49] G. F. Boudreaux-Bartels, On the use of operators vs warpings vs axiomatic derivations of new time-frequency-scale (operator) representations, in *Proceedings Twenty-Eighth Asilomar Conference on Signals, Systems and Computers*, (Pacific Grove, California), October/November, 1994.

[50] H. I. Choi and W. J. Williams, Improved time-frequency representation of multicomponent signals using exponential kernels, *IEEE Transactions on Acoustics, Speech, and Signal Processing*, vol. 37, pp. 862–871, June 1989.

[51] A. Papandreou and G. F. Boudreaux-Bartels, Generalization of the Choi–Williams distribution and the Butterworth distribution for time-frequency analysis, *IEEE Transactions on Signal Processing*, vol. 41, pp. 463–472, January 1993.

[52] P. Flandrin, Scale-invariant Wigner spectra and self-similarity, in *Proceedings European Signal Processing Conference, EUSIPCO–90*, (Barcelona, Spain), pp. 149–152, September 1990.

[53] C. Braccini and G. Gambardella, Form-invariant linear filtering: Theory and applications, *IEEE Transactions on Acoustics, Speech, and Signal Processing*, vol. 34, pp. 1612–1628, December 1986.

[54] J. Bertrand and P. Bertrand, Affine time-frequency distributions, in *Time-Frequency Signal Analysis—Methods and Applications* (B. Boashash, ed.), ch. 5, pp. 118–140, Melbourne, Australia: Longman-Cheshire, 1992.
[55] F. Hlawatsch and R. L. Urbanke, Bilinear time-frequency representations of signals: The shift-scale invariant class, *IEEE Transactions on Signal Processing*, vol. 42, pp. 357–366, February 1994.
[56] J. P. Sessarego, J. Sageloli, P. Flandrin, and M. Zakharia, Time-frequency Wigner–Ville analysis of echoes scattered by a spherical shell, in *Wavelets, Time-Frequency Methods and Phase Space* (J. M. Combes, A. Grossman, and P. Tchamitchian, eds.), pp. 147–153, Berlin: Springer-Verlag, December 1989.
[57] G. C. Gaunaurd and H. C. Strifors, Signal analysis by means of time-frequency (Wigner-type) distributions—Applications to sonar and radar echoes, *Proceedings of the IEEE*, vol. 84, pp. 1231–1248, September 1996.
[58] L. R. Dragonette, D. M. Drumheller, C. F. Gaumond, D. H. Hughes, and B. T. O'Connor, The application of two-dimensional signal transformations to the analysis and synthesis of structural excitations observed in acoustical scattering, *Proceedings of the IEEE*, vol. 84, pp. 1249–1263, September 1996.
[59] I. Gohberg and S. Goldberg, *Basic Operator Theory*. Boston, Massachussets: Birkhäuser, 1980.
[60] A. Papandreou, F. Hlawatsch, and G. F. Boudreaux-Bartels, A unified framework for the Bertrand distribution and the Altes distribution: The new hyperbolic class of quadratic time-frequency distributions, in *Proceedings IEEE Symposium on Time-Frequency and Time-Scale Analysis*, (Victoria, Canada), pp. 27–30, October 1992.
[61] A. Papandreou-Suppappola and G. F. Boudreaux-Bartels, Distortion that occurs when the signal group delay does not match the time-shift covariance of a time-frequency representation, in *Proceedings 30th Annual Conference on Information Sciences and Systems*, (Princeton, New Jersey), pp. 520–525, March 1996.
[62] A. Papandreou-Suppappola and G. F. Boudreaux-Bartels, The effect of mismatching analysis signals and time-frequency representations, in *Proceedings IEEE International Symposium on Time-Frequency/Time-Scale Analysis*, (Paris, France), pp. 149–152, June 1996.
[63] A. Papandreou, F. Hlawatsch, and G. F. Boudreaux-Bartels, Quadratic time-frequency distributions: The new hyperbolic class and its intersection with the affine class, in *Proceedings Sixth Signal Processing Workshop on Statistical Signal and Array Processing*, (Victoria, Canada), pp. 26–29, October 1992.
[64] R. A. Altes and E. L. Titlebaum, Bat signals as optimally Doppler tolerant waveforms, *Journal of the Acoustical Society of America*, vol. 48, pp. 1014–1020, October 1970.
[65] P. Guillemain and P. White, Wavelet transforms for the analysis of dispersive systems, in *Proceedings IEEE UK Symposium on Applications of Time-Frequency and Time-Scale Methods*, (University of Warwick, Coventry, UK), pp. 32–39, August 1995.
[66] D. E. Newland, Time-frequency and time-scale analysis by harmonic wavelets, in *Signal Analysis and Prediction* (A. Prochazka, ed.), ch. 1, Boston, Massachusetts: Birkhäuser, 1998.
[67] D. E. Newland, Practical signal analysis: Do wavelets make any difference? in *Proceedings ASME Design Engineering Technical Conferences, 16th Biennial Conference on Vibration and Noise*, (Sacramento, California), 1997.
[68] M. J. Freeman, M. E. Dunham, and S. Qian, Trans-ionospheric signal detection by time-scale representation, in *Proceedings IEEE UK Symposium on Applications of Time-Frequency and Time-Scale Methods*, (University of Warwick, Coventry, UK), pp. 152–158, August 1995.

[69] A. Papandreou-Suppappola and L. T. Antonelli, Use of quadratic time-frequency representations to analyze cetacean mammal sounds, Tech. Rep. 11,284, Naval Undersea Warfare Center, Newport, Rhode Island, December 2001.

[70] K. G. Canfield and D. L. Jones, Implementing time-frequency representations for non-Cohen classes, *Proceedings Twenty-Seventh Asilomar Conference on Signals, Systems and Computers*, (Pacific Grove, California), November 1993.

[71] V. S. Praveenkumar, Implementation of hyperbolic class of time frequency distributions and removal of cross-terms, Master's thesis, University of Rhode Island, Kingston, Rhode Island, May 1995.

[72] R. L. Murray, A. Papandreou-Suppappola, and G. F. Boudreaux-Bartels, New higher order spectra and time-frequency representations for dispersive signal analysis, in *Proceedings IEEE International Conference on Acoustics, Speech and Signal Processing*, vol. 4, (Seattle, Washington), pp. 2305–2308, May 1998.

[73] R. L. Murray. A. Papandreou-Suppappola, and G. F. Boudreaux-Bartels, A new class of affine higher order time-frequency representations, in *Proceedings IEEE International Conference on Acoustics, Speech and Signal Processing*, vol. 3, (Phoenix, Arizona), pp. 1613–1616, March 1999.

[74] B. Iem, A. Papandreou-Suppappola, and G. F. Boudreaux-Bartels, Classes of smoothed Weyl symbols, *IEEE Signal Processing Letters*, vol. 7, pp. 186–188, July 2000.

[75] B. G. Iem, A. Papandreou-Suppappola, and G. F. Boudreaux-Bartels, Wideband Weyl symbols for dispersive time-varying processing of systems and random signals, *IEEE Trans. on Signal Processing*, vol. 50, pp. 1077–1090, May 2002.

[76] B. G. Iem, A. Papandreou-Suppappola, and G. F. Boudreaux-Bartels, New concepts in narrowband and wideband Weyl correspondence time-frequency techniques, in *Proceedings IEEE International Conference on Acoustics, Speech and Signal Processing*, vol. 3, (Seattle, Washington), pp. 1573–1576, May 1998.

[77] B. G. Iem, A. Papandreou-Suppappola, and G. F. Boudreaux-Bartels, A wideband time-frequency Weyl symbol and its generalization, in *Proceedings IEEE-SP International Symposium on Time-Frequency and Time-Scale Analysis*, (Pittsburgh, Pennsylvania), pp. 29–32, October 1998.

11

Self-Similarity and Intermittency

Albert Benassi
Serge Cohen
Sébastien Deguy
Jacques Istas

ABSTRACT In this paper, we propose a class of stochastic processes having an extended self-similarity property as well as intermittency. These notions are characterized with two parameters, and we propose statistical estimators for them.

11.1 Introduction

Since the appearance of Mandelbrot's book *Fractal Geometry of Nature* [1], it is admitted that fractal geometry is a good tool for modeling complex media, with a special mention of random media. Various examples can be found for instance in the book, *Fractals in Engineering* [2]. Roughly speaking, a fractal is an object that looks like, after rescaling, any part of itself.

Let us recall that a real-valued stochastic process $(X(x); x \in \mathbb{R}^d) d \geq 1$, is self-similar with scaling factor $H > 0$ $(SS(H))$ if

$$Law\left(\frac{X(\lambda .)}{\lambda^H}\right) = Law(X(\cdot)) \qquad \forall \lambda > 0. \tag{11.1.1}$$

In numerous situations, we do not have (11.1.1) but a generalized notion. We shall say that a stochastic process X is semi-self-similar with scaling factor H $(SSS(H))$ if we have (11.1.1) for λ's belonging to a multiplicative subgroup $\mathcal{G}_H$ of $\mathbb{R}^{+*}$.

Another class of models, the locally asymptotically self-similar processes, is obtained by localizing the property (11.1.1), as in [3].

In turbulence theory, use of the self-similarity property is justified by the Kolmogorov theory; we refer to Frisch [4]. Numerous textures of images can also be characterized by using the concept of self-similarity (see Lévy Véhel [5], Rao [6], for example).

Another possibility of characterization of complex media is to use the intermittency property. Let us recall that in a fluid the existence of coherent structures is related to intermittency (see Frisch [4]).

For a quantity F, physicists use the dimensionless ratio

$$\frac{\langle F^4 \rangle}{\langle F^2 \rangle^2} \tag{11.1.2}$$

to evaluate its intermittency. The bracket $\langle\ \rangle$ may have various meanings; for example, an integral over $\mathbb{R}^d$, an expectation, or the mean over a large sample.

In this paper, we present a stochastic model with the semi-self-similar property and intermittency. This model is parameterized with a scaling factor H and a factor δ measuring the intermittency.

We present statistical estimators for both H and δ. The estimation of δ involves quantities like (11.1.2), justifying *a posteriori* the use of the word "intermittency."

To introduce our model, the use of the wavelet decomposition of the fractional Brownian motion is required. After having noticed the quasi-tree structure of the index set of a Lemarié–Meyer orthonormal basis of $L^2(\mathbb{R}^d)$, we introduce our model of intermittency by using subtrees of the initial encoding tree, the parameter δ being then the "growing" number of this tree.

Then, after recalling the wavelet decomposition of the $SS(H)$ Gaussian processes given in [3], we present the encoding of the dyadic cells of $\mathbb{R}^d$ with trees. To ensure a good understanding of our model, we recall some basic facts about the geometry of trees. We also propose statistical estimators for H and δ, and prove their strong consistency. Finally, in an appendix, we give some simulations of such processes as well as the estimation of their characteristic parameters.

11.2 Wavelet decomposition of self-similar Gaussian processes

We recall the wavelet decomposition of self-similar Gaussian processes with stationary increments [3].

11.2.1 Self-similar Gaussian processes

Let us consider a centered, real-valued Gaussian process X, with stationary increments (SI). There exists a spectral measure $\mu(d\xi)$, and the covariance $K(x,y) = \mathrm{e}(X_x X_y)$ can be represented as

$$K(x,y) = \int (e^{ix.\xi} - 1)(e^{-iy.\xi} - 1)\mu(d\xi)$$

(see [7] or [8]). Moreover, if X is SS(H) by applying Dobrushin's result (Theorem 3.2 in [8]) to the derivative of X in a distribution sense, the spectral measure is easily characterized in spherical coordinates. More precisely, if χ is the one-to-one correspondence

$$\begin{aligned}\chi : \Sigma_{d-1} \times (0, +\infty) &\to \mathbb{R}^d \\ (\theta, r) \quad &\to r\theta,\end{aligned}$$

where Σ_{d-1} is the unit sphere in $\mathbb{R}^d$, there exists a unique measure μ_Σ on the sphere such that

$$\chi_\star^{-1}(\mu) = r^{(d+2H)} dr \mu_\Sigma(d\theta).$$

Actually we assume as in [3] that the measure μ_Σ is equivalent to the Haar measure on Σ_{d-1}, which entails that there exists a positive constant C such that

$$C|x-y|^{2H} \le \mathbb{E}|X(x) - X(y)|^2 \le \frac{1}{C}|x-y|^{2H}, \qquad x, y \in \mathbb{R}^d.$$

Moreover, there exists a function S satisfying

$$C' \le S\left(\frac{\xi}{|\xi|}\right) \le \frac{1}{C'} \qquad \forall \xi \in \mathbb{R}^d$$

with C' a positive constant, such that

$$\mu_\Sigma(d\theta) = S^2(\theta) d\theta.$$

Then one can classically derive the reproducing kernel Hilbert space (RKHS) K of the Gaussian process as in pages 24 and 25 of [3],

$$K = \left\{ f \in L^2(\mathbb{R}^d),\ f(0) = 0,\ \int |\xi|^{d+2H} S^2\left(\frac{\xi}{|\xi|}\right) |\hat{f}(\xi)|^2 d\xi < +\infty \right\},$$

where $\hat{f}$ is the Fourier transform of $f \in L^2(\mathbb{R}^d)$. Note that functions in K are smooth enough to define $f(0)$ (see [9]). From a probabilistic point of view, we are interested in orthonormal bases of K with respect to the scalar product

$$\forall f,\ g \in K \qquad a(f, g) = \int |\xi|^{d+2H} S^2\left(\frac{\xi}{|\xi|}\right) \hat{f}(\xi)\bar{\hat{g}}(\xi) d\xi$$

since they yield Karhunen–Loeve expansions of X. In that direction, we can remind the reader of Lemma 2.1 in [3], which proposes an isometry between $L^2(\mathbb{R}^d)$ and K:

$$\phi \to f_\phi(x) = \int \frac{e^{ix.\xi} - 1}{|\xi|^{d/2+H} S\left(\frac{\xi}{|\xi|}\right)} \bar{\hat{\phi}}(\xi) d\xi.$$

Hence, orthonormal bases of K are easily deduced from orthonormal bases of $L^2(\mathbb{R}^d)$. We provide a convenient example in the following section.

11.2.2 Lemarié–Meyer basis

Let $d \geq 1$ and consider $\mathbb{R}^d$ endowed with the canonical Euclidean basis $(e_1, \dots, e_d)$. If $C = [0,1)^d$ is the semiclosed unit cube of $\mathbb{R}^d$, we can encode all the faces (of any dimension $\leq d-1$) containing $(0, \dots, 0)$, with the set $E_d = \{0,1\}^d$ in the following way.

Let $\underline{u} \in E_d$, if $u = \sum_{i=1}^d \underline{u}^i e_i$, $u/2$ is the center of a face of C (including C itself if $\underline{u} = (1, \dots, 1)$). Let us set $E_d^* = E_d - (1, \dots, 1)$ and let us consider the set $\Lambda_d = \{\lambda = (j, k, \underline{u}) \in \mathbb{Z} \times \mathbb{Z}^d \times E_d^*\}$. There exists a family $\{\Psi^u; u \in E_d^*\}$ of functions $L^2(\mathbb{R}^d)$ such that, letting $\Psi_\lambda(x) = 2^{j/2d}\Psi^u(2^j x - k)$, $\lambda = (j, k, \underline{u})$, the family $(\Psi_\lambda; \lambda \in \Lambda_d)$ is an orthonormal basis of $L^2(\mathbb{R}^d)$ (a Lemarié–Meyer basis [10]).

11.2.3 Wavelet decomposition

Let $(\Psi_\lambda; \lambda \in \Lambda_d)$ be a Lemarié–Meyer orthonormal basis of $L^2(\mathbb{R}^d)$. We define a family $\{\Phi_\lambda(x); \lambda \in \Lambda_d\}$ by

$$\Phi_\lambda(x) = 2^{-jH}\Phi^{(\lambda)}(x) := 2^{-jH}\Phi^u(2^j x - k); \tag{11.2.1}$$

where $\lambda = (j, k, \underline{u})$, and

$$\Phi^u(x) = \frac{1}{C_{H,d}} \int_{\mathbb{R}^d} \frac{(e^{ix.\xi} - 1)}{\rho(\xi)} \bar{\widehat{\Psi^u}}(\xi) d\xi, \tag{11.2.2}$$

where $\rho(\xi) = |\xi|^{1/2+H} S(\xi/|\xi|)$.

Proposition 11.2.1 ([3]). *Let X be an $SS(H)$ Gaussian process fulfilling the assumptions of Subsection 11.2.1. Then there exists an i.i.d. family $\{\xi_\lambda; \lambda \in \Lambda_d\}$ of Gaussian $N(0,1)$ random variables such that*

$$X(x) = \sum_{\lambda \in \Lambda_d} 2^{-jH}\Phi^{(\lambda)}(x)\xi_\lambda. \tag{11.2.3}$$

The proof is based on the fact that $a(\Phi_\lambda, \Phi_\mu) = \delta_{\lambda\mu}$, $\lambda, \mu \in \Lambda_d$, and is based on a Karhunen–Loeve expansion of X.

11.2.4 Estimation of H

Let us recall now a well-known way of estimating H. For that purpose, let us introduce a discrete operator D, defined in the following way:

$$Df(x) = \sum_{k \in \mathbb{Z}} a_k f(x+k). \tag{11.2.4}$$

In this paper, we mainly use

$\bigtriangledown$ for which $a_0 = 1$, $a_1 = -1$, $a_k = 0$ if $k \in \mathbb{N} - \{0, 1\}$

$\triangle$ for which $a_{-1} = 1$, $a_0 = -2$, $a_1 = 1$, $a_k = 0$ if $k \in \mathbb{N} - \{-1, 0, 1\}$.

For a function $f : \mathbb{R} \to \mathbb{R}$ we define, for $n \in \mathbb{N}$,

$$V_n^r(f, D) = \sum_{k=0}^{2^n-1} \left(Df\left(\frac{k}{2^n}\right)\right)^{2r}. \tag{11.2.5}$$

Now for functions $f : \mathbb{R}^d \to \mathbb{R}$, for $v \in \Sigma_{d-1}$, we define

$$f_v(t); t \in \mathbb{R} \quad \text{by} \quad f_v(t) = f(tv). \tag{11.2.6}$$

For estimating H, we define the estimator $\widehat{H_n}$ by

$$\widehat{H_n} = -\frac{1}{n 2 \log 2} \log \left[\frac{V_n^1(X_v, \bigtriangledown)}{2^n}\right] \qquad v \in \Sigma_{d-1}. \tag{11.2.7}$$

Theorem 11.2.1. *Let X be an $SI, SS(H)$ Gaussian process fulfilling the assumptions of Subsection 11.2.1. If $\widehat{H}_n$ is defined as in (11.2.7), for any $v \in \Sigma_{d-1}$ we have*

$$\lim_{n\to\infty} \widehat{H_n} = H \qquad a.s. \tag{11.2.8}$$

Proof: In the paper [11], Theorem 11.2.1 is proved in dimension $d = 1$. Because X_v is, for any v, a one-dimensional (1D) $SI, SS(H)$ Gaussian process, the result of [11] applies, proving the theorem. □

11.2.5 A generalized model

As the estimator of H is very robust with respect to the errors in computing the functions $\{\Phi^{\underline{u}}; \underline{u} \in E_d^*\}$, it is natural to consider stochastic processes $\{Y(x); x \in \mathbb{R}^d\}$ built by substituting some functions $\{G^{\underline{u}}; \underline{u} \in E_d^*\}$ to the initial functions $\{\Phi^{\underline{u}}; \underline{u} \in E_d^*\}$ and by setting

$$Y(x) = \sum_{\lambda \in \Lambda_d} 2^{-jH} G^\lambda(x) \xi_\lambda. \tag{11.2.9}$$

Since the functions $G^{\underline{u}}$ are put instead of vaguelette G^λ we need some estimates on discrete samples of the functions $G^{\underline{u}}$. More precisely, let us write that a function g verifies A1(M) if

A1(M) $\exists\, 0 < C < \infty$ and $0 < j < \infty$ such that

$$\left|\sum_k g(2^L k - m)^r g(2^{L'} k - m')^s\right| \leq C e^{-j|m-m'|} 2^{-\inf(|L|,|L'|)}$$

$m, m' \in \mathbb{Z}^d$, $L, L' \in \mathbb{Z}$, $r, s \leq M$

is fulfilled for some integer M. In what follows we will assume for each $u \in E_d^*$ that $\{G^u; u \in E_d^*\}$ satisfies A1(8) and that $G^u(0) = 0 \ \forall u \in E_d^*$. Then we claim the following lemma.

Lemma 11.2.1. *For functions $(G^u; u \in E_d^*)$ as above, the sum*

$$Y(x) = \sum_{\lambda \in \Lambda_d} 2^{-jH} G^\lambda(x) \xi_\lambda$$

is convergent a.s. and Y is semi-self-similar on the multiplicative group $\{2^{-jH}; j \in \mathbb{Z}\}$, with scale factor H.

Proof: Let us check briefly that the process Y is semi-self-similar

$$\begin{aligned} Y(2x) &= \sum\nolimits_{(j,k,u)} 2^{-jH} G^u(2^j(2x) - k)\xi_\lambda \\ &= 2^H \sum\nolimits_{(j,k,u)} 2^{-(j+1)H} G^u(2^{j+1}x - k)\xi_{(j,k,u)} \\ &\overset{(d)}{=} 2^H \sum\nolimits_{(j,k,u)} 2^{-(j+1)H} G^u(2^{j+1}x - k)\xi_{(j+1,k,u)} \\ &\overset{(d)}{=} 2^H Y(x), \end{aligned}$$

where $\overset{(d)}{=}$ stands for equality in distribution. □

Under assumption A1(8), the estimation of the scaling factor H for the process Y is almost as easy as for processes of Subsection 11.2.1. We postponed the proof of this estimation of H after the presentation of the model with intermittency.

11.3 Intermittency for stochastic processes

We now present a model for intermittency based on the choice of infinite subset of the encoding set Λ_d. Actually, since we can endow Λ_d with a tree-like structure, we shall select an infinite subtree of it.

11.3.1 Encoding the dyadic cells of $\mathbb{R}^d$ with a tree

11.3.1.1 The 1D case

To see the tree structure of Λ_d, let us start with $d = 1$. The regular tree T_2 is an infinite graph without cycle such that from every vertex, three edges are emanating (each vertiex has degree 3). We give T_2 a root, denoted $*$ (or the common ancestor). We assume that each father, except eventually the common ancestor, has 2 sons. Now let us denote T_2^* a maximal subtree of T_2 for which $*$ has degree 2 (there are three such subtrees). It is known that T_2^* encodes the dyadic cells of $[0, 1)$. This coding corresponds to the dyadic (base 2) decomposition of the real numbers of $[0, 1)$. Let D_2^* be the set of dyadic cells of $[0, 1)$ and let us set $D_2 := \{2^n(D_2^* - k); n \in \mathbb{N},$

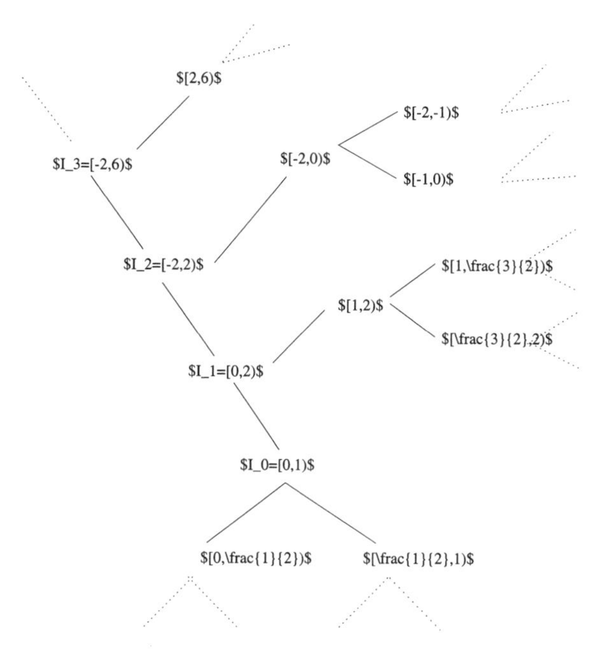

FIGURE 11.1. Encoding the dyadic cells of $\mathbb{R}$.

$k \in \mathbb{Z}\}$, the set of dyadic cells of $\mathbb{R}$. Let us see how to encode D_2 with T_2. First of all, an increasing sequence $(I_k)_{k\in\mathbb{N}}$ of dyadic cells is constructed such that $\bigcup_{k\in\mathbb{N}} I_k = \mathbb{R}$. Actually, $I_0 = [0,1)$ and I_k is inductively defined so that I_k is the first half of I_{k+1} if k is even and the second half of I_{k+1} if k is odd. It is easy to show that, if $I_k = [a_k, a_k + 2^k)$, then

$$\begin{aligned} a_{k+1} &= a_k \quad \text{if } k \text{ is even} \\ a_{k+1} &= a_k - 2^k \quad \text{if } k \text{ is odd.} \end{aligned}$$

Since $\lim_{k\to+\infty} a_k = -\infty$ and $\lim_{k\to+\infty}(a_k + 2^k) = +\infty$, $\mathbb{R} = \bigcup_{k\in\mathbb{N}} I_k$. To encode dyadic cells of $\mathbb{R}$, we consider $* = [0,1)$ and we have D_2^* encoded in the usual way by T_2^*; then $(I_k)_{k\in\mathbb{N}}$ is encoded into an infinite branch starting from the only edge emanating from I_0 that is still free. At last the dyadic cells included in $I_{k+1} \setminus I_k$ are encoded into maximal subtrees isomorphic to T_2^* emanating from I_k (see Figure 11.1).

11.3.1.2 Encoding the dyadic cells of $\mathbb{R}^d$

The dyadic cells of $\mathbb{R}^d$ are easily encoded into T_{2^d} in a way analogous to the previous section. As in the 1D case, $[0,1)^d$ is encoded into $T_{2^d}^*$, then $(I_k^d)_{k\in\mathbb{N}}$ yields the infinite branch starting from $[0,1)^d$ and $I_{k+1}^d \setminus I_k^d$ is encoded into $T_{2^d}^*$. Hence, dyadic cells of $\mathbb{R}^d$ are encoded into T_{2^d} in such a way that each dyadic cell has $2^d - 1$ sons included in it and one father that contains it. Hence, the Lemarié–Meyer wavelet basis of $L^2(\mathbb{R}^d)$ is endowed with a labeling $T_{2^d} \times E_d^*$ for which dyadic cells are encoded by a tree.

11.3.2 Intermittency

Let $\mathcal{T}$ be an infinite subgraph of $T_{2^d} \times E_d^*$. We can define a stochastic process $(Y_\mathcal{T}(x); x \in \mathbb{R}^d)$ by setting

$$Y_\mathcal{T}(x) = \sum_{\lambda \in \mathcal{T}} 2^{-|\lambda| H} G^\lambda(x) \xi_\lambda \tag{11.3.1}$$

for $0 < H < 1$, $(G^u; u \in E_d^*)$ fulfilling A1(8), $(\xi_\lambda; \lambda \in \Lambda_d)$ an i.i.d. family of normal centered Gaussian random variables.

Remark 11.3.1. Let w be an open set of $\mathbb{R}^d$ with no infinite branch of $\mathcal{T}$; then on w $X_\mathcal{T}$ has the smoothness of $(G^u; u \in E_d^*)$. On the contrary, if in w there are enough infinite branches of $\mathcal{T}$, $X_\mathcal{T}$ may be more irregular than the functions $(G^u; u \in E_d^*)$. Such sets w are called "active zones."

To better characterize the intermittency of $X_\mathcal{T}$, a more precise description of the geometry of trees is given in the following subsection.

11.3.3 Geometry of trees

One of the main parameters used to investigate the geometry of trees in relation to intermittency is the *growing number*. Let us recall its definition.

Let T be an infinite tree with root $*$. For simplicity, we shall say $\forall \sigma \in T - \{*\}$ degree$(\sigma) \geq 2$. Let us denote $|F|$ the cardinal of the set F. The *growing number* $\delta(T)$ of T is defined by

$$\delta(T) = \limsup_{n \to \infty} |T_n|^{1/n} \tag{11.3.2}$$

if $T_n = \{\sigma \in T / |\sigma| \leq n\}$. Naturally, $\delta(T) \geq 1$. We refer the reader to Lyons [12] for a discussion comparing the *growing number* and the *branching number* of a tree. In this article, we focus on the growing number since it appears in our estimations.

11.3.4 Model and main result

11.3.4.1 Main results

Let us suppose that the dyadic cells of $\mathbb{R}^d$ are encoded by T_{2^d}. Let T be a subtree of T_{2^d} with growing number δ. Let $(G^u, u \in E_d^*)$ be some functions fulfilling A1(8). Let us then set

$$Y_T(x) = \sum_{\lambda \in T \times E_d^*} 2^{-|\lambda| H} G^\lambda(x) \xi_\lambda. \tag{11.3.3}$$

From a statistical point of view, without loss of generality, we can consider summations along T only, since the cardinal of E_d^* does not depend on λ.

Let $q = \lfloor d/2 \rfloor + 1$, where $\lfloor y \rfloor$ denotes the integral part of y, and let us choose for D the operator $D = \triangle^q$, the qth power of the operator $\triangle$. Following the notation previously introduced in (11.2.7), let us define the estimators of H and of δ:

$$\widetilde{H_n} = \frac{1}{n 2 \log 2} \log \left[\frac{V_n^2(Y_T, \Delta^d)}{V_n^1(Y_T, \Delta^d)} \right] \tag{11.3.4}$$

$$\log(\widehat{\delta_n}) = \frac{1}{n} \log \left[\frac{V_n^2(Y_T, \Delta^d)}{[V_n^1(Y_T, \Delta^d)]^2} \right]. \tag{11.3.5}$$

Theorem 11.3.1. *Let $(G^u; u \in E_d^*)$ be a family of functions fulfilling the assumption* A1(8). *If $0 < H < 1$, if $\delta(T) = \lim_{n \to \infty} |T_n|^{1/n}$ for the subtree T of T_{2^d}, and if Y_T is the stochastic process defined in* (11.3.3), *we have*

(i) $\lim_{n \to \infty} \widetilde{H_n} = H$ *a.s.*

(ii) $\lim_{n \to \infty} \delta_n = \delta$ *a.s.*

11.3.4.2 The Bernoulli trees

In this part, we consider a random procedure to select the subtree T in (11.3.3). Let $p \in (2^{-d}, 1]$, and $(\mathcal{B}_\lambda; \lambda \in T_{2^d})$ be an i.i.d family of Bernoulli variables with parameter p. A subset V of vertices of T_{2^d} is then called connected if for every pair of vertices in V there exists a path, connecting the two vertices, of edges such that $\mathcal{B}_\lambda = 1$. We then consider the maximal connected subset $T_p(w)$ of T_{2^d} containing the root $*$. With a positive probability T_p is infinite.

$$Y_{T,p}(x, w, w') = \sum_{\lambda \in T_p(w)} 2^{-|\lambda| H} G^\lambda(x) \xi_\lambda(w') \tag{11.3.6}$$

In order to identify the parameter of $Y_{T,p}$, we have the following corollary.

Corollary 11.3.1. *Let $(G_\lambda; \lambda \in \Lambda)$ fulfill* A1(8). *Then, if the Bernoulli variables are independent of the Gaussian variables, we have*

1. $\delta = 2^d p$.
2. *The limit (ii) of Theorem 11.3.1 yields a consistent estimator of p.*

Let us give a sketch of the proof of Theorem 11.3.1.

If, for $n \in \mathbb{N}$, $|x - y| \simeq 2^{-n}$, we have $\Delta_n^d Y_T(k/2^n) \times 2^{nH} \simeq \eta_k^n$; η_k^n being a random Gaussian variable, we can easily check the two following properties:

1. $\Delta_n^d Y_T(k/2^n) 2^{nH}$ and $\Delta_n^d \ Y_T(l/2^n) 2^{nH}$ are asymptotically independent as $|k - l| \to \infty$,

2. $\sharp\{k; |k| \leq 2^n; \Delta_n^d Y_T(k/2^n)2^{nH} \simeq \mathcal{N}(0,1)\} \simeq \sharp\{k/2^n \in T^n\} \simeq \delta^n$, where # denotes the cardinal of a set.

Let us now consider the summations $(1/\delta^n)\sum_{|k|\leq 2^n} |\eta_k^n|^{2r}$. In that sum, if we only consider the η_k^n's attached to a branch of T, we should have, if the generalized increments are independent, $(1/\delta^n)\sum_{k/2^n\in T^n}(\eta_k^n)^{2r} \to C_r$ as $n \to \infty$.

To make this argument more precise, we have to take into account two facts:

1. As in [3], as far as the generalized increments contribute to multibivariation in V_n^r, we can consider that they are independent.
2. The contribution of the η_k^n's far from the tree T is of order $2^{n(2H-4q+d)}$. Moreover, if $q = \lfloor d/2 \rfloor + 1$, $2H - 4q + d < 0$, this contribution is asymptotically negligible.

This leads to the following

Lemma 11.3.1. $(2^{2rnH}/\delta^n)V_n^r(Y_T, \Delta^q) \to C_r$ *a.s. when* $n \to \infty$.

The proof of Theorem 11.3.1 is now a direct consequence of Lemma 11.3.1.

Let us now introduce the variations $V_n^r(Y_T, \Delta, L)$, $L \in \mathbb{N}$,

$$V_n^r(Y_T, \Delta, L) = \sum_{k=0}^{[2^n/L]} \left(Y_T\left(\frac{k+1}{2^n}.L\right) - Y_T\left(\frac{k}{2^n}.L\right)\right)^{2r}. \tag{11.3.7}$$

Corollary 11.3.2. $(2^{2rnH}/L^{2rH})(1/\delta^{n-\log_2 L})V_n^r(Y_T, \Delta, L) \to_{n\to\infty} C_r$, *where* C_r *is the constant of Lemma 11.3.1.*

11.4 Appendix

11.4.1 Simulations

In this appendix we show simulations of processes Y_T and we proceed to the estimation of their characteristic parameters. For that we only consider dimension $d = 1$, and we choose for function G the following one:

$$G(x) = e^{-\frac{(x-1/2)^2}{2}}.1_{[0,1]}. \tag{11.4.1}$$

In this situation, the tree T is a percolation tree T_p determined by its Bernoulli parameter p. As we already have noticed, $\delta(T) = 2p$, so we only have to estimate p.

For the simulation of Y_T, we have generated the following process $\widetilde{Y}$:

$$\widetilde{Y}(p) = \sum_{\substack{0\leq|\lambda|\leq 18\\ \lambda\in T(p)}} 2^{-|\lambda|H}G(2^{|\lambda|}x - k_\lambda)\xi_\lambda; \qquad x \in E_{18}, \tag{11.4.2}$$

where $E_{18} = \{k/2^{18}, 0 \leq k < 2^{18}\}$.

Figure 11.2 shows $\widetilde{Y}(0.8)$, and $\widetilde{Y}(0.9)$ and Figure 11.3 shows $\widetilde{Y}(0.95)$ and $\widetilde{Y}(1)$, for $H = 0.2$.

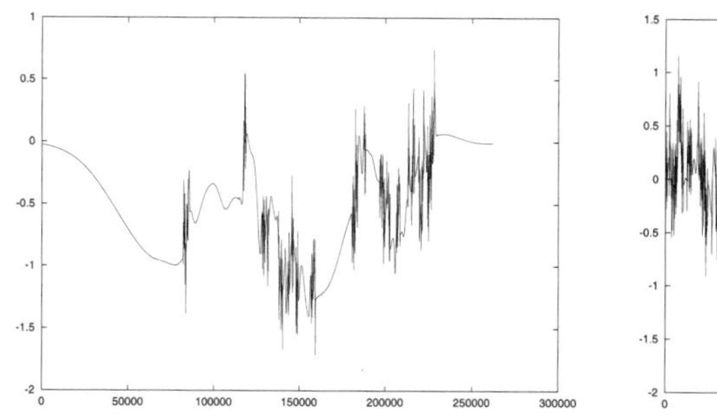

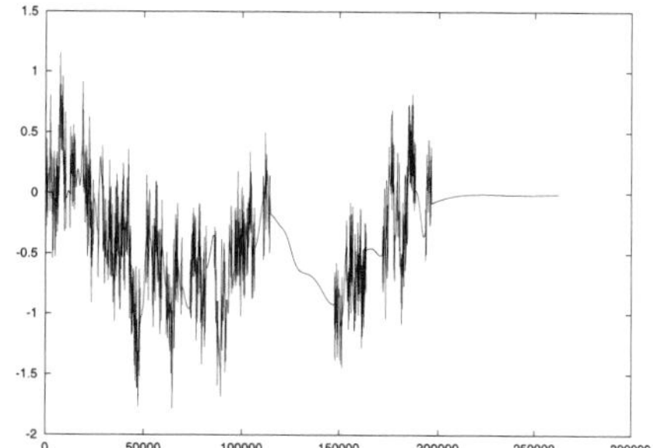

FIGURE 11.2. Simulations of the process with $H = 0.2$ and $p = 0.8, 0.9$.

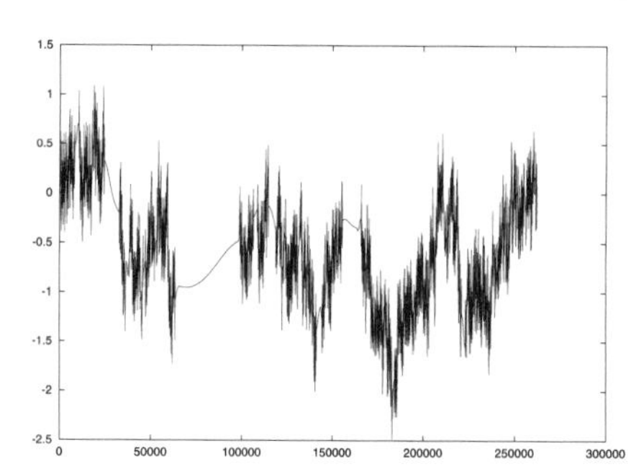

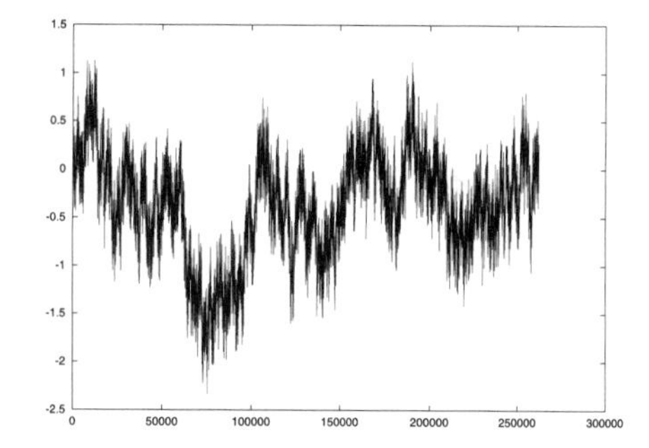

FIGURE 11.3. Simulations of the process with $H = 0.2$ and $p = 0.95; 1$

Remark 11.4.1. The function G in (11.4.1) does not fulfill $G(0) = 0$, but as $\widetilde{Y}(p)$ in (11.4.2) is not defined with negative scales, it does not matter.

11.4.2 Estimations

Let us consider the variations $V_n^r(Y_T, \Delta, L)$ defined in (11.3.7).

11.4.2.1 Estimation of p

Let us denote $Q_n(p, L)$ the quotient as follows:

$$Q_n(p, L) = \frac{(V_n^1(Y_T, \Delta, L))^2}{V_n^2(Y_T, \Delta, L)}.$$

If C_r are the constants defined in Lemma 11.3.1, from Corollary 11.3.2 we get:

$$\frac{(2p)^{n-\log_2 L}}{Q_n(p, L)} \simeq \frac{C_2}{(C_1)^2}.$$

But C_r does not depend on p. So,

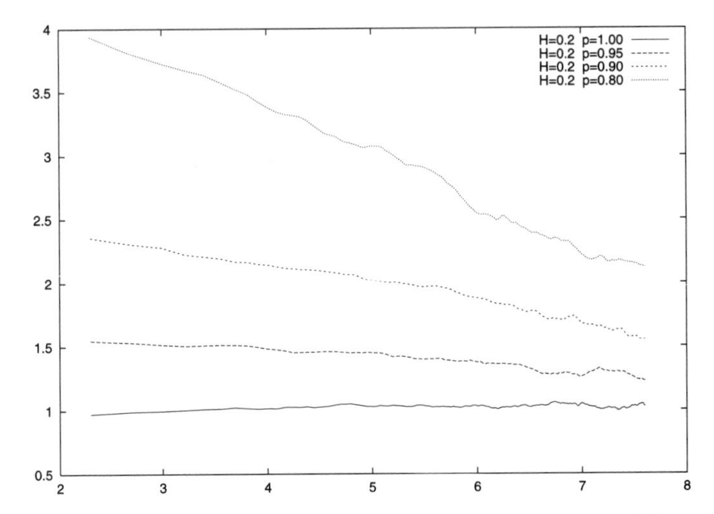

FIGURE 11.4. Results of estimation of p for $H = 0.2$ and $p = 1$, 0.95, 0.9, and 0.8. The observed slopes give us $\widehat{p} = 1.00$, 0.94, 0.84, and 0.70.

$$p^{n-\log_2 L}\left(\frac{Q_n(p)}{Q_n(1)}\right) \simeq 1$$

and, taking a logarithm,

$$(n - \log_2 L)\log_2 p - \log\left(\frac{Q_n(p)}{Q_n(1)}\right) \simeq 0.$$

So the slope of the curve $(\log_2(Q_n(p, L)); \log_2(L))$ is equal to $\log_2 p$.

In Figure 11.4 we have plotted the curves $(\log_2(Q_n(p, L)); \log_2(L))$ for $p = 1$, 0.95, 0.9, 0.8.

11.4.2.2 Estimation of H

From Lemma 11.3.1 and Corollary 11.3.2, the slope of the curve

$$\left(\log_2\left(\frac{V_n^2(Y_T, \Delta, L)}{V_n^1(Y_T, \Delta, L)}\right); \log_2(L)\right)$$

is equal to $\widetilde{H}_n$.

In Figures 11.5 and 11.6, we have plotted the curves

$$\left(\log_2\left(\frac{V_n^2(Y_T, \Delta, L)}{V_n^1(Y_T, \Delta, L)}\right); \log_2(L)\right)$$

for $H = 0.2$ and $p = 1$, 0.95, 0.9, and 0.8.

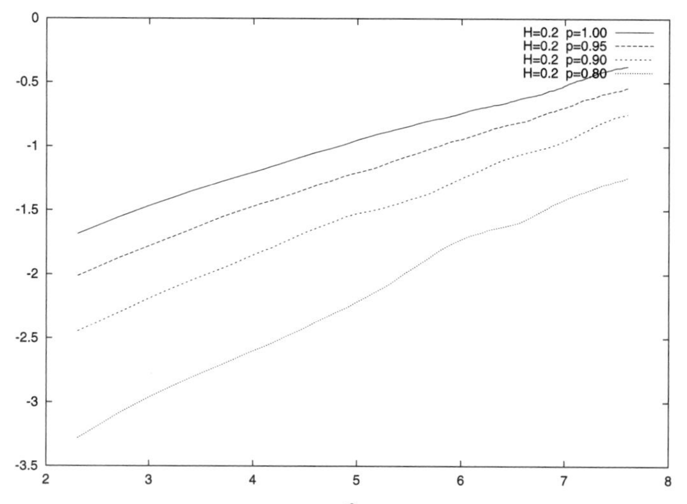

FIGURE 11.5. Results of estimation of H by $\widehat{H}$ for $H = 0.2$ and $p = 1, 0.95, 0.9$, and 0.8. The observed slopes give us $\widetilde{H} = 0.23, 0.26, 0.31$, and 0.37.

11.4.2.3 Results of estimations

Let us give the results of 1000 simulations/estimations of the parameters H and p with the help of the estimators described in the previous sections (see Table 11.1). In this table, μ and σ, respectively, represent the mean and the observed standard deviation of the estimated quantities over the sample of 1000 elements.

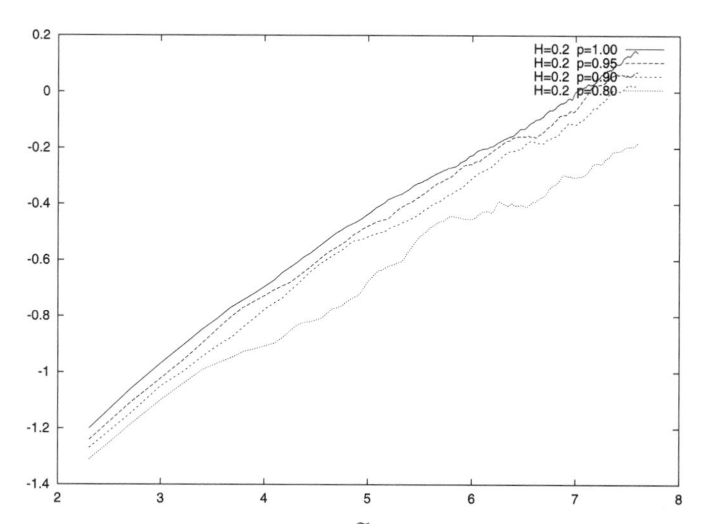

FIGURE 11.6. Results of estimation of H by $\widetilde{H}$ for $H = 0.2$ and $p = 1, 0.95, 0.9$, and 0.8. The observed slopes give us $\widetilde{H} = 0.23, 0.23, 0.22$, and 0.19.

	$H = 0.2$							
	$p = 1.0$		$p = 0.95$		$p = 0.9$		$p = 0.8$	
	μ	σ	μ	σ	μ	σ	μ	σ
$\widehat{H}$	0.236495	0.010632	0.272687	0.062869	0.325054	0.180294	0.461795	0.335543
$\widetilde{H}$	0.237282	0.010958	0.233452	0.062183	0.243806	0.185560	0.280588	0.364497
$\widehat{p}$	1.001623	0.009777	0.926431	0.041691	0.852885	0.049961	0.706511	0.098023

	$H = 0.5$							
	$p = 1.0$		$p = 0.95$		$p = 0.9$		$p = 0.8$	
	μ	σ	μ	σ	μ	σ	μ	σ
$\widehat{H}$	0.527317	0.014743	0.571475	0.097759	0.624274	0.164591	0.743743	0.256816
$\widetilde{H}$	0.524475	0.014828	0.529753	0.099513	0.538740	0.174246	0.563881	0.291960
$\widehat{p}$	0.994479	0.017093	0.920245	0.023601	0.846221	0.054737	0.706091	0.089533

	$H = 0.8$							
	$p = 1.0$		$p = 0.95$		$p = 0.9$		$p = 0.8$	
	μ	σ	μ	σ	μ	σ	μ	σ
$\widehat{H}$	0.833783	0.017033	0.873275	0.043683	0.921597	0.126598	1.030810	0.233935
$\widetilde{H}$	0.803925	0.034912	0.796961	0.059558	0.796199	0.151863	0.790308	0.306825
$\widehat{p}$	0.943634	0.053466	0.860972	0.063718	0.783095	0.076761	0.633597	0.125976

TABLE 11.1. Results of simulations/estimation.

As an ultimate remark, we can note that:

- The smaller p is, the less the quality in estimating H and p.
- The smaller p is, the more the estimator $\widetilde{H}$ prevails over the estimator $\widehat{H}$ in quality.

References

[1] B. Mandelbrot, *Fractal Geometry of Nature*. W.H. Freeman and Company, New York, 1983.

[2] Lévy Véhel, Lutton, and Tricot, editors, *Fractals in Engineering*. Springer-Verlag, berlin and New York, 1997.

[3] A. Benassi, S. Jaffard, and D. Roux, Gaussian processes and pseudo-differential elliptic operators. *Revista Mathematica Iberoamericana*, 13 (1):19–89, 1997.

[4] U. Frisch, *Turbulence*. Cambridge University Press, Cambridge, England, 1995.

[5] J. Lévy Véhel, Texture analysis using fractal probability functions. *INRIA*, 1992.

[6] A. R. Rao, *A Taxonomy for Texture Description and Identification*. Springer-Verlag, Berlin and New York, 1999.

[7] A. M. Jaglom, Some classes of random fields in n-dimensional space related to stationary random processes. *Theor. Probability Appl.*, 2:292–337, 1957.

[8] R. L. Dobrushin, Gaussian and their subordinated self similar random fields. *Ann. Proba.*, 7:1–28, 1979.

[9] G. Bourdaud, Réalisation des espaces de Besov homogènes. *Arkiv. Math.*, 26:41–54, 1988.

[10] Y. Meyer, *Ondelettes et Algorithmes Concurents*. Hermann ed, Paris, 1992 .

[11] A. Benassi, S. Cohen, J. Istas, and S. Jaffard, Identification of filtered white noises. *Stoch. Processes and their Applications*, 75:31–49, 1998.

[12] R. Lyons, Random walk and percolation on tree. *Ann. Proba.*, 18 (3):937–958, 1990.

12

Selective Thresholding in Wavelet Image Compression

Prashant Sansgiry
Ioana Mihaila

ABSTRACT Image compression can be achieved by using several types of wavelets. Lossy compression is obtained by applying various levels of thresholds to the compressed matrix of the image. This paper explores nonuniform ways to threshold this matrix and discusses the effects of these thresholding techniques on the quality of the image and the compression ratio.

12.1 Introduction

Images saved on the computer take up a fairly good amount of disk space. On the world wide web, transmitting image files also takes a large amount of time due to the size of these files. Thus, image compression is a big issue. In this paper, we will present work related to compressing images via wavelet transforms.

For image compression we will use matrices representing digitized versions of pictures in which the colors are expressed into 256 levels of gray. These matrices are transformed using standard wavelet decomposition. Since applying the wavelet transform represents a change of basis for the matrix that needs to be compressed, this can be accomplished by matrix multiplication [1]. At each multiplication step we will create a lower version resolution of the original matrix and store the details in certain areas of the transformed matrix. A matrix transformed using wavelets might normally have more zeros than the original. To make the transformed matrix more sparse, we drop all its elements less than or equal in absolute value compared with a chosen threshold. If a threshold was applied after compression, the uncompressed matrix will not be identical to the original. We will decide on the level of thresholding by considering a certain percent of the maximum of the absolute value of certain elements in the compressed image. We will use B-spline wavelets of degrees 0, 1, 2, and 3 for compression [2]. B-splines of degree 0 are also known as Haar wavelets. The Haar wavelets are the only ones that are orthogonal. For orthogonal wavelets, details of the picture at the same resolution level are kept in distinct areas; for the others, they are not.

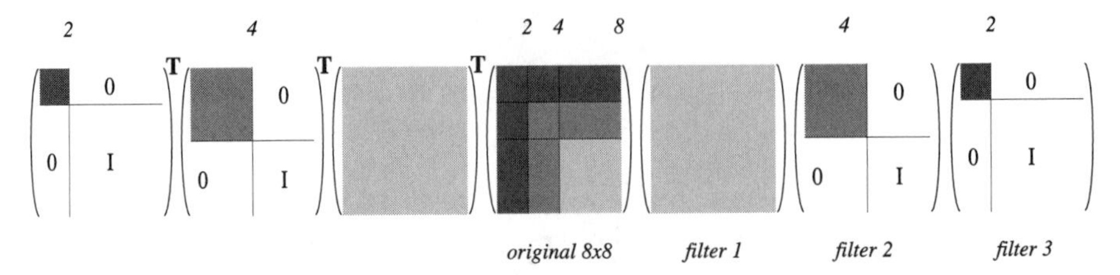

FIGURE 12.1. Compression of an 8 ×8 matrix via multiplication by filter matrices and their transposes.

12.2 Selective thresholding

Before we consider an image made up of a certain number of rows and columns, we present this idea for a matrix of size 8×8 compressed using Haar wavelets. We follow the notation introduced in the paper by C. Mulcahy [1], where the wavelet matrix W is given by a product of A_1, A_2, and A_3. The A_j's are called filter matrices. Each matrix A_j consists of a nontrivial block of size $2^j \times 2^j$ in the upper left corner, 1's along the diagonal, and zeros everywhere else. In Figure 12.1 the nontrivial areas of the filter matrices are represented by shades of gray. The original matrix is denoted P, and the compressed image is stored in $T = W^T PW$. The original image can be recovered by reversing the above matrix calculations, i.e., $P = (W^T)^{-1}W^{-1}$. To make the compressed matrix T more sparse, resulting in reducing the number of nonzero entries stored, one drops all the compressed matrix's elements less than or equal in absolute value compared to a chosen threshold value. This type of thresholding is called *uniform thresholding*. If the thresholded matrix is called U, then the image can be recovered by the calculation $Q = (W^{-1})^T UW^{-1}$. This does not give the original picture, hence the name *lossy compression*. An analysis of this procedure has led us to propose another way of thresholding the compressed matrix, which we explain below.

Multiplication by A_j and $(A_j)^T$, respectively, affects the first $2^j + d$ rows and columns of P, where d is the degree of the B-splines used. Therefore, we can partition the matrix U into blocks according to the number of row and column multiplications that they have been subject to. For an 8×8 matrix compressed using Haar wavelets, this is presented in Figure 12.1.

Figure 12.2 shows the number of multiplications that affects each area of the compressed matrix.

Dropping a coefficient in the lower right corner of the compressed matrix affects fewer of the entries of the original matrix than dropping a coefficient in one of its adjacent blocks, and so on. The bottom right corner of the compressed matrix ends up containing mainly the fine details of the original picture, since they are represented in matrix form by local changes in value from one entry to the next one, while the upper left corner contains overall averages of big blocks of the original matrix.

Hence, our idea was to eliminate more of the entries of the compressed matrix that influence fewer coefficients of the original matrix; that is, to threshold the de-

6	5	4
5	4	3
4	3	2

FIGURE 12.2. Partition of an 8 ×8 matrix into blocks showing the number of times they were multiplied by the filter matrices.

16	15	14	13	12	11	10	9
15	14	13	12	11	10	9	8
14	13	12	11	10	9	8	7
13	12	11	10	9	8	7	6
12	11	10	9	8	7	6	5
11	10	9	8	7	6	5	4
10	9	8	7	6	5	4	3
9	8	7	6	5	4	3	2

FIGURE 12.3. Partition of a 256 × 256 matrix into blocks showing the number of times they were multiplied by the filter matrices.

tails at a higher level. The level of thresholding will thus be computed as a percent of the maximum of the absolute values of the entries in a certain block of the compressed matrix. The effects of this procedure will be slightly different when using Haar wavelets than when using B-splines of higher degree, because the higher degree wavelets are not orthogonal, and hence there is some overlapping in the way the details are stored in the various areas of the compressed matrix. With selective thresholding, we can decide how much to threshold the coefficients at each level of resolution.

12.3 Results

We implemented the preceding ideas using MATLAB software [3]. We first extracted a 256×256 matrix of a gray picture in MATLAB. This corresponds to a set of 8 matrices in each filter bank, creating 15 areas in the matrix, to whicht one can apply selective thresholding. These areas, sorted by the number of matrix multiplications by which they were affected, are presented in Figure 12.3. The thresholding percentages of the various blocks used for the images presented in the examples below were going from the small detail blocks to the overall average ones) 85, 65, 65, 50, 50, 30, 10, 0, 0, 0, 0, 0, 0, 0, 0. Hence, the coarse details were not thresholded at all, while the fine details in the bottom right corner were thresholded at 85% of the element

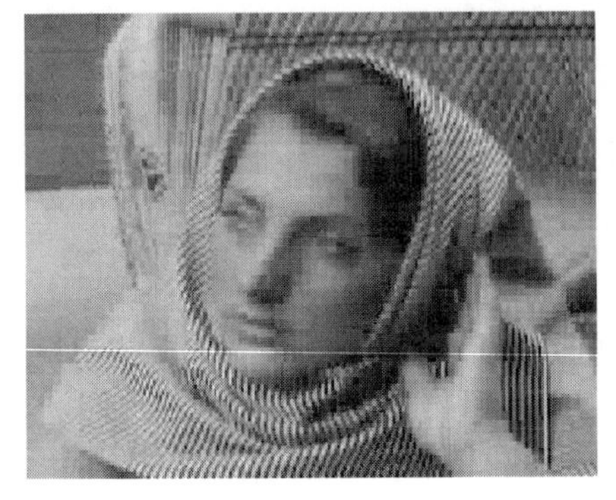

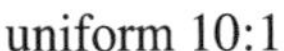

uniform 10:1 modified 10:1

uniform 15:1 modified 15:1

FIGURE 12.4. Comparison of image compression done with B-splines of degree 1 using a modified versus a uniform thresholding procedure.

with the maximum absolute value in this block. We have decided on this level of thresholding after running numerous experiments on various pictures and trying to achieve a compression ratio between 10:1 and 15:1.

Thresholding the blocks containing the details more than the others produces a brighter, but blurry, picture. Pictures with soft lines look better under the modified thresholding, for the same compression ratio. Pictures with clear sharp changes in color look better under uniform thresholding. Examples of these cases are presented in Figure 12.4.

Another issue we addressed was using the selective thresholding technique to improve an image that has noise overlapped on the original data. In order to do this, we added to the picture a matrix of the same size containing randomly generated numbers between 0 to 80. By using modified thresholding and removing more of

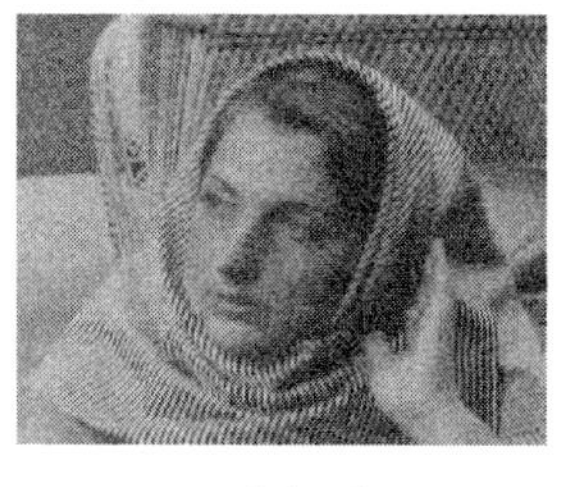

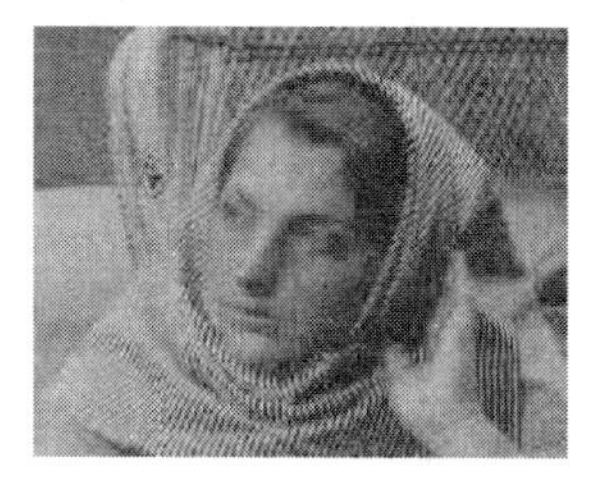

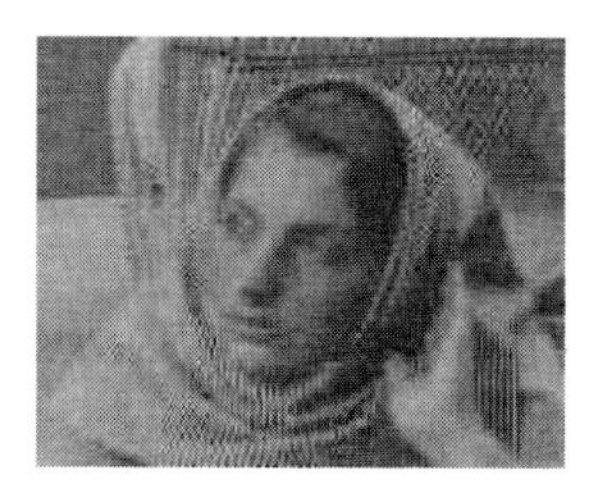

original uniform 11.7:1 modified 12.1:1

FIGURE 12.5. Comparison of image compression/de-noising done with B-splines of degree 2 using a modified versus a uniform thresholding procedure.

the details, we were able to eliminate more of the noise than we could via uniform thresholding. An example of this process is presented in Figure 12.5.

12.4 Conclusions

Listed below are several of our observations for the selective thresholding process.

- Modified thresholding allows us to choose the details to be modified by their level of resolution.
- For the same compression procedure, the compression ratio varies from picture to picture. Selective thresholding gives a more consistent compression ratio for different pictures.
- When the two methods were applied to removing noise from an image, the selective thresholding method worked better as the noise manifests itself mainly in the details.

Acknowledgments. The authors would like to thank Colm Mulcahy for exposing them to the topic of wavelets. The authors also acknowledge the help students Sheila Page and Heidi Zeleznik provided in various stages of this project.

References

[1] Mulcahy, C., *Plotting and Scheming with Wavelets*, Mathematics Magazine 69, December 1996.

[2] Stollnitz, E., DeRose, T., and Salesin, D., *Wavelets for Computer Graphics*, Morgon Kaufmann, San Francisco, CA, 1996.

[3] MATLAB, The MathWorks, Inc., Natick, MA, 1984–1994.

13

Coherent Vortex Simulation of Two- and Three-Dimensional Turbulent Flows Using Orthogonal Wavelets

Alexandre Azzalini
Giulio Pellegrino
Jörg Ziuber

ABSTRACT This article reviews a new method to extract and compute coherent structures of turbulent flows, called coherent vortex simulation (CVS). It is based on the orthogonal wavelet decomposition of the vorticity field, which can be highly compressed applying wavelet filters. Here, we introduce in detail an iterative algorithm for the wavelet filtering and show results for the analysis of artificial and turbulent flow fields. Numerical simulations using CVS are performed for turbulent flows in two and three dimensions; i.e., the integration of the Navier–Stokes equations is carried out in a compressed adaptive basis. In comparison to direct numerical simulation results, we find that the nonlinear dynamics of the flow are well retained with CVS, though most of the modes are discarded. CVS appears promising as an efficient method for the simulation of turbulent flows at high Reynolds number.

13.1 Introduction

The numerical simulation of turbulent flows still represents a challenging problem in fluid mechanics. Although it is now widely accepted that the Navier–Stokes equations describe turbulent phenomena completely, their numerical solution remains extremely demanding. Direct numerical simulation (DNS) of fully developed turbulent flows involves a large number of degrees of freedom which are proportional to Re in two dimensions and $Re^{9/4}$ in three dimensions. Therefore, when the Reynolds number increases, the computational cost becomes prohibitive, and thus DNS cannot be used for practically relevant problems with the available computational technology. To overcome this obstruction, it is necessary to simplify the problem by introducing some turbulence models [5, 20]. The techniques developed represent a compromise between the accuracy of the simulation and its cost.

Reynolds averaged Navier–Stokes (RANS) simulations calculate only the mean flow by averaging in time. This method is widely used because its results are accurate enough for many engineering applications at a relatively low cost. For those flows where unstationary phenomena are relevant, an unsteady version of RANS (URANS) exists [33]. The performance of these methods depends on the choice of the turbulence model (mixing length, k-ε, and so on) and of its adjustable parameters, which must be adapted for each flow configuration.

More detailed and more accurate results can be achieved with large eddy simulations (LESs), which perform a deterministic computation of the largest turbulence scales while the influence of the smaller ones is modeled. The computational cost of these simulations is higher than that of RANS, but the turbulence models are more general; thus a LES adapts more easily to different flow configuration.

In this article, we review a new semideterministic method to compute turbulent flows called coherent vortex simulation (CVS). This method, introduced by Farge et al. in 1999 [18], can be placed in terms of performance between LES and DNS. The general idea is to split a flow into two parts, a coherent part which retains the main multiscale structures (vortices) developed by turbulence, and an incoherent background which contains those small unpredictable oscillations that are essentially responsible for the turbulent diffusion [17]. We show that this proposed separation can be achieved by thresholding the coefficients of the wavelet representation of vorticity. The filtering procedure employed is inspired by a de-noising technique developed by Donoho and Johnstone in the context of signal processing [9,10]. We observe that the coherent part contains most of the flow energy and is mainly responsible for the nonlinear dynamics of the flow. Therefore, we compute its evolution deterministically. On the other hand, the incoherent background is noise-like and contains little energy. Moreover, the analysis of its statistical properties, such as the probability density function (PDF) and energy spectrum, suggests that it reaches a sort of thermal equilibrium. These considerations lead us to conjecture that its influence on the coherent flow should be very limited and can be easily statistically modeled or just neglected, as in the examples that we will discuss later. Performing the CVS filtering, it turns out that the coherent part can be described by only a few wavelet modes. This makes CVS promising for computing turbulent flows at high Reynolds number. It is in fact possible to exploit the space localization of the wavelet functions to implement an adaptive space discretization and thus reduce the computational cost of a numerical simulation in terms of memory requirements and CPU time.

This article is divided into three parts. First, we introduce the filtering technique, we test it on artificially noised benchmark signals, and we show how it can be modified for our application.

Then, we perform some *a priori* tests by applying the wavelet filter to one instance of two-dimensional (2D) and three-dimensional (3D) simulations of homogeneous isotropic turbulence. We compare then the results obtained with those obtained by using for the separation a classical low-pass Fourier filter, typically used for LES. In this way we quantify the gain of performance of the nonlinear CVS wavelet filter used with respect to a linear LES-type low-pass filter.

The third part is devoted to an *a posteriori* test: in order to check the potential of the CVS approach, we compare a DNS and CVS (where the wavelet filter is applied at each time step) in both two dimensions and three dimensions, and we analyze the effect of the filtering procedure on the dynamical evolution of the flow.

13.2 Wavelet filtering for coherent vortex extraction

The CVS method was proposed by Farge, Schneider, and Kevlahan [18]. It is a method for numerically simulating turbulent flows which limits the computational cost by reducing the number of degrees of freedom used for the integration. It consists in projecting the vorticity field of a turbulent flow onto a wavelet basis, splitting the wavelet coefficients into two subsets. From the inverse transform we obtain the "coherent" part of the vorticity, which contains most of the enstrophy and the energy [30]. It shows nearly the same statistics as the total flow, and it is described by a very small number of wavelet coefficients. The other part, called "incoherent," corresponds to the large number of remaining coefficients. It contains only a negligible amount of energy and enstrophy, and it is very close to Gaussian noise. The method for selecting the coefficients for both parts is derived from a process proposed by Donoho and Johnstone for de-noising of signals [10].

The following sections provide for the reader basic information about the wavelet transform (Subsection 13.2.1) and about the de-noising procedure (Subsection 13.2.3). Subsection 13.2.4 then introduces a new method for setting a crucial parameter used in Donoho and Johnstone's technique.

13.2.1 Wavelet projection

Wavelets are families of oscillating and fast decaying functions in $L^2(\mathbb{R}^n)$ that are obtained by dilation and translation of a single zero-mean function ψ. ψ can be either real or complex valued and is designed for being well localized in both spectral space and physical space. ψ has to comply with the following conditions (see [7] and [29] for more details):

- the admissibility condition:

$$C_\psi = \int_{\mathbb{R}^n} |\hat{\psi}(\boldsymbol{k})|^2 \frac{d^n \boldsymbol{k}}{|\boldsymbol{k}|^n} < \infty \qquad \text{where } \hat{\psi}(\boldsymbol{k}) = \int_{\mathbb{R}^n} \psi(\boldsymbol{x}) e^{-i\boldsymbol{k}\cdot\boldsymbol{x}} d^n \boldsymbol{x} \tag{13.2.1}$$

- vanishing moments up to a high order m:

$$M_i = \int_{\mathbb{R}^n} x_1^{\alpha_1} x_2^{\alpha_2} \cdots x_n^{\alpha_n} \psi(\boldsymbol{x}) d^n \boldsymbol{x} = 0$$
$$(\alpha_1 + \alpha_2 + \cdots + \alpha_n) = i,\ 0 \le i < m. \tag{13.2.2}$$

The wavelet family is then given by

$$\psi_{l,\boldsymbol{x}',\theta_1,\dots,\theta_{n(n-1)/2}}(\boldsymbol{x}) = l^{-n/2}\psi\left[\Omega^{-1}(\theta_1,\dots,\theta_{n(n-1)/2})\left(\frac{1}{l}(\boldsymbol{x}-\boldsymbol{x}')\right)\right], \tag{13.2.3}$$

where $l \in \mathbb{R}$ is the scale parameter, and $\Omega(\theta_1,\dots,\theta_{n(n-1)/2})$ is the rotation operator in $\mathbb{R}^n$ associated to angles $\theta_1,\dots,\theta_{n(n-1)/2}$. In Fourier space, this reads

$$\hat{\psi}_{l,\boldsymbol{x}',\theta}(\boldsymbol{k}) = l^{n/2}\hat{\psi}\left[l\Omega^{-1}(\theta)(\boldsymbol{k})\right]e^{-i\boldsymbol{k}\cdot\boldsymbol{x}'}. \tag{13.2.4}$$

The wavelet family is then equivalent to a bank of band-pass filters. Figure 13.1 shows three 1D and 2D wavelets of different scales and position parameters.

The wavelet coefficients of a signal are obtained by computing the inner product between the signal and the wavelets. It defines a continuous wavelet transform [24] that is reversible, provided that the admissibility condition (13.2.1) is fulfilled.

As ψ is well localized in physical space, its inner product with the signal is only sensitive to variations of the signal that are in a close neighborhood of its center of mass. Because of the simultaneous localization of ψ in spectral space, its convolution with the signal is equivalent to a band-pass filtering around its main wavenumber. The other members of the family inherit exactly the same properties but with different wavenumbers or locations, according to the dilation and translation (rotation in multidimensional domains) l,x' (and θ) parameters.

For some specially designed real-valued mother wavelet ψ, the 1D discrete family

$$\psi_{j,i}(x) = 2^{j/2}\psi(2^j x - i), \qquad (j,i) \in \mathbb{Z}^2,\ x \in \mathbb{R}$$

is an orthonormal basis of $L^2(\mathbb{R})$ [31]. The sequence of closed embedded subspaces $(V^J)_{J\in\mathbb{Z}}$ of $L^2(\mathbb{R})$ respectively generated by the $\{\psi_{j,i}\}_{j<J,i\in\mathbb{Z}}$, forms a 1D multiresolution analysis (MRA) [28]. The projection of a signal f onto V^j gives an approximation of f up to the resolution level J. The orthogonal complement between V^{J+1} and V^J, $W^J = V^{J+1} \ominus V^J$ is thus spanned by $\psi_{J,i}$ and contains the

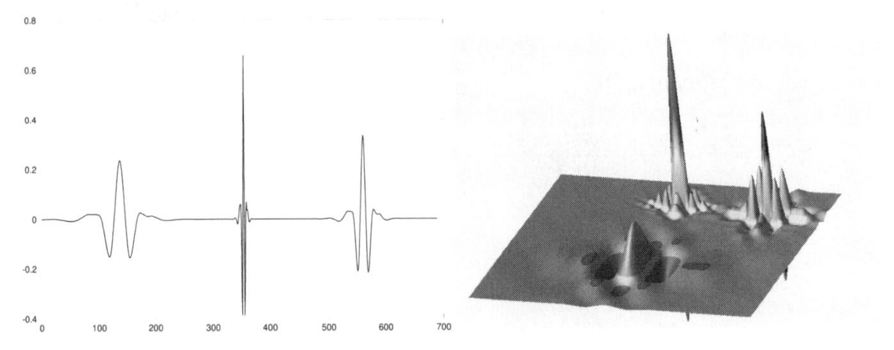

FIGURE 13.1. Example of 1D and 2D wavelets with different scale, position, and orientation (in 2D) parameters.

information needed for a doubling of the resolution of the approximation of f. The MRA framework can be extended to n-dimensional domains. An example in a 3D domain will be given in Subsection 13.2.2.

With little adaption, an orthonormal basis of the interval or periodic domain can be built. For a signal f discretely sampled over a grid of size N, a fast discrete wavelet transform (FWT) is available, and its numerical complexity is $O(N)$, which is small compared to $O(N \log N)$ obtained for FFT.

The signal f can be decomposed to

$$f(x) = \overline{f}\phi(x) + \sum_{j=0}^{J-1} \sum_{i=0}^{2^j-1} \tilde{f}_{j,i}\psi_{j,i}(x), \qquad \text{with } 2^J = N, \tag{13.2.5}$$

$$\text{and with } \tilde{f}_{j,i} = \langle f, \psi_{j,i} \rangle, \tag{13.2.6}$$

where ϕ is the *scaling function* (see Subsection 13.2.2) corresponding to the wavelet ψ. Thus, there are $N - 1$ wavelet coefficients $\tilde{f}_{j,i}$ plus one scaling coefficient $\overline{f}$ balancing the number of N samples in physical space.

13.2.2 *Multiresolution analysis*

The standard 1D MRA can be extended to the n-dimensional case by using a construction based on the tensor product of 1D spaces (e.g., [8]). In this section, we show how an MRA can be built for a 3D domain. We start from a series of 1D embedded spaces $V_{1D}^j \subset V_{1D}^{j+1}$ with $V_{1D}^{j+1} = V_{1D}^j \bigoplus W_{1D}^j$, and we build the 3D spaces V_{3D}^j by using a tensor product of the 1D ones V_{1D}^j:

$$V_{3D}^{j+1} = V_{1D,x}^{j+1} \bigotimes V_{1D,y}^{j+1} \bigotimes V_{1D,z}^{j+1}.$$

To obtain the complementary space W_{3D}^j, we substitute the expression of V_{1D}^{j+1} and we get

$$V_{3D}^{j+1} = (V_{1D,x}^j \bigoplus W_{1D,x}^j) \bigotimes (V_{1D,y}^j \bigoplus W_{1D,y}^j) \bigotimes (V_{1D,z}^j \bigoplus W_{1D,z}^j). \tag{13.2.7}$$

Using the following definitions:

$$\begin{aligned}
V_{3D}^j &= V_{1D}^j \bigotimes V_{1D}^j \bigotimes V_{1D}^j \\
W_{3D}^{j,1} &= V_{1D}^j \bigotimes V_{1D}^j \bigotimes W_{1D}^j \\
W_{3D}^{j,2} &= V_{1D}^j \bigotimes W_{1D}^j \bigotimes V_{1D}^j \\
W_{3D}^{j,3} &= W_{1D}^j \bigotimes V_{1D}^j \bigotimes V_{1D}^j \\
W_{3D}^{j,4} &= V_{1D}^j \bigotimes W_{1D}^j \bigotimes W_{1D}^j \\
W_{3D}^{j,5} &= W_{1D}^j \bigotimes V_{1D}^j \bigotimes W_{1D}^j
\end{aligned}$$

$$W_{3D}^{j,6} = W_{1D}^{j} \bigotimes W_{1D}^{j} \bigotimes V_{1D}^{j}$$
$$W_{3D}^{j,7} = W_{1D}^{j} \bigotimes W_{1D}^{j} \bigotimes W_{1D}^{j}, \qquad (13.2.8)$$

we obtain the following expression for V_{3D}^{j+1}:

$$V_{3D}^{j+1} = V_{3D}^{j} \bigoplus W_{3D}^{j} = V_{3D}^{j} \bigoplus_{\mu=1}^{7} W_{3D}^{j,\mu},$$

where the wavelet space W_{3D}^{j} is the direct sum of the spaces $W_{3D}^{j,\mu}$ obtained by the 7 possible combinations of the tensor product between the approximation space V_{1D}^{j} and the wavelet space W_{1D}^{j}. The extra parameter μ is called the orientation or wavelet direction.

The spaces V_{3D}^{j} and W_{3D}^{j} are generated by 3D scaling functions

$$\phi_{j,(i_x,i_y,i_y)}(x,y,z)$$

and wavelets $\psi^{\mu}_{j,(i_x,i_y,i_y)}(x,y,z)$, respectively, defined as follows from the 1D scaling functions $\phi_{j,i}(x)$ and wavelets $\psi_{j,i}(x)$:

$$\phi_{j,(i_x,i_y,i_z)}(\boldsymbol{x}) = \phi_{j,i_x}(x)\,\phi_{j,i_y}(y)\,\phi_{j,i_z}(z)$$

$$\psi^{\mu}_{j,(i_x,i_y,i_z)}(\boldsymbol{x}) = \begin{cases} \phi_{j,i_x}(x)\,\phi_{j,i_y}(y)\,\psi_{j,i_z}(z); & \mu = 1 \\ \phi_{j,i_x}(x)\,\psi_{j,i_y}(y)\,\phi_{j,i_z}(z); & \mu = 2 \\ \psi_{j,i_x}(x)\,\phi_{j,i_y}(y)\,\phi_{j,i_z}(z); & \mu = 3 \\ \phi_{j,i_x}(x)\,\psi_{j,i_y}(y)\,\psi_{j,i_z}(z); & \mu = 4 \\ \psi_{j,i_x}(x)\,\phi_{j,i_y}(y)\,\psi_{j,i_z}(z); & \mu = 5 \\ \psi_{j,i_x}(x)\,\psi_{j,i_y}(y)\,\phi_{j,i_z}(z); & \mu = 6 \\ \psi_{j,i_x}(x)\,\psi_{j,i_y}(y)\,\psi_{j,i_z}(z); & \mu = 7. \end{cases} \qquad (13.2.9)$$

The FWT introduced in (13.2.5) can be generalized in the n-dimensional case as follows:

$$f(\boldsymbol{x}) = \bar{\omega}_{0,\mathbf{0}}\,\phi_{0,\mathbf{0}}(\boldsymbol{x}) + \sum_{j=0}^{\log_2(N)-1} \sum_{d=1}^{n} \sum_{i_d=0}^{2^j-1} \sum_{\mu=1}^{2^n-1} \tilde{f}^{\mu}_{j,\boldsymbol{i}} \psi^{\mu}_{j,\boldsymbol{i}}(\boldsymbol{x}).$$

13.2.3 Donoho's theorem

One of the main applications of wavelets is de-noising. Donoho and Johnstone pioneered this field of application [9, 10]. They showed that nonlinear filtering of the wavelet coefficients of a noisy signal sampled on N grid points in a suitable way was more efficient than other methods, especially when no further information on the noise is available.

Donoho's filtering is based on threshold estimators as follows: Let X be the signal to be filtered; then its filtered version is

$$F_\varepsilon(X) = \sum_{j,i} \rho_\varepsilon(\langle f, \psi_{j,i}\rangle)\psi_{j,i}, \tag{13.2.10}$$

where

$$\rho_\varepsilon(a) = \begin{cases} a & \text{if } |a| > \varepsilon \\ 0 & \text{if } |a| \leq \varepsilon. \end{cases} \tag{13.2.11}$$

ρ_ε is called the *hard thresholding filter* implemented with threshold ε. It retains only the coefficients that are stronger than the threshold ε. Weaker coefficients are discarded. F_ε is called the *hard threshold estimator.* F_ε is called a diagonal estimator because it operates on each wavelet coefficient independently from the others. The idea of the method is the following: Let f be a signal whose wavelet coefficients form a sparse sequence. Let $X = f + W$, where W is an uncorrelated Gaussian random variable of standard deviation σ_W. Provided that f is well localized within the wavelet basis, its energy will be concentrated on very few coefficients of large amplitude. On the contrary, Gaussian uncorrelated noise will be spread all over the basis. Therefore, in wavelet space, the main coefficients of f will emerge from a swamp of coefficients dominated by the noise. Thresholding is thus a suitable way to extract the noise from the signal.

Following Donoho, there exists an optimal threshold value for ε that minimizes the error of the estimation. Donoho and Johnstone proposed the following formula as an asymptotic and easily accessible approximation of this optimal threshold:

$$\varepsilon_D = \sigma_W \sqrt{2\log_e(N)}. \tag{13.2.12}$$

Figure 13.2 gives an example of the de-noising of an artificial 1D signal superimposed with Gaussian white noise. The plot in Figure 13.2(d) shows the evolution of the error for the variation of the threshold value ε. One can check that the threshold given by (13.2.12) (the vertical line) is very close to the optimal one.

An interpretation for the choice of this value is given by Mallat [29]. His main arguments are summarized as follows.

The optimal threshold value should correspond to the maximum of the wavelet coefficients of W. Indeed, when the number of elements of the basis N is large, one can prove that the probability that this maximum is inside the interval

$$p(N) = P\left(\max_{j,i}(\langle W, \psi_{j,i}\rangle) \in \left[\varepsilon_D - \frac{\sigma_W \log_e(\log_e(N))}{\log_e(N)}, \varepsilon_D\right]\right)$$

is tending to 1:

$$\lim_{N\to\infty} p(N) = 1.$$

This means that it is almost impossible to find any sample of W of magnitude greater than $\varepsilon_D = \sigma_W \sqrt{2\log_e(N)}$. Because of the shape of the Gaussian probability density, the number of coefficients of a given magnitude increases very fast as soon as

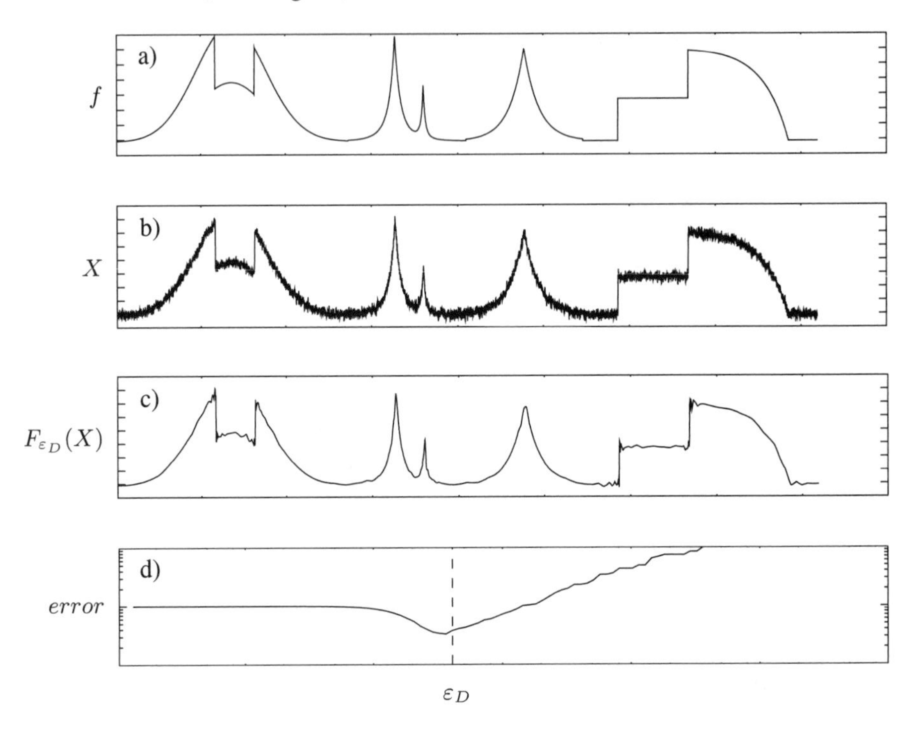

FIGURE 13.2. Example of de-noising in 1D space using Donoho's threshold. (a) original signal, (b) noisy signal, (c) de-noised signal. The error of the approximation when varying the threshold value is shown in (d), where the vertical line corresponds to the value following (13.2.12).

this magnitude is taken slightly smaller than ε_D. Therefore, if a threshold $\varepsilon < \varepsilon_D$ is chosen, a large number of noisy coefficients would be kept in the estimated signal $F_\varepsilon(X)$, and the error of the estimation would then increase. On the contrary, if $\varepsilon > \varepsilon_D$, the error also increases. The reason is that the wavelet representation of f is a sparse sequence of coefficients much stronger than the noise level. As a consequence, the additionally removed coefficients would at worst contain a noisy part equal to ε_D and most of the error would come from the loss of strong coefficients of the original signal f. This explains the increase of the curve of the error for values larger than ε_D in Figure 13.2.

In Figure 13.3 we compare the plot of ε_D to the histograms of the maximum values of a large number of realizations of Gaussian white noise. For each N, 100 realizations of Gaussian white noise exposing $\sigma_W = 1$ are sampled over N grid points. Their 100 maximum magnitudes are taken, and the histogram is computed. The dark area on the gray-scale image corresponds to the most frequently found values for the maxima of the noise realizations. The histogram corresponding to $N = 2^{12}$ is shown as an example on the right-hand figure; the horizontal line corresponds

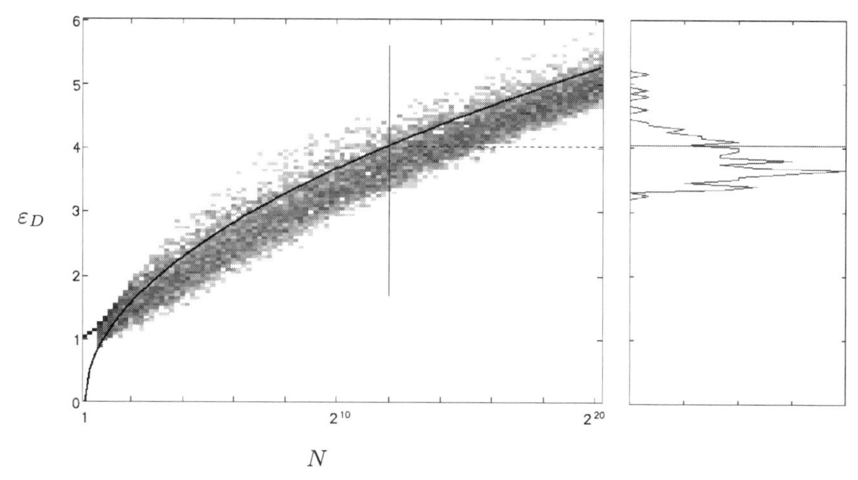

FIGURE 13.3. ε_D (solid curve) compared to the histograms of the observed maximum values of Gaussian white random variable sampled over N grid points. The histogram for $N = 2^{12}$ is shown as an example on the right-hand side.

to ε_D. One observes that formula (13.2.12) is a suitable upper bound of the maximal amplitude of the noise for large N.

13.2.4 Iterative threshold estimation

In Subsection 13.2.3, we have seen that formula (13.2.12) gives an almost optimal value for the choice of a threshold ε. However, this value is not directly accessible in practice, because the variance of the noisy incoherent part generally is unknown. In this section, we present an iterative algorithm that provides an estimated value for ε as the limit of a recursively defined sequence. The idea is to start with an estimated initial threshold value ε_T built on the standard deviation of the noisy signal:

$$\varepsilon_T = \sigma_X \sqrt{2 \log_e(N)}. \tag{13.2.13}$$

The signal is then split into its first estimate $F_{\varepsilon_T}(X)$ and a noisy remaining part $X - F_{\varepsilon_T}(X)$ whose standard deviation is taken as an improved estimation of the noise standard deviation. By iterating the algorithm, one can expect to get better and better estimates of the unknown threshold by approaching the magnitude of the strongest noise coefficient. This section discusses the convergence and the efficiency of this algorithm.

Let X be the noisy signal of vanishing mean defined in Subsection 13.2.3. Let $(\varepsilon_n)_{n\in\mathbb{N}}$ and $(\sigma_n)_{n\in\mathbb{N}}$ be two sequences of $\mathbb{R}^{\mathbb{N}}$ defined such that:

- $\sigma_0 = \sqrt{\mathrm{E}\{|X|^2\}}$ is the standard deviation of the noisy signal X
- $$\left.\begin{aligned} \varepsilon_n &= \sigma_n \sqrt{2 \log_e(N)} \qquad &(13.2.14)\\ \sigma_{n+1} &= \sqrt{\mathrm{E}\{|X - F_{\varepsilon_n}(X)|^2\}} \qquad &(13.2.15) \end{aligned}\right\} \quad \forall n \in \mathbb{N}$$

where $\mathrm{E}\{x\}$ denotes the statistical expected value of x.

Merging (13.2.14) and (13.2.15), we get

$$\varepsilon_{n+1} = \sqrt{2\log_e(N)}\sqrt{\mathrm{E}\{|X - F_{\varepsilon_n}(X)|^2\}} = I_{X,N}(\varepsilon_n), \qquad \forall n \in \mathbb{N} \quad (13.2.16)$$

We call $\varepsilon \mapsto I_{X,N}(\varepsilon)$ the iteration function associated to the noisy signal X sampled over N grid points. The behavior of the sequences ε_n and σ_n totally depends on $I_{X,N}$.

Remark 13.2.1.

$$(I_{X,N}(\varepsilon))^2 = 2\log_e(N)\sum_{j,i}\frac{\rho_\varepsilon(\tilde{X}_{j,i})^2}{N} = 2\log_e(N)\int_{x=0}^{\varepsilon} x^2 \sum_{j,i}\frac{\delta(|\tilde{X}_{j,i}| - x)}{N}dx :$$

the iteration function is the square root of the second order moment of the function

$$x \mapsto \sum_{j,i}\frac{\delta(|\tilde{X}_{j,i}| - x)}{N}.$$

This function is the limit of the histogram of $|\tilde{X}|$ when the width of each bin used for empirically computing this histogram tends to zero. Thus, $I_{X,N}$ is a function of the empirical PDF of $|\tilde{X}|$.

Remark 13.2.2. $I_{X,N}$ is increasing. Indeed, let ε^a and ε^b be two threshold values such as $\varepsilon^a \leq \varepsilon^b$. Then one has

$$\begin{aligned}\varepsilon^a \leq \varepsilon^b &\Rightarrow \sqrt{\mathrm{E}\{|X - F_{\varepsilon^a}(X)|^2\}} \leq \sqrt{\mathrm{E}\{|X - F_{\varepsilon^b}(X)|^2\}} \\ &\Rightarrow I_{X,N}(\varepsilon^a) \leq I_{X,N}(\varepsilon^b). \end{aligned} \qquad (13.2.17)$$

Remark 13.2.3. Because of the form of $I_{X,N}(\varepsilon)$ in Remark 13.2.1, $\varepsilon \mapsto I_{X,N}(\varepsilon)$ is piecewise constant on a finite number of intervals. The bounds of these intervals are the threshold above which new coefficients are taken into account in the sum. The consequence is that $\varepsilon \mapsto I_{X,N}(\varepsilon)$ is not continuous.

In order to study the behavior of ε_n, we study fixed points of $I_{X,N}$. Fixed points would usually be defined as the intersections between the curve of $I_{X,N}$ and the line $y = x$. When sequences are defined recursively using a function h and a formula of the form $U_{n+1} = h(U_n)$, their convergence depends on the existence of fixed points for the graph of h. The recursive mechanism in the definition of U_n can be followed graphically. Figure 13.4 shows an example of such a situation. The initial value of the sequence is U_0. By definition, $U_1 = h(U_0)$ corresponds to the ordinate of the intersection I between the vertical dashed line at abscissa U_0 and the graph of h, namely the curve C. The horizontal dashed line starting from I hits $y = x$ at the coordinate (U_1, U_1) and a new vertical line corresponding to abscissa U_1 is drawn. It hits C at the point (U_1, U_2). The process goes on and a "stair" is drawn between C and $y = x$. If C is continuous and intersects $y = x$, U_n converges to this intersection.

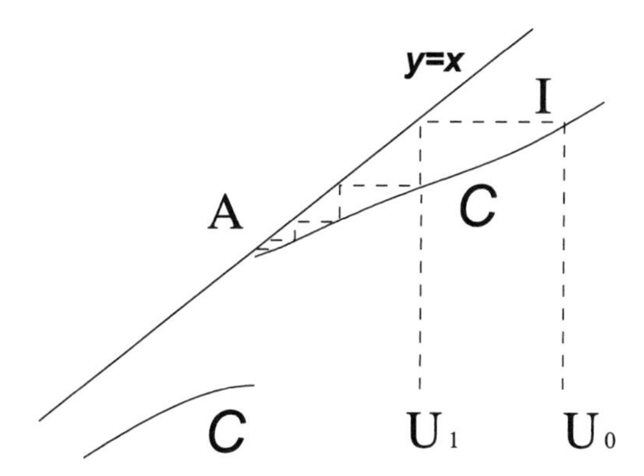

FIGURE 13.4. Example of fixed point for a discontinuous function.

As $I_{X,N}$ is not continuous, fixed points should be defined more generally as points where the two graphs approach infinitely close. In Figure 13.4, when followed from the right to the left, the curve C breaks just before the point A and so does not intersect the line $y = x$. However, if one uses h for computing a recursive sequence U_n with U_0 as initial value, U_n will converge to the abscissa of A. On the contrary, if h has no fixed points, U_n will never converge to any limit U_∞.

Now, we can consider the following property illustrated in Figure 13.5 for the discontinuous increasing function $I_{X,N}$ and the sequence ε_n.

Proposition 13.2.1. *Let $[\varepsilon^a, \varepsilon^b]$ be an interval of $\mathbb{R}+$. Suppose that the curve of $I_{X,N}$ is above the line $y = x$ on the left border of $[\varepsilon^a, \varepsilon^b]$ (i.e., $I_{X,N}(\varepsilon^a) \geq \varepsilon^a$) and is below $y = x$ on the right border of $[\varepsilon^a, \varepsilon^b]$ (i.e., $\varepsilon^b \geq I_{X,N}(\varepsilon^b)$). Then, if the sequence $(\varepsilon_n)_{n \in \mathbb{N}}$ once hit the interval $[\varepsilon^a, \varepsilon^b]$, there exists a value $\varepsilon_{it} \in [\varepsilon^a, \varepsilon^b]$ such as $\lim_{n\to\infty} \varepsilon_n = \varepsilon_{it}$.*

Indeed:

- Suppose that there exists $\varepsilon_{n_0} \in [\varepsilon^a, \varepsilon^b]$ and $I_{X,N}(\varepsilon_{n_0}) \leq \varepsilon_{n_0}$. Then $\varepsilon_{n_0+1} \leq \varepsilon_{n_0}$ and $I_{X,N}(\varepsilon_{n_0+1}) \leq I_{X,N}(\varepsilon_{n_0})$. Finally $\varepsilon_{n+1} \leq \varepsilon_n \quad \forall n \geq n_0$, so ε_n decreases.
- Now, as $\varepsilon^a \leq \varepsilon_n$, then $\varepsilon^a \leq I_{X,N}(\varepsilon^a) \leq I_{X,N}(\varepsilon_n) = \varepsilon_{n+1}$. Thus ε_n converges to a limit between ε^a and ε_{n_0}.
- On the contrary, if $\varepsilon_{n_0} \leq I_{X,N}(\varepsilon_{n_0})$, then the proof is similar with ε_n increasing and bounded.

Even being discontinuous, the increasing function has at least one fixed point of abscissa ε^{fp}. Moreover, any sequence hitting the interval $[\varepsilon^a, \varepsilon^b]$ will converge to the abscissa of one of these fixed points (in the case of the figure, one only is present for abscissa ε^{fp}, so one will have $\varepsilon_{it} = \varepsilon^{fp}$).

Proposition 13.2.1 gives a sufficient condition on $I_{X,N}$ and ε_0 to test the convergence of the algorithm to a limit ε_{it}. We now want to know the condition for ε_{it} to be a good estimate of Donoho's threshold ε_D. We will first focus on the case where

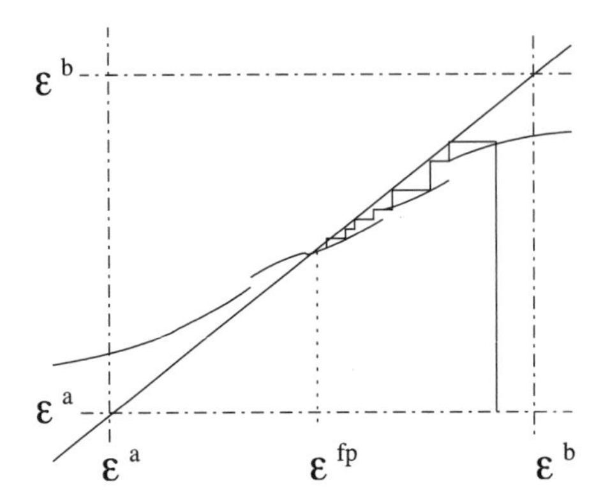

FIGURE 13.5. Illustration of Proposition 13.2.1 for sequences ε_n defined recursively with a discontinuous and increasing function. Although $I_{X,N}$ is piecewise constant, it does not need to be so for satisfying the hypothesis of the proposition. Therefore, the curve is drawn as a general example.

X is only white noise (i.e., $f = 0$). In this case, the following proposition shows that the algorithm converges exactly to ε_D. The error made by the wavelet estimator F_ε is then equal to zero.

Proposition 13.2.2. *If X is white noise (i.e., $f = 0$):*

- $\varepsilon_0 = \sigma_0\sqrt{2\log_e(N)} = \sqrt{\mathrm{E}\{|X|^2\}}\sqrt{2\log_e(N)} = \varepsilon_D$ *is a fixed point. Indeed,* $\varepsilon_0 = \varepsilon_D$ *estimates the greatest magnitude of the wavelet coefficients of X, then* $F_{\varepsilon_0}(X) = \sum_{j,i}\rho_{\varepsilon_0}(\tilde{X}_{j,i})\psi_{j,i} = X$ *and* $I_{X=W,N}(\varepsilon_0) = \varepsilon_0$.
- *On the left-hand side of ε_0, there exists an interval $[\varepsilon^a, \varepsilon_0]$ on which the curve of $I_{X,N}$ bumps over the line $y = x$ (see example given in Figure 13.6). Proposition 13.2.1 then implies that, if a sequence ε'_n has once a value inside $[\varepsilon^a, \varepsilon_0]$, it will converge to ε_0.*

The next step consists in determining if the algorithm would converge to a value close to ε_D for the case of a clear signal superimposed with noise (i.e., $f \neq 0$). Two conditions are needed:

- The curve of $I_{X=f+W,N}$ must be close to the one of $I_{W,N}$ around ε_D and must have a fixed point **A** close to the fixed point **B** of the noise.
- The curve of $I_{X=f+W,N}$ is below the line $y = x$ between **A** and its right-hand extremity **C** (which corresponds to $\varepsilon_T = \sqrt{\mathrm{E}\{|X|^2\}}\sqrt{2\log_e(N)}$).

Now, suppose that $\sqrt{\mathrm{E}\{|W|^2\}} \ll \sqrt{\mathrm{E}\{|f|^2\}}$ and that f is well approximated in the wavelet basis. Then most of its coefficients have a magnitude far below ε_D. The remaining rare coefficients bear almost all the variance of f, and are much stronger than ε_D. Therefore, the rare remaining coefficients are not taken into account in $I_{X,N}(\varepsilon)$ for $\varepsilon \leq \varepsilon_D$. On the other hand, $I_{X,N}(\varepsilon)$ is not perturbed too much by the majority of coefficients of f whose mean magnitude is very small compared to ε_D.

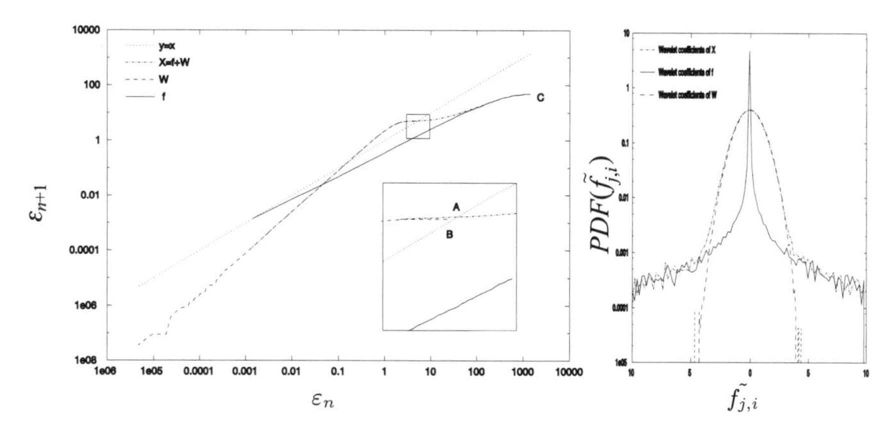

FIGURE 13.6. Left: Plot of the curves of $I_{X=f+W,N}$, $I_{f,N}$, and $I_{W,N}$. The unknown threshold ε_D is the abscissa of the intersection **B** between the curve of $I_{W,N}$ and the line $y = x$. The actual limit ε_{it} for the algorithm corresponds the intersection **A** between the curve of $I_{X,N}$ and the line $y = x$. The abscissa of the point **C** corresponds to ε_T. If $\varepsilon_0 = \varepsilon_T$, the ordinate of **C** is equal to ε_1 (threshold obtained after a single iteration). Right: PDFs of the wavelet coefficients for X, f, and W.

It should then remain very close to the curve of $I_{W,N}$ for $\varepsilon \sim \varepsilon_D$. This corresponds to the first condition. For $\varepsilon \geq \varepsilon_D$, the rareness of the strong coefficients of f forces $\varepsilon \mapsto I_{X,N}(\varepsilon)$ to increase slower than $\varepsilon \mapsto \varepsilon$. The curve of $I_{X,N}$ thus stays below the line $y = x$. The second condition is thus also fulfilled.

The consequence is illustrated in Figure 13.6, which is an example of the curve of $I_{X,N}$ for an artificial noisy signal whose further description will be given in Section 13.3. In this plot, $I_{X,N}(\varepsilon)$ is approximately identical to $I_{W,N}(\varepsilon)$ for $\varepsilon \leq \varepsilon_D$. The zoomed plot shows that the points **A** and **B** are almost superimposed. The unknown threshold ε_D needed for the best de-noising corresponds to the intersection **B** between the curve of $I_{W,N}(\varepsilon)$ corresponding to the noise and the line $y = x$. The actual limit for the iterative algorithm ε_{it} corresponds to the point **A**, which is the intersection between the curve of the noisy signal X and the line $y = x$.

One can check that ε_{it} is a far better estimator for ε_D than the threshold ε_T built on the standard deviation of the whole field. In this example, $\sqrt{\mathrm{E}\{|W|^2\}}/\sqrt{\mathrm{E}\{|f|^2\}} = 0.1$. According to Proposition 13.2.1, the sequence ε_n with initial value $\varepsilon_0 = \varepsilon_T$ will naturally converge to $\varepsilon_{it} \simeq \varepsilon_D$.

A much longer and more rigorous discussion about the behavior of $I_{X=f+W,N}$ would focus on the following aspects: It would require a precise statistical model for the modulus of the wavelet coefficients of f. Considering Remark 13.2.1 would allow us to determine the form of $I_{X=f+W,N}$ and its sensitivity to the model parameter. In the example of Figure 13.6, the PDF of the wavelet coefficients of f is sharply peaked around zero. As a consequence, $I_{f,N}$ increases slower than $\varepsilon \mapsto \varepsilon$ (see Figure 13.6, left). On the contrary, because of the shape of the Gaussian distribution, the curve of $I_{W,N}$ bumps over the line $y = x$ on the left side of **A** and **B**, as described in

Proposition 13.2.2. Around ε_D, $I_{W,N}$ dominates $I_{f,N}$. The consequence is that the curve of $I_{X=f+W,N}$ follows $I_{f,N}$ for large ε but is deviated toward $I_{W,N}$ for $\varepsilon \sim \varepsilon_D$.

13.3 Application of wavelet filtering to 2D and 3D data fields

The application of wavelet techniques to turbulence has been studied for more than a decade [3, 8, 11–14, 31]. The filtering method introduced in Section 13.2 will now be applied to the analysis of 2D and 3D data fields. We first decompose an artificial academic 2D field (Subsection 13.3.1). We then apply it to a turbulent flow field (Subsection 13.3.2). In Subsection 13.3.3, we extend the method to 3D vector fields.

13.3.1 Artificial academic 2D field

We first test the filtering method by considering an academic case. We construct a 2D scalar field, with a resolution of 512^2, which mimics a typical 2D vorticity field as encountered in 2D turbulent flows. For this, we created a superposition of several isotropic cusps, having a local scaling in $r^{1/2}$ (r being the distance to the core of the considered vortex) and different amplitudes, and whose positions are randomly distributed in space (Figure 13.7); this field has a standard deviation of 10. We then superimposed a Gaussian white noise of standard deviation 1, which results in a reference field. Furthermore, we show in Figure 13.7 a 1D cut through the cusps (solid line), the noise (dotted line), and their superposition (dashed line). The corresponding enstrophy spectra

$$Z(k) = \frac{1}{2} \sum_{k-1/2<|\boldsymbol{k}|\leq k+1/2} \hat{\omega}(\boldsymbol{k})^2 \tag{13.3.1}$$

are shown in Figure 13.8: the cusps and the total field have the same power-law scaling in k^{-3} (k being the modulus of the wavenumber), while for the smallest scales we recover the k^{+1} scaling of the white noise. The corresponding PDFs show that the cusps and the total field present the same non-Gaussian PDF, whereas the noise exhibits a Gaussian PDF with a very reduced variability. We also plot the PDFs of the wavelet coefficients for the three fields, which show very different behaviors for the cusps and the noise: furthermore, the wavelet coefficients of the total field whose modulus is smaller than 5 are dominated by the Gaussian noise, whereas the coefficients whose modulus is larger are dominated by the cusps.

We now apply the filtering procedure to this academic signal. The noniterated algorithm gives an estimated threshold ε_T of 50.2. The coherent contribution (Figure 13.9), namely the field reconstructed from the wavelet coefficients whose modulus is larger than ε_T, corresponds to 0.2% of the coefficients, which retain 95.39% of the variance of the original field. In Figure 13.9, we observe that the coherent structures, thus extracted, are not as isotropic as those of the original field. This is due to the

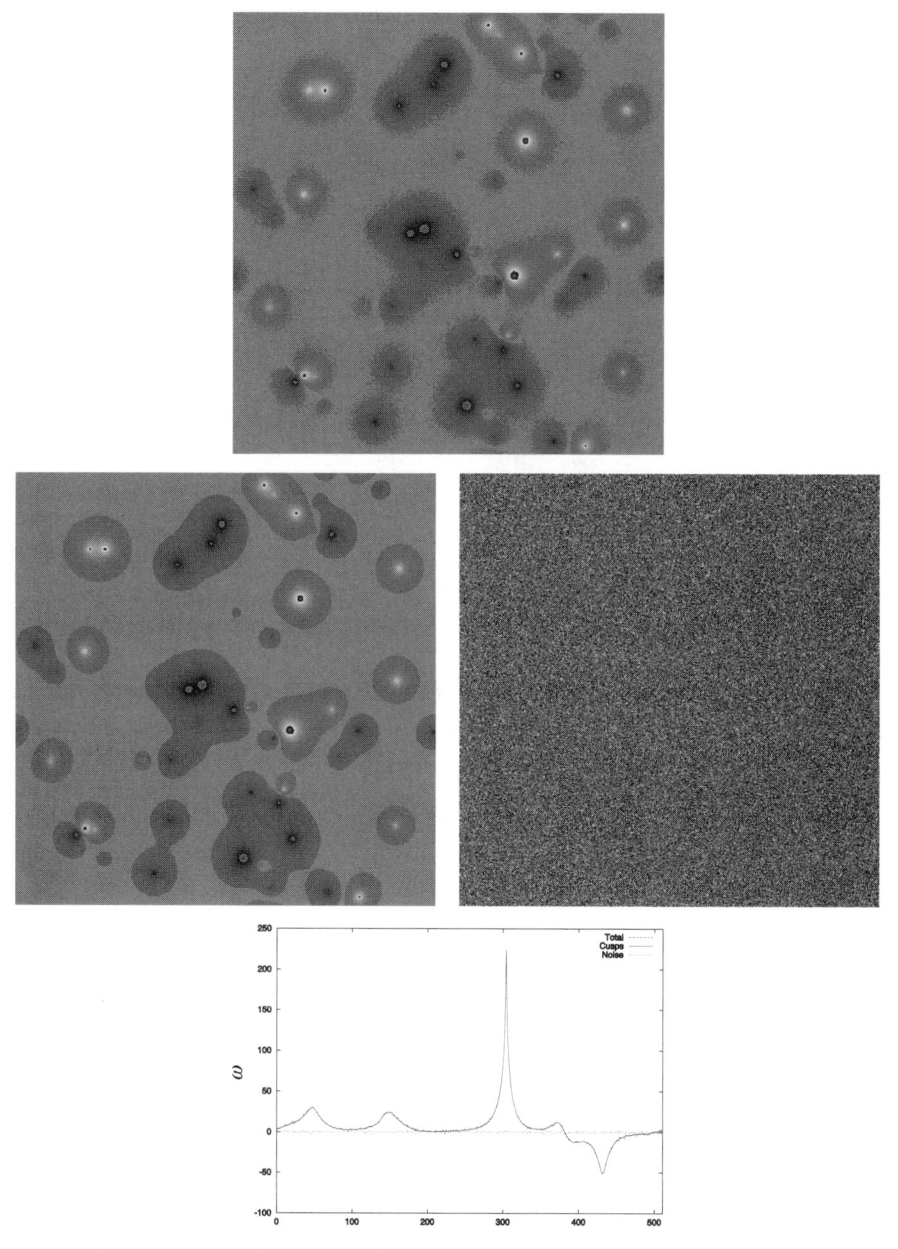

FIGURE 13.7. Artificial 2D scalar vorticity field (top) made of isotropic cusps (middle, left), and white Gaussian noise (middle, right). Cuts across the total field and its two components (bottom).

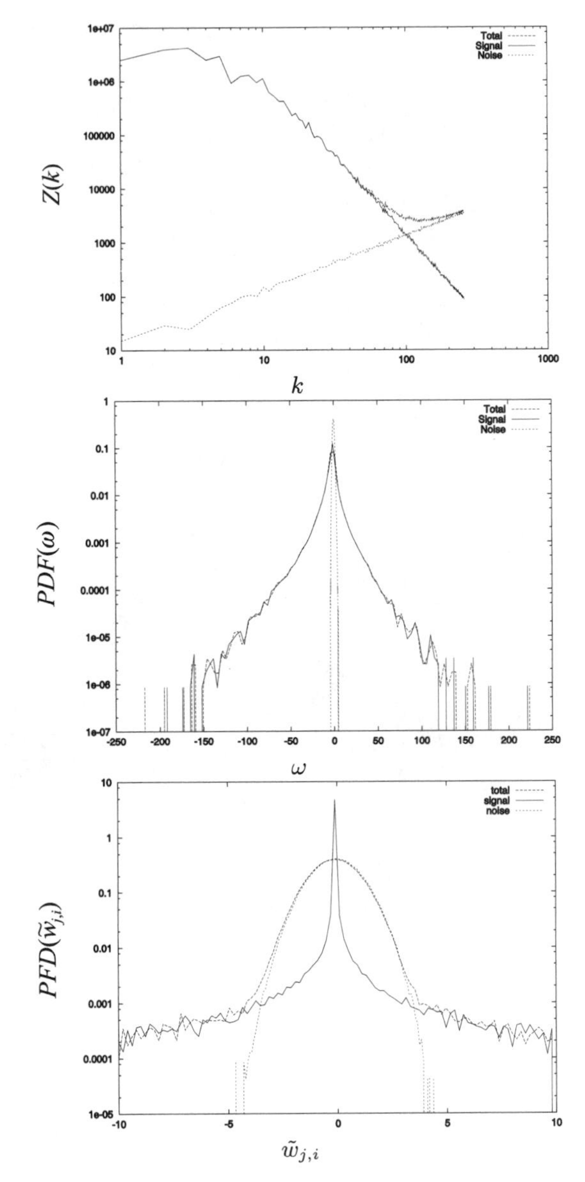

FIGURE 13.8. Spectra $Z(k)$ (top), PDFs of $\tilde{\omega}$ (middle), and PDFs of the wavelets coefficients (bottom) for total field, isotropic cusps, and white Gaussian noise.

lack of isotropy of the 2D wavelets constructed by the tensor product of two 1D wavelets. Moreover, the wavelet basis is invariant only with respect to a discrete set of translation. This induces an asymmetry in the representation of vortices that are not perfectly located at some particular grid points. This lack of symmetry is normally compensated by recovering several wavelets. For the strong compression ratio

$$C = \frac{\text{\# coefficients in the coherent part}}{N} \tag{13.3.2}$$

obtained with the noniteratively evaluated threshold, the number of retained wavelets at small scales is so small that they cannot compensate the lack of isotropy of the larger scale wavelets. They are also unable to compensate the lack of symmetry of wavelet representation. This can be observed when comparing 1D cuts from the filtered (Figure 13.9) and unfiltered (Figure 13.7) fields. Another effect of the very strong compression ratio is that the incoherent field presents some weak organized structures (see Figure 13.9) which are also observed in the 1D cut. These two problems, namely the lack of isotropy of the coherent field, and of homogeneity of the incoherent field, are solved when using the iterated threshold procedure. In this case, 1.1% of the coefficients are kept and 98.94% of the variance is retained, with many more coefficients in the small scale to ensure a good isotropy of coherent structures. Therefore, the coherent structures extracted in the coherent field are as isotropic as the cusps present in the original field and the incoherent field is perfectly homogeneous (Figure 13.9). The compression is weaker for the iterated threshold, which has converged after seven iterations.

Considering the PDFs in Figure 13.10, we find that filtering with both thresholds ε_T and ε_{it}, the coherent and original field have exactly the same non-Gaussian distributions. In contrast, the incoherent fields present very different distributions. For ε_T, the PDF of the vorticity is non-Gaussian, which is consistent with the fact that some coherent structures have not been fully removed in this case. For ε_{it}, the vorticity is Gaussian in Figure 13.10, being consistent with the perfect homogeneity of the incoherent field.

Finally, we compare the spectra $Z(k)$ corresponding to the fields obtained using both methods (Figure 13.11). For ε_T, we observe that the scaling of the coherent contribution becomes steeper in the small scales, since a very small number of wavelets are retained. The incoherent contribution presents a spectrum very different from the k^{+1} scaling of the Gaussian white noise to be removed. In contrast, the filtering with ε_{it} perfectly removes the white noise.

We can conclude that filtering with ε_{it} has successfully removed the white noise from the academic signal. In particular, it is striking to remark that ε_{it} is 5.16, very close to the value 5 that we have previously evaluated from the wavelet coefficient PDF. This value is also consistent with the theoretical one, $\varepsilon_D = \sqrt{2\log_e(512 \times 512)} = 4.995$ based on the standard deviation of the noise ($\sigma_W = 1$).

13.3.2 2D turbulent field

Having tested both filtering methods on an academic signal, we apply them to a 2D turbulent vorticity field (Figure 13.12). This field was computed by direct numerical simulation with a resolution of 512^2, using hyperviscosity. We used a pseudo-spectral code written by Marie Farge and modified by Alexandre Azzalini for solving

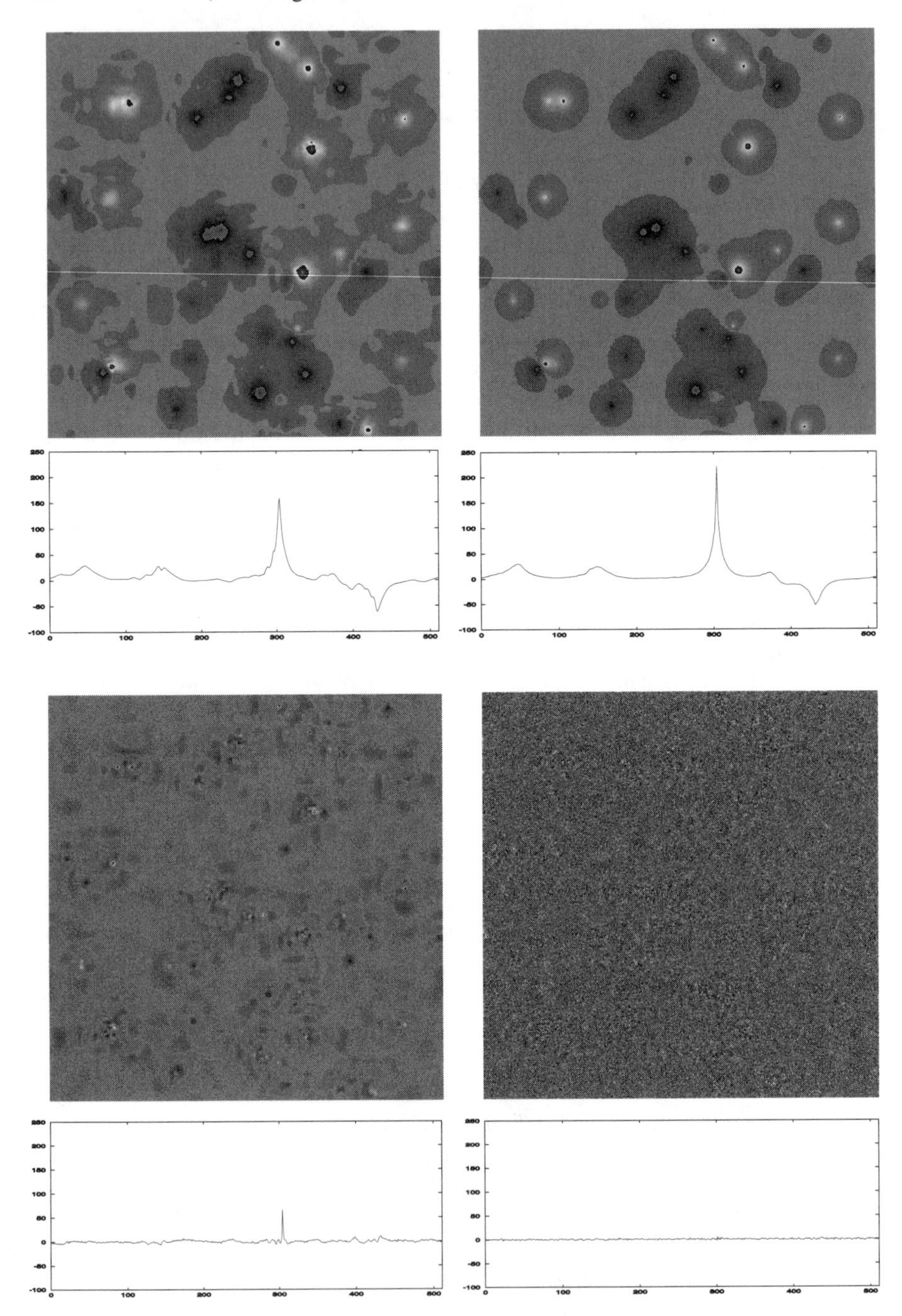

FIGURE 13.9. Split of an artificial 2D scalar vorticity field using ε_T (left column) and ε_{it} (right column). Top row: coherent parts; Bottom row: incoherent parts.

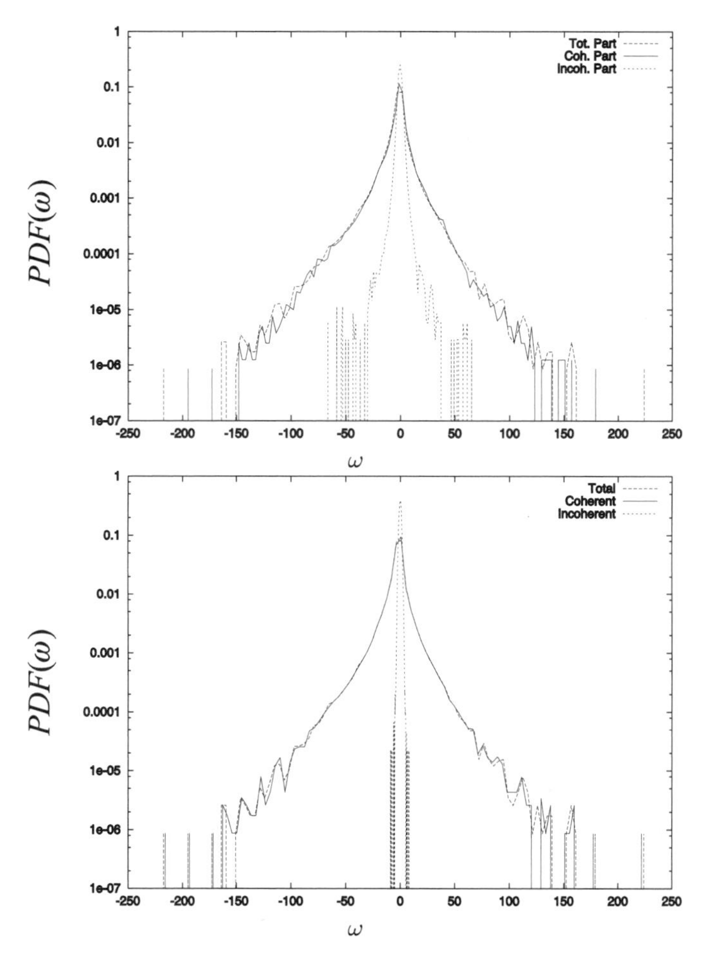

FIGURE 13.10. PDFs obtained for the decomposition of an artificial 2D scalar vorticity field using ε_T (top) and ε_{it} (bottom).

rotating shallow water equations. The ageostrophic contribution of the initial condition was taken equal to zero, and the parameters were chosen so that the flow behaves closely like an uncompressible nonrotating turbulent flow (deformation radius equal to twice the lengthscale of the initial energy peak, Froude number of order 10^{-1}). Using the threshold ε_T, we find that $C = 0.32\%$ of the modes are retained as the coherent contribution, which contains 82.31% of the variance of the vorticity. Using ε_{it} gives a weaker compression ratio, with 3.22% of the modes retaining 95.44% of the enstrophy. As we have already found with the academic signal test case, filtering with ε_T does not preserve the isotropy of the coherent vortices (Figure 13.13). Moreover, considering 1D cuts of the incoherent field in Figure 13.13, we observe that some vortices can be found. In contrast, the isotropy of the vortices is perfectly respected

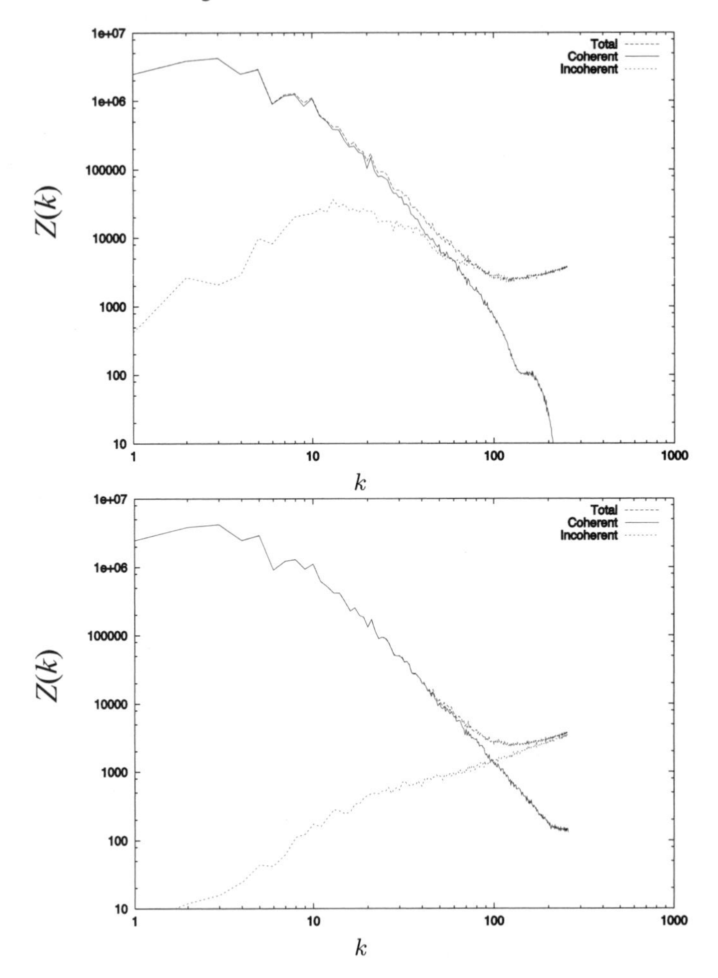

FIGURE 13.11. Spectra obtained for the decomposition of an artificial 2D scalar vorticity field using ε_T (top) and ε_{it} (bottom).

using ε_{it} and all vortices have been removed from the incoherent background flow. This fact is confirmed when we look at the PDFs of the vorticity (Figure 13.14), since we have a Gaussian distribution for the incoherent background obtained when filtering with ε_{it}, while we have a non-Gaussian distribution when using ε_T, which is due to the presence of small vortices still present in the incoherent field. Nevertheless, both filtering methods show a very good fit between the PDFs of the total and coherent fields. Finally, if we consider the enstrophy spectra (Figure 13.15), we observe that the incoherent background obtained using ε_{it} shows a tendency towards enstrophy equipartition, which corresponds to a k^{+1} scaling. Moreover, we observe that in the large scales, up to wavenumber $k = 30$, all the contribution to the enstrophy comes from the coherent modes. Concerning the filtering using ε_T, the behavior

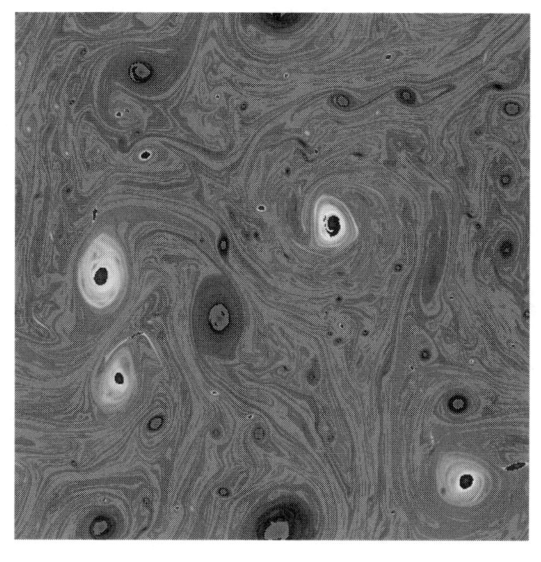

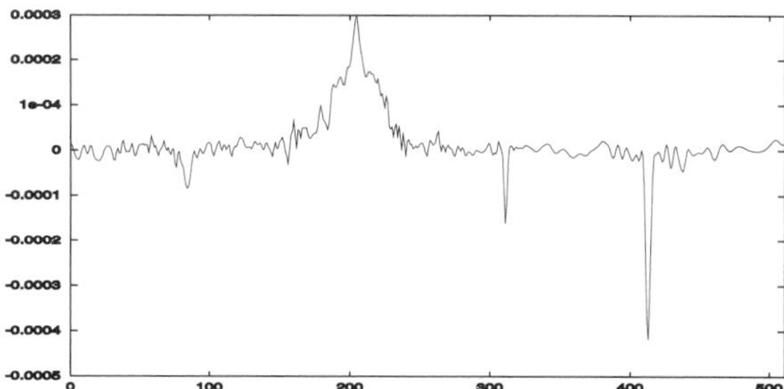

FIGURE 13.12. 2D turbulent vorticity field and cut.

of the enstrophy spectrum confirms that we did not properly separate coherent and incoherent contributions, since the coherent and total enstrophy spectra begin to depart for scales smaller than wavenumber $k \simeq 10$. The incoherent enstrophy does not present the same equipartition as we have obtained using ε_{it}.

13.3.3 Wavelet filtering of 3D turbulent flows

Here, we present the extension of wavelet filtering for 3D turbulent flows. In contrast to 2D flows, where the vorticity is a pseudo scalar, in 3D flows the vorticity is a vector field. Hence, the filtering techniques previously presented have to be adapted to the vector-valued case. Based on the 3D MRA shown in Subsection 13.2.2, we present the extension of the nonlinear wavelet filtering to vector-valued wavelet coefficients and then show results of CVS filtering of a 3D homogeneous isotropic turbulent flow [15]. We also compare the CVS filtering and the classical Fourier low-pass filtering, which is frequently used for LESs [27].

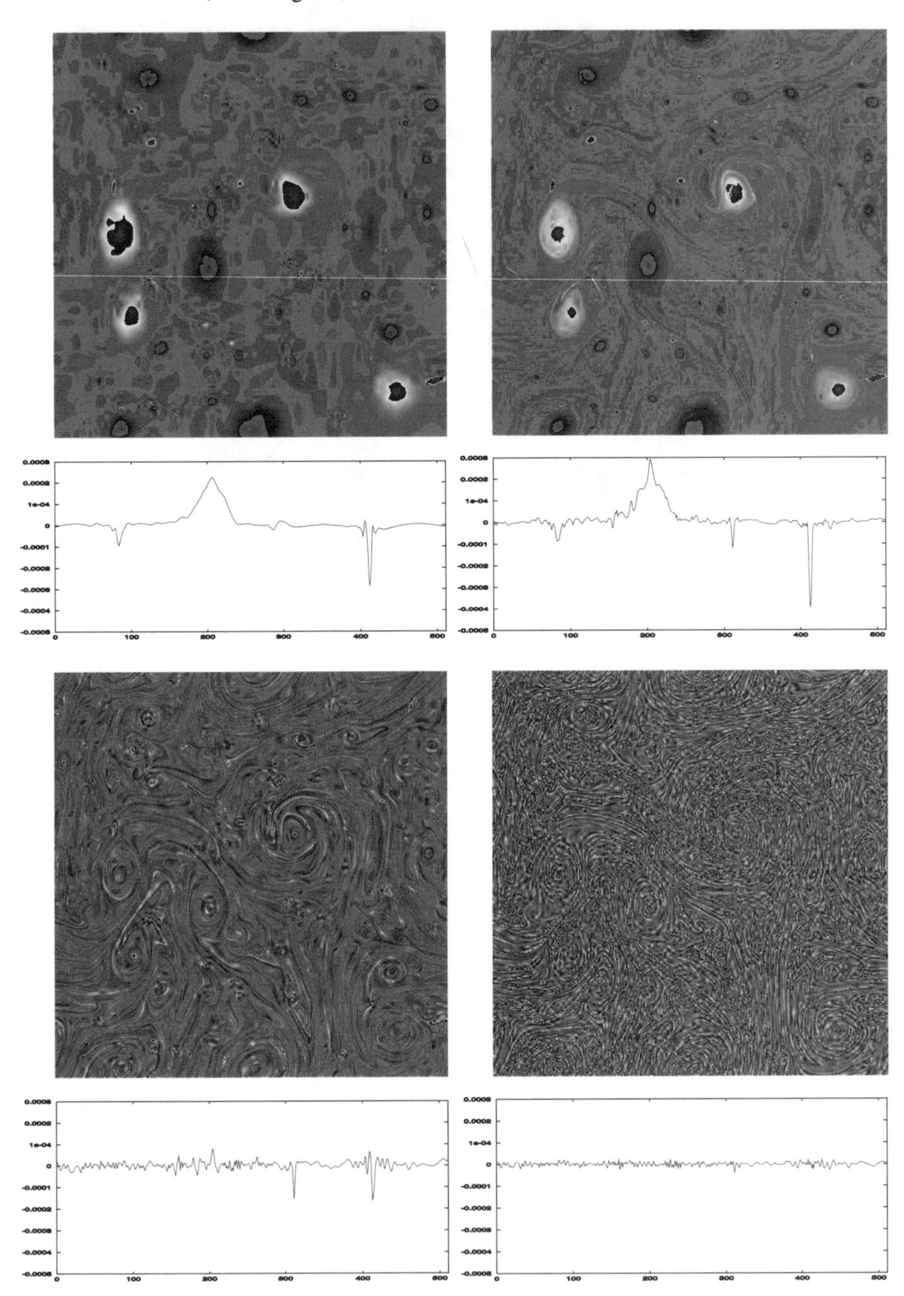

FIGURE 13.13. Split of a turbulent 2D scalar vorticity field using ε_T (left column) and ε_{it} (right column). Top row: coherent parts; Bottom row: incoherent parts.

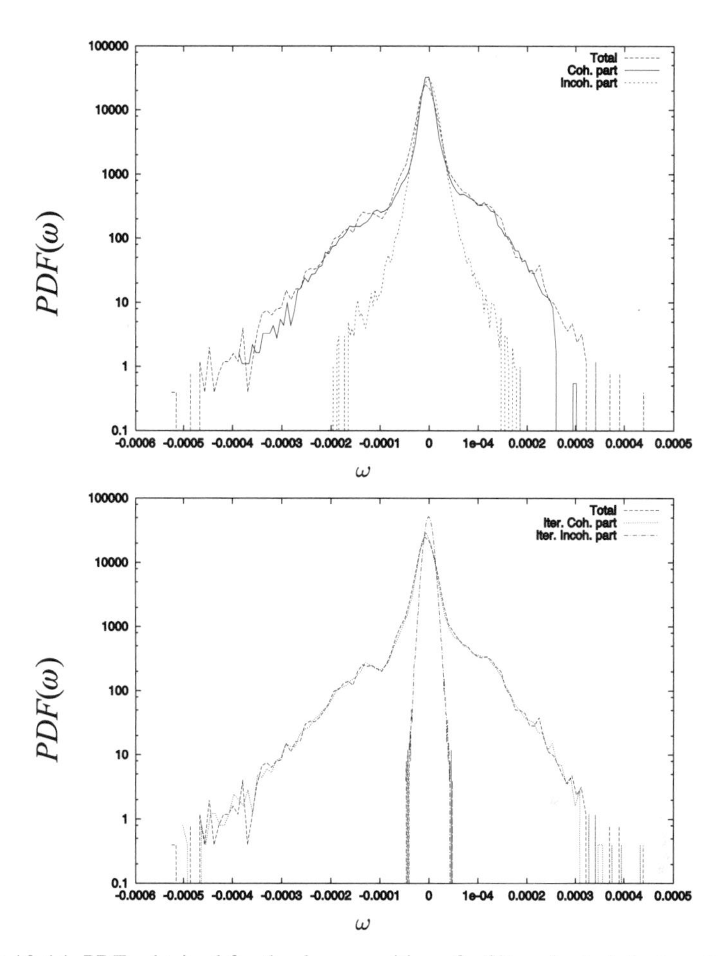

FIGURE 13.14. PDFs obtained for the decomposition of a 2D scalar turbulent vorticity field using ε_T (top) and ε_{it} (bottom).

13.3.3.1 3D vector-valued wavelet decomposition

A vector-valued function $\boldsymbol{\omega}$ of V_{3D}^J can be expressed in terms of the 3D wavelet basis defined in Subsection 13.2.2:

$$\boldsymbol{\omega}(\boldsymbol{x}) = \bar{\boldsymbol{\omega}}_{0,(0,0,0)}\ \phi_{0,(0,0,0)}(\boldsymbol{x}) + \sum_{j=0}^{J-1} \sum_{i_x=0}^{2^j-1} \sum_{i_y=0}^{2^j-1} \sum_{i_z=0}^{2^j-1} \sum_{\mu=1}^{7} \tilde{\boldsymbol{\omega}}^{\mu}_{j,(i_x,i_y,i_z)} \psi^{\mu}_{j,(i_x,i_y,i_z)}(\boldsymbol{x}).$$

Here the scaling function and wavelet coefficients ($\bar{\boldsymbol{\omega}}_{0,(0,0,0)}$ and $\tilde{\boldsymbol{\omega}}^{\mu}_{j,(i_x,i_y,i_z)}$ respectively) are vectors since $\boldsymbol{\omega}(\boldsymbol{x})$ is a vector field.

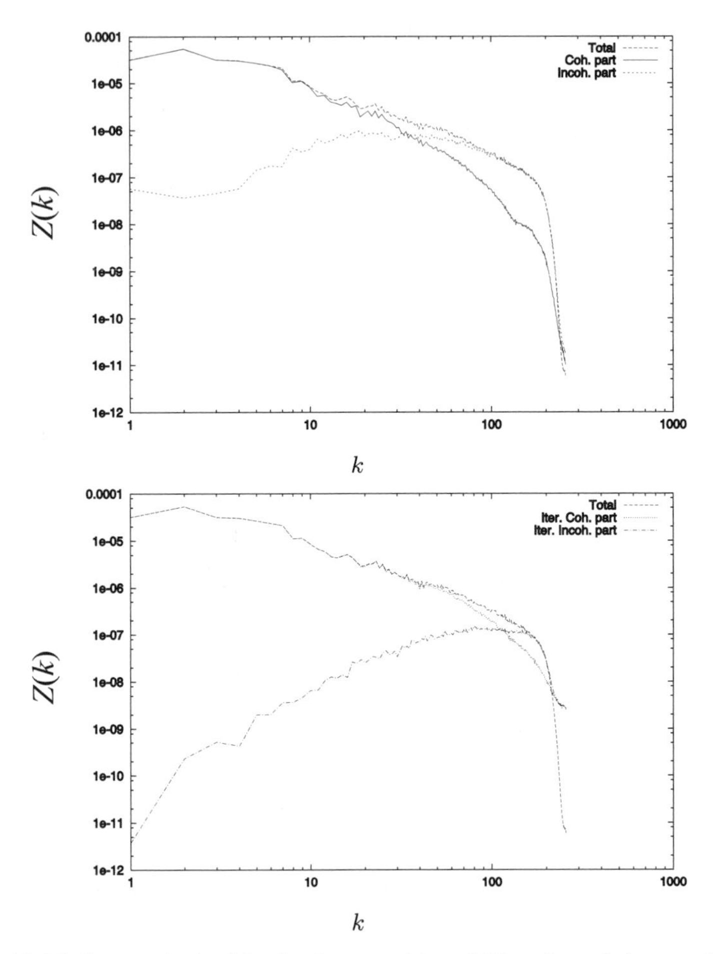

FIGURE 13.15. Spectra obtained for the decomposition of 2D scalar turbulent vorticity field using ε_T (top) and ε_{it} (bottom).

13.3.3.2 Wavelet filtering of vector-valued data

Once the 3D wavelet transform is defined, the extension of the Donoho filtering technique is straightforward for scalar fields but needs some adaptation for vector fields like the vorticity. In this case, we decided to consider the three scalar components ω_1, ω_2, and ω_3 of $\boldsymbol{\omega}$ as three independent fields. For each of them we compute the corresponding Donoho threshold $\varepsilon_{D,i} = \sqrt{2\langle\omega_{i<}\rangle \log_e(N)}$ (where $\langle\omega_{i<}\rangle$ is the background variance). Assuming that the background flow is homogeneous, we can define the threshold as the quadratic mean of the three thresholds previously obtained:

$$\varepsilon_D = \sqrt{\frac{\varepsilon_{D,1}^2 + \varepsilon_{D,2}^2 + \varepsilon_{D,3}^2}{3}} = \sqrt{2\frac{\langle\boldsymbol{\omega}_<, \boldsymbol{\omega}_<\rangle}{3} \log_e(N)}.$$

With this threshold, we reconstruct the coherent vorticity

$$\omega_C(x) = \tilde{\omega}_{>0,(0,0,0)}\ \phi_{0,(0,0,0)}(x) + \sum_{j=0}^{J-1} \sum_{i_x=0}^{2^j-1} \sum_{i_y=0}^{2^j-1} \sum_{i_z=0}^{2^j-1} \sum_{\mu=1}^{2^n-1} \tilde{\omega}^{\mu}_{>j,(i_x,i_y,i_z)}\ \psi^{\mu}_{j,(i_x,i_y,i_z)}(x), \tag{13.3.3}$$

where

$$\tilde{\omega}^{\mu}_{>j,(i_x,i_y,i_z)} = \begin{cases} \tilde{\omega}^{\mu}_{j,(i_x,i_y,i_z)}, & \|\tilde{\omega}^{\mu}_{j,(i_x,i_y,i_z)}\| > \varepsilon_D \quad , \\ 0, & \|\tilde{\omega}^{\mu}_{j,(i_x,i_y,i_z)}\| \leq \varepsilon_D, \end{cases}$$

and the incoherent vorticity ω_I which is reconstructed from the remaining weak coefficients. As described in the previous sections, this filtering procedure leads to an implicit definition of the background vorticity ω_I, which can be calculated by using the iterative threshold estimation.

13.3.3.3 Coherent vortex extraction in 3D homogeneous isotropic turbulence

In the following paragraphs we show the application of the CVS filtering technique previously described to extract coherent vortices out of a 3D flow. We then point out the differences between a nonlinear wavelet filter and a linear low-pass Fourier filter, classically used in LES [19]. Here, we consider a homogenous isotropic turbulent flow computed with a DNS by Alan Wray [6, 25] which exhibits a self-similar behavior in the inertial range. This simulation has been performed at a resolution of 256^3 for a microscale Reynolds number $Re_\lambda = 168$. For the visualisation, subcubes of size 64^3 are used in order to observe the small scale structures in detail.

The application of wavelet and Fourier filters to the vorticity ω leads to the definition of the complementary fields ω_C/ω_I (coherent/incoherent vorticity) and ω_L/ω_S (large-scale/small-scale vorticity). After this splitting, we use the Biot–Savart operator to reconstruct the corresponding velocities $V = \nabla^{-2}\nabla \times \omega$. In this way we can compute the corresponding velocity fields V_C, V_I, V_L, V_S and study their respective properties. It is important to point out that, as the wavelet basis employed is orthonormal, the fields ω_C and ω_I are by construction orthogonal, which ensures the separation of the total enstrophy into $Z = Z_C + Z_I$ because the interaction term $\langle \omega_C, \omega_I \rangle$ vanishes. The same consideration also holds for the splitting in Fourier space. Furthermore, as the Biot–Savart operator is diagonal in the Fourier space, also the velocity fields V_L and V_S are orthogonal, and hence the total energy decomposes into $E = E_L + E_S$. This property also holds approximately for the wavelet separation because the Biot–Savart operator is almost diagonal in the wavelet space. Therefore, the interaction term remains small, although it does not vanish. For our example we obtain $E = E_C + E_I + \epsilon$, where $\epsilon = \langle V_C, V_I \rangle$ and $\epsilon \leq 0.5\% E$.

13.3.3.3.1 CVS filtering

Figure 13.16 shows the isosurfaces of the vorticity modulus of the total field. To extract the coherent vortices, we proceed as follows: first we project the vorticity

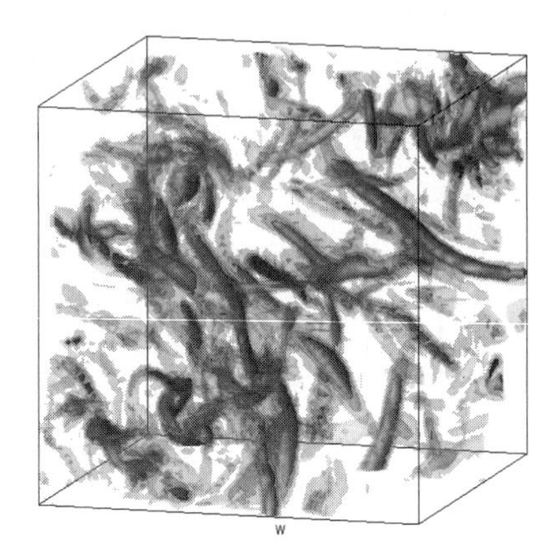

FIGURE 13.16. Isosurfaces of modulus of vorticity (total field).

field onto an orthogonal wavelet basis, then we compute the threshold for the wavelet coefficients using the iterative method (note that for this example only one iteration is used), and finally we filter. From the wavelet coefficients whose modulus is larger than the threshold we obtain the coherent vorticity $\boldsymbol{\omega}_C$, and from the remaining weak wavelets coefficients we obtain the incoherent vorticity $\boldsymbol{\omega}_I$. We find that only 3% of the wavelet modes describe the coherent flow, which retains the major part of the total enstrophy (75%) and nearly all the energy (98%). Figure 13.17 (left) shows isosurfaces of $\boldsymbol{\omega}_C$. We observe that all structures of the original field are also present in the filtered one. The incoherent background flow, represented by the remaining 97% of wavelets modes, contains 25% of the enstrophy and only 2% of the energy. Figure 13.17 (right) shows the modulus of the incoherent vorticity in physical space. Here the structures present in the original field are completely lost.

13.3.3.3.2 LES filtering

To perform the LES-type filtering we have employed an isotropic sharp cutoff filter in Fourier space, which is optimal in the proper orthogonal decomposition (POD) sense for homogeneous turbulence [2]. We have chosen a cutoff frequency $k_c = 47$ to obtain the same compression as obtained for the wavelet filtering. Figure 13.18 shows the large-scale vorticity $\boldsymbol{\omega}_L$ and the small-scale vorticity $\boldsymbol{\omega}_S$. Looking at these figures, we observe two facts: first, the large scales are much smoother than in the original field and second, that the small scales still present well-organized structures that correspond to the cores of the vorticity tubes.

13.3.3.3.3 Statistical analysis

We now present some statistical analyses of the Fourier and wavelet filtered fields that we have just described. In Figure 13.19 we plot the isotropic energy spectrum,

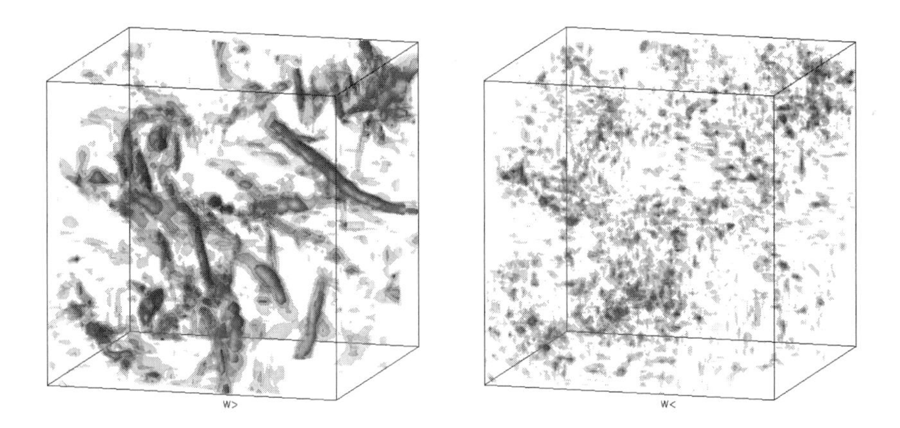

FIGURE 13.17. Wavelet separation (left: coherent part, right: incoherent part).

which is defined by

$$E(k) = \frac{1}{2} \sum_{k-1/2<|\boldsymbol{k}|\leq k+1/2} |\hat{\boldsymbol{v}}(\boldsymbol{k})|^2 \tag{13.3.4}$$

and correlated to the enstrophy spectrum (13.3.1) by

$$Z(k) = k^2 E(k), \tag{13.3.5}$$

of the total flow, the coherent and the incoherent parts as well as the cutoff frequency used for the Fourier filtering. We can remark that both, the coherent and incoher-

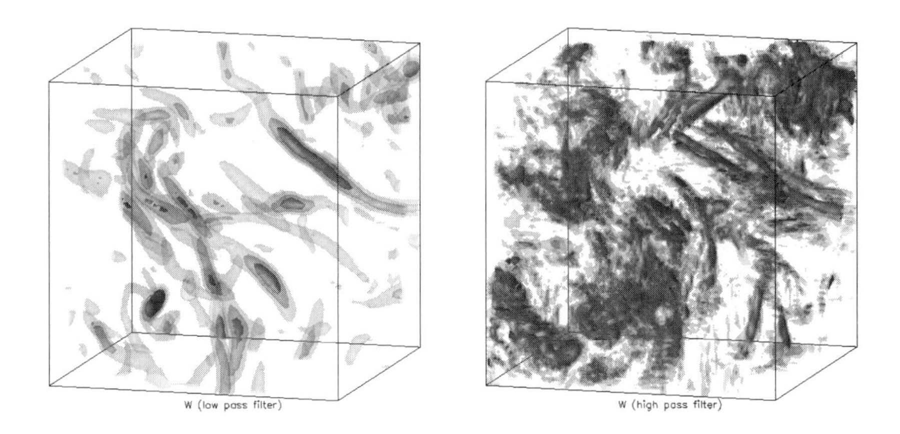

FIGURE 13.18. Fourier separation (left: low-pass filtered part, right: high-pass filtered part).

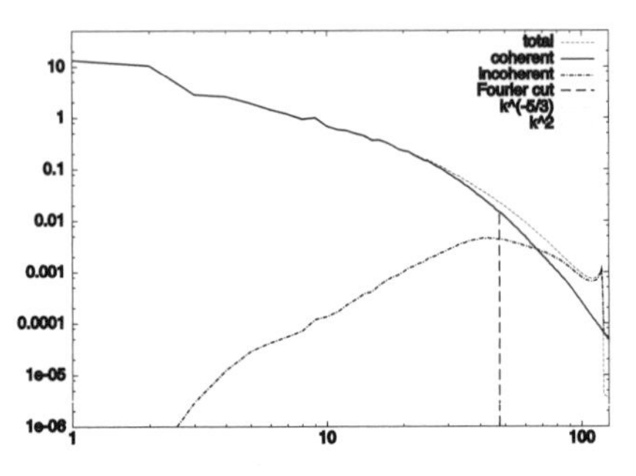

FIGURE 13.19. Energy spectra.

ent flows exhibit a multiscale behavior. In particular, the former has the same $k^{-5/3}$ scaling as the total flow throughout the whole self-similar inertial range. We conjecture that the fall-off observed in the dissipative range corresponds to some coherent energy that has been transferred into incoherent energy before being dissipated. This is confirmed by the fact that there is an equipartition of incoherent energy up to the dissipative scales, since its spectrum scales about like k^2 (which in 3D corresponds to white noise because the isotropic spectrum is integrated over spherical shells) followed by a decay due to the dissipation of incoherent energy at the small scales.

Figure 13.20 shows the PDF of the vorticity field for wavelet and Fourier filtering methods. We observe that the PDF of coherent vorticity exhibits the same stretched exponential behavior as the total vorticity, while the PDF of the incoherent vorticity is exponential with a much smaller variance. Considering the Fourier filtered vorticity fields, we find that the PDF of the large-scale vorticity has an exponential shape and looks very different from the PDF of the total flow. The corresponding background vorticity PDF (small scales) is instead of the stretched exponential type, and its tails present values even bigger than those present in the large-scale vorticity. This confirms that in the small scales there remain regions where the local enstrophy is intense which are probably responsible for energy transfers from the small to the large scales (backscatter).

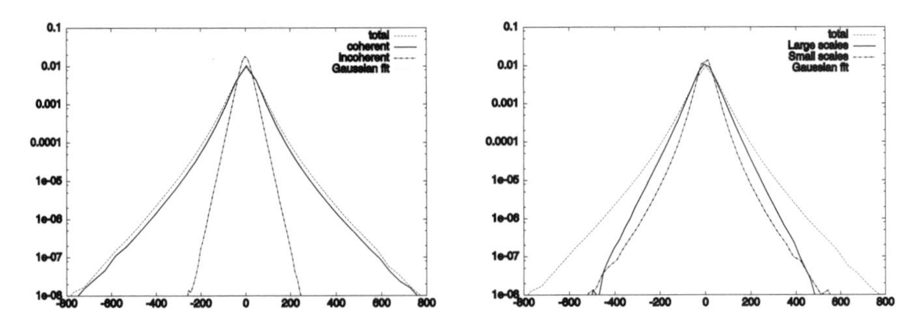

FIGURE 13.20. PDF of vorticity (left: wavelet separation, right: Fourier separation).

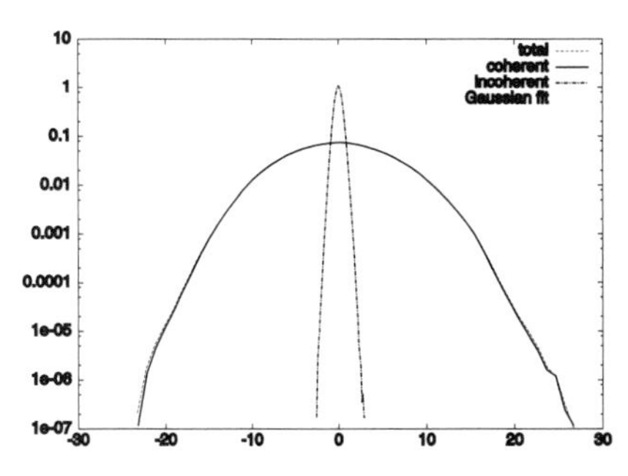

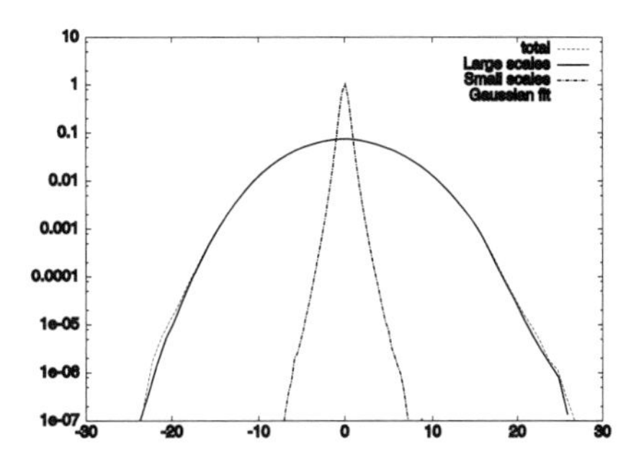

FIGURE 13.21. PDF of velocity (left: wavelet separation, right: Fourier separation).

Wavelet and Fourier filtering techniques are both able to represent the PDF of the total velocity quite well (Figure 13.21). The difference between the two types of filters becomes clear when we consider the background flows. The PDF of the small-scale velocity is exponential, while the incoherent velocity presents a Gaussian behavior, which suggests that a sort of thermal equilibrium is reached for the incoherent velocity.

13.4 CVS of evolving turbulent flows

In this section, we present the application of the previously introduced wavelet filtering technique to turbulent flow simulations, performing CVSs [18]. For the statistical analysis of single turbulent flow fields, it has been demonstrated in Section 13.3 that wavelet filtering is very effective for the extraction of coherent structures from an incoherent background flow. Although the filtered coherent flow part is highly compressed in wavelet space, it conserves most of the total energy and enstrophy of the original flow field [12, 16]. This property makes it possible to successfully simulate dynamically evolving flows using only the reduced wavelet-filtered basis instead of the total number of degrees of freedom, which will reduce the overall computational cost.

We first introduce the adaptive basis concept of CVS and explain the implementation of the basis selection. We then show results for the CVS of 2D and 3D turbulent flows. The effect of CVS on the flow will be quantified by comparing the calculated evolution to DNSs, which, by definition, are model-free [23].

13.4.1 Implementation of CVS: adaptive basis selection

Performing numerical simulations using CVS implies the integration of the governing equations with a reduced set of degrees of freedom. This is achieved by filtering the flow represented in wavelet space at every time step, and thus defining the basis for the integration [34]. The filtered wavelet basis of each integration step corresponds to a locally refined computational grid in physical space and scale. Because the coherent vortices are moving during the evolution, and due to their successive extraction at each time step, the simulation is finally carried out in a locally and temporally adaptive compressed basis [21]. The CVS integration basis is composed of the filtered wavelet coefficients obtained from threshold application. Furthermore, a *security zone* is added to the basis. It consists of the direct neighbor coefficients of the former selection, and it is intended to take into account the nonstationary behavior of the evolving flow.

13.4.1.1 Security zone

When applying the CVS filtering technique to a turbulent flow field, it has been shown in Section 13.3 that coherent structures are extracted from the background flow, which apparently is negligible from the energy point of view. Employing the filtering in a simulation, the integration is performed only on the grid points corresponding to the filtered flow part. The time evolution of a turbulent flow results in the movement of its structures in space and scale. With respect to the reduced integration domain, this includes movements to regions that are not captured by the initially filtered wavelet basis. This effect is illustrated for a single 1D structure in Figure 13.22. For the case of filtering the wavelet coefficient field at the beginning of a time step by applying only a threshold ε, a structure, e.g., a vortex, is represented by the coefficients $|\tilde{\omega}| > \varepsilon$ (white area under the solid-line curve). All coefficients $|\tilde{\omega}| \leq \varepsilon$ are neglected, and their corresponding wavelets are not considered for the integration. The structure moves in space and scale within the determined integration step (indicated by the arrow), and then it can be represented by the coefficients under the dashed-line curve. We recognize that some of these coefficients (hatched area) are not captured by the initial filtering with ε. Therefore, they would be ignored for the integration, and their corresponding dynamical information would be irreversibly lost. To avoid this effect, we retain for the integration basis, together with the coefficients $|\tilde{\omega}| > \varepsilon$, their local neighbors in wavelet space, which constitute the *security*

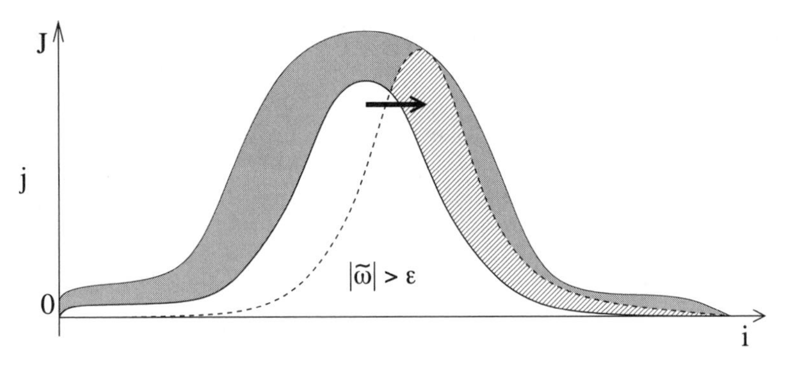

FIGURE 13.22. Effect of the security zone extension. The wavelet coefficient space of a 1D signal is shown, spanned by the scale j and the position i.

zone (gray area). It is the function of the security zone to provide room for the solution to evolve in space and scale. In Figure 13.22 it is illustrated that the movement of the structure within the determined integration step is fully captured by the enlarged basis and thus no dynamical information is lost.

It is obvious that the set of filtered wavelet coefficients extended by the security zone reduces compression. The challenge is to find a minimal security zone capturing all degrees of freedom necessary to catch the flow dynamics. The security zones chosen for the simulations in two and three dimensions will be presented in detail.

13.4.1.2 Security zone in two dimensions

For CVS the filtering of the flow field is done in wavelet space, and therefore the applied filter is also defined in wavelet space. Figure 13.23 illustrates in detail how the security zone is selected for 2D flows, similar to [21]. For every coefficient fulfilling the threshold filtering condition $|\tilde{\omega}| > \varepsilon$, the coefficient itself (•) and automatically all its neighbours ($\times$) are retained in the filtering procedure. This neighborhood includes coefficients on the same scale (light grey) as well as the corresponding finer (dark grey) and coarser (white) scales. It was found, that the filter varies very slightly for the chosen time step width, so we use a reasonable procedure considering direct neighbors only. Because wavelet coefficients are localized in both phase and space,

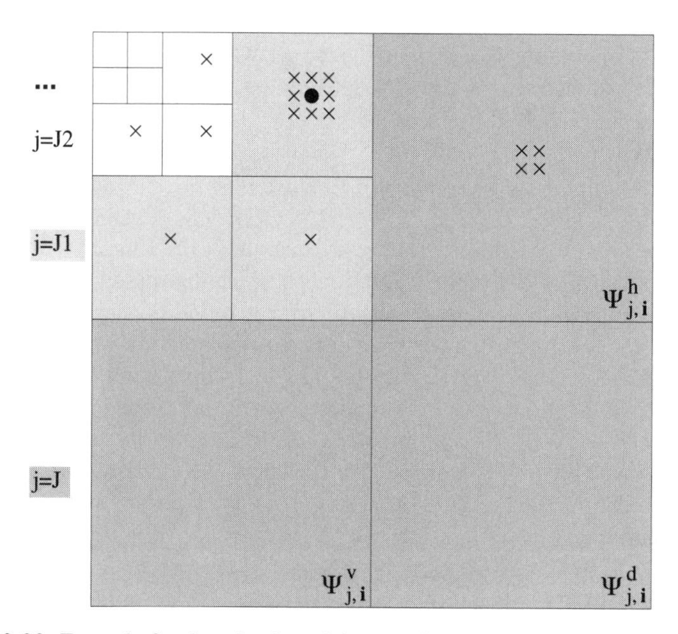

FIGURE 13.23. Example for the selection of the security zone in the 2D wavelet space. For every scale j each of the three remaining quarters represents 2^{2j} coefficients corresponding to the wavelets $\Psi^{\mu=h,v,d}_{j,i}$ oriented in the horizontal (h), vertical (v), and diagonal (d) directions, respectively [8].

we include proximity information locally as well as in phase space. The security zone presented in Figure 13.23 is a proposal, offering to the solver the possibility of evaluating a solution in all directions and scales. Many different types of security zones are conceivable, affecting compression and the capability of conserving dynamical information of the filtered flow.

13.4.1.3 Security zone in three dimensions

The security zone for 3D simulations is built almost the same way as that in two dimensions. The increase of dimension involves an increase in size of the wavelet coefficient field, i.e., in a 3D domain, many more wavelets must be considered. Compared to the 2D wavelet representation in Figure 13.23, in three dimensions each wavelet in fact has 27 neighbors (instead of 9), and we find 7 wavelets (instead of 3) that have the same localization with different orientations in space. It is obvious that the 2D security zone evaluation scheme scaled up for 3D coefficient fields causes a strong degradation of compression. A further reduction of the number of retained coefficients is desirable but difficult to realize, because then it must be known *a priori*, in which direction new wavelets will be needed in the next time step. Therefore, we introduce the concept of "neighbor orientation." It is based on the supplementary reasonable hypothesis that a wavelet with a particular orientation most probably will not generate wavelets of totally different orientation in the next time step, but will continue to evolve in almost the direction of their orientation. For example, if a wavelet with orientation x is retained, then the wavelets with orientation xy and xz are included in the security zone (discarding those with orientation y, z, yz, and xyz).

13.4.2 *Simulation of 2D turbulent flows*

In this section, we present results for the application of CVS to 2D turbulent flows. First, we show the DNS of a flow evolution, representing the reference case. Considering different ways of threshold selection for wavelet filtering, we present simulations using the estimated threshold ε_T and the iteratively evaluated threshold $\varepsilon_{it} \approx \varepsilon_D$, both introduced in Subsection 13.2.4. In both cases the security zone extension (see Subsection13.4.1.2) is applied. Two further simulation are stressed to illustrate by comparison the effect and necessity of the security zone, one with and the other one without it. As in the analysis section, we focus on the compression properties of each method as well as on its capability of conserving flow dynamics in the simulation.

13.4.2.1 DNS of 2D turbulence

In DNS the governing equations are solved directly without any subgrid-scale model. The presented DNS of a 2D incompressible turbulent flow yields the model-free reference case for comparisons with CVS. The Navier–Stokes equations in the velocity-vorticity formulation in two spatial dimensions read as

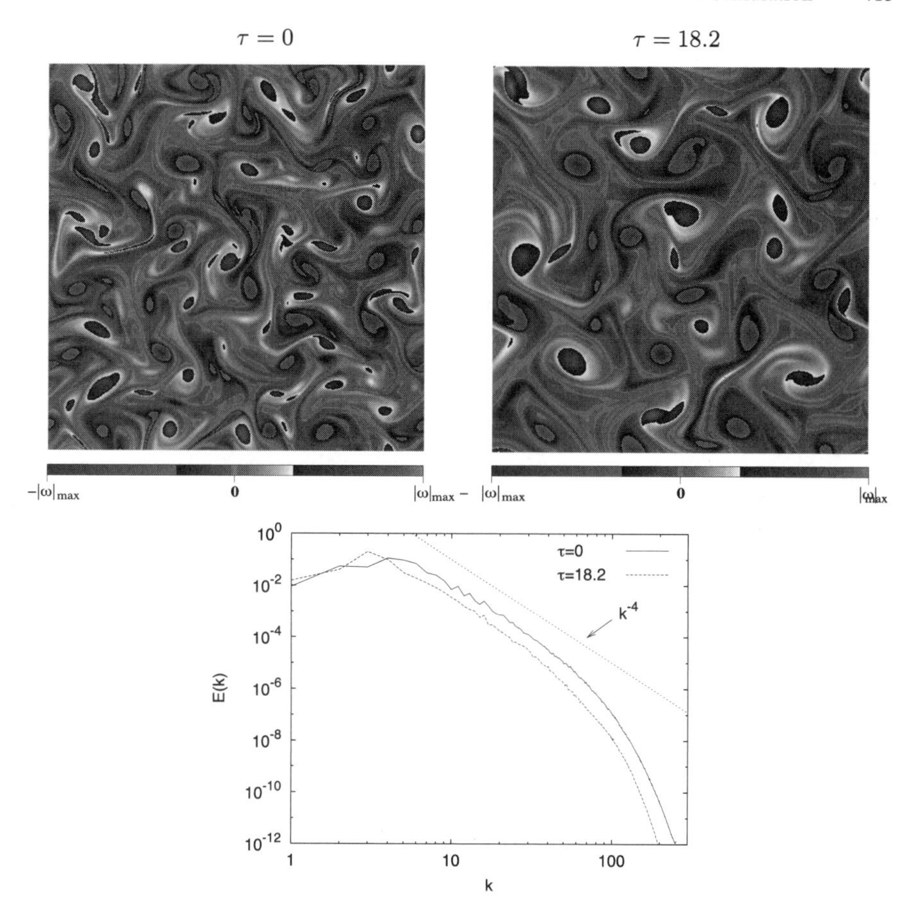

FIGURE 13.24. DNS of a turbulent flow. Vorticity plots for the initial condition ($\tau = 0$, $|\omega|_{\max} = 232.5$) and data field at $\tau = 18.2$ ($|\omega|_{\max} = 156.3$) are given with corresponding energy spectra.

$$\partial_t \omega + \boldsymbol{v} \cdot \nabla \omega - \frac{1}{Re} \nabla^2 \omega = 0 \tag{13.4.1}$$

$$\nabla \cdot \boldsymbol{v} = 0, \tag{13.4.2}$$

where $\boldsymbol{v} = \binom{v_x}{v_y}$. The equations are normalized by the initial root-mean-square velocity v_0 and vorticity ω_0 and the computational domain length L_0. Because the upcoming CVSs imply a model by themselves, it is necessary to compare them to a reference noninfluenced by any additional model, and thus decaying turbulence without forcing is considered. Equations (13.4.1) and (13.4.2) are completed by the given initial condition (see Figure 13.24) and periodic boundary conditions. With the selected kinematic viscosity ν the resulting initial Reynolds number for the presented flow becomes $Re = v_0 L_0/\nu \approx 8722$. The equations are integrated on a domain

$\Omega = [0;1[^2$ with a resolution of 512^2 degrees of freedom. A pseudo-spectral Fourier method is used, resulting in high order and accuracy [4].

For DNS simulations of 2D turbulence, a code, basically developed by K. Schneider and enhanced by J. Ziuber, has been used. In Figure 13.24 the vorticity fields are depicted for the initial condition and the end of the simulation. The given time scales are normalized with $\tau = 1/\sqrt{2Z_0}$, where Z_0 is the total enstrophy at $t = 0$. From the vorticity plots, the typical behavior of 2D turbulent flows can be identified: The number of vortices decreases as a consequence of the merging of co-rotating vortices. Fine-scale structures like elongated filaments are formed by vortex interaction, i.e., local fluid elements are compressed in one direction and simultaneously stretched in another. The overall vorticity amplitude decreases due to viscous dissipation.

The energy spectra are plotted in Figure 13.24 as well. From statistical theory, a k^{-3} inertial range would be expected for 2D turbulence [1, 26]. For the presented simulation we find spectra with a k^{-4} slope, which is almost consistent with results of other authors for numerical simulations in two spatial dimensions [20]. Because decaying turbulence is considered without energy injection by a forcing term, we find a decay of total energy, caused by energy dissipation at the molecular level. The shape of the spectrum remains very similar during the evolution. The approximately constant decay of the inertial range with a slope of k^{-4} indicates a fully developed turbulent flow evolution.

13.4.2.2 CVS of 2D turbulent flows

In Section13.3 we have seen that wavelet filtering a single turbulent flow field preserves almost the whole energy and most of the enstrophy, coinciding with a very high compression. Compared to the DNS run in Subsection 13.4.2.1, we now determine if this holds for the CVS of turbulent flows. As in the analysis section, we focus on the compression properties of the presented filtered simulations as well as on their capability of preserving flow dynamics by reproducing the reference flow deterministically and statistically.

For these purposes, the code used for the simulations presented in Subsection 13.4.2.1 is modified. At the beginning of each time step, from the vorticity field ω its coherent part ω_C is extracted by filtering. The choice of filter differs depending on the simulation runs described below. The integration for one time step is performed for ω_C only using the pseudo-spectral scheme, which yields a solution $\omega_C^\star$. The time step is completed by reprojecting $\omega_C^\star$ to the basis of ω_C by applying the filter again. Based on this scheme, Eq. (13.4.1) and (13.4.2) are consequently integrated as

$$\partial_t \omega_C + (\boldsymbol{v}_C \cdot \nabla \omega_C)_C - \frac{1}{Re}\nabla^2 \omega_C = 0 \tag{13.4.3}$$

$$\nabla \cdot \boldsymbol{v}_C = 0. \tag{13.4.4}$$

For the following studies, the Coiflet12 wavelet family is used for filtering, because it is known to be suitable for extracting the major physical information from a single flow field. The simulation parameters are identical with those of the DNS run.

To study two general aspects, four simulation runs are presented:

- The threshold evaluation techniques based on Donoho's theorem were applied in Section 13.3 to the statical analysis of flow fields. Here, we compare flow evolutions obtained from CVS using the threshold ε_T (run labeled **CV-TS**) and ε_D (**CV-DS**), respectively. In both cases the security zone is applied.
- The necessity and effect of the security zone are demonstrated by comparing two simulation runs, one with the security zone (labeled **CV-1S**) and the other one without it (labeled **CV-6**), starting both simulations with about the same compression. Using the iterative threshold estimation algortithm introduced in Subsection 13.2.4, here the number of iterations is limited in both cases, as indicated by the particular label, i.e., one iteration for CV-1S and six iterations for CV-6. Only in the case of the CV-1S run is the security zone added. The number of iterations chosen for these two runs is justified with the approximately identical compression ratio for their resulting initial condition.

The starting point for all these simulations is the vorticity initial condition of the DNS run in Figure 13.24. After the initial wavelet filtering, we obtain the initial conditions for each simulation. The corresponding compression ratios and the percentages of energy and enstrophy conserved are listed in Table 13.1.

The vorticity fields for the CV-TS and the CV-DS simulation run are depicted in Figure 13.25. For the CV-TS run, the vorticity field at $\tau = 0$, though it contains most of the energy compared to the unfiltered DNS reference (see Table 13.1), looks quite fragmented, and many structures are not completely resolved. For $\tau = 18.2$ we recognize that, compared to the reference case, the vortices are blurred and deformed, and their positions are out of phase. Even their sizes and number differ from the reference. Fine-scale structures are not sufficiently resolved, which is quantified by the spectrum plot in Figure 13.26 for $\tau = 18.2$. Starting at about $k = 10$ in the CV-TS spectrum, the large-scale energy develops far below the reference, indicating non-resolved fine structures. The increase of the CV-run spectra over the DNS spectrum at high wavenumbers is due to an artifact of the wavelet transform, which is of negligible importance for the presented studies. The global enstrophy evolution plot depicts an increasing difference, compared to DNS. Representing with its initial condition 96.2% of the reference's enstrophy, only 88% of the enstrophy is conserved for the final vorticity field. These effects, due to the high compression, starting at $\tau = 0$ with a ratio of 2.2% (see Table 13.1). The relative compression ratio plot in

Run	*Compression ratio [%]*	*Energy [%]*	*Enstrophy [%]*
CV-DS	46.9	100	100
CV-6	7.0	100	99.9
CV-1S	6.7	99.9	99.4
CV-TS	2.2	99.6	96.2

TABLE 13.1. Compression ratio (defined in (13.3.2)) and conservation properties of the initial conditions ($\tau = 0$) for the CVS-filtered simulations compared to DNS.

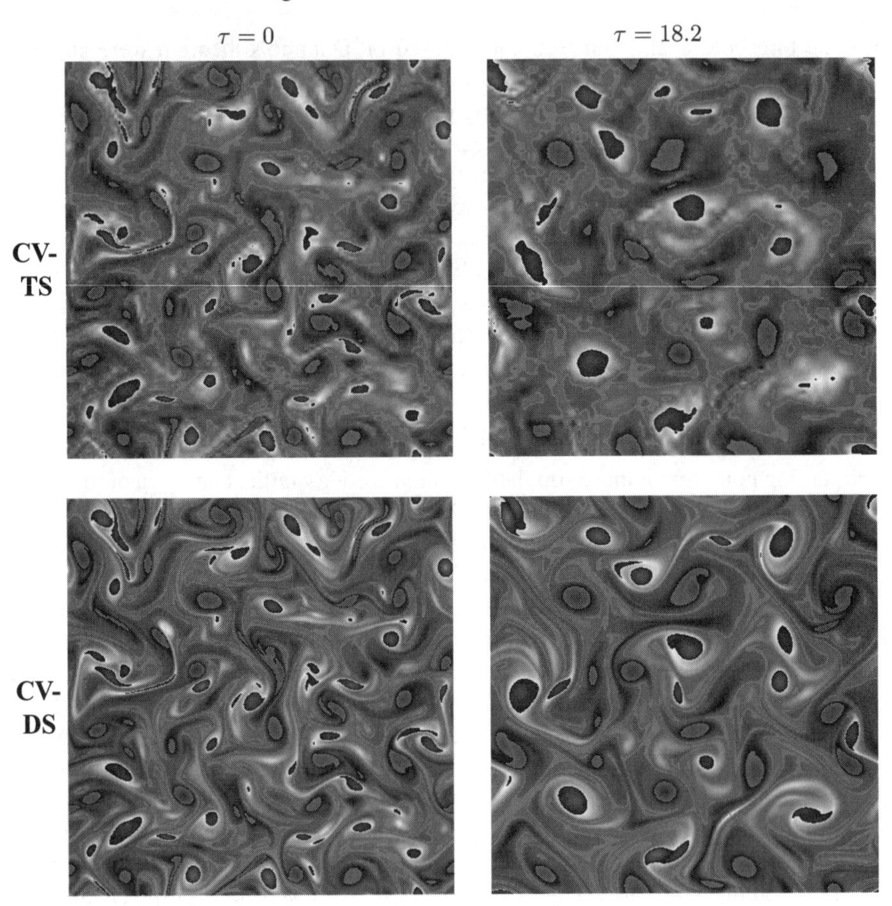

FIGURE 13.25. Vorticity plots for initial conditions ($\tau = 0$) and data fields at $\tau = 18.2$ of simulation runs CV-TS using an estimated threshold ε_T, and CV-DS based on the iteratively evaluated Donoho threshold ε_D. The security zone is applied for both runs. For the color palette, see Figure 13.24.

Figure 13.26 illustrates a further decay to even about 1.3% at $\tau = 18.2$, i.e., we find a concentration of the adaptive basis to structures of increasing coherence shifting more and more structures into the neglected incoherent part. Furthermore, a comparison of the PDF of the vorticity at $\tau = 18.2$ illustrates a significant deviation for the vorticity maxima. Large amplitudes become very rare in CV-TS. So, we can conclude that the compression resulting from the CV-TS run obviously is too high to reproduce the reference flow accurately with all structures resolved.

When we compare the vorticity fields of the CV-DS (Figure 13.25) and DNS (Figure 13.24) runs, no difference can be detected by eye. The plots for the total enstrophy evolution, the PDFs of vorticity, and the energy spectra yield the same result: there is approximately no difference between CV-DS and the reference. This

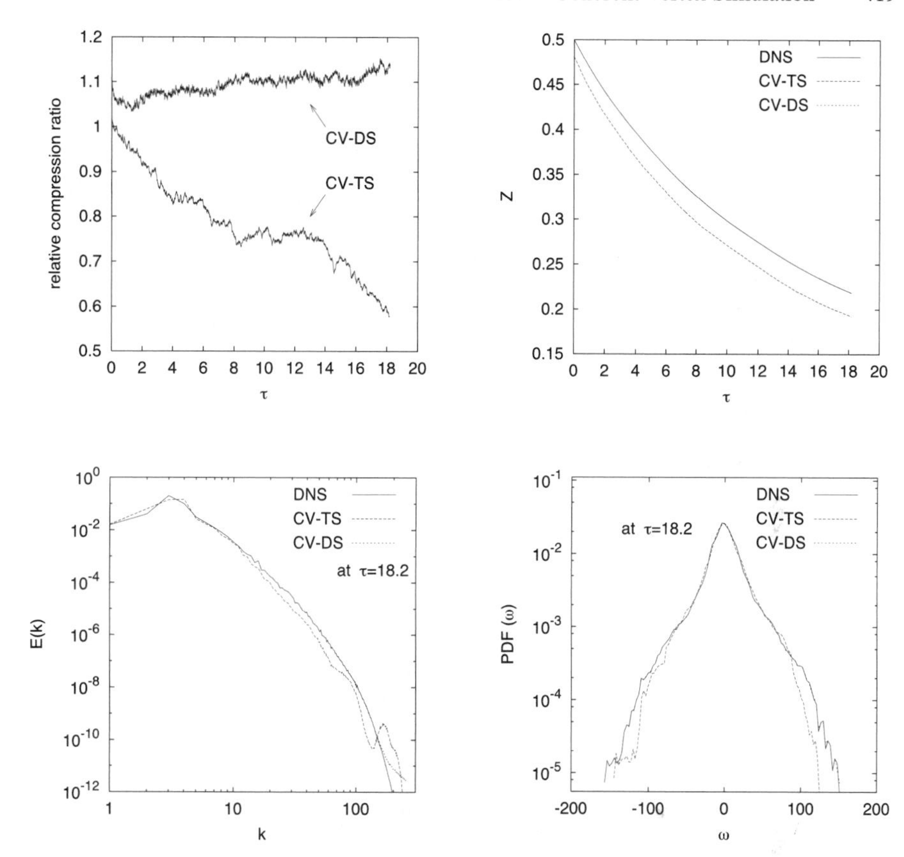

FIGURE 13.26. Statistical information on the CV-TS and CV-DS runs compared to the DNS reference. Top-left: Relative compression ratio, i.e., compression normalized by the particular initial compression at $\tau = 0$. Top-right: Total enstrophy evolution. Bottom-left: Energy spectra at $\tau = 18.2$. Bottom-right: PDF of vorticity at $\tau = 18.2$. In some plots only two curves can be recognized because DNS and CV-DS are approximately identical then.

high degree of conformity becomes obvious considering the compression that goes along with CV-DS (cf. Figure 13.25). The filtered initial vorticity field is compressed to only 46.9% of degrees of freedom. In addition, the relative compression ratio evolution plot in Figure 13.26 illustrates a global increase to 53.3%, i.e., the compression is much too low to benefit from in CVS.

From the results of the CV-DS and CV-TS simulation runs, we propose for the practical application of CVS filtering in numerical turbulent flow simulations a modification of the threshold estimation algorithm. Instead of using one of the presented extremes, i.e., a threshold based on the total vorticity's variance or the threshold obtained from a converging iteration process, the threshold should be determined by the known iterative threshold estimation procedure performing only a small fixed number of iterations. The number of iterations can be well adjusted to the desired

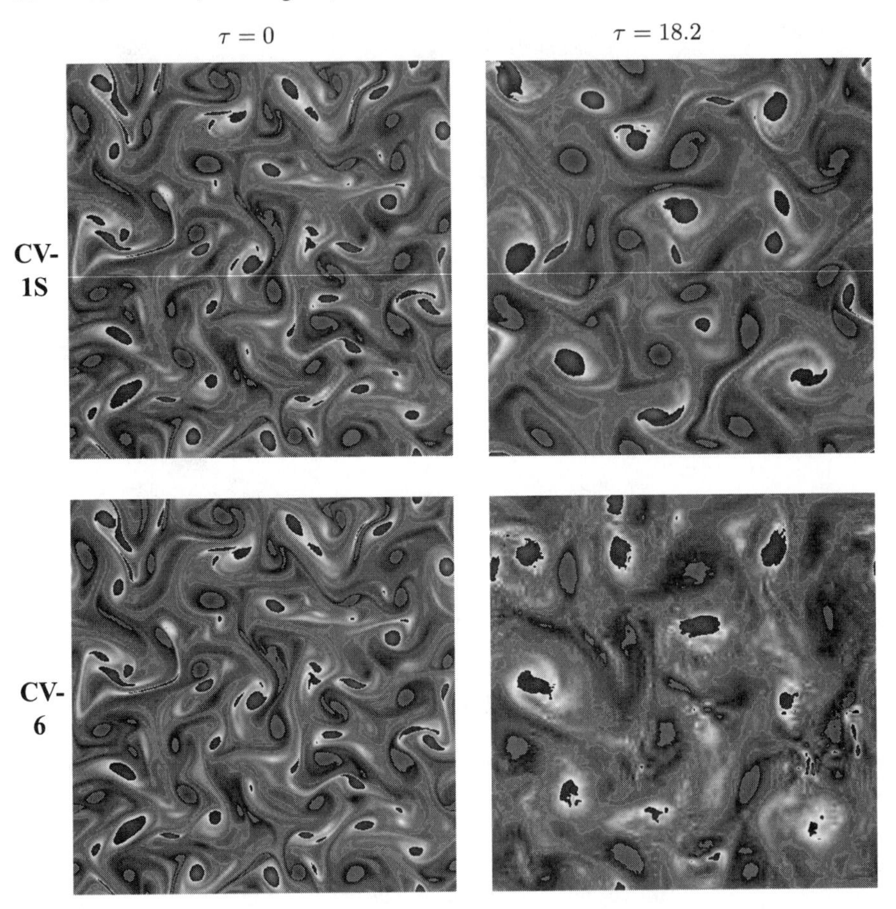

FIGURE 13.27. Vorticity plots for initial conditions ($\tau = 0$) and data fields at $\tau = 18.2$ of simulation runs CV-1S using a Donoho-based threshold with one iteration plus the security zone, and CV-6 using six iterations without the security zone. For the color palette, see Figure 13.24.

compression and properties to conserve the energy of the particular physical problem. With the following CV-1S run, we demonstrate that, for the considered 2D flow, one iteration step for the threshold estimation already yields a satisfactory flow reproduction in accordance with an acceptable compression.

The vorticity fields of the CV-1S and the CV-6 run are depicted in Figure 13.27. Run CV-1S starts with an initial condition reproducing the reference (cf. Figure 13.24) slightly fragmented, but globally no vortices are missing and even finer structures like filaments can be found in the filtered field. At $\tau = 18.2$, the vortices appear a little mutated in shape and some smaller vortices as well as small-scale filaments are missing, compared to the reference. However, the vortices, number and their positions in space confirm quite well the reference case. Certainly, compared

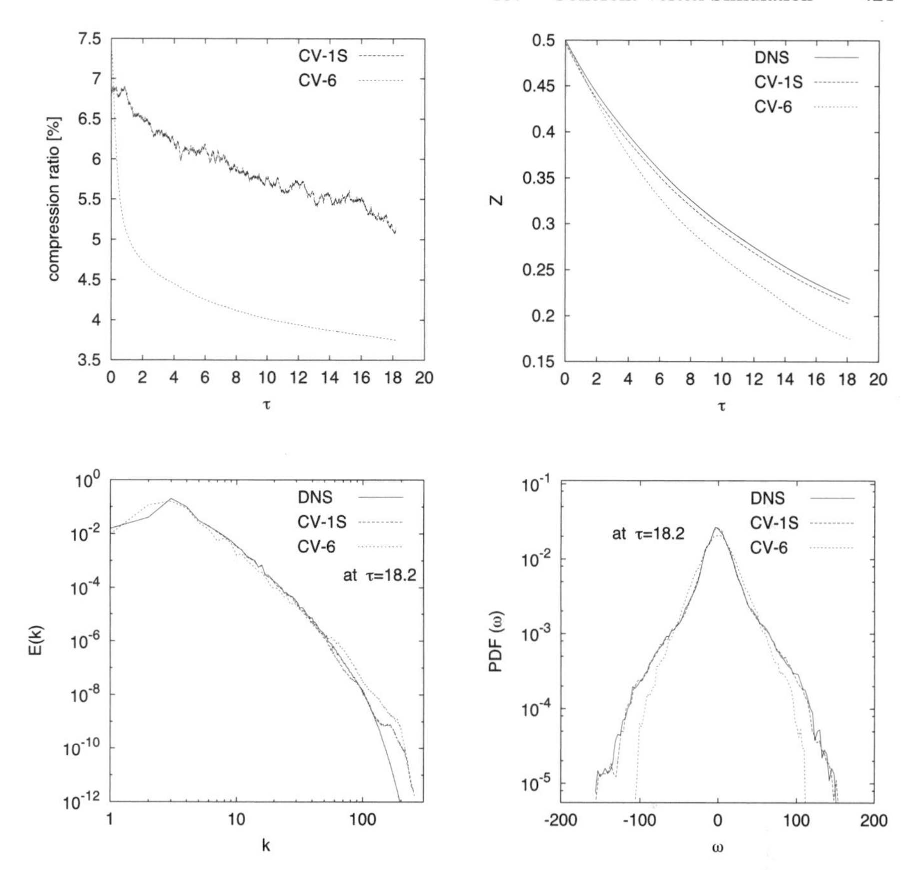

FIGURE 13.28. Statistical information on the CV-1S and CV-6 runs compared to the DNS reference. Top-left: Compression ratio evolution. Top-right: Total enstrophy evolution. Bottom-left: Energy spectra at $\tau = 18.2$. Bottom-right: PDF of vorticity at $\tau = 18.2$.

to the highly compressed CV-TS run (see above), the flow field for both time steps is better reproduced in the CV-1S case. The total enstrophy for this run does not differ very much from the reference (cf. Figure 13.28). Containing 99.4% of the enstrophy of the reference at the beginning, a small decrease to 97.6% at the end of the simulation is found. Simultaneously, the compression ratio decays from 6.7% to about 5.1% during the evolution. The spectra of CV-1S and DNS at $\tau = 18.2$ are almost identical. Solely at medium wavenumbers ($\sim k > 50$) the CV-1S spectrum slips distinctly under the reference, indicating the lack of corresponding fine-scale structures. Finally, the PDF of the vorticity for CV-1S shows no significant deviation from the reference, i.e., the final vorticity field is statistically well reproduced. As a summary, we can conclude that, the major reference dynamical flow informations are conserved well in this filtered simulation run. The proposed threshold is affirmed to be a reasonable choice.

The CV-6 run starts with an initial compression ratio of 7%, being only slightly larger than in the CV-1S case. Because no security zone is added in CV-6, the whole set of selected wavelet coefficients yielding the coherent vorticity field represent coherent structures. The CV-1S initial condition is represented by about the same number of wavelet coefficients. But one must kept in mind that, unlike CV-6, only a portion of these coefficients represent coherent structures—the rest are only neighbors of coherent structures in wavelet space, chosen to be part of the coherent vorticity ω_C by no "natural" criterion, i.e., the coefficients' magnitude being larger than a threshold. The result of this advantage for the CV-6 initial condition is, that it evidently looks more similar to the reference's initial condition than the one of the CV-1S run (Figure 13.27). But the obvious advantage turns into an enormous disadvantage as soon as the simulation is started and dynamics evolve within the selected basis. Because of the lack of "space" for the solution to evolve within, the effect described in Subsection 13.4.1.1 and illustrated in Figure 13.22 for a too small basis takes place: within each time step some dynamically relevant information, e.g., structures of nonnegligible energy partly leave the integration basis and therefore are cut off. As a consequence, the dynamics of the considered flow are successively destroyed, resulting in a heavily fragmented vorticity field at the end of the simulation for the CV-6, as shown in Figure 13.27. Only the positions of the largest vortices roughly fit the DNS reference. It is not possible to assign smaller vortices to vortices in the reference flow. They are completely out of phase and their number does not correspond to the reference. Taking into account that CVS filtering is effective in extracting coherent structures from a single flow field, the CV-6 compression ratio decay (cf. Figure 13.28) indicates a fast reduction of the coherent structure quantity. Furthermore, the monotonous decay of the compression ratio exhibits that, in the case of a missing security zone, the simulation cannot emerge new coherent structures in the reduced integration basis, which normally is a natural and therefore necessary effect. The steeper decay of the total enstrophy, the energy spectrum, and the fact that the PDF describing vorticity fields is very different from the DNS or the CV-1S run are subsequent results of the loss of dynamics in the CV-6 simulation run.

There are two main results from this consideration of the CV-1S and CV-6 simulation runs. First, for the application to simulations of 2D turbulent flows, we proposed a CVS filtering based on the iterative threshold estimation process performing only one iteration. Applied with an additional security zone, this choice was found to be a good compromise between high compression and flow reproduction comparable to DNS. The respective extent of information loss that is acceptable for any application has to be decided in each individual case, and it can be controlled by the number of iterations for the threshold estimation and the size of the security zone. Second, the security zone concept was shown to be a reliable way to make possible the evolution of dynamics in both space and scale.

13.4.3 Simulation of 3D flows

In this section, we present the analysis of the performance of the CVS for 3D flows, studying a 3D temporally developing mixing layer. This flow is chosen because its time evolution can be efficiently computed with a pseudo-spectral code. Furthermore, mixing layers are very interesting test cases because they present nonlinear behaviors that occur in almost all 3D flows.

(a) *Numerical method.* For the 3D simulation, we integrate the Navier–Stokes equations in terms of the primitive variables, velocity $\boldsymbol{v}$ and pressure P. The vorticity $\boldsymbol{\omega}$ is then derived from the velocity to compute the nonlinear term:

$$\partial_t \boldsymbol{v} + \boldsymbol{\omega} \times \boldsymbol{v} + \nabla P - \frac{1}{Re} \nabla^2 \boldsymbol{v} = 0$$
$$\nabla \cdot \boldsymbol{v} = 0$$
$$\nabla \times \boldsymbol{v} = \boldsymbol{\omega}.$$

We consider periodic boundary conditions in the two horizontal directions and a mirror symmetry in the vertical direction. To solve these equations, we employ a pseudo-spectral code developed by Pierre Comte [6]. The method uses Fourier transforms in the horizontal planes and sine/cosine transforms in the vertical direction. The nonlinear term is computed by collocation in physical space. A low-storage Runge–Kutta scheme of third order is used for the time integration.

(b) *Flow configuration.* As an initial condition, we have imposed a hyperbolic tangent profile for the stream-wise velocity, and we perturb it with a Gaussian noise $\boldsymbol{W}(\boldsymbol{x})$:

$$\boldsymbol{v}(\boldsymbol{x}, t_0) = U \tanh\left(\frac{2x_3}{\delta_i}\right) \boldsymbol{e_1} + \boldsymbol{W}(\boldsymbol{x}),$$

where $\boldsymbol{e_1}$ is the streamwise unity vector and δ_i the initial vorticity thickness.

This flow has been widely studied using linear theory, and some results on its stability properties are available in the literature. For inviscid fluids, it has been shown that the most amplified mode is the 2D Kelvin–Helmholtz instability with wavelength $\lambda_a = 7.066\delta_i$ [32]. Viscous flows present the same behavior asymptotically for $Re_\delta = \delta_i U/\nu > 100$. We have employed this linear theory to generate the initial condition, and the value of the initial vorticity thickness $\delta_i = L_0/4\lambda_a$ (where $L_0 = 2\pi$ is the size of the computational domain) has been determined to correspond to the formation of four Kelvin–Helmholtz rollers.

The perturbation has been designed in such a way that four rollers are rapidly excited, by taking $\boldsymbol{W}(\boldsymbol{x})$ as a Gaussian noise with a spectrum proportional to k^4 at large scales with a peak at the wavenumber $k_a = 4L_0/\lambda_a$ and an exponential decay for higher wavenumbers. This noise is also modulated in physical space by a Gaussian filter $f(x_3) = \exp(-0.5(x_3/\delta_i)^2)$, so that the perturbations affect only the unstable rotational region and vanish outside.

We study the time evolution of such a flow with a DNS at resolution 128^3. This corresponds to a Reynolds number $Re_{\delta_i} = 448$ based on the initial vorticity thickness, which ensures that turbulence develops.

13.4.3.1 DNS of 3D mixing layer

In this section, we present the DNS of the 3D mixing layer previously defined, which will then be the reference simulation we will use to evaluate the performances of the CVS (see Subsection 13.4.3.2).

Figure 13.29 shows the isosurfaces of the vorticity modulus computed by DNS at three significant time steps. The beginning of the simulation is characterized by the transition to turbulence. In this phase, the vortex stretching mechanism is negligible; thus, there is no production of enstrophy, which actually weakly dissipates (see Figure 13.30). The 2D Kelvin–Helmholtz instability starts to develop in the streamwise direction, creating four rollers which are then deformed by spanwise 3D instabilities ($t = 9\delta_i/U$). As these instabilities are in phase opposition for each vortex pair, there are regions where the rollers get close and thus merging occurs. After this first pairing, hairpin vortices are generated and stretched by velocity gradients, and the production of vorticity becomes important ($t = 18\delta_i/U$). More and more small vortices are generated around the main rolls, and the mixing layer becomes turbulent. This first pairing produces two turbulent rollers bigger than the original ones. Once created, these two vortices start merging with a mechanism close to the previous one ($t = 25\delta_i/U$). The simulation was stopped when the vorticity approached the upper boundary because in that situation the mirror boundary condition employed for the simulation is no longer realistic.

13.4.3.2 CVS of 3D mixing layer

In the Subsection 13.4.2, we applied the CVS filter to 2D flows and we showed that a CVS using the iterative threshold estimation for the nonlinear wavelet filter and with the addition of a security zone allows us to simulate turbulent flows with a reduced numbers of modes, although the accuracy remains comparable to DNS. Furthermore, we showed that, using a small number of iterations to estimate the optimal threshold value for the filter, it is possible to reach a good compromise between the compression properties of the CVS and the quality of the simulation. This filtering strategy has been kept for the CVS of the 3D mixing layer. We have chosen to use 3 iterations for the estimation of the threshold and a security zone defined as explained in Subsection 13.4.1.3. The time integration has been performed with the same code used for the reference DNS simulation.

13.4.3.2.1 Initial condition

It does not make much sense to start the CVS at $t = 0$ because as the filtering is designed to eliminate the random part of the flow, it causes big changes in the random perturbation field and thus in the further evolution of the flow. From a physical point of view, this is not a major problem because, as only the perturbation is affected, the resulting simulation will be another realization of the same flow and thus statistically equivalent to DNS. In practice, this makes the comparison with the reference DNS extremely difficult because all instabilities (Kelvin–Helmholtz rolls and hairpin

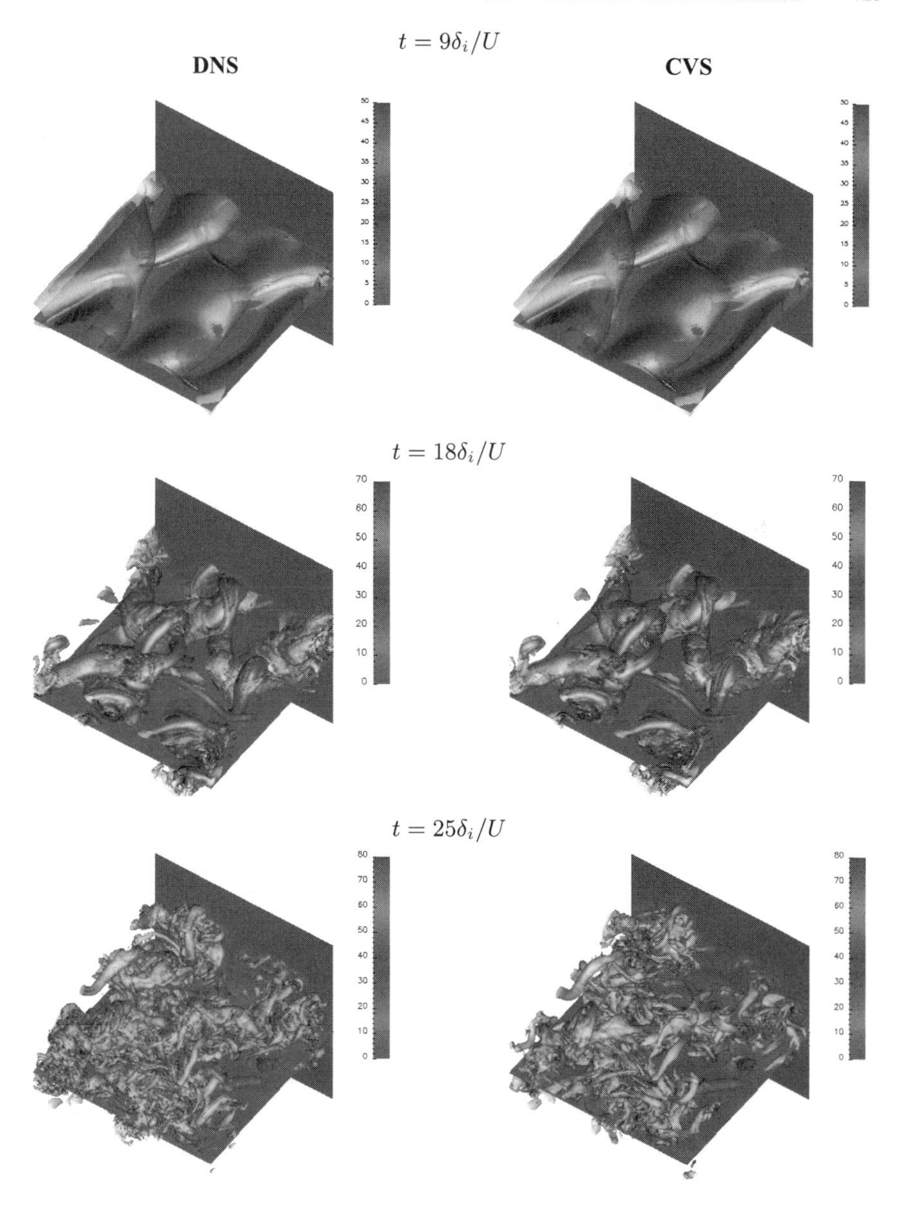

FIGURE 13.29. Isosurfaces of vorticity modulus colored by spanwise vorticity.

vortices) will not develop at the same time and space location for the DNS and for the CVS. Therefore, we have decided to use as an initial condition for the CVS the flow computed by the DNS at time $t = 4\delta_i/U$. At this time the random perturbation has already excited the deterministic perturbation that will determine the further evolution of the main structures of the flow.

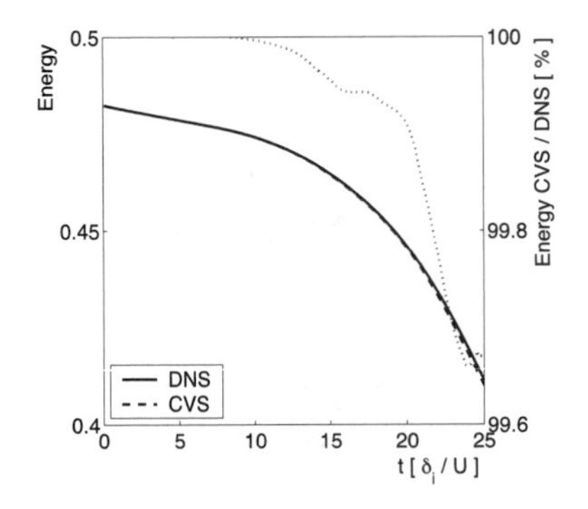

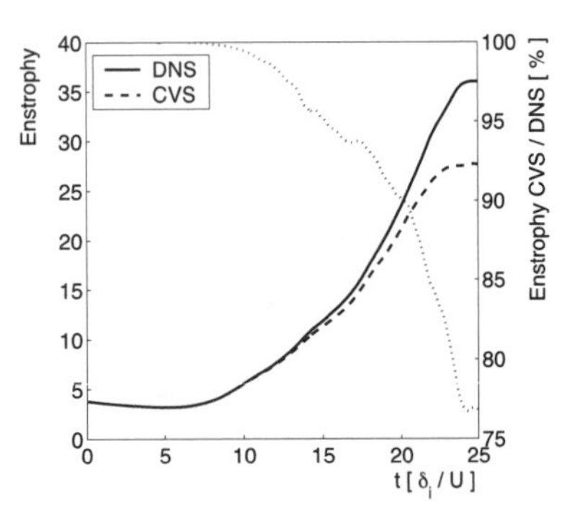

FIGURE 13.30. Time evolution of energy (left) and enstrophy (right). Solid line: DNS, dashed line: CVS, dotted line ratio CVS/DNS.

13.4.3.3 Comparison between CVS and DNS

Before performing any quantitative analysis of the difference between DNS and CVS, it is fruitful to look at the vorticity fields in physical space as presented in Figure 13.29. At $t = 9\delta_i/U$ no differences can be observed between DNS and CVS. The flow is not turbulent; thus, it is deterministically computed, and the wavelets modes retained describe the flow completely. During the transition to turbulence, this situation changes progressively. Around $t = 18\delta_i/U$ and until the end of the simulation ($t = 25\delta_i/U$), the CVS calculated flow appears slightly simpler than the one computed with DNS, because the random fluctuations of the incoherent vorticity have been filtered out. Notwithstanding, all the main structures of the vorticity are well described.

These observations are confirmed by the time evolution of integral quantities like energy and enstrophy. Figure 13.30 shows that, during the transition phase, both energy and enstrophy are identical for CVS and DNS. When the turbulent regime is attained, after the mixing transition, incoherent enstrophy is produced, which corresponds to turbulent dissipation. CVS automatically dissipates the incoherent background, and this explains why, CVS flow contains less enstrophy than the DNS one. For DNS in fact the only possible mechanism to dissipate the energy is the molecular viscosity, which is much smaller than the turbulent viscosity. On the other hand, the time evolution of energy of CVS stays very close to the one of DNS.

In Figure 13.31 we present the number of wavelet modes retained during the simulation which varies between 8% and 15%. The plot shows that the wavelet approximation basis evolves in time. In particular, when rolls pairing occurs, more wavelet modes are needed, which corresponds to the two peaks in Figure 13.31.

13.4.3.3.1 Statistical analysis

We present now some statistical analysis of the flow simulation at two time steps, $t = 18\delta_i/U$ and $t = 25\delta_i/U$. Figure 13.32 shows that the mean velocity profile

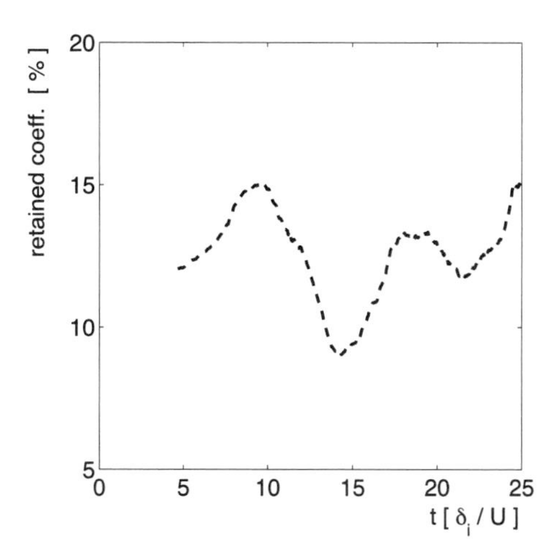

FIGURE 13.31. Number of retained wavelet coefficients during the CVS.

obtained using CVS does not present any relevant differences from the one obtained using DNS.

In Figure 13.33 we analyse the profiles of the second order moments of velocity and vorticity fluctuations, i.e., their variances in each horizontal plane. Here again, for the velocity we find a very good accordance between CVS and DNS. If we consider the vorticity field (Figure 13.33), we notice that, when the flow becomes more turbulent a part of the vorticity fluctuations is filtered out and the profile obtained with CVS is slightly more narrow than that of the reference. However, the maximal value of the vorticity variance is almost the same for both simulations.

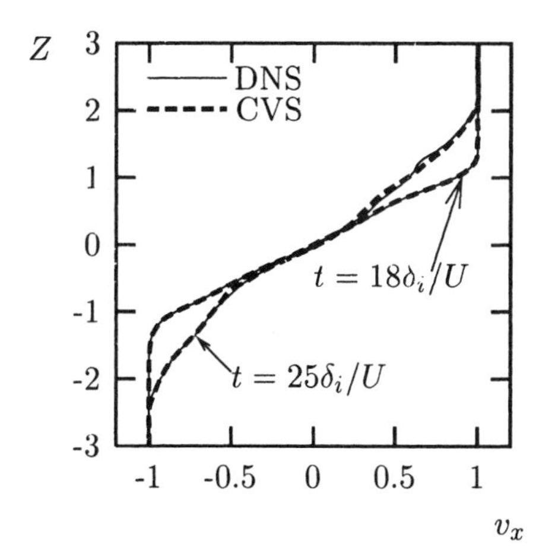

FIGURE 13.32. Profiles of the mean streamwise velocity.

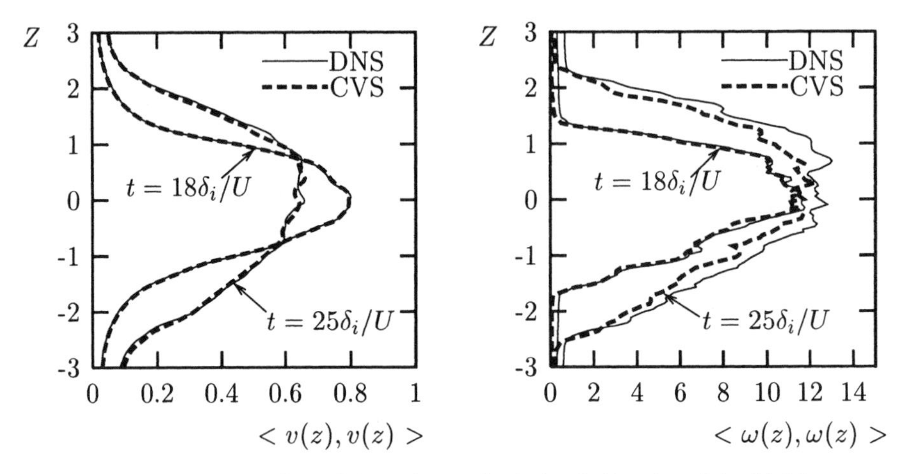

FIGURE 13.33. Profiles of the variance of velocity (left) and vorticity (right).

In Figure 13.34 we present 1D streamwise energy spectra for each of the three velocity components averaged over vertical planes. During the simulation, the energy at large scales is transferred to the smaller ones, and a self-similar inertial range with a $k^{-5/3}$ develops. This is particularly evident for the span-wise and vertical component, while the streamwise component presents a stronger large-scale correlation. The comparison between CVS and DNS confirms the result of the *a priori* test in Subsection 13.3.3.3: the energy spectra in the CVS case are the same as the reference ones at large scales, while at small scales the filtering procedure dissipates the incoherent energy (equivalent to turbulent dissipation).

Focusing on higher order statistics, we consider the PDF of the vorticity. At $t = 18\delta_i/U$ the PDF of the span- and streamwise vorticity appears slightly skewed due to the Kelvin–Helmholtz instabilities and ribs, respectively. The increase of turbulence enhances the mixing of the vorticity, thus at $t = 25\delta_i/U$ the PDFs become more symmetric. We also observe that the maximal values of vorticity increase due to the production of enstrophy. Comparing CVS and DNS, we also find a good accordance for the PDFs, even at the end of the simulation. This, together with the visualisation (Figure 13.29), demonstrates that CVS represents all the structures present in the flow also from a dynamical point of view.

13.5 Summary and conclusion

In this chapter, we presented a survey of the CVS, which is a new method for the simulation of turbulent flows. The extraction of dynamically relevant structures from an uncorrelated background flow is done by filtering the flow representation in wavelet space. The threshold selection is inspired by Donoho's theorem for optimal de-noising used in signal processing. An iterative algorithm for the threshold estima-

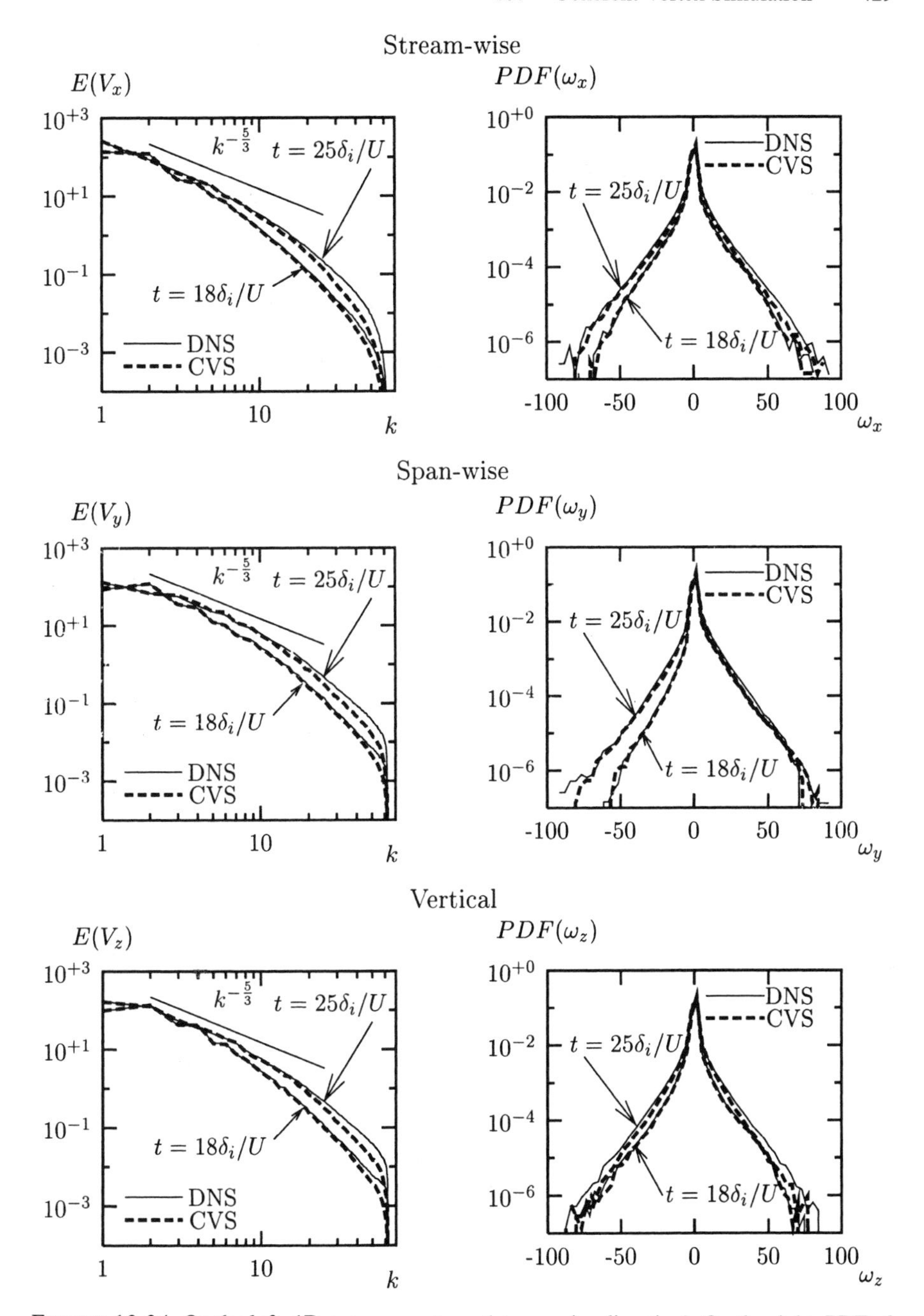

FIGURE 13.34. On the left: 1D energy spectrum (streamwise direction). On the right: PDF of vorticity. From top to bottom: streamwise, spanwise, and vertical components.

tion was introduced, and it was shown that it converges on the presence of a Gaussian white noise. Filtering artificially generated noisy signals with the calculated optimal threshold, we obtain an approximately perfect separation into the de-noised signal and the noise itself.

We applied this filtering technique to single turbulent flow fields, in particular, the pseudo-scalar vorticity field in two dimensions and the velocity vector field in three. The flow part represented by the great majority of wavelet coefficients contains a negligible amount of energy. Since it has a spectrum indicating approximately white noise and a PDF of Gaussian shape, it represents the decorrelated background flow. The extracted flow part is highly compressed as it is reconstructed from only a few wavelet coefficients, though it contains most of the flow field's energy. It is dominated by coherent structures like vortices and can hardly be distinguished from the unfiltered reference. The conformity with the reference was quantified by comparing the PDFs and the spectra. From these results, it was assumed that the coherent part contains the dynamically relevant flow information.

This assumption was verified by the CVS of 2D and 3D turbulent flows. For these purposes, the Navier–Stokes equations were integrated for the coherent part of the flow using a pseudo-spectral Fourier method. The background flow was completely discarded, and no further turbulence model was added. The coherent flow part was obtained by filtering the vorticity field in a wavelet basis at every time step of the flow simulation. The threshold estimation was optimized in terms of compression and conservation properties, while the filtering procedure was extended by the security zone concept, which is necessary for an evolving flow. As the main result, we found that the simulated coherent flow part presents almost the same evolution as the reference flow, calculated with DNS. This was quantified by comparing spectra and PDFs indicating statistical equivalence.

For the simulation of turbulent flows, the presented CVS method yields results similar to DNS calculations, despite the lack of a turbulence model. The nonlinear procedure of the CVS filtering provides an adaptive computational grid with a compression ratio depending on the dynamics. Even though, up to now, CVS demands more computational resources than other methods like LES, it offers the possibility of enhancing flow simulations by its advantages. With increasing Reynolds number, the compression of CVS should exceed LES significantly, corresponding to an intense reduction of the number of degrees of freedom for the simulation. Furthermore, some applications can take advantage of the CVS property to preserve some active small-scale structures, which are neglected in LES. This becomes important, e.g., in the mixing of chemically reacting species, where small-scale structures in the scalar field are a prerequisite for molecular transport and chemical reactions to take place [3, 22].

Acknowledgments. We are grateful to Marie Farge and Kai Schneider for providing many ideas developed and realized in this article. We thank them, and Henning Bockhorn as well, for supervising and supporting our work. We acknowledge Pierre Comte for giving us the 3D spectral code to compute the mixing layer. To

LMD (Paris), ICT (Karlsruhe), L3M (Marseille), and IDRIS (Orsay) we are thankful for providing computing facilities. We acknowledge the French–German PROCOPE program (contract 01220ZE), the European Community's TMR project "Wavelets in numerical simulation" (contract FMRX-CT98-0184), and the French–German DFG-CNRS program "Computational Fluid Dynamics" for partial support of this work. A. Azzalini is thankful to CNRS-SPI for providing him a BDI Ph.D. grant.

References

[1] Batchelor, G., Computation of the energy spectrum in homogeneous two–dimensional turbulence. *Phys. Fluids Suppl. II*, 12:233–239, 1969.

[2] Berkooz, G., Holmes, P., and Lumley, L., The proper orthogonal decomposition in the analysis of turbulent flows. *Ann. Rev. Fluid Mech.*, 25:539–575, 1993.

[3] Bockhorn, H., Gerlinger, W., Schneider, K., and Ziuber, J., Simulation and analysis of mixing in two–dimensional turbulent flows using Fourier and wavelet techniques. In Keil, F., Mackens, W., Vo, H., and Werther, J., editors, *Scientific Computing in Chemical Engineering II*, pages 344–351. Springer, Berlin and New York, 1999.

[4] Canuto, C., Hussaini, M., Quarteroni, A., and Zang, T., *Spectral Methods in Fluid Dynamics*. Springer, Berlin, 1988.

[5] Chen, C., and Jaw, S., *Fundamentals of Turbulence Modeling*. Taylor & Francis, Washington, 1998.

[6] Comte, P., Lesieur, M., and Lamballais, E., Large- and small-scale stirring of vorticity and passive scalar in a 3-d temporal mixing layer. *Phys. Fluids A*, 4(12):2761–2778, 1992.

[7] Daubechies, I., *Ten Lectures on Wavelets*. Soc. Ind. Appl. Math. (SIAM), Philadelphia, 1992.

[8] Do-Khac, M., Basdevant, C., Perrier, V., and Dang-Tran, K., Wavelet analysis of 2D turbulent flows. *Physica D*, 76:252–277, 1994.

[9] Donoho, D., Unconditional bases are optimal bases for data compression and statistical estimation. *Appl. Comput. Harmon. Anal.*, 1(1):100–115, 1993.

[10] Donoho, D. and Johnstone, I., Ideal spatial adaptation via wavelet shrinkage. *Biometrica*, 81:425–455, 1994.

[11] Farge, M., Wavelet transforms and their applications to turbulence. *Ann. Rev. Fluid Mech.*, 24:395–457, 1992.

[12] Farge, M., Goirand, E., Meyer, Y., Pascal, F., and Wickerhauser, M., Improved predictability of two-dimensional turbulent flows using wavelet packet compression. *Fluid Dyn. Res.*, 10:229–250, 1992.

[13] Farge, M., Kevlahan, N., Perrier, V., and Goirand, E., Wavelets and turbulence. *Proceedings of the IEEE*, 84(4):639–668, 1996.

[14] Farge, M., Kevlahan, N., Perrier, V., and Schneider, K., Turbulence analysis, modeling, and computing using wavelets. In J. van den Berg, editor, *Wavelets in Physics*, pages 117–200. Cambridge University Press, London, 1999.

[15] Farge, M., Pellegrino, G., and Schneider, K., Coherent vortex extraction in 3D turbulent flows using orthogonal wavelets. *Phys. Rev. Lett.*, 87(5):45011–45015, 2001.

[16] Farge, M., and Rabreau, G., Transformée en ondelettes pour détecter et analyser les structures cohérentes dans les écoulements turbulents bidimensionnels. *C. R. Acad. Sci. Paris, Série II*, 307, 1988.

[17] Farge, M., and Schneider, K., Coherent vortex simulation (CVS), a semi-deterministic turbulence model using wavelets. *Flow, Turbulence and Combustion*, 66(4):393–426, 2001.

[18] Farge, M., Schneider, K., and Kevlahan, N., Non–gaussianity and coherent vortex simulation for two-dimensional turbulence using an adaptive orthogonal wavelet basis. *Phys. Fluids*, 11(8):2187–2201, 1999.

[19] Farge, M., Schneider, K., Pellegrino, G., Wray, A., and Rogallo, B., CVS decomposition of three-dimensional homogeneous turbulence using orthogonal wavelets. In *Proceedings of the 2000 Summer Program, Center for Turbulence Research*, pages 305–317, NASA Ames and Stanford University, 2000.

[20] Frisch, U., *Turbulence*. Cambridge University Press, London, 1995.

[21] Fröhlich, J. and Schneider, K., Computation of decaying turbulence in an adaptive wavelet basis. *Physica D*, 134:337–361, 1999.

[22] Gerlinger, W., Schneider, K., Falk, L., and Bockhorn, H., Numerical simulation of the mixing of passive and reactive scalars in two-dimensional flows dominated by coherent vortices. *Chem. Eng. Sci.*, 55:4255–4269, 2000.

[23] Givi, P., Model-free simulations of turbulent reactive flows. *Prog. Energy Combust. Sci.*, 15:1–107, 1989.

[24] Grossmann, A. and Morlet, J., Decomposition of hardy functions into square integrable wavelets of constant shape. *SIAM J. Math. Anal.*, 15(4):723–736, 1984.

[25] Jiminez, J. and Wray, A., The structure of intense vorticity in isotropic turbulence. *J. Fluid Mech.*, 255:65–90, 1993.

[26] Kraichnan, P., Inertial ranges in two-dimensional turbulence. *Phys. Fluids*, 10:1417–1423, 1967.

[27] Lesieur, M., *Turbulence in Fluids*. Kluwer, Dordrecht, 3rd rev. and enlarged edition, 1997.

[28] Mallat, S., Multiresolution approximations and wavelet orthonormal bases of $L^2(\mathbb{R})$. *Trans. Amer. Math. Soc.*, 315(1), 1989.

[29] Mallat, S., *A Wavelet Tour of Signal Processing*. Academic Press, New York, 1998.

[30] McWilliams, J. C., The emergence of isolated coherent vortices in turbulent flows. *J. Fluid Mech.*, 146:21–43, 1984.

[31] Meneveau, C., Analysis of turbulence in the orthonormal wavelet representation. *J. Fluid Mech.*, 232:469–520, 1991.

[32] Michalke, A., On the inviscid instability of the hyperbolic-tangent velocity profile. *J. Fluid Mech.*, 19:543–556, 1964.

[33] Pope, S. B., *Turbulent flows*. Cambridge University Press, Cambridge, London, 2000.

[34] Schneider, K., Kevlahan, N., and Farge, M., Comparison of an adaptive wavelet method and nonlinearly filtered pseudo-spectral methods for two-dimensional turbulence. *Theoret. Comput. Fluid Dyn.*, 9:191–206, 1997.

Index

admissibility condition, 5, 385
affine class, 179, 315, 334
affine group, 181
affine modulation, 212
affine operator, 190
affine operators, 180
ambiguity function, 146
analytic signal, 138, 141
AND decision fusion, 127
approximation space, 106
asymptotic reassignment point, 258
autocorrelation function, 44

B-spine wavelet, 377
bandwidth, 7
Bernoulli tree, 369
Beurling weight, 43
bilinear TF representations, 179
biorthogonal wavelet family, 42
branching number, 368

canonical Hamiltonian, 62
canonical quadratic form, 99
chirp detection, 259
chirp signal, 244, 283
coadjoint orbit, 208
cocycle, 182
Cohen's class, 142, 179, 199, 200, 218, 239, 314, 328
coherent vortex simulation (CVS), 383, 411
conditional distribution, 9
conditional moments, 9
conditional variances, 125
conjugate operators, 187
continuous wavelet transform, 178
correlation function, 44
correlation matrix, 45
covariance, 195
covariance property, 177
covariant, 195

decision function, 124
differential reassignment, 256
dilation matrix, 26
direct numerical simulation (DNS), 383, 414
dispersive warping operator, 318
dispersively warped α-form WD, 327, 333
dispersively warped Bertrand P_0-distribution, 333
dispersively warped classes, 336
dispersively warped scalogram, 334
dispersively warped WD, 327
displacement function, 177, 179, 201
displacement operator, 190, 193
distribution of energy, 143
Donoho's theorem, 388, 428
dual modulation, 212
dual operators, 180, 187
duration, 7
dyadic cells, 367

energy distributions, 220, 221
energy of the signal, 4
enstrophy spectrum, 409
exponential class, 179, 335

Fourier transform, 189
Fourier transforms, 4

Fréchet derivative, 95
frequency, 7
frequency moments, 18, 21
frequency-shift operator, 178

Gabor wavelet, 12
Gaussian envelope, 244
Gaussian fixed point, 66, 73
Gaussian logon, 244
Gaussian noise, 423
generalized impulse, 319
general quadratic form of TFDs, 143
generalized self-similar function, 24
geometric phase, 254
groups, 180
group delay, 141
group Fourier transform, 227
growing number, 368

Hamburger moment problem, 113
hard thresholding, 110
hard threshold estimator, 389
hard thresholding filter, 389
Hardy space, 190
Heisenberg's uncertainty relation, 142
hyperbolic class, 179, 241, 328
hyperbolic wavelet transform, 178
hyperbolically warped affine class, 334

influence cone, 249
intermittency, 368
Isis variables, 63
isotropic energy spectrum, 408

joint spectral radius, 30

kernel filter, 145
Klauder wavelet family, 249
Kullback–Leibler distance, 256
κth power class, 335
κth power exponential class, 336

Lemarié–Meyer orthonormal basis, 364
level curves, 254
linear chirp, 244
local centroids, 239
localization function, 220, 221

Mallat's wavelet, 114
marginal distribution, 9
marginal properties, 219, 221
marginals, 5
MATLAB software, 379
maximum resolution, 106
mean frequency, 7, 8
mean time, 7, 16
mean time squared, 16
Mellin transform, 321
minimal accuracy, 32
minimax thresholding, 110
modulation function, 183
modulation operator, 183, 211
modulation operators, 179
moments, 6
momentum-slicing, 56
Morlet wavelet, 11
Morlet wavelet, modified, 12
mother wavelet, 178, 386
multiresolution analysis, 35, 106, 386

Navier–Stokes equation, 383, 414, 415
normalized displacement vector, 242

OR decision fusion, 125
orthonormal basis, 107, 386
Osiris wavelets, 56

Parseval's formula, 45
positive frequencies, 138
power class, 241
power wavelet transform, 178, 197
projection mapping, 205
projective representation, 182

quadratic form, 59, 81
quincunx dilation matrices, 27
quincunx sublattice, 27

reassigned distribution, 239
reassigned spectrogram, 237
reassignment principle, 234
recursion formula, 73
refinable function, 26
refinement equation, 25
refinement operator, 26
Riesz basis, 45, 47
robustness, 118

scale covariance, 316
scale moments, 18

scaling function, 26, 35, 106, 387
scaling operator, 178
scalogram, 4, 236
security zone, 412
self similar Gaussian process, 362
self-similar, 24
sensor-angle distributions (SADs), 297, 301
separable kernel filter, 150
shift invariant, 43
short-time Fourier transform, 178, 197, 235
signal kernel, 144
signal-to-noise ratios (SNRs), 105, 117
SNR enhancement, 276
spatial ambiguity functions, 296
spatial time-frequency distributions (STFDs), 270, 271, 273
soft thresholding, 110
spectrogram, 236
stationary signals, 234
Stein's unbiased risk estimator, 111
supervised reassignment, 255
symmetric ambiguity function, 146

TF representations, 201
TFD kernel filter, 145
time delay, 141
time moments, 18
time-frequency distribution (TFD), 136, 140, 271
time-shift operator, 178
topological group, 181
topologically isomorphic, 181
two-scale equation, 106

unitary equivalence, 179
unitary operator, 194
unitary representation, 181

warped exponential class, 339
warped hyperbolic class, 338
warping function, 184
warping operator, 180, 183, 211, 212, 318
warping operators, 179
wavelet, 4
wavelet basis, 107
wavelet family, 386
wavelet filtering, 403
wavelet shrinkage, 108
wavelet transform, 4, 178, 197, 235
Wiener filter, 111
Wiener–Khintchine theorem, 144
Wigner distribution, 7
Wigner–Ville distribution, 144, 236
window, 178